American Institute of Chemical Engineers

6th Topical Conference on Natural Gas Utilization 2006

AA003161

Held at the 2006 AIChE Spring National Meeting

April 23-27, 2006
Orlando, Florida, USA

Printed from e-media with permission by:

Curran Associates, Inc.
57 Morehouse Lane
Red Hook, NY 12571
www.proceedings.com

ISBN: 978-1-60423-526-5

Some format issues inherent in the e-media version may also appear in this print version.

American Institute of Chemical Engineers

6th Topical Conference on Natural Gas Utilization
2006

Table of Contents

Session 26 - LNG I - Plant & Operation

Papers on LNG plants and Their Operation

CoChair: Yu-Nan Liu

Chair: Chen-Hwa Chiu

Session 60 - LNG II - Technology & Equipment

Issues of LNG Technology and Equipment

CoChair: Yoshitsugi Kikkawa

Chair: Chen-Hwa Chiu

Session 128 - Advanced Gas Conversion and Gasification Processes

The symposium will primarily deal with the recent advances in production of syngas from natural gas, as the first step to add value to the resources. Papers related to gasification processes for refinery use of petroleum coke, heavy oil, oil sands bitumen, kerogen from shale oil and other hydrocarbons for making hydrogen, fuel components, or fuels are appropriate for this session.

CoChair: Rameshwar Srivastava

Chair: Venkat Venkataraman

277 I T M Syngas: Ceramic Membrane Technology for Lower Cost Conversion of Natural Gas
Christopher M. Chen, Douglas L. Bennett, Michael F. Carolan, Christopher F. Miller, James J. Steppan, William E. Waldron

283 Metal Promoted Binary Oxides of Ceria and Zirconia for Low Temperature Water-Gas Shift
Gary Jacobs, Sandrine Ricote, Yaying Ji, Patricia M. Patterson, Burtron H. Davis

289 Efficiencies in Small Scale Hydrogen Production by Methane Steam Reforming in High Temperature Catalytic and Low Temperature Plasma Processes
Trung Pham, Trung Hoang, Richard G. Mallinson, Lance Lobban

297 Oxidative Coupling of Methane Enhanced by Thermally Optimized Reactors and Intermediate Separation
Ames Kulprathipanja, Alexander S. Lindquist, Joseph C. Poshusta, Jerry L. Martin

304 Development and Evaluation of the Gasification Technology as Alternative Power Generation
Emilio Manzanares-Papayanopoulos, Manuel Fernández-Montiel, Jorge A. Altamirano-Bedolla, Agustín Moisés Alcaraz-Calderón, César A. Romo-Millares

318 Optimization of Amins Mixture `S Formulation by Neural Networks in Gas Sweeting
iman Farasat, Mohammad Samipoor giri, Mojtaba Shariati Niasar

Session 138 - LNG IV - Control & Simulation

Papers on Controlling and Simulating LNG Processes

CoChair: Hans Kimmel

Chair: Chen-Hwa Chiu

319 CFD Study on Natural Gas Fluidized Bed Combustors
Felix Severino Farias Junior, Rubens Maciel Filho, Sergio Lucena, Jornandes Dias Silva

325 Rasgas Makes Extensive Use of Process Operator Training Simulators in LNG Operations
Ahmet Pekediz, James M. Dawson, John W. Womack

Session 181 - Fischer-Tropsch and Alternative Syngas Conversion Processes

Gas Conversion to Liquid Fuels and Related Advance Fuels symposium will primarily deal with the conversion of natural gas, associated gas and/or synthesis gas to liquid fuels and related advanced products that both add value to the gas resource and improve refinery operations.

Cochairs: Rameshwar Srivastava
Christopher M. Chen

Session 183 - LNG V - Cost & Facilities

LNG Costs and Facilities

CoChair: Joseph Cho

Chair: Chen-Hwa Chiu

Session 222 - LNG VI - Risk & Safety

Issues of Risk and Safety in LNG

CoChair: Harry West

Chair: Chen-Hwa Chiu

Session 228 - Synthetic Fuels from Oil Shale, Coal and Natural Gas

The symposium will primarily deal with recent advances in the production of gaseous and liquid hydrocarbons from unconventional resources, including oil shale. Papers dealing with modeling approaches, R&D activities aimed at commercialization, and current commercial operations are appropriate for this session.

Cochairs: Venkat Venkataraman
Rameshwar Srivastava

Chair: Hugh D. Guthrie

New Advances of LNG in China

Anzhong Gu, Xuesheng Lu, Wensheng Lin
Shanghai Jiaotong University
Shanghai, China

AIChE Spring Meeting, April 2006
6th Topical Conference on Natural Gas Utilization
Orlando, Florida, April 23-27, 2006

Abstract

There is rich natural gas source in China. The Chinese government put great effort to exploit and utilize the natural gas for improving energy structure, reducing pollution, and developing economy in China. The industry of liquefied natural gas (LNG) has already started and shaped. In recent 10 years, every link in LNG chain has advancement. Especially in the recent years, some aspects in the LNG chain make great progress.

Besides the LNG plants in Shanghai, Zhongyuan of Henan, and Guanghui of Xinjiang, China is building or has already built LNG plants in Gudong, Hainan, Guangxi, Inner Mongolia, Gansu, and Sichuan, respectively.

To meet the every increasing energy demand, China plans to construct a series of LNG terminals. The first LNG terminal of China was built in Shenzhen, Guangdong Province, and the second LNG terminal of China was projected in Fujian Province. Moreover, the LNG terminals in Zhejiang, Shanghai, Qingdao, and Dalian have made substantive progresses.

The Chinese government has planned to build LNG carriers at shipyard in Shanghai. The LNG road tankers made by manufacturers in Sichuan, Jiangsu and Hebei have already put into practice; the LNG storage containers made by manufactures in Jiangsu have been launched to the market. In some cities in Shandong, Jiangsu, Henan, Zhejiang, and Jiangxi provinces, there are satellite stations, where the LNG is re-gasified and the natural gas is supplied to residents and industries by pipeline.

The regulations and standards of LNG are been establishing now in China.

In this paper, the technical data and flow diagram about LNG plants, LNG receiving terminals, LNG road tankers and storage containers, and LNG re-gasification stations are introduced.

Introduction

LNG plants in China include the peakshaving LNG plant in Shanghai, Zhongyuan LNG plant at Puyang of Henan province. A third one is being built in Xinjiang province. At the mean time, an LNG receiving terminal is being built for import at shenzhen of Guangdong province and the second receiving terminal will be built in Putian, Fujian province. LNG receiving terminals are also planned in Shanghai, Guangdong Zhuhai, Zhejiang Ningbo, Shandong Qingdao. Regarding to LNG transportation, LNG road tankers made by domestic manufacturers have already been put into use while tank containers used to deliver liquefied natural gas have also been developed. The first LNG carrier in China being made in Shanghai now heralded in new technological breakthroughs in the Chinese shipyard industry. As for LNG application, in some cities of Shandong, Jiangsu, Henan, Zhejiang and Guangdong provinces, there are LNG satellite stations, where LNG is re-gasified and supplied to residents and industries by pipeline. To meet the rapid progress of LNG demand in China, related standardization work had already been started. The basic research projects have been carried out at universities and institutes.

The course of advances of LNG industrial chain in China, including LNG plant, receiving terminal, transportation and satellite station is as follows.

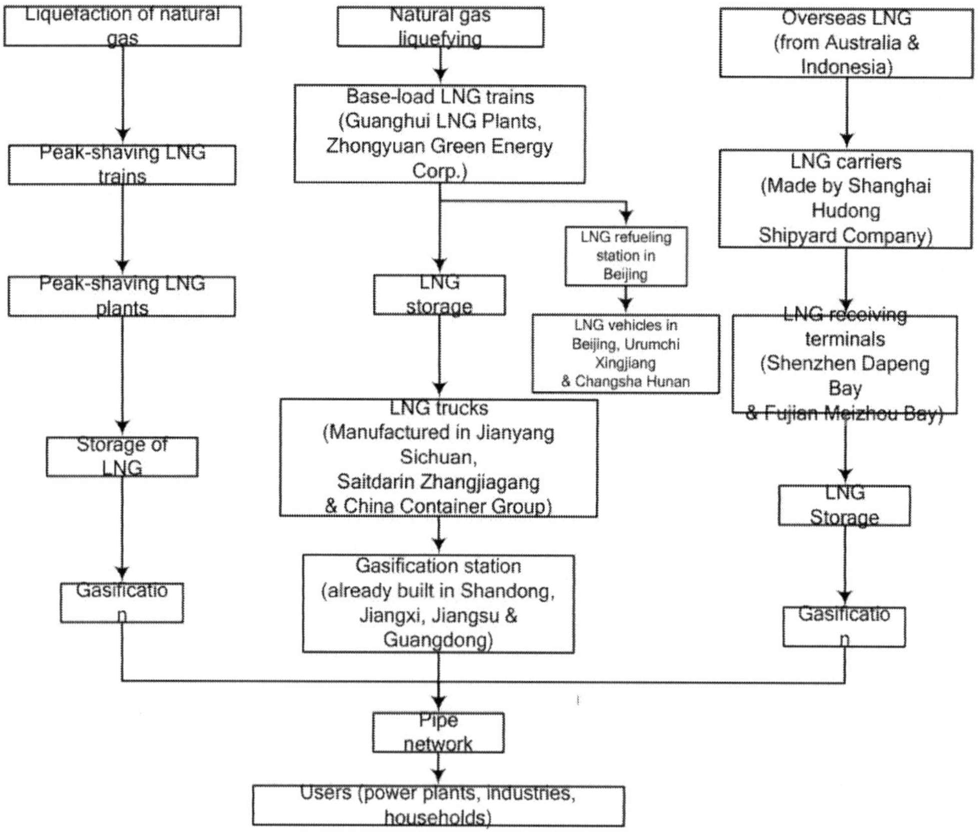

Fig.1. LNG Industrial Chain in China

LNG Industrial Chain

LNG Plant

China began its LNG plant research in the late 1980s. In the following part, a survey was done into the research and development of small-scale LNG plants which may provide useful insights into LNG liquefaction technology in China.

1) LNG Plants Developed by CAS Cryogenic Center, etc.

Two LNG plants had been developed jointly by Chinese Academy of Science (CAS) cryogenic center, Sichuan Mianyang Gas Group, the exploration bureau of China Petroleum and Natural Gas Group and Jilin Oil Administration Bureau. One has a capacity of 0.3 m^3 of LNG per hour and applied natural gas expansion cycle; while the other used Nitrogen expansion closed cycle, with a capacity of 0.5 m^3 of LNG per hour.

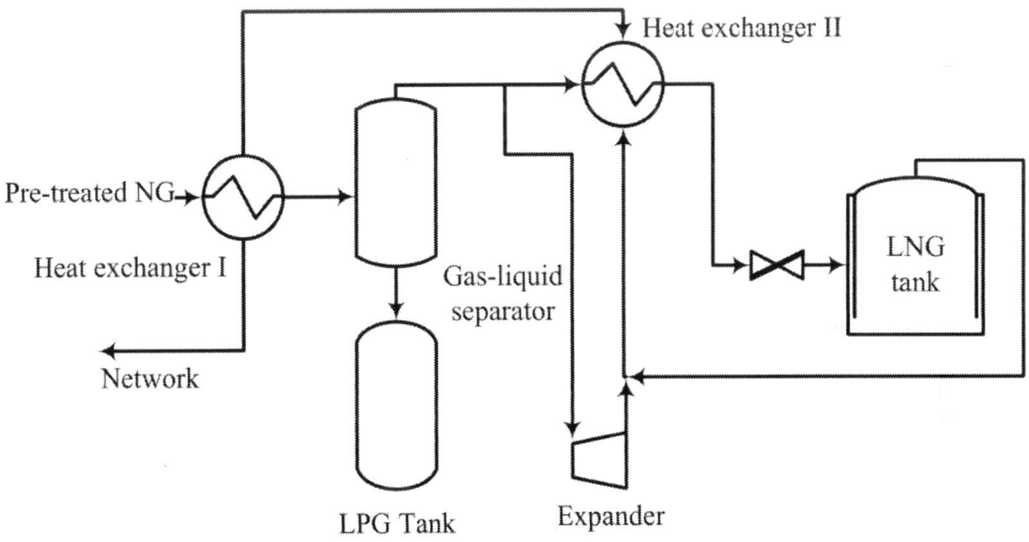

Fig.2. Flow Diagram of Expansion Refrigeration Cycle

Fig. 2. shows using natural gas expansion cycle at an inflow pressure of 4 Mpa, and outflow pressure of 0.8 Mpa. High-pressure gas is pre-cooled in heat-exchanger No. 1 and it enters the gas-liquid separator to get rid of heavy Hydrocarbon. One part of the gas produced by the separator enters the expander to lose some temperature and then enters heat exchanger No. 2 and cools the natural gas waiting to be liquefied. The rest goes to heat exchanger No. 2 to be cooled and kept in LNG tank after being cooled in the throttle valve.

2) Zhongyuan LNG Plant

In 2001, the successful start-up of the first small-scale LNG plant at Zhongyuan oil field symbolizes an important progress of Chinese LNG plants.

There, LNG produced is supplied to Shandong, Jiangsu, and other provinces by LNG tank trucks.

Zhongyuan LNG plant has a production capacity of 15×104 Nm^3/d. Its gas resource is 12 MPa, 30□, and with a methane mole percentage 93.35~95.83%. It uses cascade refrigeration cycle with Propane and Ethene as refrigerant. The flowchart of the liquefaction process is shown in Fig. 3.

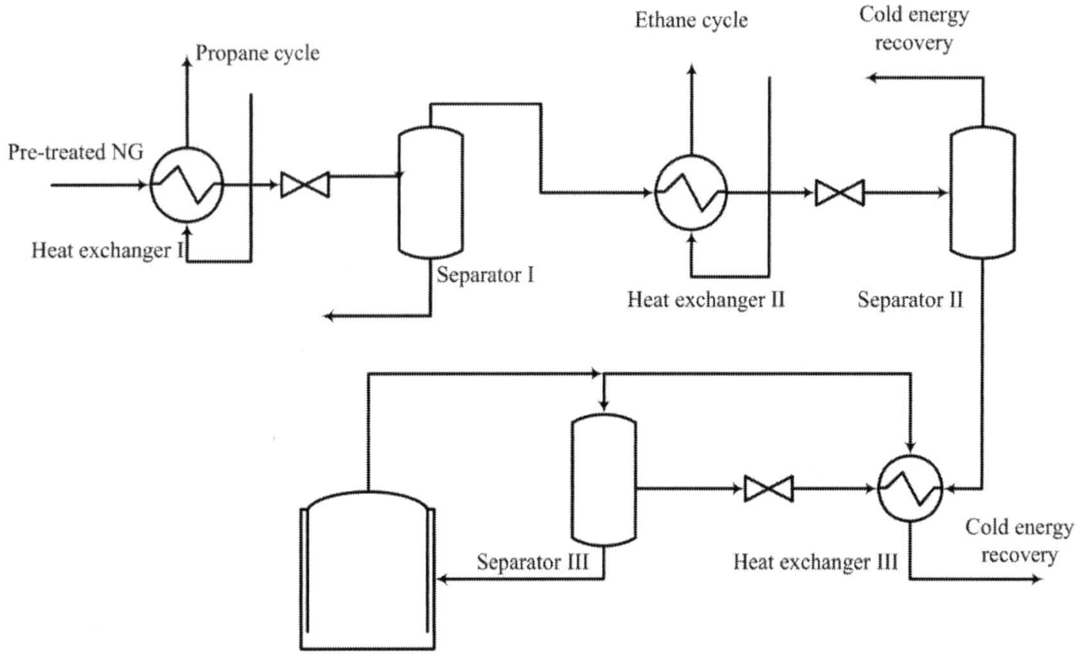

Fig.3. Process of Natural Gas Liquefying Unit

After the gas stock enters the plant, firstly it goes to the separation unit to remove the liquid and then enters the filter to remove large-sized liquid and solid. The filtered natural gas enters the de-carbon process and removes the CO2 by ethylate amine (MEA). Then the natural gas is dehydrated with molecule salvage. There are two dryers in the pre-treatment process for alternative use, one for drying, and the other for regeneration.

Natural gas purified is pre-cooled by the propane refrigeration cycle in heat-exchanger 1, then is throttled and the temperature of the gas decreases, and is divided into liquid and gas stream in separator 1. The liquid phase return to the separator; and the gas phase is cooled by the ethylene heat-exchanger 2, chilled via throttle valve, and then enters separator 2, then liquefied natural gas is throttled and the pressure decrease. LNG is stored in a storage tank at normal pressure.

LNG produced by liquefaction plant was stored in tanks. Each tank contains seven small tanks and one outer shell. The insulating material is filled between the outer shell and small tanks. Nitrogen is pressed into the insulation layer to prevent air and water penetrating into the insulation layer.

3) Shanghai Peak-shaving LNG Plant

Shanghai LNG Plant is the first peak-shaving LNG plant in China. It is the primary part of the east sea natural gas project. It supplies natural gas to Shanghai from the east sea in case of typhoon and pipeline problem etc.

The natural gas at 1.5 MPa from the main line is filtrated, gauged, and compressed to 5 MPa. CO2 and H2S in the natural gas are removed by MEA, H2O is removed by zeolite, then purified natural gas is liquefied and stored in two $104m^3$ tanks. LNG re-gasified, odorized, regulated to the city gas network as demand. The evaporated gas in LNG tank streams into the main line after BOG (Boiling of Gas) treatment.

Main parameters of this plant are as follows: (1) volume of tank, $2\times104m^3$. (2) daily evaporation rate of LNG tank, 0.08%/d. (3) LNG production, $165m^3$/d. (4) gasification capacity, $120m^3$/h in LNG.. (5) the refrigerant components, N_2: 7.145, CH_4: 35.17, C_2H_6: 45.588, C_3H_8: 7.259, n-C_4H_{10}: 2.34 , i-C_4H_{10}: 0.923, n-C_5H_{12}: 0.361, i-C_5H_{12}: 1.215.

The refrigeration cycle of the process is integral incorporated cascade cycle (CII) designed by GAZ DE France. The main equipments of this cycle are compressors, fractionation unit and cold box (see Fig.4). The whole liquefaction process can be divided into two parts: natural gas liquefaction system and mixed refrigerant cycle. CII is characterized by the following points: 1) concise processes and require less equipment; 2) the cool box uses fined-plate heat exchanger which is small in size and convenient to install; 3) the patterns of compressor and driver are both simple and reliable.

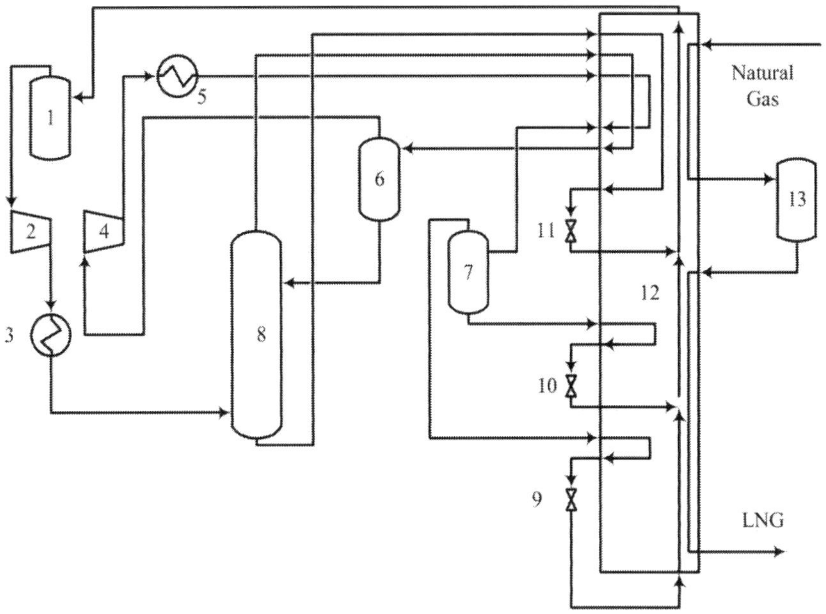

1,6,7,13—separator 2,4—compressor 3,5—water cooler
8—fractionation tower 9,10,11—throttle 12—cold box

Fig 4 CII Cycle Diagram

4) Xinjiang LNG Project

At present, the Xinjiang LNG Project is under construction. The prospective markets are: (1) Southeast of Fujian province, the 14 cities along the railway from Fuzhou to Xiamen. (2) Hubei, Hunan, and other provinces in eastern China. (3) Northern China and Tianshan Economic Zone in Xinjiang.

The natural gas resource is located in Tuha gas field. The output of Xinjiang LNG plants is 1.5×106 Nm^3/d in natural gas and it is the largest LNG production base in China at present. The annual LNG output is about 4.56×105 t. A LNG tank of 3×104m^3 will be built which can provide enough LNG needed for regular production time of ten days. At present, phase II of Guanghui LNG project had been launched and it is supposed to reach a capacity of above 12×106 m^3/d and a supplying capacity of more than 4 billion m^3 after phase IV be completed in June, 2008.

The above LNG plant is comprised of five major parts: purification, liquefaction, LNG tank, LNG export and flare. Figure 5 is the sketch of the mixed refrigerant cycle for the Xinjiang LNG plant.

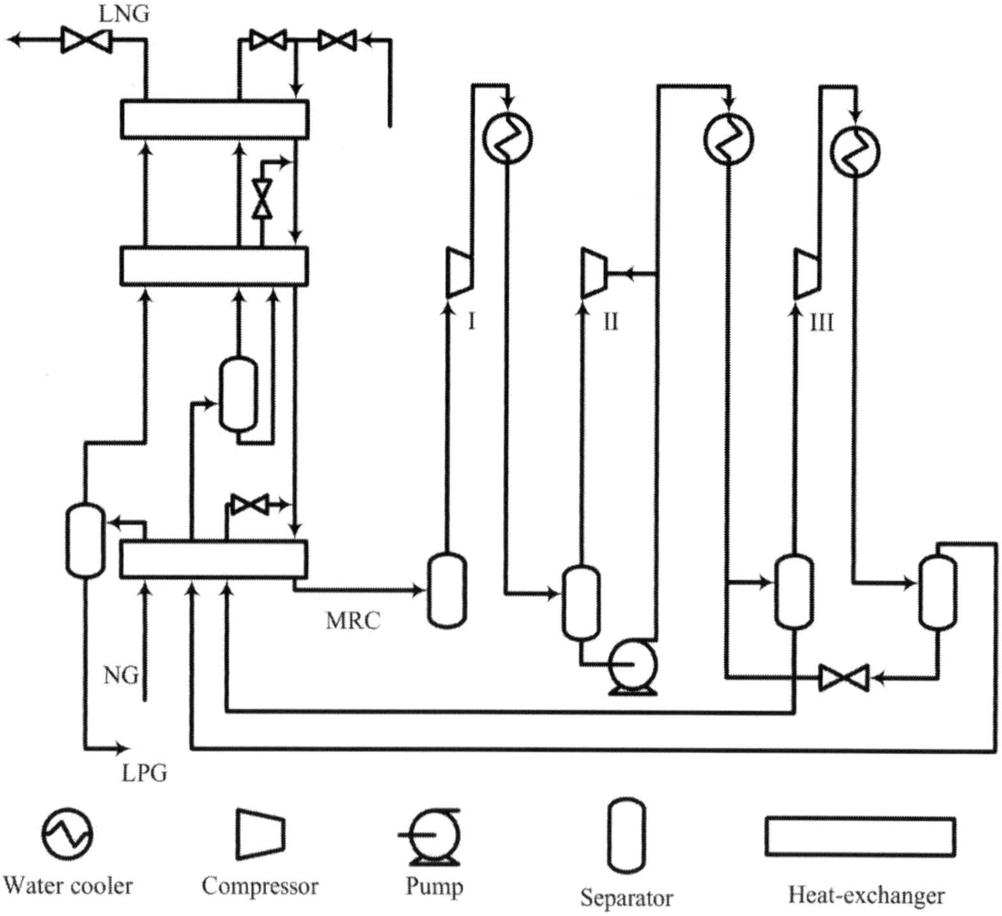

Fig. 5 Sketch of the Mixed Refrigerant Cycle for the Xinjiang LNG Plant

In addition, there are LNG plants of different scales already built or being built in Hainan, Guangxi Beihai, Inner Mongolia Wu Shenqi and Lanzhou.

LNG Receiving Terminal

China has planned to import LNG in its Zhujiang River Delta, Yangtze River Delta and Fujian Province. Shenzhen will have the first LNG receiving terminal ever built in China with its phase I project put into production in 2006 and having an import capacity of 3 million tons per year. The area it covers will include Shenzhen, Dongguan, Guangzhou, Foshan and part of Hongkong. Phase II of the project will be completed in 2008 and increase the annual import capacity to 5 million tons. Guangdong Province has already began to build six LNG power stations converting the current oil power plants into LNG plants. Huizhou LNG power plant had an investment of $8.5 billion and it's a major part in the project of LNG Integration. It is designed to have a capacity of 2 million KW and will provide enough power for Huizhou and the petrol-chemical project jointly sponsored by China Ocean Petroleum Group and Shell Company after it is put into production. Australian LNG companies had signed LNG supply contracts with Guangdong Province with contracting fees ranging from 20 billion to 25 billion Australian Dollars. Beginning in the year 2005 to 2006, LNG will be imported from Australia and the contract includes 3×106t/a LNG for a period of 25 years.

The land-filling project for Fujian LNG receiving terminal had begun in Xiuyu Bay of Putian in 2004 with a total scale of 5 million tons per year. The investment into phase I was above 22 billion CNY including some 6 billion CNY invested into the receiving terminal project which was composed of a receiving terminal, main pipelines, three gas power plants and five urban gas projects in Fuzhou, Putian, Quanzhou, Xiamen and Zhangzhou. The project will have a receiving capacity of 2.6 million tons of LNG and will be put into use in April, 2007.

LNG receiving terminal project in Shanghai has entered the critical period till now. The project was designed to receive 6 million tons of LNG annually and will be completed in two phases which include LNG receiving terminal, LNG shipping docks and the bottom sea transmitting pipelines. Phase I will be put into production in June, 2008 with a receiving capacity of 3 million tons of LNG per year and capable of providing Shanghai with 4 billion m^3 of natural gas annually.

The second LNG receiving terminal in Guangdong Province will be built in Zhuhai, and the first choice of site is the Pingpai Mountain of Gaolan Bay. The receiving terminal mainly provides some power plants and urban industry and households in the west bank of Zhujiang River Delta with reliable fuel. The designed receiving capacity of phase I is 3 million tons per year and the final capacity will be 9 million tons per year.

LNG receiving terminal project in Ningbo has gained fundamental progress and will be put into production in the year 2008. The project comprises of a receiving terminal, docks, gas pipelines and an LNG power plant. Phase I of the project scales 3 million tons a year while phase II increases to 6 million tons per year. Total investment of phase I is about 4.3 billion yuan including a loading dock capable of receiving 80,000 to 165,000 m^3 of LNG carriers.

And LNG receiving terminal to be constructed in Qingdao, Shandong Province will have a receiving capacity of 3 million tons of LNG per year.

LNG Tank Truck and Tank Container

At present, the transportation of LNG from production site to users takes the form of LNG tank trucks or tank containers which had been made by manufacturers in Sichuan, Jiangsu and Hebei Provinces.

1) LNG tank truck

There are two kinds of tank trucks in China, 27m3 and 40m3. In light of the cryogenic character of LNG, the tanks are thermally insulated by vacuumed fiber. The inner shell and pipeline is made of austenitic stainless steel-0Cr18Ni9; the outer shell is made of low alloy steel-16MnR. The support between the inner and outer shell is epoxy glass fiber reinforced plastics, which is suitable for thermal insulation at cryogenic temperature.

The truck contains both loading and unloading system, a self-contained pressuring system, a cleaning system, an system controlling instrument, an emergency shut valve, a safety system, a vacuuming unit, and a sampling and full load analysis device.

Since LNG is combustible liquid, the truck has such safety facilities as an emergency shut-down switch, a quick melting plug, fireproofing equipment, an nitrogen flashing and filling system, grounding and a fire hydrant.

2) LNG tank container

The sizes of LNG tank containers fall into three types: 17.5 m^3, 40 m^3 and 43.9 m^3. They have high vacuum multi-layer insulation thus provide good insulation. They are also capable of long time harmless storage while boasting of very light weight. LNG tank containers provide flexible ways of transportation, either by road, water, railway or the combination of the three, and can be used directly as storage tanks.

LNG Carrier

Two LNG carriers worth $400 million are under construction by Shanghai Hudong Zhonghua Ship Making Group which symbolizes the Chinese ship manufacturing industry's entry into a brand-new technological field. Both carriers are GTT NO.96 with a capacity of 147,200 m^3, and are 292 m in length, 43.35 m in width. The cryogenic inner wall supported directly with a double layer shell is composed of two layers of films made of the same material and two independent insulating layers. The inner wall is made of 0.7mm non-expansion steel(36% Ni alloy steel). The first carrier was supposed to come into use in October, 2007 while the second will be in 2008.

LNG Applications

1) LNG satellite station

LNG satellite stations, which receive LNG from Zhongyuan Oil Field and Xinjiang Guanghui LNG plant, have been built at Zhibo in Shandong province, Suzhou, and Jiangyan in Jiangsu province etc.

The Suzhou LNG satellite station has following functions: (1) LNG unloading and storage; (2) re-gasification, regulation, blending with water gas and coal gas, and transported to city pipelines; (3) LNG loading and transported to community gas stations and LNG vehicle fueling stations. So far the first two functions have been realized.

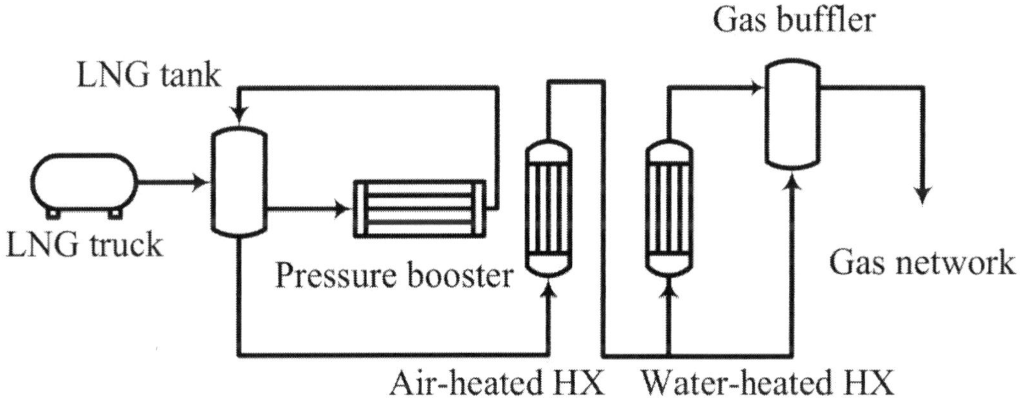

Fig.6. LNG Gasification Process

The station has a current gas supply ability of 20000 Nm³/d. The future gas supply ability will be 30000 Nm³/d and will cover the community gas and LNG vehicle refueling. The peak supply volume is 1200 Nm³/h.

LNG is re-gasified by the combination of both air and steam evaporators. The air evaporator mainly works in summer, and the steam evaporator mainly works in winter. Air heated evaporator is a fined tube heat exchanger and LNG flows in the tube. The process of LNG gasification is shown in Fig. 6.

The main facilities of the satellite station are as follows, (1) two cryogenic tanks, which is 100m³ each. The space between inner shell and outer shell is vacuumed and filled with pearlite, (2) two pressuring evaporators which help to pressurize and discharge fluids in the tank, the output gas rating is 150kg/h, (3) two air-heated evaporators and one steam-heated evaporator, each of which has a production rating of 600 Nm³/h, (4) A BOG heater which can produce 400 Nm³/h of natural gas.

2) LNG vehicle

To reduce the air pollution brought by waste emission from fuel cars, China is positively boosting the research of LNG vehicle. The Clean Vehicle plan initiated by the National Science and Technology Ministry put forward the demonstration project of LNG-powered buses. The project is supposed to enhance the application of LNG cars in China and to boost the development of domestic LNG vehicle technology.

Zhongyuan Green energy Science and Technology Company is researching into LNG-powered buses jointly with Beijing municipality. The results showed that LNG vehicle superiors over others in the following seven points: (1) the construction of LNG fueling stations is not affected by the natural gas system and easy for networked sites; (2) longer driving distance, each refueling can support 400-600 km's of driving which solved the problem of CNG vehicle; (3) LNG fueling stations require less land and large scale power equipments are not needed, noise level is low thus suitable for urban areas; (4) convenience of storage and transportation, economical; (5) easy loading and the same driving requirements with conventional vehicle; (6) LNG had been removed of the possible solid stuff produced during deep cryogenic process, therefore is more purified and environment-friendly; (7) LNG can be evaporated directly into CNG and used to refuel CNG vehicle.

Sponsored by the Shanghai Science and Technology department, a prototype bus powered by LNG has been made by Shanghai Jiaotong University and Shanghai Sunwin Bus Corporation. LNG bus is robust in power and conforms to the European III Emission Standards.

R&D of LNG-powered bus has also been started in Xinjiang Urumchi and Hunan Changsha.

Theoretic Research of LNG

Fundamental researches on the heat transfer and flow characteristics of LNG are of great significance to the developing of LNG plants and receiving terminals based on domestic technology.

Elementary research about LNG is as follows: (1) thermal property and transfer characteristic of natural gas and refrigerant, (2) equipment simulation of natural gas liquefaction process, (3) simulation and optimization of natural gas liquefaction process, (4) numerical simulation on LNG storage, stratification and rollover, (5) forecasting the instability of LNG storage.

During the last two decades, Shanghai Jiaotong University has carried out comprehensive research of LNG sponsored by National Natural Science funds, the National Education Ministry as well as by LNG projects in various enterprises. More than 80 papers have been published both home and broad. The research results also include the first monograph on LNG in China, and the journal papers has been accepted and published by leading LNG conferences in the world consecutively for five years.

3 Prospect

In China, LNG is still an emerging industry. There are big gaps with international LNG importers and exporters in many aspects of LNG production and application. There's still a long way to go before China can catch up. The following is a list of LNG techniques to be initiated by the industry.

Speed up R&D on Small-Scale LNG Plants Suitable for China

Small-scale LNG plants can be used to explore distant stranded gas reserves. Together with the non-pipeline transportation of natural gas, small-scale LNG plant breaks the technological barrier to tap into scattered and distant gas fields. In urban areas small-scale LNG plant can be used as a peak-shaving alternative. And in natural gas refueling stations it can be used to produce LNG and refuel LNG-powered vehicle. Therefore, the research into small-scale LNG plant is of great practical value.

At present many countries in the world like America and Finland are stepping up R&D into small-scale LNG plants, which is also very worthwhile for China to carry on developing. Shanghai Jiaotong University has already carried out R&D on small-scale LNG plants jointly with some companies in the world.

The Technology of LNG Cold Utilization

LNG is a mixture of cryogenic liquid (-162□) produced by cryogenic refrigeration and liquefaction process. It takes 850kWh of electricity on power plant and public facility to produce one ton of LNG; while at LNG receiving terminals, it needs to be evaporated through evaporators before put into use. During the evaporation process, it gives out large amount of cold (about 830KJ/Kg). Cold from LNG can be utilized through certain processes to save energy and enhance economic performance. There have already been wide and fundamental research into the application of LNG cold energy which has been used in power supply, food refrigeration, air liquefaction, dry ice production and cryogenic pulverization. These have provided enormous benefit to the economy and society.

During China's current effort to build LNG receiving terminals, the technology of LNG cold utilization should be applied.

Domestic Designing and Production of Equipment in LNG Industrial Chain

There are various equipment involved in LNG industrial chain, for example, compressor, expander, pump, heat-exchanger and valve etc. Advancing the domestic designing and production of such equipment is critical to further lower cost and enhance LNG technology in China.

The compressor is the key equipment for pressure boosting and transmitting gas used in small-scale LNG plant. It falls into three types: reciprocating cycle, centrifugal cycle and axial-flow compressor used in LNG plant should pay enough attention to the combustible character of gases. No leakage is allowed and anti-blast facility is required in electrical and driving parts. Compressors worked under deep cryogenic temperatures should also take into account the effect of coldness on the performance of component materials. In addition, the freezing of grease oil should not be ignored.

Heat-exchanger is an important part of LNG plant in the liquefaction process or in liquid-gas transfer. It also has many types and when used in LNG it takes the main form of winding pipe and fined-plate heat-exchangers which had been widely used in cryogenic liquefaction and air separation units.

In LNG system, there is another kind of heat-exchanger specially used in the transportation of liquid LNG into gas LNG. It varies in types with different work conditions, heating methods and the scale of gasification. For example, the heating method can be divided into air heating, sea water heating and combustion heating etc.

Beside pressure transportation, LNG transportation process must use LNG pump under the condition of mass transportation and high pipe pressure loss.

LNG pump not only requires to be used at cryogenic temperatures just as conventional pumps, but also requires high sealing performance and electrical safety guarantee. There are special requirements on the structure, material of the pumps to ensure the transportation of combustible cryogenic media. There's a kind of submergible electrical pump installed in the sealed tank which is widely used in LNG system. In addition, some conventional centrifugal pumps can also be applied in LNG transportation after improvement in sealing structure and material.

Expander is the key equipment to obtain low temperature in LNG plant. In general, turbo-expander is the most commonly used in LNG plants.

Attention must be paid to the piping parts like cryogenic valves, safety valves, automatic break valves and rupture disk etc. in order to ensure the reliability and safety of the system.

4 Conclusion

This paper presented an overview of LNG industry in China, covering LNG plants, receiving terminals, tank trucks and satellite stations.

LNG industry is relatively new in China. In light of the enormous LNG market and the future demand increase, China needs to make efforts on the following aspects in order to gain further progress.

First, holistic solution with Chinese characteristics is needed in LNG industry. To achieve overall development of LNG industry, it is important for China to learn from foreign countries while sticking to its own national conditions in the process of advancing and applying LNG technology. Chinese LNG industry needs balanced solutions and processes which cover LNG plants, LNG terminals, LNG transportation and satellite stations. Only when these critical industrial components advance proportionately can benefit be obtained in the whole industry.

Secondly, LNG regulations and standards should be established to better adapt to the global context. There's a set of international regulations regarding each aspect of LNG industry, esp. LNG plant, LNG receiving terminal and LNG transportation. On one hand, China should conform to these regulations in the exploration and utilization of LNG; on the other it is also very important for China to take the chance and establish our own LNG standards and regulations.

Lastly, global integration should be promoted. As a systematic and enormous industry, LNG development needs global sourcing and coordination. Regarding to each aspect of LNG industry, either it be technology, project construction or supply of end-product, large amount of international cooperation and sourcing is inevitable. Therefore, China should take more initiative in the international integration process and make procedures easier and more reliable for this end.

References

1. Gu AZ, Lu XS, Wang RS, et al. (2004), "*Liquefied Natural Gas Technology*", Beijing, China Machine Press.
2. Colleen Taylor Sen, (2002), "The World LNG Industry", *Hydrocarbon Engineering*
3. "*World LNG Source Book 2005*", (2005), Gas Technology Institute
4. Finn A J, Johnson G L, Tomlinson T R, (1999), "Development in natural gas liquefaction", *Hydrocarbon Processing*, No. 4, pp47-59
5. Johnson G L, Finn A J, Tomlinson T R, (1999), "Offshore & smaller-scale liquefiers", *LNG Journal*, No. 4, pp19-22
6. Terry Lavin, (1998), "Comparison of Liquefaction Process", *LNG Journal*, No. 4, pp28-33
7. Yoshitsugi Kikkawa, Moritaka Nakamura, (1997), "Development of Liquefaction Process for Natural Gas", *Journal of Chemical Engineering of Japan*, vol.30, No.4, pp626-630

Biography

Anzhong Gu

Institute of Refrigeration and Cryogenics
Shanghai Jiaotong University
1954 Huashan Road
Shanghai, 200030
China

Phone Number: +86-21-62932602
Fax Number: +86-21-62932602
Email: azgu@sjtu.edu.cn

Born 1939
B.Sc. 1962, Shanghai Jiaotong University, Marine Power Engineering
1993-current, Professor of Refrigeration and Cryogenics,
Shanghai Jiaotong University, Institute of Refrigeration and Cryogenics

Honors/Distinctions (Partial list):
Member of Commission A3, International Institute of Refrigeration (1992-1999)
Council Member, Chinese Association of Refrigeration (1996-current)

Principal Research, Education and Consulting Activities:
Liquefaction, separation, storage and transportation of gases; LNG technology; Enhanced heat transfer at cryogenic temperature; Adsorption of natural gas and hydrogen; Fuel cell; Refrigeration and air-conditioning.

Published about 150 refereed papers, chapters in books and books.

Integrated Approach for the Design of Refrigeration and Power Systems

Del Nogal, F.L.[1], Kim, J., Perry, S.J. and Smith, R.

Centre for Process Integration
School of Chemical Engineering and Analytical Science
The University of Manchester
Sackville Street, Manchester M60 1QD, U.K.

[1] Currently with M.W. Kellogg Limited
Kellogg Tower, Greenford Road
Greenford, Middlesex UB6 0JA, U.K.

AIChE Spring National Meeting 2006
6th Topical Conference on Natural Gas Utilization.
Orlando, Florida, April 23-27, 2006.

ABSTRACT

The use of vapour-compression refrigeration in low temperature processes creates significant demands for mechanical power. Therefore, the power system supplying this energy will have an impact on the performance and economics of these processes. This is particularly true for LNG.

This paper follows previous work on the optimal design of refrigeration systems (Del Nogal *et al.*, 2005a) and power systems (Del Nogal *et al.*, 2005b) individually. In the first case, the optimal compositions and operating conditions in mixed refrigerant cycles were obtained by using a genetic algorithm approach and a process model that considers multiple compression stages and full enforcement of temperature feasibility. In the latter, a mathematical programming formulation was presented for the optimal driver and power plant selection in power-demanding industrial processes with proper consideration of discrete gas turbine and power plant options as well as availability and energy efficiency implications.

As an effort to improve the overall design of low temperature processes, a systematic methodology for the integrated design of refrigeration and power systems was recently produced as an extension to the previous work above. Additional key degrees of freedom such as stage pressure ratios and plant capacity are optimised, alongside other design variables, in order to provide greater flexibility in the matching of power supply and demands. The optimisation framework is genetic algorithm-driven and features an inner optimisation for the design of the power system. This strategy is applied to an LNG case study with cascaded mixed refrigerant cycles and shows the convenience of this approach as the interactions between the refrigeration and power systems are systematically exploited.

INTRODUCTION

Power Systems

Designing a power system involves discrete decisions, not only because of the discrete options available for gas turbines and power plants, but also because of the discrete power demands and arrangement alternatives of the components in the system. Maximum utilisation of the installed capacity is a key issue as well. The start-up motors of gas turbines are often used as helper drivers to boost shaft power. The optimal power system must satisfy a set of mechanical, electrical and heating demands by employing the most cost-efficient number, type and model of mechanical drivers, compressor stage arrangement, helper motors or generators and power plants, either in simple or combined cycle. A proper optimisation model for this problem must allow for the flexible arrangement alternatives of driven equipment on different mechanical shafts in order to consider the different compressor casing alternatives. It must also be able to consider the use of helper motors or generators as balancing elements. The main power system interactions and measurable implications are illustrated in Figure 1.

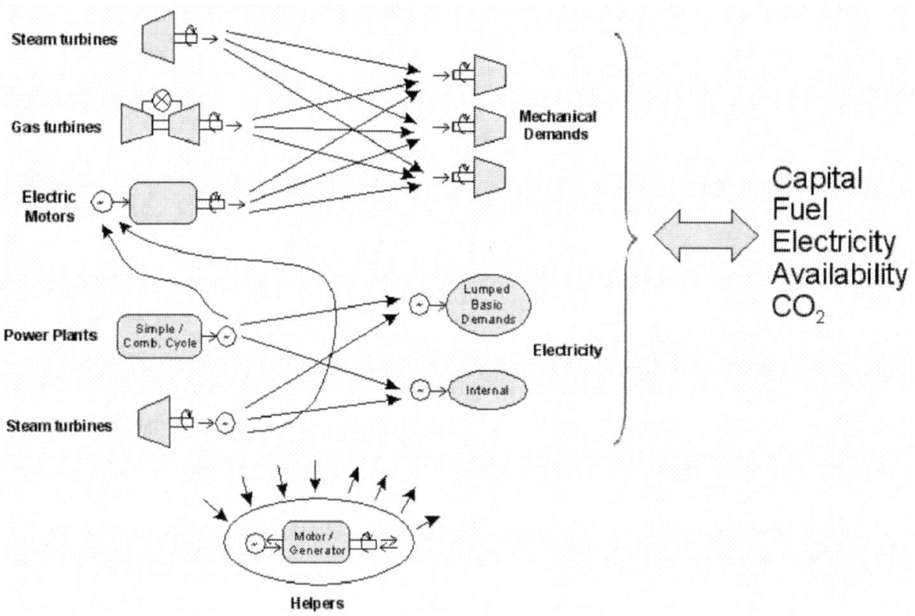

Figure 1. Power system interactions and implications.

A mathematical programming methodology for the optimal design of power systems has been previously published (Del Nogal *et al.*, 2005a), addressing the above considerations. The model is based on a superstructure approach and the designer is able to influence the final design by modifying the system constraints, equipment choices and economic parameters.

Figure 2 shows, in two parts, the superstructure on which the power system optimisation model is based. Figure 2a shows all the elements not requiring a customised steam system (steam may still be present in combined cycle power plants). That is, gas turbine drivers, electric motors, mechanical demands, power plants and the relevant electrical connections. Helper motors or generators are allowed on each gas turbine shaft. Each compressor stage is allowed to run on any of the gas turbines or motor shafts, hence allowing each compressor to be split into different casings. Figure 2b shows the steam-based elements and how they are linked to the non-steam superstructure through potential heat recovery steam generators at each gas turbine driver and simple cycle power plant. This multilevel steam superstructure includes steam generators, such as boilers and supplementary fired waste heat recovery units, which may produce steam at any one of the existing levels, extraction steam turbine drivers and generators, letdown valves, vents, deareator and connections to provide/receive steam to/from the process at each level. Heat may be provided to the process through steam or through an implicit hot oil system. This makes the overall superstructure flexible enough to represent systems from no steam-at-all to pure-steam ones. The optimisation model was applied to an LNG case study and proved to be appropriate, robust and flexible.

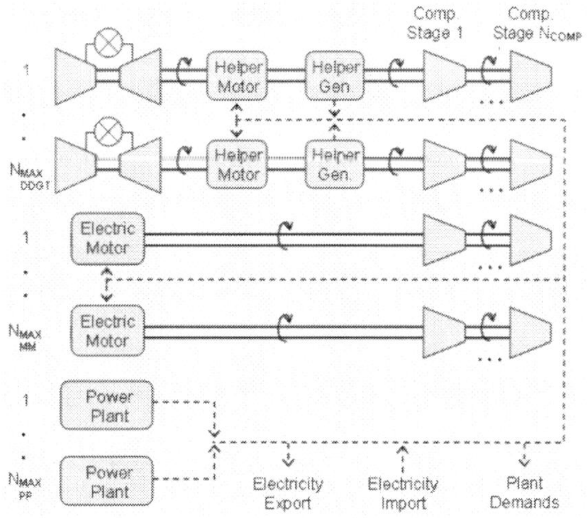

a) No-steam superstructure

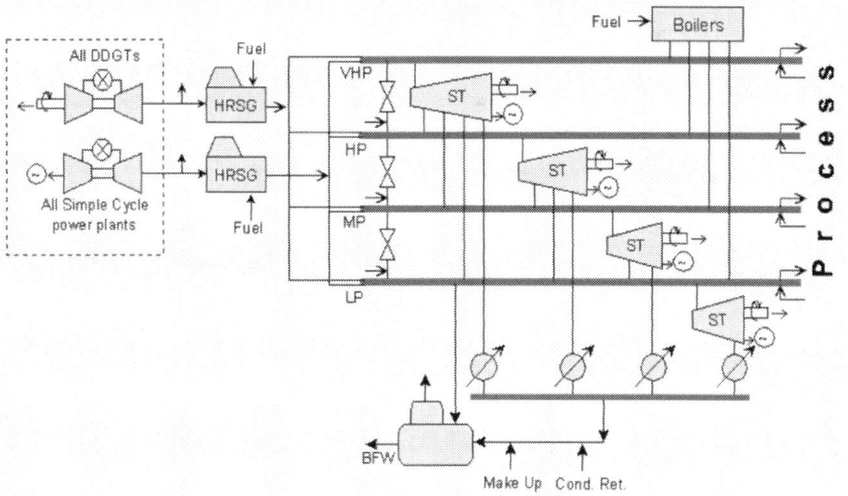

b) Steam superstructure.

Figure 2. Superstructures for the synthesis of power systems.

Mixed Refrigerant Systems

A major limitation of vapour compression cycles that make use of pure components as refrigerants is that refrigeration is provided at a constant temperature while the cold refrigerant is evaporating. For a chosen refrigerant, the temperature at which cooling is provided is a consequence of the evaporator pressure, i.e. the saturation temperature. The advantage of mixed refrigerants (MRs) over pure refrigerants is that they undergo isobaric phase change along a range of temperatures within the dew and bubble temperatures of the mixture. Given the correct refrigerant pressures, flowrate and composition, this would allow a good match between the process and refrigerant temperature profiles (an indicator of thermodynamic efficiency) with a simpler configuration than multilevel pure refrigerant cycles. Refrigerant composition becomes an important degree of freedom both at the design stage and during operation. The integrated design approach for refrigeration and power systems presented in this paper can be applied to any kind of refrigeration cycles (e.g. pure refrigerants, mixed refrigerants, expander cycles). However, the initial focus has been on mixed refrigerants.

An approach for the optimal design of mixed refrigerant cycles has been previously described (Del Nogal *et al.*, 2005b). It considers multistage refrigerant compression, full enforcement of the minimum temperature difference along the temperature profiles, simultaneous optimisation of variables, incorporation of capital costs in the objective function and the use of stochastic optimisation (genetic algorithm, or GA in short) to overcome local optima. The effectiveness of the method was illustrated by revisiting previously published LNG case studies, for which better and feasible solutions were produced. It has also proved the importance of considering multistage compression and capital costs during optimisation and that the application of genetic algorithms in these applications permits having greater confidence in the optimality of the results.

The base model by Del Nogal *et al.* (2005b) has been recently extended to support generic mixed refrigerant cycles in cascade. An example diagram is shown in Figure 3. The variables subject to optimisation include the refrigerant compositions and flowrates, the intermediate temperatures between cryogenic heat exchanger sections, the suction pressures, discharge pressures and stage compression ratios in the refrigerant compressors and the partition temperature (T_P, which defines the share of process heat duty for each cycle). Flowsheet options such as the number of cryogenic heat exchanger sections, or stages, for each cycle, whether the hot streams are precooled against the warmer cycle or not, subcooling or not subcooling the high pressure liquid refrigerant streams and the use of either liquid expanders or Joule-Thompson valves can be set by the designer prior to optimisation.

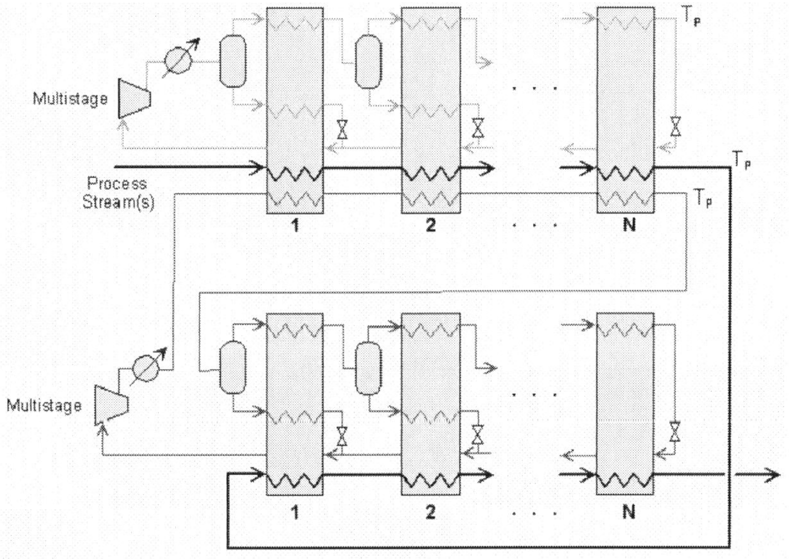

Figure 3. Two generic MR cycles in cascade.

The optimisation strategy and the flow of information within the framework are illustrated in Figure 4. The optimiser (GA in this case) proposes a series of candidate solutions and relies on the simulator for their assessment. How the optimisation task evolves is decided by the GA based on the 'fitness' of the candidate solutions as dictated by the objective function. The interactions between the GA and the simulator result in a set of the best solutions found over a discretised solution space. Standard Non-Linear Programming (NLP) optimisation(s) can be carried out afterwards having the best discretised solution(s) as initial guess in order to fine-tune and finally report the optimal solution on the basis of a continuous solution space.

AN INTEGRATED DESIGN APPROACH

Although the methods discussed above for the design of refrigeration systems and the design of power systems have been successful, they only focus on solving one part of a bigger problem at one time. For instance, the power system design approach departs from a given set of energy demands (heat, electricity and mechanical power) and assumes that the refrigeration system has already been designed, without challenging the convenience of its features. On the other hand, although the approach for optimising MR systems is able to consider some capital costs (including compressors), it does not foresee the economic and thermal performance implications that the resulting solution will have on the driver and power plant selection. See Figure 5.

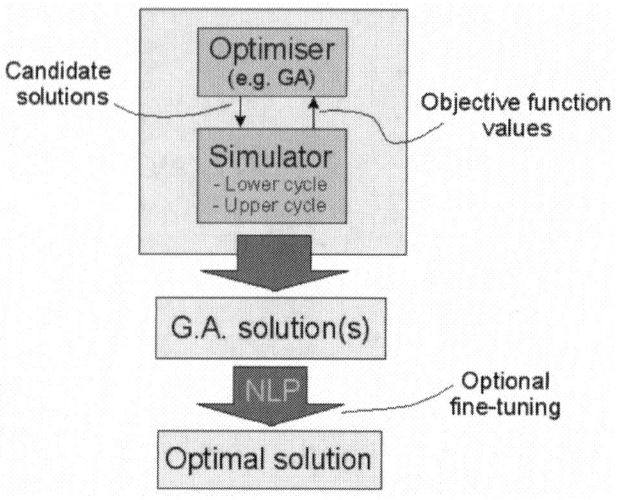

Figure 4. Optimisation strategy for refrigeration system design.

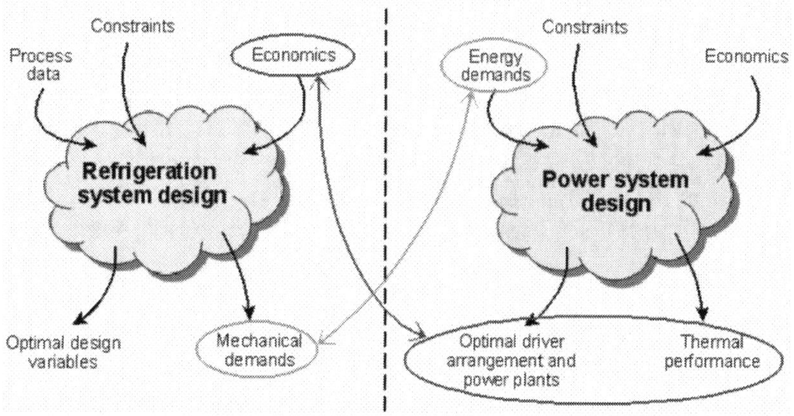

Figure 5. Interactions between refrigeration system design and power system design.

The interactions between refrigeration and power systems are strong and must be exploited in order to achieve a full and efficient capital utilisation. Usually these systems are designed in sequence; refrigeration system first and then the power system, in a sort of 'manual' iteration loop, until the designer is satisfied with the outcome. Such a non-systematic procedure is unlikely to produce truly optimal solutions (or near-optimal, for practical purposes), unless a very significant amount of effort and time are invested. Ideally, both systems should be optimised simultaneously as one single design problem.

However, these are systems of a very different mathematical nature. The refrigeration system design problem is a highly non-linear one (mainly due to physical properties) with relatively few variables, all of which are continuous. Conventional NLP methods are likely to get trapped in local optima in these circumstances and therefore a stochastic optimiser has been proposed (Del Nogal *et al.*, 2005a). On the other hand, in the synthesis of power systems the majority of the many variables subject to optimisation are binary, due to discrete equipment choices. Although the formulation for power system design is in principle Mixed Integer Non-Linear (MINLP), a Mixed Integer Linear formulation (MILP) can be derived from it without a significant loss of accuracy, thus improving the robustness of the solver employed and ensuring global optimality. If these two formulations (i.e. refrigeration and power systems) were merged in a single optimisation problem, the resulting formulation would be unavoidably highly non-linear and with many variables, most of them binary, which would seriously compromise the robustness of the solver and the quality of the solution.

Given the above considerations, Figure 6 illustrates the optimisation framework adopted in this work. A GA-based approach is still applicable. In this case, for each candidate solution the power system is optimised (in GAMS) after the refrigeration cycle(s) is(are) simulated (in FORTRAN). The results form both steps contribute to a unique objective function which is then evaluated and its value sent as feedback to the optimiser. Although not a simultaneous optimisation, this strategy still permits the master optimiser to realise and exploit the interactions between both subsystems. Furthermore, it permits solving each subproblem with the most appropriate formulation and optimiser, as well as keeping them in their original software platform, provided the feasibility of a two-way communication. An optional initial guess can be provided to the Master Optimiser to speed up convergence, although this is not strictly necessary. This initial guess can be obtained from standalone optimisation of the refrigeration system as in Figure 4 with an objective function set to either minimum power or minimum capital cost.

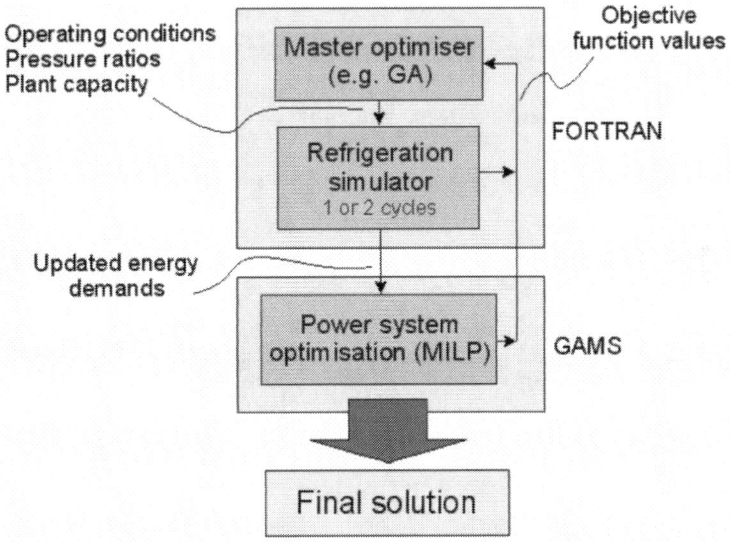

Figure 6. Integrated optimisation strategy.

The manipulation of key degrees of freedom such as compression ratios will ultimately facilitate a feasible match between a set of mechanical and electricity demands and a particularly attractive set of drivers and/or power plants. At this moment, it is also useful to introduce a new optimisation variable such as plant capacity. Allowing the plant capacity to vary, at least within a restricted space, would permit the master optimiser to ensure a full utilisation of the equipment capacity for a given machinery selection. If the objective function is targeting costs, there is now the need to ensure these are defined on a specific basis (i.e. cost per unit of plant capacity). As absolute costs are expected to increase at higher capacities, there would be an implicit bias towards lower capacities if costs were expressed on an absolute basis.

CASE STUDY

A natural gas stream is to be liquefied using two mixed refrigerant cycles in cascade to produce 6 MTPA of LNG. The temperature-enthalpy profile of the natural gas stream is given in Figure 7. 462.8 MW of combustion heat (LHV) are available from the end flash system. Any amount of fuel demand over this figure would be at the cost of 1 \$/MMBtu of LHV. The electricity demand is 39.9 MW excluding any helper motors. Power plants are subject to N+1 sparing philosophy.

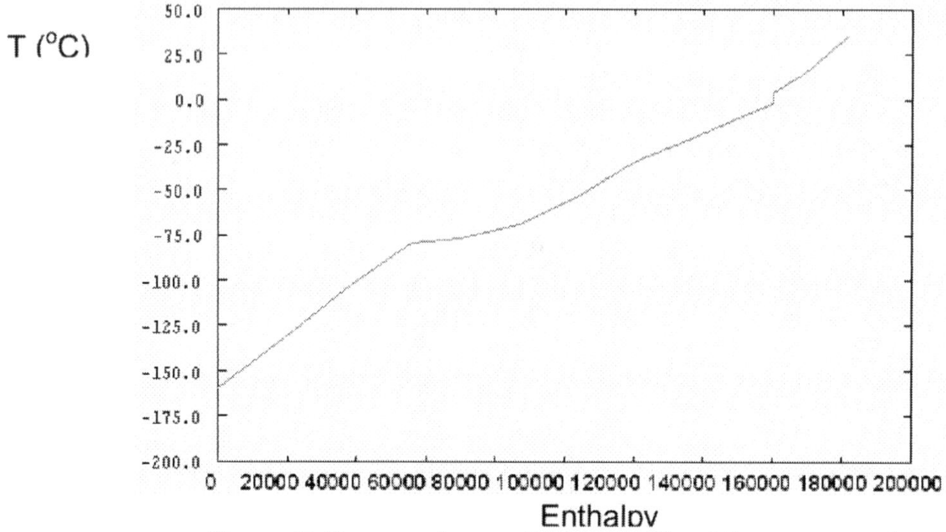

Figure 7. Temperature-enthalpy profile of the process stream.

An initial non-integrated design was obtained by optimising only the refrigeration system with minimum power as objective function and a ΔT_{min} of 3 °C, followed by synthesis of the power system. It has to be said that targeting capital plus fuel costs in the refrigeration optimisation step would not be appropriate as first, the actual fuel demand will not be known until the power system optimisation step and second, assuming any kind of trend regarding fuel consumption would be considerably inaccurate as this is subject to the decisions on driver and power plant selection. Also note that minimising power will have, in practice, the same effect as minimising capital cost as compressor costs are dominant in this subproblem.

The features of the refrigeration system with minimum compression power are summarised in Table 1. The total mechanical demand from the refrigerant compressors is 199.44 MW, or 33.24 MW per metric ton/yr of LNG, with eight compression stages in total, and the partition temperature is -56.3 °C. The power system, once optimised for this input data, makes use of four gas turbine drivers; three LM6000 and one frame 7 units, with a Trent gas turbine as simple cycle generator plus the respective spare unit. For this 'sequential', or non-integrated, design of refrigeration and power system, the total cost is 299.93 MM\$, or 49.99 \$ per metric ton of annual capacity, which includes the installed cost of major equipment (e.g. compressors and drivers, power plants, main heat exchangers, at 295.37 MM\$ or 49.23 \$/t·yr LNG) and the present value of the fuel costs along the plant lifetime (4.56 MM\$ or 0.76 \$/t·yr LNG).

In order to illustrate the influence of plant capacity and the discrete equipment options over the overall system design, a sensitivity analysis was carried out. All the variables in the refrigeration system were kept the same as in the previous solution and the plant capacity was varied between 4 and 7 MTPA. For each capacity, the optimal power system was obtained. The results are summarised in Figure 8. As a result of the economy of scale, the overall specific cost trend is to decrease as the plant capacity increases. However, there are numerous and sudden upsets in the trend due to the discrete decisions in equipment selection and configuration. These sudden changes sometimes pull in the same direction of the economy of scale and help in reducing the specific cost, but sometimes go in the other direction and are big enough to overcome the effect of the economy of scale at least locally. Compression ratios have a similar effect as plant capacity on the total specific cost. This is why these variables are key in achieving minimum total cost through a fine match between power demand and supply.

Table 1. Minimum power refrigeration system.

	Lower cycle	Upper cycle
Refrigerant flowrate (kmol/s)	11.90	18.70
Refrigerant composition (mole %)		
N_2	11.09	0.08
CH_4	40.08	9.78
C_2H_6	35.33	41.06
C_3H_8	13.03	11.80
n-C_4H_{10}	0.47	37.28
Compressor inlet pressure (bar)	1.99	4.00
Compressor outlet pressure (bar)	42.21	37.29
Compression ratios		
Stage 1	3.00	2.91
Stage 2	3.00	1.43
Stage 3	1.56	1.51
Stage 4	1.51	1.48
Compression power (kW)		
Stage 1	31204	58804
Stage 2	39850	15842
Stage 3	16398	13514
Stage 4	14435	9308

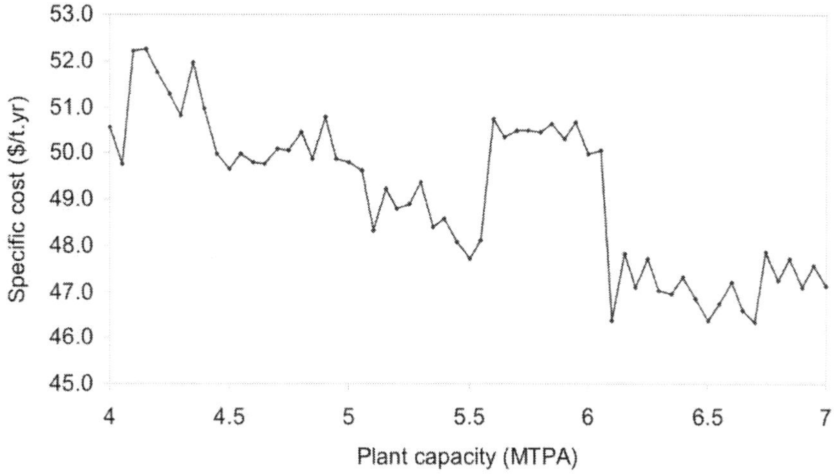

Figure 8. Sensitivity analysis results.

The integrated optimisation strategy in Figure 6 was applied in this case study, aiming for minimum specific cost. The plant capacity, within a range of ±10%, and the compression ratios were subject to optimisation. The other variables were kept the same as in the minimum power solution (i.e. refrigerant compositions, suction and discharge pressures and partition temperature), as they provide an acceptable match between the composite curves, except refrigerant flowrates, which were pro-rated according to the plant capacity. The resulting integrated solution has a capacity of 5.92 MTPA LNG, which is not too far from the original 6 MTPA. The optimisation framework allowed a reduction of the specific cost by 8.7% to $45.66 per metric ton of annual capacity. This integrated solution features four Trent gas turbine drivers and two frame 5 generators in simple cycle plus one spare unit. As expected, the surplus power is nearly zero, which indicates that the capacity of this power system configuration is being fully utilised.

Table 2 shows the compression ratios and compression powers of the integrated solution. The total compression power in this case is 197.62 MW. Note that, although the specific power has increased slightly to 33.41 MW per metric ton/yr of LNG, the savings resulting from a better match between energy demands and supply are large enough to overcome this. From Table 2, the main difference between the integrated and the non-integrated solutions is in the compression ratios of stages 1 and 2 of the upper (i.e. pre-cooling) MR cycle. With the integrated solution compression ratios, and the consequent stage powers, it is now possible to accommodate the first stage of the upper cycle compressor in one aeroderivative gas turbine driver, more efficient than the previous frame 7. Furthermore, all the compression stages can now be grouped in such a way that they are all feasibly driven by Trent gas turbine drivers, slightly more efficient than the LM6000 units in the previous solution. The specific fuel cost was reduced by 59.9% to 0.30 $/t·yr LNG and the specific capital cost by 7.9% to 45.36 $/t·yr LNG. This was possible thanks to the adaptation of the refrigeration systems, through the rearrangement of the energy demands, in order to exploit a more cost-effective driver and power plant configuration with minimum impact on the main process features.

Table 2. Integrated solution.

	Lower cycle	Upper cycle
Compression ratios		
Stage 1	2.99	2.47
Stage 2	2.96	1.68
Stage 3	1.59	1.53
Stage 4	1.50	1.47
Compression power (kW)		
Stage 1	30738	48884
Stage 2	38750	26366
Stage 3	16962	13281
Stage 4	14007	8548

This case study has centred in minimising costs. A further assessment would be required in order to ensure that the reduction in profit resulting from the decreased production capacity do not overcome the cost savings. Also, different solutions could be obtained depending on the driver and power plant options and the relative size of helper motors allowed by the designer.

CONCLUSIONS

Although there is clearly a good potential for producing integrated refrigeration and power systems in an LNG context, doing so proves difficult without a systematic approach. This paper presented an optimisation framework for the integrated design of refrigeration and power systems. The framework is based on a combination of stochastic and MILP optimisation methods and relies on recently developed approaches for the design of refrigeration and power systems individually. The robustness of the integration approach and the magnitude of the impact were proven in a case study. The systematic optimisation of key degrees of freedom, such as compression ratios and plant capacity, allowed a better exploitation of the interactions between both systems, achieving an 8.7% of reduction on the specific cost of fuel and major capital equipment.

Planned extensions to this work at the University of Manchester's Centre for Process Integration include the incorporation of pure refrigerant and expander cycles as well as the consideration of variable ambient conditions and their effect on equipment performance.

ACKNOWLEDGEMENTS

The main author would like to thank M.W. Kellogg Ltd. for their valuable technical advice and the Centre for Process Integration at the University of Manchester and the Overseas Research Student Award Scheme for the financial sponsorship of this project.

REFERENCES

Del Nogal, F.L., Kim, J., Perry, S.J. and Smith, R., Improved Design of Mixed Refrigerant Cycles Using Genetic Algorithm, 5th Topical Conference on Natural Gas Utilization, AIChE Spring National Meeting, Atlanta, 2005(a).

Del Nogal, F.L., Kim, J., Perry, S.J. and Smith, R., Systematic driver and power plant selection for power-demanding industrial processes, Energy Efficiency for a Sustainable Economy, AIChE Spring National Meeting, Atlanta, 2005(b).

THE SPEAKER

Frank Del Nogal obtained his first degree in Chemical Engineering from the *Universidad Simón Bolívar* in Caracas, Venezuela, in 1997. He performed as Technical Consultant on process simulation with Aspen Technology for nearly five years. In 2001 he joined the Centre for Process Integration at UMIST, UK, and obtained an MSc degree with distinction on Advanced Process Design. He recently completed his PhD research in the same institution, focusing on mixed refrigerant cycles and driver selection in an LNG context and winning the GPA Europe's Aungier Award in 2003 with an earlier publication. Frank joined M.W. Kellogg Ltd. at Greenford, UK, as a Senior Process Engineer in late 2005.

Frank Del Nogal
M.W. Kellogg Limited
Kellogg Tower, Greenford Road
Greenford, Middlesex UB6 0JA, U.K.
Tel. +44-(0)208-8727258
E-mail: frank.delnogal@mwkl.co.uk

Processes for High C2 Recovery from LNG -
Part I: Schemes based on Refluxed Demethanizer

Stanley Huang, Dennis Cook, Jame Yao, Douglas Elliot*

IPSI LLC
3000 Post Oak Boulevard
Houston, TX 77056
* To whom all correspondences be addressed

AIChE Spring Meeting, April 2006
6th Topical Conference on Natural Gas Utilization
Orlando, Florida, April 23-27, 2006

1. Introduction

Due to the declining domestic natural gas production, the LNG import to the U.S. is projected to play an important role in the overall natural gas balances. It is generally accepted that imported LNG will account for 10% of overall domestic gas consumption by 2010, and 16% by 2025 [Adair and Starr, 2004]. The re-gasified LNG must meet gas quality requirements and satisfy the gas interchangeability before it can be accepted in the U.S. pipeline grid. Toward this goal, many existing and prospective LNG terminal owners are considering C2+ extraction or nitrogen injection [Huang et al., 2004 and 2005].

The declining domestic natural gas production also implies decreasing NGL output. Currently, the U.S. is the largest NGL producer and accounts for about 30% of global production. Overall, the U.S. produces about 2 MMBPD of NGL, which is mostly used as gasoline blends and chemical feedstock [True, 2005]. Although increasing oil import to make up the anticipated NGL shortfall is viable, another route would be to extract maximum amounts of NGL from the imported rich LNG.

This work is a continuation to a previous article [Huang et al., 2004] in which schemes for extracting C2+ components from LNG were discussed. The previous article focused on lowering the heating values of the re-gasified LNG to meet pipeline specifications. The C2 recovery level was non-restrictive and may go below 50%. For example, some schemes were designed to by-pass half of the inlet LNG and use it to absorb the other processed half.

This work will focus on improving the C2 recovery levels. A minimum recovery of 95% is set as a criterion. In particular, the conventional approach of using refluxed demethanizer (DeC1) design is investigated. The idea is to utilize the refrigeration in the LNG to perform vapor condensation which, in turn, purifies the residue gas and prevents the C2+ components from leaving the column. The C2+ components are recovered from the bottom effluent. Conclusions of this work will serve as guidelines for future development.

Extracting C2+ components from natural gas is a well-established practice. Many innovative processes are described in the U.S. patents. Of particular importance is the expander-based technology [Lee et al., 1999]. In Part II of this work, the alternative of utilizing expander-based technology for recovering C2+ components will be explored.

2. Technical Background

Following the structure of the previous article, the C2 extraction schemes are classified into three categories: (A) Residue Compression, (B) Residue Compression and Condensing, and (C) Residue Condensing. Table 1 lists the categories and their respective representative figures.

The schemes presented in the figures are selected for further development into high C2 recovery applications. Several criteria were used to screen out less suitable ones in the previous article. Among other factors, the following two criteria are crucial:

1. The scheme should be able to handle the entire LNG inlet. The overall C2 recovery level is proportional to the facility capability. If a facility can only handle 50% of inlet LNG, its C2 recovery level cannot exceed 50%.
2. Separation by flash drums is undesirable because severe loss of C2 would result.

The DeC1 presented in the Figures 1, 2, and 3 are not rectified by any conditioned streams, i.e. streams lean in C2. Although the scheme in Figure 1 shows a reflux stream, it is a side-draw from the raw LNG inlet. Consequently, all three schemes are constrained by the thermodynamic phase equilibrium at the top tray and unable to achieve high C2 recoveries.

The upper limit of using an un-refluxed DeC1 is presented in Figure 4. The lower curve in the figure is generated by directing the entire LNG inlet to the top tray in Figure 1. Since this operation mode assures the lowest temperature at the top tray, the C2 recovery thus acquired is the maximum level achievable. When the column operating pressure decreases, the thermodynamic equilibrium favors the separation between methane and other components. Consequently, the C2 recovery can reach 97% if the column pressure is reduced to 17 bar. However, the compression duty becomes excessive at this pressure. In the previous article the DeC1 pressure was set at 35 bar and the maximum achievable C2 recovery was 90%.

The Category B scheme as depicted in Figure 2 is forgiving in DeC1 operating pressure because the compressor only provides a differential pressure to facilitate condensing. Compared with the lower curve of Figure 4, slight deterioration in performance is anticipated because the inlet LNG is heated to the two-phase region before entering the top tray. The maximum achievable C2 recovery is about 93% if the column is operating at or below 17 bar.

The Category C scheme as depicted in Figure 3 demands a minimum operating pressure of about 35 bar in order to fully condense the residue gas. Due to this relatively high operating pressure, the maximum achievable C2 recovery is about 86%.

3. Scope and Approach

This work will convert the DeC1 in each category to the refluxed design in order to overcome the aforementioned thermodynamic constraint. Here the term "reflux" specifically implies a conditioned stream which is lean in C2. The target recovery level is 95%.

4. Study Basis

Simulations in this work are performed using HYSYS and simulation parameters are listed in Table 2. The LNG inlet condition represents the holding mode (no ship-unloading) which is considered as the normal operation scenario. The flexibility of each category in handling ship-unloading mode will be commented briefly.

5. Results

The results are presented in subsections following the sequence of Categories A, B, and C. Initially, a restraint will be imposed that the conversion to a refluxed DeC1 should not include additional compressors. Adding compressors may change the basic features of a scheme. This
restraint will be lifted after the first three subsections. Four more schemes will be presented in the fourth subsection.

5.1 Category A: Residue Compression

Figure 5 depicts a refluxed DeC1 with residue compression. The residue from the top of column was partially condensed by the incoming LNG and the condensed liquid goes back to the column as reflux. The remaining vapor is re-compressed and delivered to the pipeline.

Since all the refrigeration in the incoming LNG is solely used to generate reflux, the achievable C2 recovery level can reach a level of 99.5% at a column operating pressure of 35 bar. The upper curve in Figure 4 shows the results. The required compression duty is mainly determined by the DeC1 operating pressure. The limitation for achieving even higher C2 recovery level is the temperature approach in the LNG Pre-heater/Condenser. Some enhancements based on the refluxed DeC1 were described elsewhere [Wilkinson and Hudson, 2004].

During ship unloading, the temperature of inlet LNG will rise after the absorption of additional boil-off-gas (BOG). The operation of the system will be insignificantly impacted because the majority of refrigeration in the LNG remains available for condenser operations.

Figure 6 presents a variation in this category. Instead of using a reflux drum and pumps (not shown in Figure 5) to complete the reflux loop, a side-draw from the re-compressor discharge is taken, chilled by the inlet LNG, followed by a pressure let-down, before entering DeC1 top tray. The working principles remain the same, with potential savings in leaving out the reflux drum.

5.2 Category B: Residue Compression and Condensing

Figure 7 shows the converted DeC1 based on the configuration of Figure 2. The residue gas from the top of column, after compression and condensing, goes into a reflux drum where it is split into two streams. One of the streams may be recycled directly to the DeC1 without pumps, depending on the boosted pressure. The other stream goes to the booster pump, vaporizer, and pipeline.

It was mentioned previously that Category B is forgiving with respect to the column operating pressure, because the compressor only provides a differential pressure to facilitate residue condensation. Also, the compression duty requirement is insensitive to the pressure levels. This observation still holds true here.

During ship unloading, the system shows good flexibility toward rising LNG temperatures. The compressor may be required to output more duty to help condensing the overhead residue. Alternatively, the product specification may be relaxed slightly (by reducing the reflux rate) in order to assure the total condensation of the overhead residue.

Figure 8 presents a variation in this category. Instead of using a reflux drum and possible pumps (not shown in Figure 7) to complete the reflux loop, a side-draw from the LNG Booster Pump discharge is taken, followed by a pressure let-down, before entering DeC1 top tray. The working principles remain the same, with potential savings in leaving out the reflux drum.

5.3 Category C: Residue Condensing

Figure 9 shows the converted scheme based on Figure 3. The basic requirement to fully condense the residue gas remains the same and there is a minimum operating pressure of about 35 bar. The small reflux stream and the elimination of subcooling for pumps slightly improve the C2 recovery level from 86% to 89%. Attempts to further increase the reflux rate for higher C2 recovery were met with incomplete condensing or a temperature cross-over in the LNG Preheater/ Condenser.

Raising the column operating pressure will allow fuller utilization of the refrigeration in the LNG. However, higher operating pressures also result in adverse thermodynamic equilibrium for separations at the top of the column. The net impact is poor C2 recovery.

Methods to enhance the recovery in Category C will be discussed in the next section, where additional compressors are allowed.

Category C schemes are relatively poor in absorbing the impact of ship-unloading operations. Since inlet LNG is the only refrigeration source in the scheme, once the ship-unloading BOG is absorbed, the raised LNG temperature would severely limit the flexibility of the scheme. Certain fraction of inlet LNG has to by-pass the DeC1 during ship-unloading operations in order to assure total condensation of the overhead residue.

5.4 Enhancements with Additional Compressors

If additional compressors are allowed for process enhancements, opportunities for novel schemes are wide open. Compressors are essential components of heat pumps, which are capable of moving heat from low to high temperatures [Elliott and Lira, 1998]. When the temperature distribution of system is altered by heat pumps, the C2 recovery level can be significantly impacted. Figure 10 shows an example of a heat pump. A recycling side-draw is taken from the overhead residue and compressed by a separate compressor. The compressor discharge is condensed by the incoming LNG before entering the DeC1 to form a lean reflux. The enhancement is a result of lean reflux and the chilling effect by moving heat from the top tray to a lower one. With the help of this recycle stream, C2 recovery to very high levels is achievable.

Since Category C can only utilize a small portion of the available LNG refrigeration for residue gas purification, the achievable C2 recovery level is only 89% even with the help of a refluxed DeC1. Figure 11 depicts a way to overcome this limitation. The reflux drum overhead is compressed by a new Overhead Recycle Compressor, and the discharge is fully condensed in an Overhead Recycle Chiller. The condensed, high-pressure residue is letdown in pressure before returning to the reflux drum. The returned stream is lean in C2 and the additional refrigeration by pressure let-down also helps the C2 recovery level to reach the 95% level. The enhancement also improves the process flexibility during ship-unloading operations.

Figure 12 depicts yet another way of enhancement. The column overhead is condensed by the incoming LNG to the maximum extent. Meanwhile the reflux ratio is also set as required to meet the C2 recovery specifications. As a result of this overload, incomplete condensation of the overhead vapor is anticipated. The non-condensed vapor is then compressed by a new compressor to the pipeline level and exported. This scheme provides excellent process flexibility to meet both C2 recovery specifications and ship-unloading operations. However, the dual provision of booster pumps and compressors will be relatively costly.

6. Cost Comparison

With the exception of the Section 5.4, the C2 recovery enhancements mainly include reflux drums and pumps. They will not result in major impacts on the cost structures as described in the previous article.

Figure 13 provides an indexed cost comparison for different categories of C_2+ extraction facilities. The abscissa is the total mechanical work, which includes both pump and compressor contributions. The cost estimate only includes key equipment items, as presented in respective figures of the schemes. The heat sources required to operate the DeC1 (furnaces and heating medium circuit) and C_2+ product handling facilities are excluded in the estimates.

The backbone of this figure (open diamonds) is the scheme presented in Figure 12, the DeC1 in which operates at 35 bar. Various levels of mechanical work represent different settings of the reflux ratios. The higher the reflux ratio, the more overhead vapor needs to be compressed. Therefore, the compression duty also increases. Meanwhile, higher reflux ratios imply improved C2 recovery levels, which range from 95 to 99.8% in Figure 12. Also presented in the figure are enhanced category schemes which are capable of achieving 95% recovery.

Figure 13 indicates that the backbone process (Figure 12) is relatively costly for the same mechanical duty requirement. The reason can be traced to the dual provision of pumps and compressors. At the top of all data points is the un-refluxed Category A design using low column operating pressure to achieve the target C2 recovery. This approach is unfavorable from cost perspective.

The total equipment cost can be correlated loosely by the total mechanical duty. As a rule of thumb, the compression duty should be minimized whenever possible. From this perspective, Category C has an advantage over the other two. The decision for the other two categories depends on economic justifications.

In the selection of optimal schemes, Life Cycle Cost analysis (LCC) is typically used for cost optimization in order to capture the impacts of both one-time capital expenditure (CAPEX) and longer-term operation expenditure (OPEX). In recent trends, increases in domestic natural gas cost have influenced the analysis results. Also, environmental regulations may significantly impact the plant design.

7. Conclusions

The conclusions of this paper are listed as follows:

- Process schemes for extracting C2+ components from LNG are classified into three categories: (A) Residue Compression, (B) Residue Compression and Condensing, and (C) Residue Condensing. Each category can be improved by converting the DeC1 to a refluxed column.
- For the first two categories, straightforward conversion is capable of achieving 95% in C2 recovery. The compressors already in the schemes provide process flexibility.
- Category C can only achieve 89% of C2 recovery by the straightforward conversion. Additional compressors must be added to achieve the target level of

95%. The compressors also improve the process flexibility during ship-unloading operations.

- Total equipment cost is roughly correlated with the total mechanical horsepower (pumping and compression). As a rule of thumb, the compression duty should be minimized.

8. References

Adair, L., Starr, S., "High NGL, natural gas prices make processing margins more volatile", Oil & Gas Journal, issue of September 27 (2004).

Elliott, J.R., Lira, C.T.; "Introductory Chemical Engineering Thermodynamics", Prentice Hall, NJ (1998).

Ertl, B., Durr, C., Coyle, D., Mohammed, I., Huang, S., "New LNG Receiving Terminal Concepts", paper presented in 1 , Johannesburg, South Africa (2005).

Huang, S., Coyle, D., Cho, J., and Durr, C., "Select the Optimal Extraction Method for LNG Regasification", Hydrocarbon Processing, Issue of July (2004).

Huang, S., Kotzot, H., De la Vega, F., and Durr, C., "Select the Optimal Scheme for N2 Injection in an LNG Terminal ", paper presented in 84th GPA Annual Convention, San Antonio, March 13 - 16 (2005).

Lee, R., Yao, J., Elliot, D., "Flexibility, efficiency to characterize gas-processing technologies", Oil&Gas Journal, Issue of December 13 (1999).

Markbreiter, S.J., Weiss, I., "Processing Liquefied Natural Gas to Deliver Methane-enriched Gas at High Pressure", U.S. Patent 3,837,172 (1974).

Marshall, Jr., W. H., "Processing Liquefied Natural Gas", U.S. Patent 2,952,984 (1960).

Rambo, C. I., Wilkinson, J. D., Hudson, H. M., W., "Liquefied Natural Gas Processing", U.S. Patent 5,114,451 (1992).

True, W., "Regional shifts in worldwide capacity, production slow", Oil & Gas Journal, Issue of June 27 (2005).

Yoshida, N., Yamaguchi, S., Ohara, S., Kumar, A., "Process and Apparatus for Separation of Hydrocarbons from Liquefied Natural Gas", U.S. Patent 2005/0218041 A1 (2005).

Wilkinson, J. D. and Hudson, H. M., "Liquefied Natural Gas Processing", International Publication Number WO 2005/035692 A2 by World Intellectual Property Organization, 2005.

9. List of Tables

Table 1 Basic schemes suitable for developing into high C2 recovery applications

Category	Description	Figure	Note
A	Residue Compression	1	The figure in the previous article is replaced by that of Markbreiter and Weiss (1974).
B	Residue Compression and Condensing	2	--
C	Residue Condensing	3	--

Table 2 Simulation parameters used in this work

LNG inlet	
Inlet temperature	-160 °C
Inlet pressure to C_2+ facilities	10 bar
Column operating pressure, unless specified otherwise	35 bar
Components	Mole Fraction
Nitrogen	0.00127
CO2	0.00008
Methane	0.90692
Ethane	0.08617
Propane	0.00552
i-Butane	0.00003
n-Butane	0.00001
i-Pentane	0.00000
TOTAL	1.00000
HHV, MJ/Sm3	40.3
Btu/SCF	1,082
Plant capacity, inlet LNG	100 MMscfd
Send-out gas pressure	100 bar
Number of trays in DeC1 column	10
C1 composition in DeC1 bottom product	1%
Pressure drop through plate-fin exchanger	0.2-0.3 bar
Pressure drop through air coolers	0.6 bar
Minimum delta T in plate-fin exchanger	2.5 ± 0.2 °C

9. List of Figures

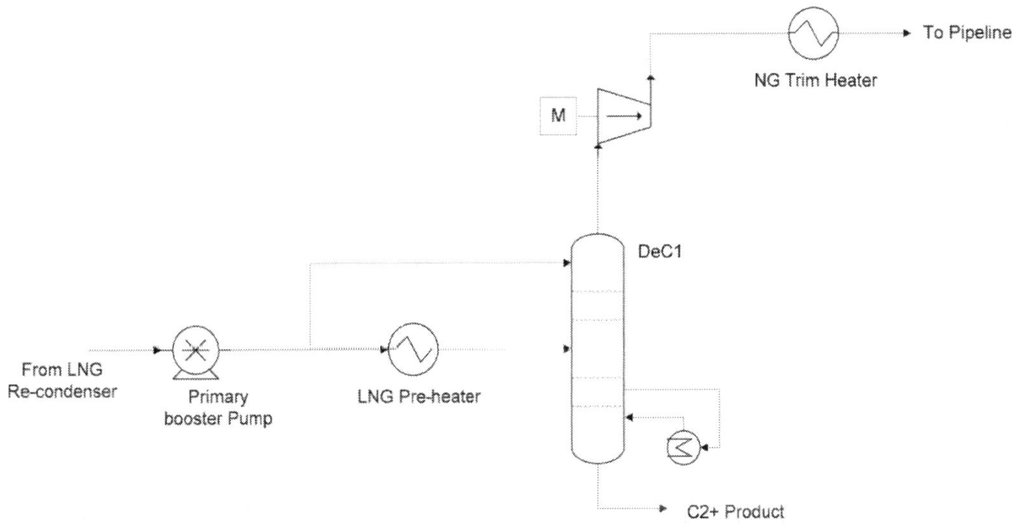

Figure 1 Residue Compression using raw LNG for rectification [Markbreiter & Weiss, 1974]

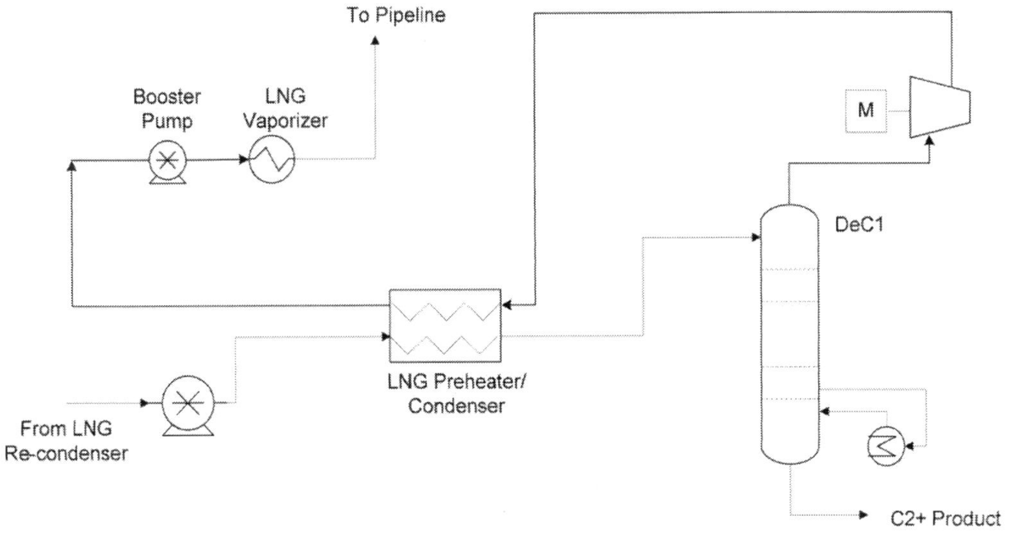

Figure 2 Residue Compression and Condensing with un-refluxed DeC1

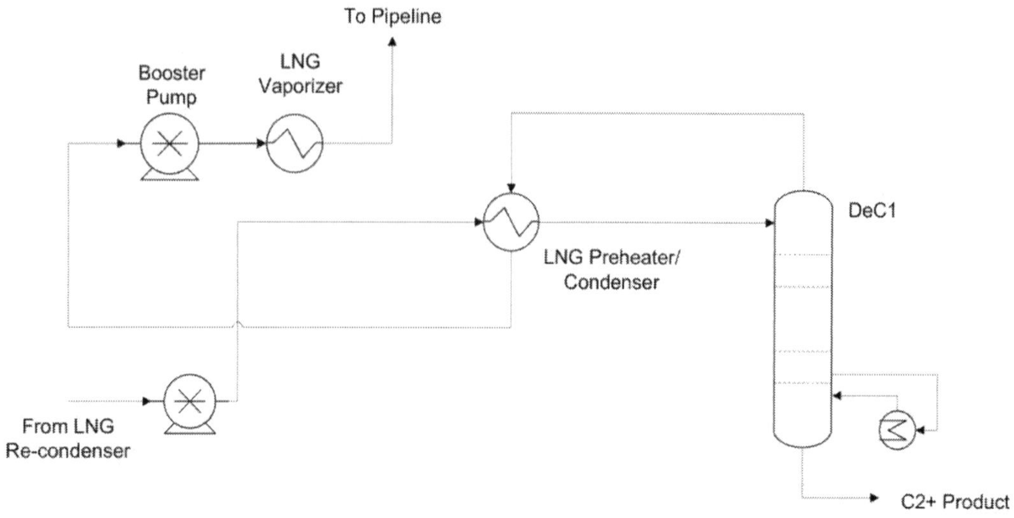

Figure 3 Residue Condensing with un-refluxed DeC1

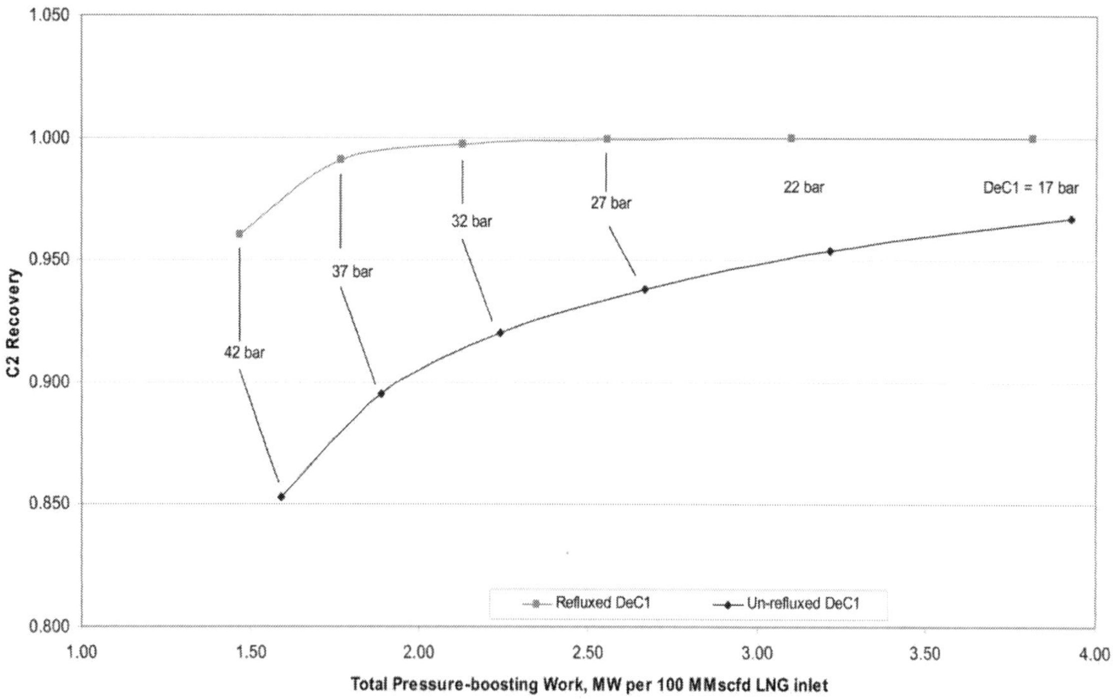

Figure 4 C2 recovery limits of refluxed and un-refluxed DeC1

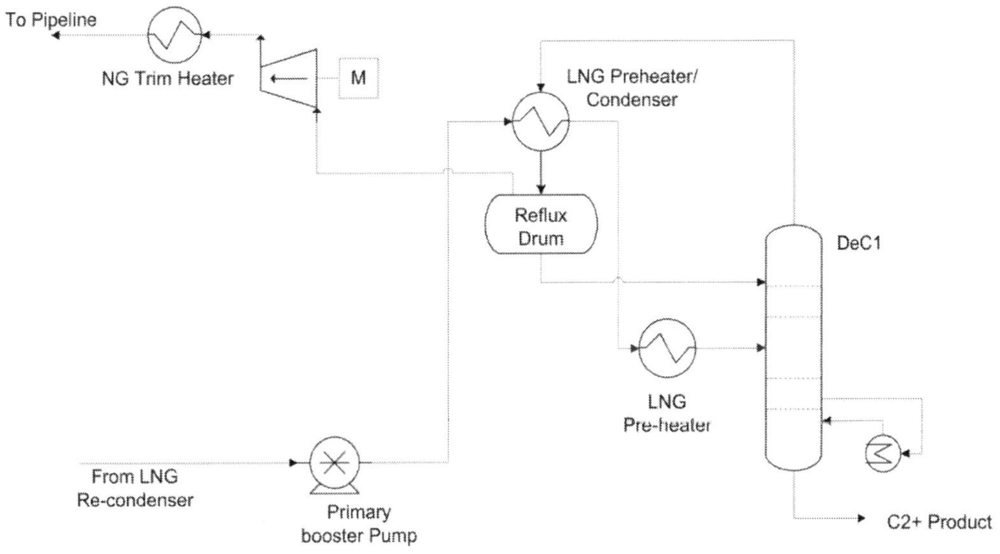

Figure 5 Residue Compression with refluxed DeC1 [Marshall, Jr., 1960]

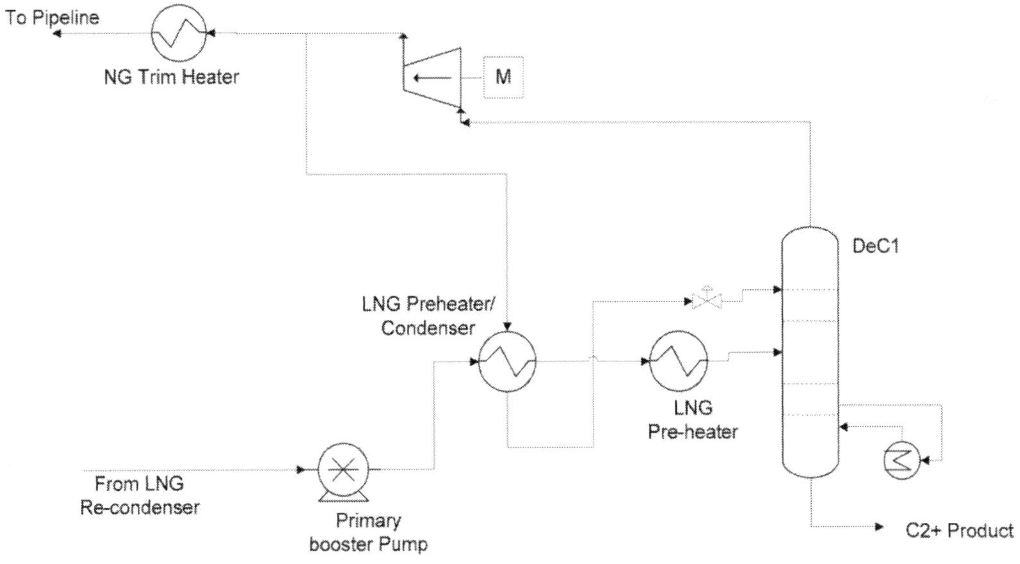

Figure 6 Residue Compression with reflux to DeC1 [Ertl et al., 2005]

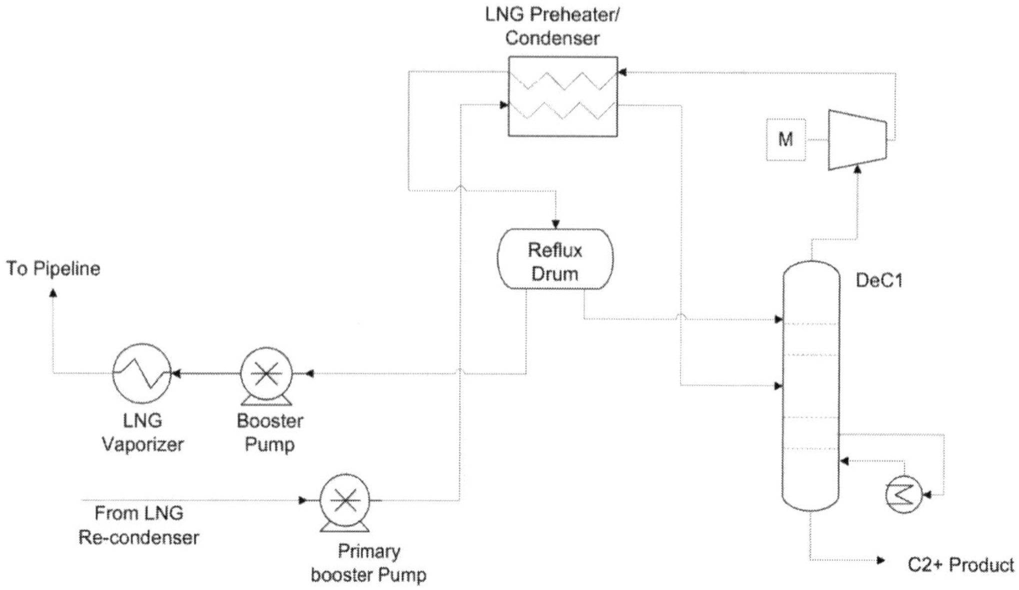

Figure 7 Residue Compression and Condensing with refluxed DeC1

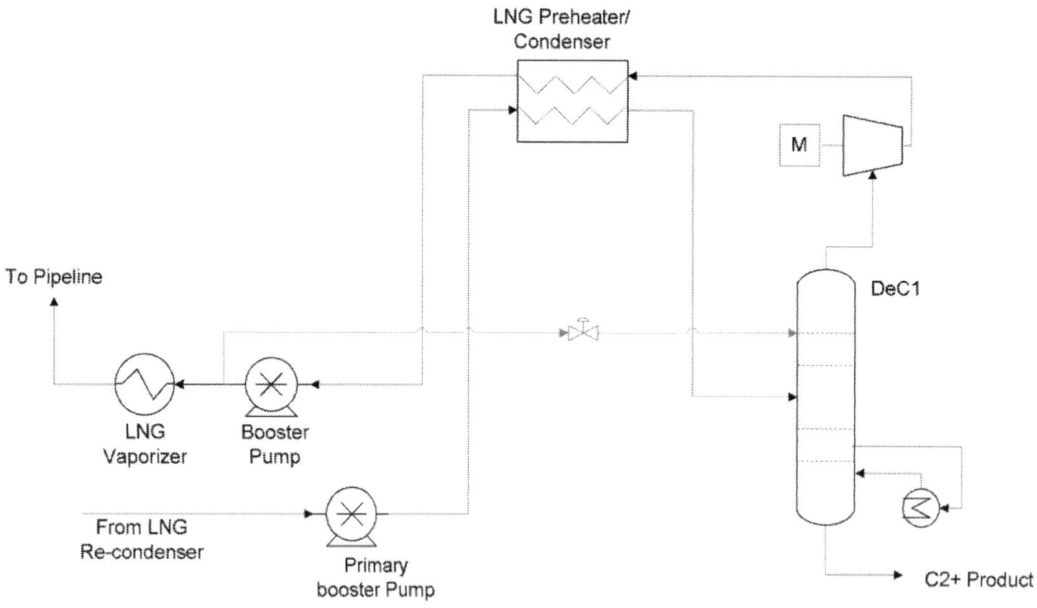

Figure 8 Residue Compression and Condensing with reflux to DeC1 [Ertl et al., 2005]

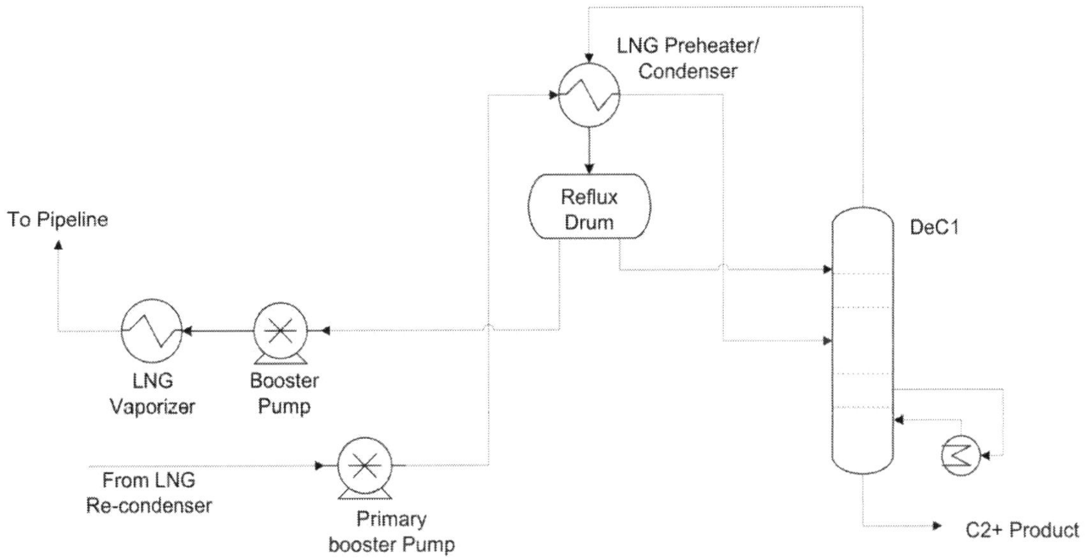

Figure 9 Residue Condensing with refluxed DeC1 [Ertl et al., 2005]

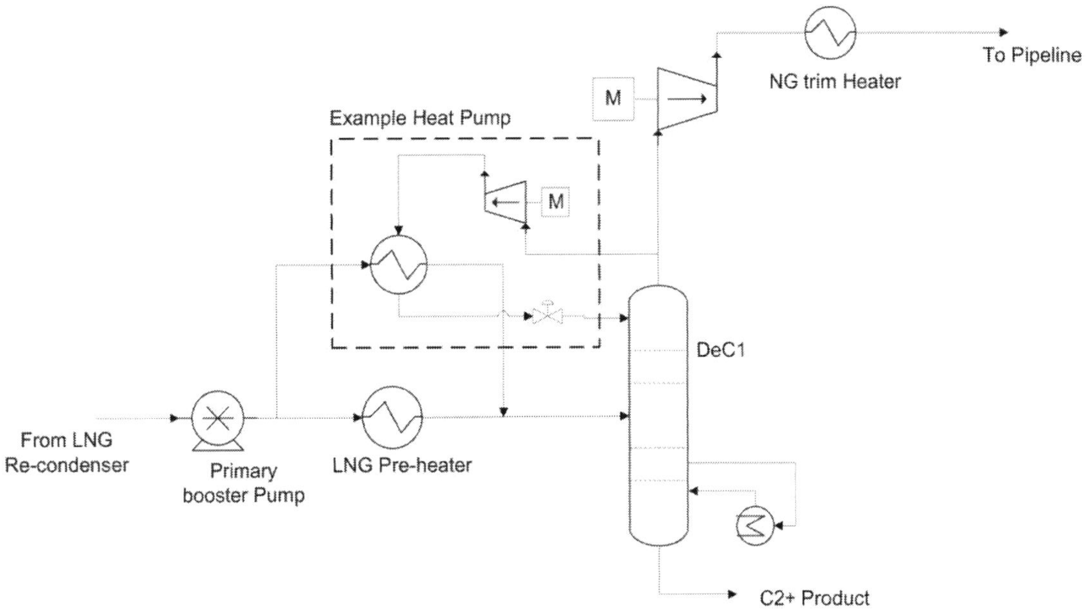

Figure 10 Residue Compression with enhancement, configuration 1 [Rambo et al., 1992]

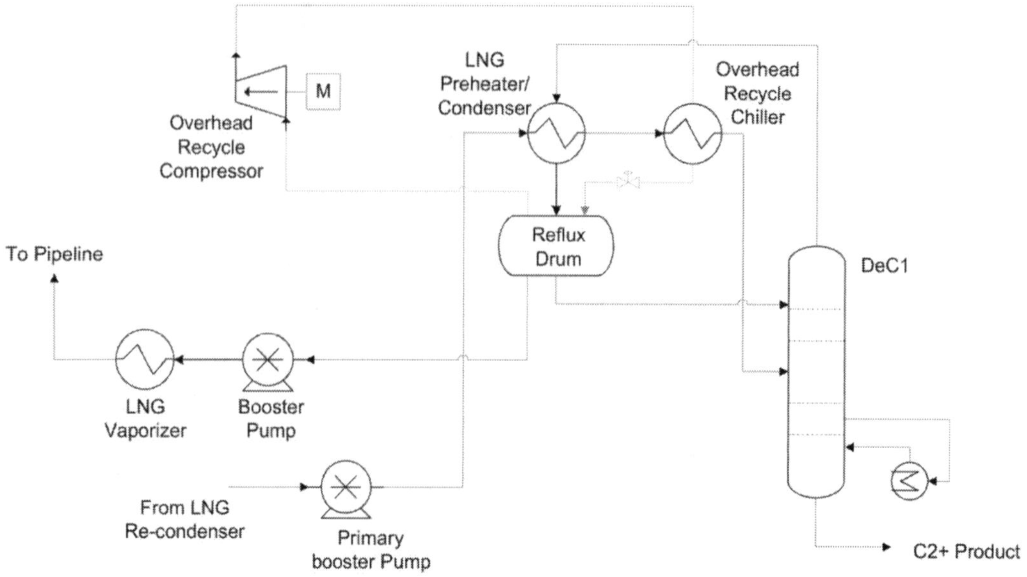

Figure 11 Residue Condensing with enhancement using compressors, configuration 1 [Ertl et al., 2005]

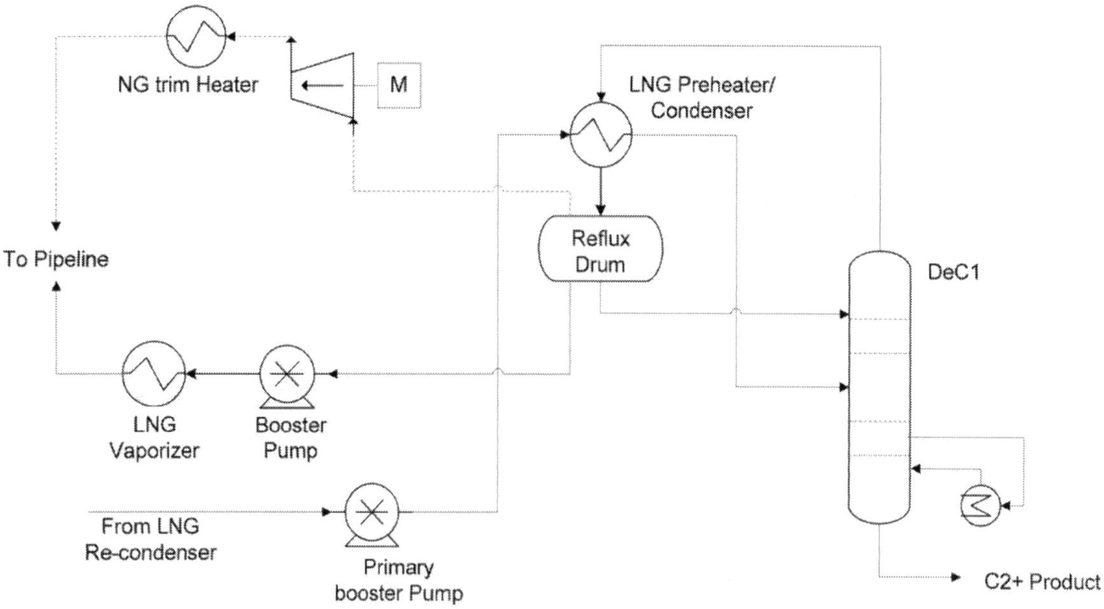

Figure 12 Residue Condensing with enhancement using compressors, configuration 2 [Yoshida et al., 2005]

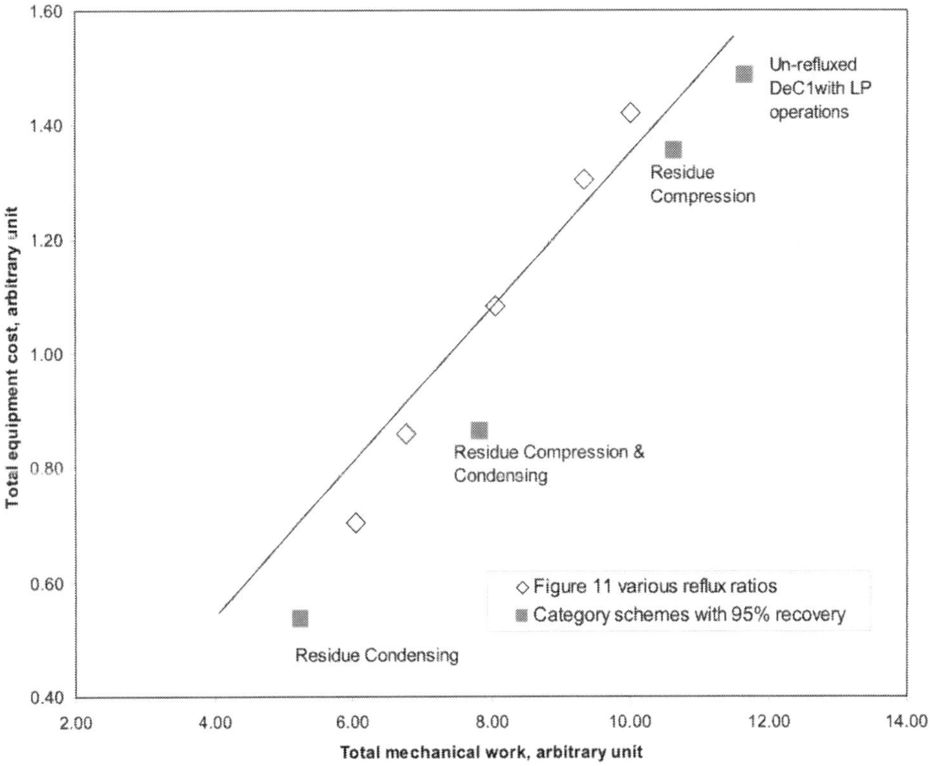

Figure 13 Cost Comparison for three categories

Biography

Stanley Huang is a Staff LNG Process Engineer in Chevron Energy Technology Company. He is a registered professional engineer and has eighteen years of experiences in the hydrocarbon industry. Stanley graduated with a Ph. D. degree in chemical engineering from Purdue University in 1986.

Validation of the Air Recirculation CFD Simulations on a Multi-Train LNG Plant

Dan Lin, Wing Kong Yee, Philip Diwakar and Vibhor Mehrotra
Bechtel Corporation, 3000 Post Oak Boulevard
Houston, TX 77056

AIChE Spring Meeting, April 2006
6th Topical Conference on Natural Gas Utilization
Orlando, Florida, April 23-27, 2006

Higher than ambient temperatures was measured at air cooler inlets in a LNG plant. The higher temperatures at the air cooler inlet can cause reductions in LNG production rate. Understanding the mechanism for the temperature rise is beneficial to better design the facility and understand the design margin.

The causes of the higher temperatures may be due to several factors: cross contamination from nearby heat sources, air cooler exhaust re-circulates to the inlet and/or other plant activities. The temperature rise also depends on wind condition. To study the effect of wind, recirculation and cross contamination, 25 temperature sensors were installed at various locations of the air cooler inlets. Measurement data together with weather data were collected every 15 minutes for a six month period. A CFD model was built to predict the air cooler inlet temperatures. The predictions were compared with measurements at the LNG plant. It was found that under wind directions parallel to the air cooler rack, CFD predictions agree reasonably well with measurement. Under cross wind directions, larger discrepancies can occur at some locations.

Reduction of Flare Loading during a Refrigerant Compressor Blocked Discharge in a LNG C3/MR process

Matthew J. Okasinski
Principal Engineer
Air Products and Chemicals, Allentown, PA, USA
Myrian A. Schenk
Principal Engineer
Air Products and Chemicals, Hersham, UK

AIChE Spring Meeting, April 2006
6th Topical Conference on Natural Gas Utilization
Orlando, Florida, April 23-27, 2006

Abstract

A refrigerant compressor blocked discharge scenario is used to size the flare systems for baseload LNG plants. This scenario is defined by closing a valve at the compressor discharge, causing the pressure in the discharge piping to rise to the pressure relief setting. At this point the pressure safety valves (PSVs) open and the refrigerant flows to the flare system. A peak relief flowrate and pressure are calculated as this event proceeds and the flowrate is used to determine the flare piping diameter, the height of the flare and its distance from the plant to protect workers from thermal radiation.

This scenario can be easily simulated in steady-state mode because the pressure relief setting is known by design constraints. However, dynamic simulation of the LNG refrigeration circuits has shown that it is possible to reduce the amount of refrigerant flared or eliminate it altogether. The key to this scenario is understanding what happens before the relief valves open and how the plant reacts to the pressure buildup. An interesting option is to bogdown the gas turbine and helper motor that drives the compressors and trip them on low speed before the PSV to flare reaches its set pressure.

The gas turbine's maximum available power can be reached faster by increasing the mass flowrate through the compressor stages, so that the compressors demand more power. The additional mass flowrate can be achieved by opening either the existing anti-surge control (ASC) valves or additional valves (e.g. a Recycle PSV), if they exist in the plant. When the compressor power demand reaches the gas turbine's maximum available power, the turbine reduces speed in order to maintain a power balance between itself and the compressors. In turn, this lower speed reduces the head achieved by the compressor and causes the pressure to rise at a slower rate. The speed reduction also reduces the maximum available power. So the gas turbine continues to reduce speed until it trips on low speed.

The gas turbine's trip could remove the refrigerant compressor blocked discharge case as the flare sizing case, allowing the flare and its header to be sized to a less severe scenario leading to important economic savings. This paper will examine this effect on a propane refrigeration circuit of a typical LNG C3/MR plant.

Background

As computing power has dramatically increased over the past two decades, an engineer's ability to solve large sets of interdependent differential algebraic equations (DAEs), as required for dynamic simulation, has also grown. (1) Specifically in the LNG industry, engineering companies have been using dynamic simulation in the various stages of an LNG plant's lifecycle. (2,3). For example, relief analyses and preliminary control studies can be done during the Engineering, Procurement and Construction (EPC) stage, operating training during the subsequent startup / commissioning stage and finally during the operation stage, process enhancements can be evaluated before being implemented in the plant.

Dynamic simulation can also give insights into the liquefaction process which are not readily seen when designing from a steady-state perspective. The accumulation and depletion of mass and energy in different sections of the flowsheet will affect how the overall closed loop refrigeration system responds to a disturbance. For example, while it is possible to obtain conservative relief flow estimates using steady-state assumptions, the time-dependent interaction between the compressors, gas turbine and other process equipment can only be studied using dynamic simulation. Through dynamic simulation of a compressor blocked discharge scenario, it was found that flare loadings could be reduced or eliminated through the use of recycle flows around compressor string to increase power demand and proper PSV set pressure selection to provide time to bogdown the gas turbine. The reduction in flare loading could lead to economic savings in the flare system.

Basis and Assumptions

The flowsheet shown in Figure 1 is the propane (C3) circuit of a C3/mixed refrigerant (MR) natural gas liquefaction process. Beginning the cyclical process at the compressor suction, propane vapor from the evaporators is compressed through one or more stages. The resulting high pressure vapor is then cooled and condensed through a series of heat exchangers before being collected in an accumulator. It is then sent through a series of evaporators operating at descending pressures. These evaporators use propane to cool various process streams in the liquefaction plant such as the warm natural gas feed to the main cryogenic heat exchanger (MCHE) and also the mixed refrigerant circulating through the MCHE. The vapor boil-offs from these evaporators are collected in the appropriate compressor stage suction drum to complete the C3 refrigeration circuit.

The flowsheet was built using APCI's proprietary dynamic models and run in Aspen Custom Modeler (ACM) 12.1.5 using APCI's proprietary thermodynamic database. Each train of evaporators are connected both on the process and propane sides so an upset is realistically propagated through the train. Relief valves are located at the discharge of the HHP compressor (K0605) and at the HHP suction drum (V0610). The compressor blocked discharge is initiated by ramping closed the block valve between the propane desuperheater and condenser in 30 seconds. The process flows to the evaporators are assumed to continue at their steady-state values during the disturbance.

A process design heat and material balance is used to initialize the simulation. Specific plant information, such as piping, equipment and vessel volumes, compressor curves, and evaporator UAs are obtained from the engineering design company.

Results and Discussion

The cases discussed below illustrate how a gas turbine can be bogged down by the combination of recycle flows and appropriate set pressure selection on the relief to flare PSVs.

The dynamic simulation starts at steady-state and the disturbance begins at time = 0.5 hours on the graphs. During the first few minutes of the disturbance, the two simulations are identical. As soon as the block valve between the desuperheater (E0608) and the condenser (E0609) begins closing, the pressure at the HHP discharge of the propane compressor begins rising. This moves the HHP compressor stage towards surge and causes the stage's anti-surge controller to open its valve. This increases the HHP suction drum pressure and moves the HP stage towards surge. This effect cascades through the MP and LP stages. As the block valve continues to close, all the anti-surge controllers open their valves to keep the compressor stages away from surge. The rise in the suction drum pressures reduces or even suppresses the propane boil-off flow from the evaporators within the first minute of the disturbance. Figure 2 shows a representative graph of the evaporator flowrates – notice the sharp drop in vapor flows into the suction drum from the evaporators. The HHP evaporator boil-off flows are reduced or suppressed first, quickly followed by the HP, MP and LP evaporators. So the dynamic simulation reveals that the compressors are isolated from the evaporators and completely dependent on their recycle flows within the first minute. However as the disturbance continues, some of the evaporators may resume their boil-off flows if the rate of evaporator pressurization due to continued process heat input overcomes the rate of suction drum pressurization due to opening the compressor anti-surge valves.

The dynamic simulation of the two cases presented in this paper deviate when the HHP Suction PSV set pressure is reached in Case 1. The description of each case from that point on is discussed below.

Case 1:

At 4.8 minutes, the HHP suction pressure exceeds the HHP Suction PSV set pressure and the relief valve opens as seen in Figure 3. When the PSV opens, it depressures the suction drums and allows the evaporators to resume their boil-off flows. The resuming boil-off flows add mass to the compressor loop faster than it can be relieved so the pressure continues to build. Note that Figure 4 shows the compressor power demand is climbing during this time.

At 9.3 minutes, the HHP discharge pressure exceeds the HHP Discharge PSV set pressure and that relief valve also opens, sending additional mass to the flare system. As shown in Figure 4, this event creates a maximum in the compressor power demand at 95% of the gas turbine's maximum available power. The opening of the Discharge PSV also causes maximums in the Suction PSV flowrate at 11% of the steady-state circulating flowrate and in the Suction PSV pressure at 15% above the set pressure as seen in Figure 5.

At 11.2 minutes, the peak total relief flowrate of 98% of the steady-state circulating flowrate is reached along with a maximum in the HHP Discharge pressure. At 11.7 minutes, the simulation is stopped because the propane accumulator is empty.

Case 2:

For this case, the HHP Suction PSV and Discharge PSV set pressures have been raised to 267% and 182% of their respective steady state pressures. So at 4.8 minute mark, the HHP Suction flare PSV has not opened yet. Pressures and temperatures continue to build around the compressor string and the compressor power demand keeps increasing.

At 9.6 minutes, the compressor power demand reaches the gas turbine's maximum available power and the compressors begin to bogdown as seen in Figure 6. In response to reaching maximum power, the gas turbine reduces speed in order to maintain the power balance between itself and the compressor string. In turn, this lower speed reduces the head achieved by the compressor and the HHP stage's discharge pressure rises at a slower rate as seen in Figure 7. The speed reduction also reduces the maximum available power, so the gas turbine continues to reduce speed.

At 14.6 minutes, the compressors trip on low speed (2877 rpm) and the simulation is stopped. Neither relief to flare PSV has opened so no mass was relieved during the event. With the higher set pressures, the HHP Suction PSV does not open, which would reduce the power demand by removing mass from the closed system. So the compressor power demand has time to reach the maximum available power of the gas turbine.

To help accelerate the onset of bogdown, a PSV valve can be placed in parallel with the ASC valve. These "Recycle" PSVs would have a set pressure lower than the relief to flare PSVs but sufficiently high enough to avoid inadvertent opening during normal plant operations. The sole purpose of the Recycle PSVs is to increase the mass flowrate through the compressors (move the compressors away from surge towards stonewall) so the compressors demand more power faster from the gas turbine. (4)

Conclusions

Dynamic simulation shows that it is possible to reduce flare loading during a compressor blocked discharge scenario. The reduction is achieved by consuming all available power from the gas turbine using recycle flows around the compressor string combined with appropriate relief to flare PSV set pressures.

Bogging down the gas turbine during a compressor blocked discharge event can reduce or eliminate the relief of refrigerant to the environment, provide an economic savings during the design and construction phase and provide a faster restart time as refrigerant inventory does not need to be replenished.

References

1 Mandler, J., 'Modeling for control analysis and design in complex industrial separation and liquefaction processes' *Journal of Process Control* **10**:167-175.
2 Cameron, D. et al., 'Integrated Distributed Dynamic Simulation of the LNG Value Chain' *Chemical Engineering Research and Design* **83**(A6): 686-691.
3 Valappil J. et al., 'Virtual Simulation of LNG Plant' *LNG Journal* January/February 2004.
4 Lucas, C. et al., 'Utilization of bogdown in single-shaft gas turbines in minimize relief flows in baseload LNG plants' *US Patent Application US20050022552 A1.*

Biography of Speaker

Dr. Matthew J. Okasinski joined the LNG group at Air Products in 2002 to execute dynamic simulation projects for external clients and work with Dr. Myrian Schenk on development of dynamic simulation capabilities. He received a BS in Chemical Engineering from Cornell University and received his doctoral degree in Chemical Engineering from the University of Massachusetts – Amherst in 1998. He is a Registered Professional Engineer in the State of Pennsylvania.

Figure 1 - Propane Refrigeration Loop

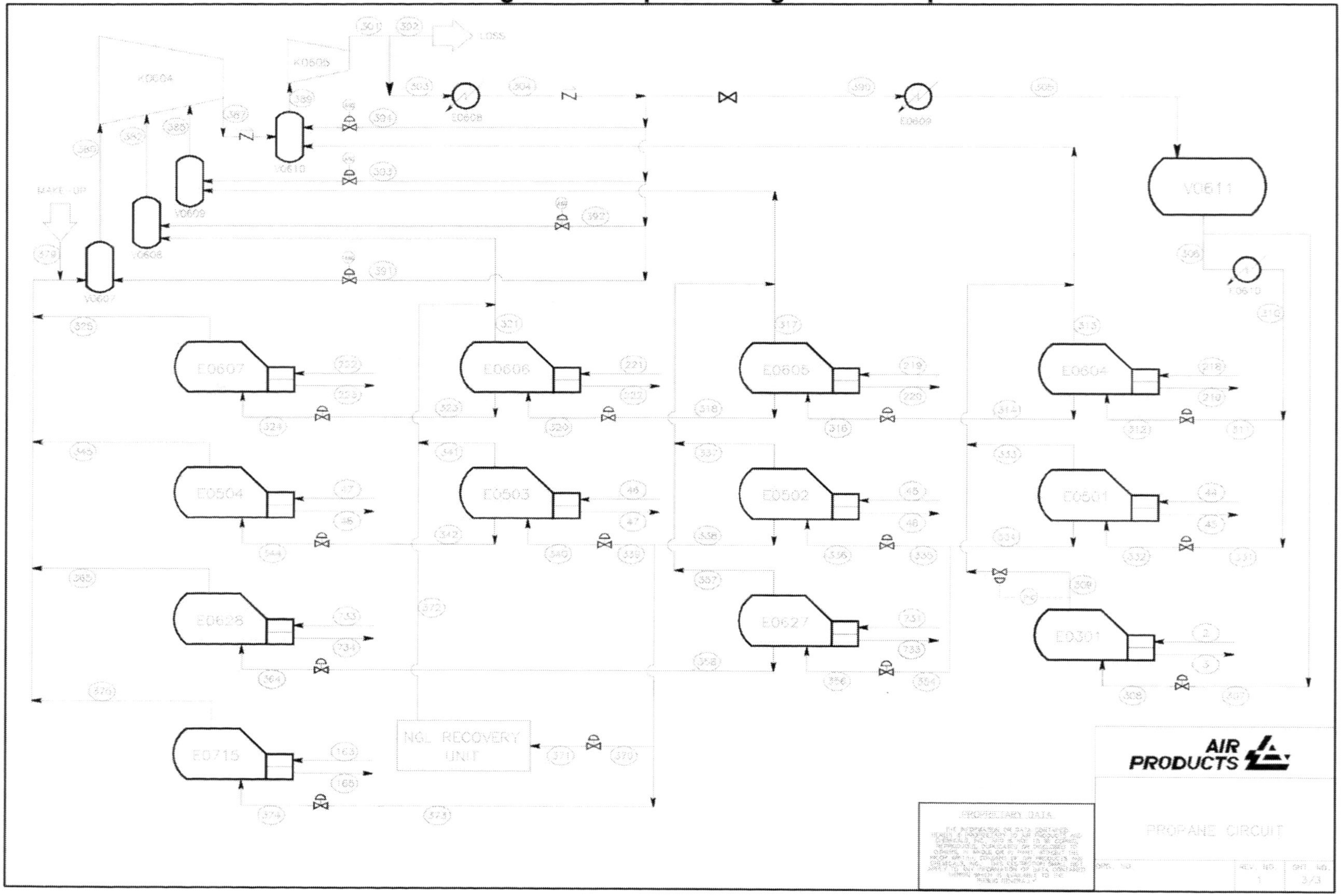

Figure 2 - Case 1 MP Evaporator Boil-off Flows

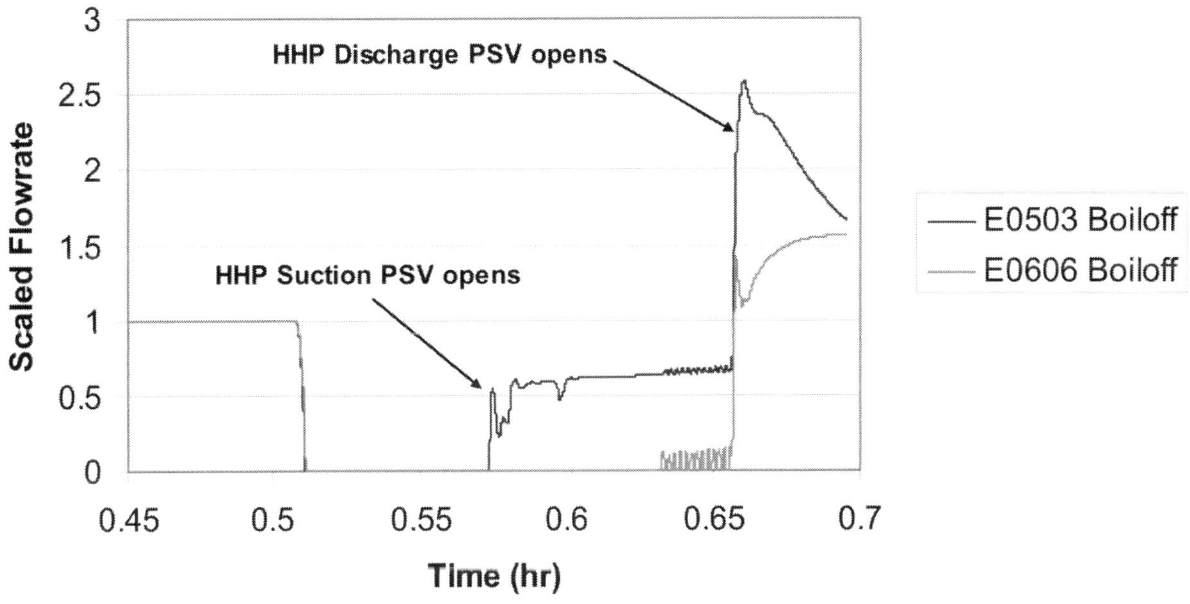

Figure 3 - Case 1 HHP Suction PSV

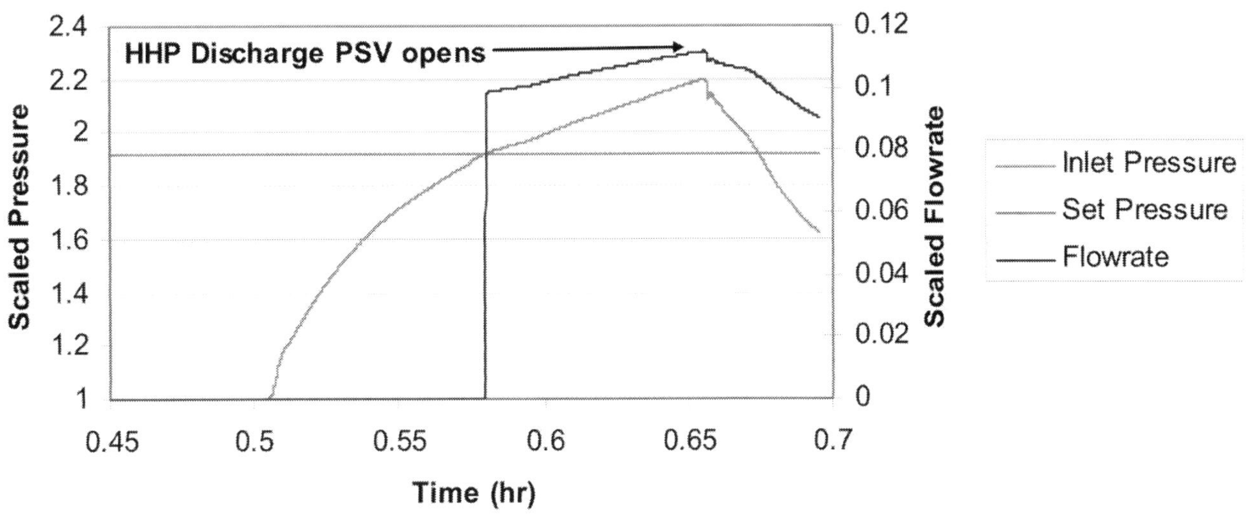

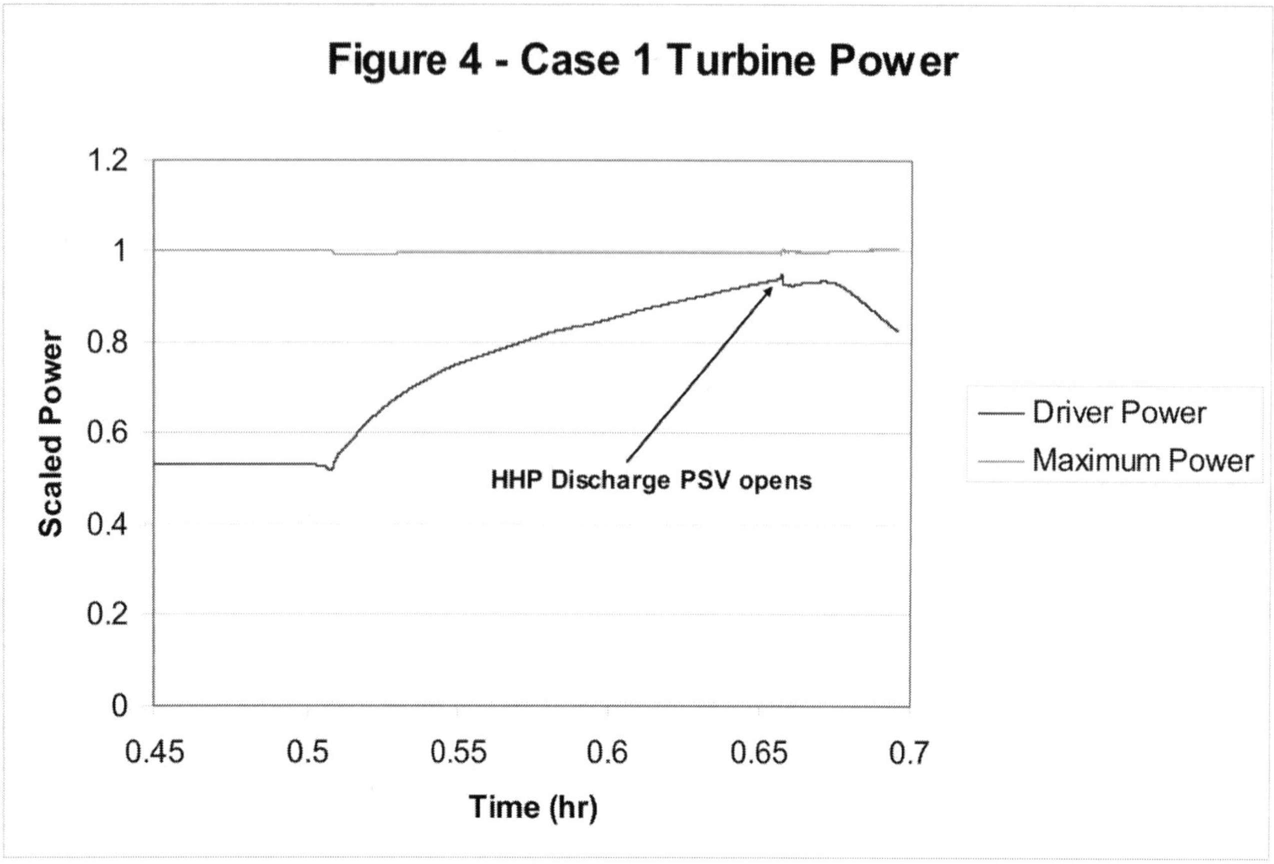
Figure 4 - Case 1 Turbine Power
1.2
1
0.8
0.6
0.4
0.2
0
Scaled Power
HHP Discharge PSV opens
0.45
0.5
0.55
0.6
0.65
0.7
Time (hr)
Driver Power
Maximum Power

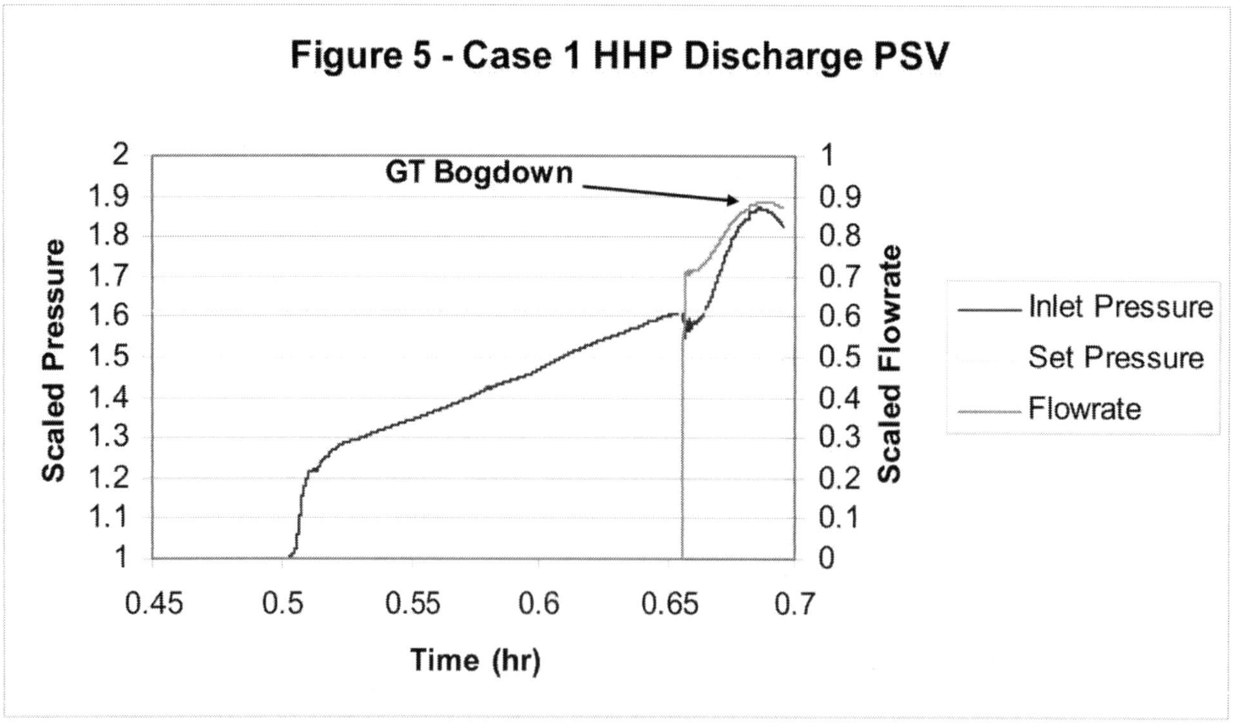
Figure 5 - Case 1 HHP Discharge PSV
GT Bogdown
2
1.9
1.8
1.7
1.6
1.5
1.4
1.3
1.2
1.1
1
Scaled Pressure
1
0.9
0.8
0.7
0.6
0.5
0.4
0.3
0.2
0.1
0
Scaled Flowrate
0.45
0.5
0.55
0.6
0.65
0.7
Time (hr)
Inlet Pressure
Set Pressure
Flowrate

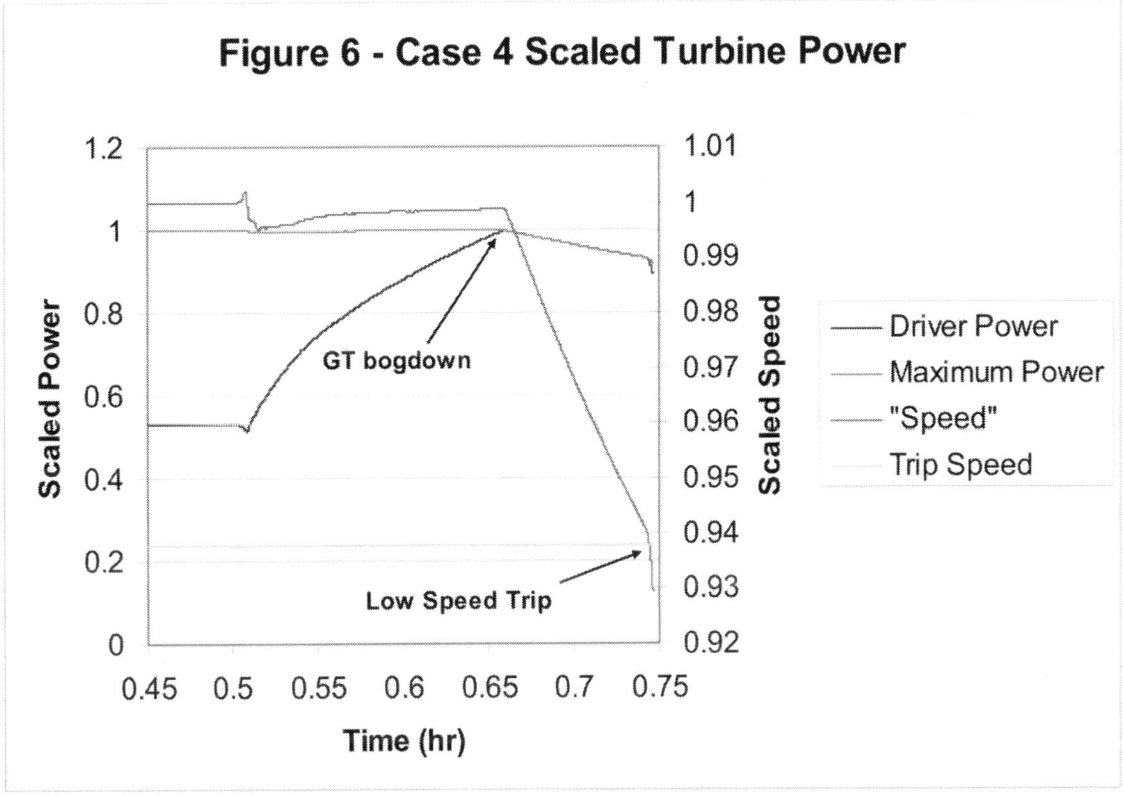
Figure 6 - Case 4 Scaled Turbine Power
Scaled Power
Scaled Speed
Time (hr)
GT bogdown
Low Speed Trip
Driver Power
Maximum Power
"Speed"
Trip Speed
1.2
1
0.8
0.6
0.4
0.2
0
1.01
1
0.99
0.98
0.97
0.96
0.95
0.94
0.93
0.92
0.45
0.5
0.55
0.6
0.65
0.7
0.75

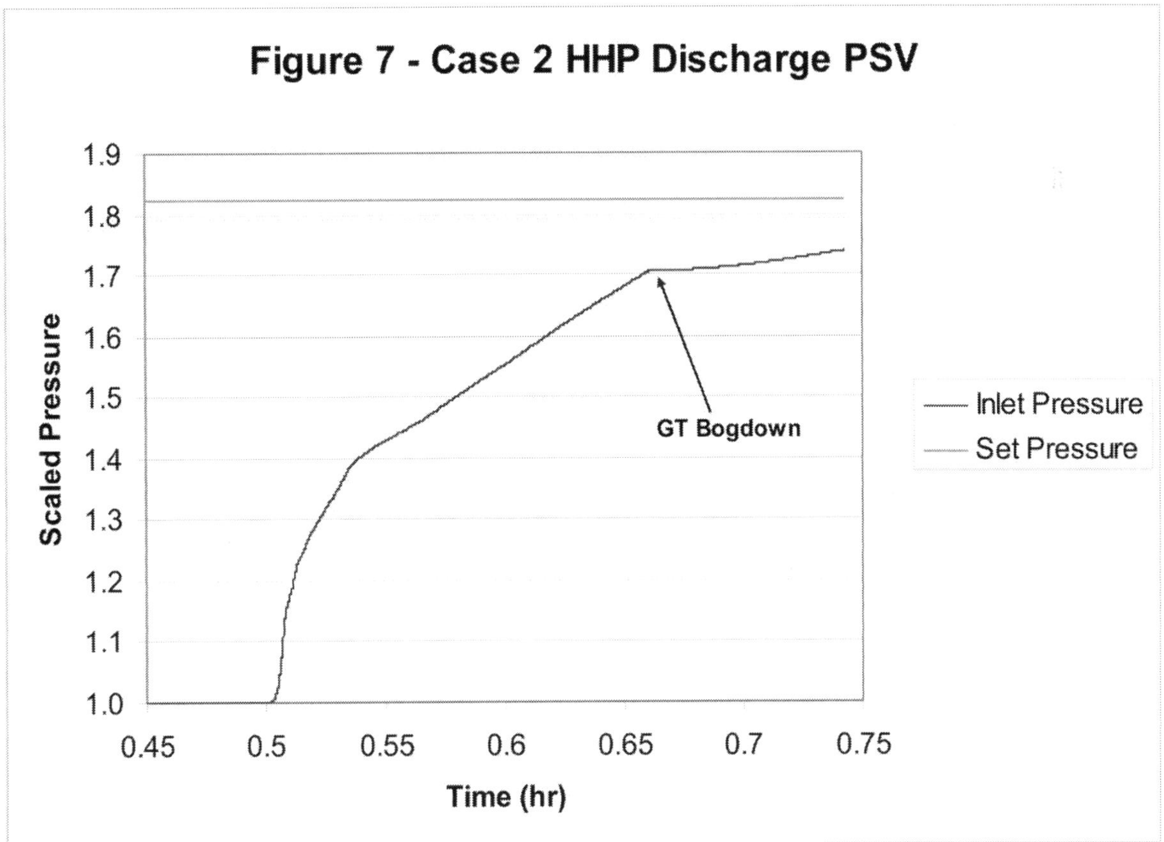
Figure 7 - Case 2 HHP Discharge PSV
Scaled Pressure
Time (hr)
GT Bogdown
Inlet Pressure
Set Pressure
1.9
1.8
1.7
1.6
1.5
1.4
1.3
1.2
1.1
1.0
0.45
0.5
0.55
0.6
0.65
0.7
0.75

LNG as a Source of Clean Energy

Chen-Hwa Chiu
Senior Technology Advisor
Chevron Energy Technology Company
Houston, Texas

AIChE Spring Meeting, April 2006
6th Topical Conference on Natural Gas Utilization
Orlando, Florida, April 23-27, 2006

Introduction

The Liquefied Natural Gas (LNG) industry has gone through dramatic changes over the years. It is now becoming an annual industry of over 100 million metric tons per annum and is closely related to the combined cycle gas turbine power generation and LNG has now becoming the source of clean energy for the household, the industry and the transportation. This paper intends to present the recent developments in the baseload LNG plant liquefaction technology and to discuss what are the considerations for the LNG receiving terminals.

The baseload LNG technology used to rely heavily on Air Products propane-precooled mixed refrigeration process. Over 85% of the present day baseload LNG plant is based on this technology. However, recent trends have made the competitions quite more interesting. The ConocoPhillips Optimized Cascade Cycle is being revitalized and used to build several plants. The Shell Double Mixed Refrigerant process is being used to some newer plants. The Linde/Statoil Mixed Fluid Cascade process is being used in the Snohvit plant and may be used in more application.

Larger baseload LNG plants also push forward technology development in equipment. Aero-derivative gas turbines are going to be used, and the single Frame-9 gas turbines. Air Products has developed a nitrogen expander combined AP-X™ process. Uses of liquid expanders are common place these days. The LNG plants of 7 to 8 million metric tons per annum are being built. Electric motors of the larger size are now being considered in the all electric driven LNG plants to cut down the greenhouse gas emissions. CO_2 is being considered to be reinjected to underground aquifers in the Snohvit and the Gorgon LNG plants.

The considerations of the LNG receiving terminals have also generated new ideas and thinking. The energy integration with neighboring facilities is considered as necessary. Larger plant for the future also requires the considerations of designing the plants with future expansions in mind. These are reflected in the facilities, LNG storage tanks and the marine facilities. Also the environmental and energy efficiency considerations are affecting the use of the submerged combustion vaporizers and the use of LNG for power generation inside the receiving terminal

Recent LNG Activities

Global LNG trade is now 146 MTPA, with more than 60 MTPA liquefaction capacities are under construction. LNG ships are now at 181 and growing with 129 on the order- book. Continuing demand growth in Asia-Pacific, Europe and US markets, especially for power generation. There are rapid depletion of reserves in US and UK. In China there are transportation and environmental bottlenecks for using coal. Furthermore there are attractive LNG pricing.

To understand the drive of using LNG as a clean energy in China, for instance. Currently there are two LNG receiving terminals under construction in Shengzhen, Guangdong and in Fujian. China is building its own LNG ships. Other LNG terminals are proposed for Liaoning (Dalian and Yingkou), Hebei (Tianjin and other), Shangdong, Jiangsu (Lianyungang, Yangcheng and Rudong), Zhejiang, Guangdong (Shangtou and Houng Kong), Guangxi and Hainan Island. Similar developments in North America, India, and UK have also seen recent booming activities in the construction of LNG receiving terminals. And to satisfy the growing demands for LNG, there are activities in developing baseload LNG projects in Africa, Australia, South America and Indonesia.

Figure 1 About 20 LNG Terminals proposed for China

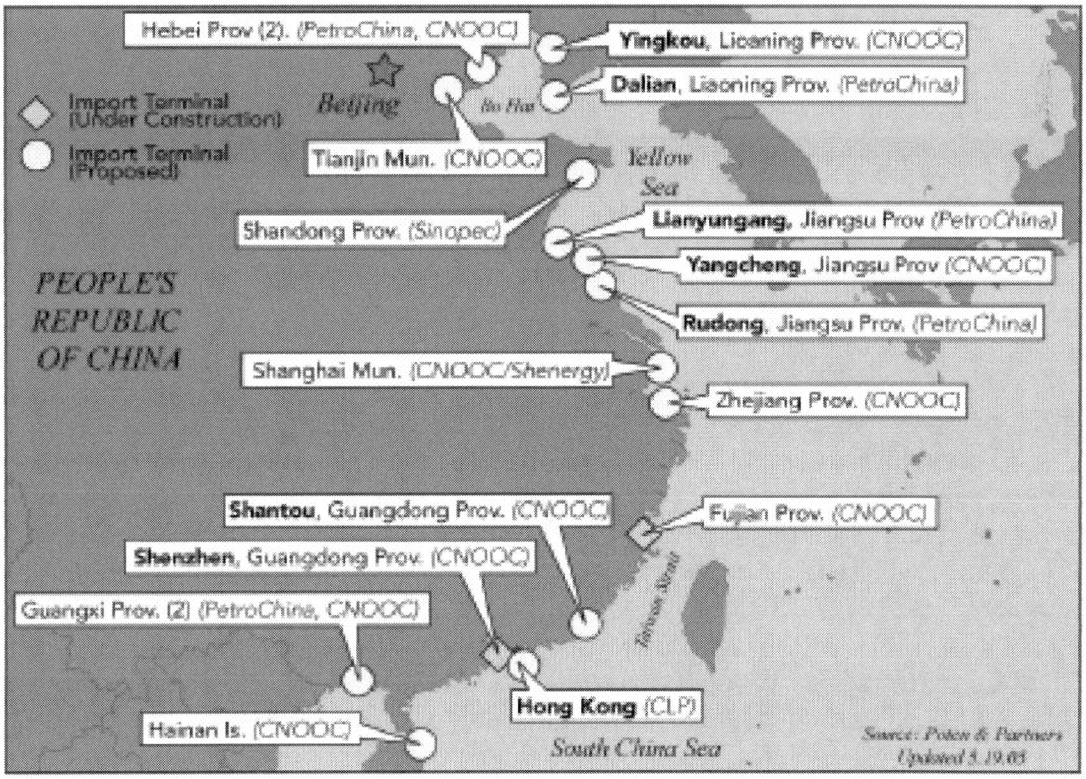

Courtesy: Poten & Partners

To understand an LNG value chain, we can roughly visualize the typical cost breakdown for a 4 MTPA (~0.5 Bcf/d) LNG project. The production will be 26% to 28%, liquefaction will be 26% to 28%, LNG ships will be 31% to 38% and receiving terminals will be 10% o 13%.

Development in Baseload LNG Technology

Baseload LNG plants have been constructed mainly using APCI C3-mixed refrigerant process (around 85% of world capacity) and Phillips optimized cascade process. Phillips optimized cascade cycle has been configured differently for driver arrangement for larger liquefaction plant size. However recently we have witnessed a series of new LNG liquefaction processes. These are the Statoil-Linde's mixed fluid cascade process (MFC™), the French Axen's Liquefin™ process, Air Products AP-X™ process and Shell's single and dual mixed refrigerant process.

The Statoil-Linde mixed fluid cascade process (MFC™) uses a 3 mixed refrigerant cycles. This process is being used for Statoil's Snøhvit LNG project near Hammersfest. It will use all electric drive design, but can be designed for direct turbine drive as well. The Snøhvit project embraces the first export facility for LNG in Norway and Europe. The refrigeration process and a cold Arctic climate help to boost the liquefaction plant's energy efficiency. This in turn cuts polluting emissions substantially by comparison with similar facilities. The main spiral wound cryogenic heat exchangers weighing 120 tons. They were installed in the cold boxes, the massive insulated towers. The four cold boxes have since been transported to an assembly yard in Belgium, where they all were mounted on one template for further transport to Melkøya in the spring of 2005. For pre-cooling the brazed aluminum plate-fin exchangers are used.

The Axen Liquefin™ process is designed to get the best use of large gas turbines and plate-fin heat exchangers. Compared to other technologies, it provides an increased capacity and a reduced investment cost. It is a dual mixed refrigerant process. It has a deep pre-cooling of -60 C with both cycles using two identical gas turbines. It can reach very high capacities, up to 6 MTPA. Several pre-FEED and FEED are under way.

APCI has developed the AP-X™ process, which can reach beyond 7.5 MTPA LNG. It combines the proven C3-MR cycle with a proven nitrogen expander cycle and retains single train proven equipment and machines. It can use GE Frame 7, Frame 9 or motor drive. It still maintains high process efficiency. The AP-X™ process is being used to build some of the LNG expansion in Qatar. Frame 9 machines are used for the large capacity LNG plants. Currently there are 5 trains in design and construction.

Shell has developed several mixed refrigerant LNG processes. The Shell single mixed refrigerant LNG process uses two parallel refrigerant compressors and part of the single MR is used to precool the natural gas and MR. The remaining MR is used to liquefy the feed gas. Electric motors are used as the drivers. It is suitable for plant capacity around 0.5 to 2.0 MTPA.

Shell's dual MR LNG process uses mixed refrigerant for precooling, replaces C3 precooling in C3-MR process. Propane/ethane mixture is generally around equal molar basis. The dual MR process satisfies the safety requirements of offshore application and offers high plant availability. The main cryogenic equipment will not be subjected to unnecessary thermal cycles. Both MR circuits are designed to contain the MR during a total shutdown to avoid flaring. It is suitable for plant capacity above 4 MTPA and offshore LNG.

The Gorgon LNG in Australia plans to re-inject the reservoir CO_2, removed from the feed gas stream, into geologic strata. Plans to install LNG facilities on Barrow Island provide a unique opportunity to re-inject CO_2 into saline reservoir. The Snøhvit LNG project in Norway also plans to re-inject its reservoir CO_2 into a formation below the producing reservoir.

Figure 2 CO_2 Re-Injection Scheme for Gorgon LNG

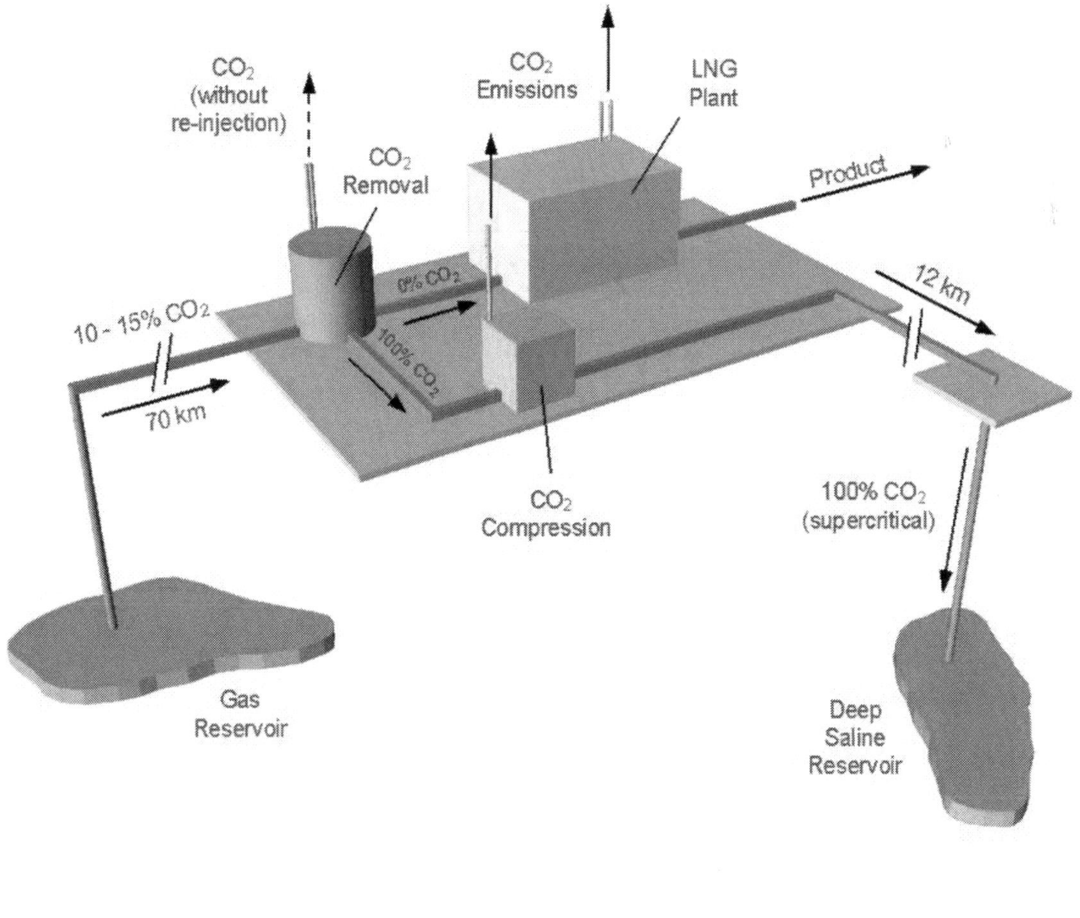

Development in LNG Receiving Terminals

Recent developments of LNG receiving terminals are such that there are emphasis on site selection and energy integration. Site selection has becoming a critical issue especially in the North America. The site selection needs to meet not only the health, safety and environmental but also to achieve any possible energy integration with neighboring facilities, such as a refinery or a CCGT (combined cycle gas turbine) plant, if possible. It also needs to comply with the regulatory requirements in order to obtain necessary permits. On the marine aspects attentions need to be focused on berth availability and LNG carrier turning basin. Capital cost, operating cost and operability are all important issues.

There are ideas on how energy integration can be achieved. For instance, precooling of the air intake to gas turbine of the CCGT plant with LNG cold to increase the efficiency and the power output of the gas turbine and to reduce the overall cost of the combined LNG/CCGT facility. Steam turbine condensers cooling water can be used for assisting the vaporization of LNG before send-out. Also, there are possibilities to reduce the thermal discharge arising from an integrated LNG import terminal and CCGT plant. The cold water can be discharged with proper dispersion mechanisms to reduce temperature rises caused by warm water discharge from the CCGT plant.

Conclusion

LNG is a clean energy source. The LNG trade is increasing at a rapid rate. LNG technology is undergoing a tremendous change. Continuous focus on efficiency and energy integration are necessary. We need to keep LNG as an economic and efficient clean energy for our future.

References

- "Commercial and Technical Considerations in the Developments of Offshore Liquefaction Plant", Chen-Hwa Chiu and Doug Quillen, to be presented at World Gas Conference, Amsterdam, June 2006.
- "Liquefied Natural Gas - A Source of Clean Energy", Chen-Hwa Chiu, China, USA, and Japan Joint Chemical Engineering Conference, Beijing, China, October 2005.
- "Comprehensive Evaluations of LNG Transfer Technology for Offshore LNG Development", David McDonald, Chen-Hwa Chiu and Dean Adkins, LNG-14, Doha, Qatar, March 2004.
- "Reduce Greenhouse Gas Emissions across the LNG Chain", Chen-Hwa Chiu, Chris Knaus, and Craig Lewis, World Gas Conference, Tokyo, June 2003.
- "New Developments in Floating LNG Technology", Chen-Hwa Chiu and L. Doug Quillen, LNG North America, Omni Hotel, Houston, March, 2002.
- "Overview of the North American LNG Market", Chen-Hwa Chiu and Doug Quillen, IBC LNG Conference, Amsterdam, November 2001, LNG Journal, Jan/Feb, 2002.
- "Turbo-Expander Technology Development for LNG Plants", Chen-Hwa Chiu and Hans Kimmel, LNG-13 Conference, Seoul, Korea, May 2001.

- "Upstream and Commercial Development of LNG Projects", Chen-Hwa Chiu and Doug Quillen, AIChE Spring Meeting, Houston, April 2001.
- "Optimization of the Integrated LNG to Independent Power Producer (IPP) Chain", Chen-Hwa Chiu, Frank W. Richardson and John R. Siegel, Twelfth International Conference on Liquefied Natural Gas (LNG-12), Perth, Australia, May 1998.

Biography

1500 Louisiana Street, Houston, TX 77002, USA.
Phone: 832-854-5305, Fax: 832-854-5325, Cell: 713-397-4097.
E-mail: Chen-Hwa.Chiu@Chevron.com.

Dr. Chen-Hwa Chiu is a Senior Technology Advisor of the Chevron Energy Technology Company. He is involved in LNG technology development in support of commercialization efforts. His expertise is in optimization of total LNG chain and offshore LNG development. Dr. Chiu has served as a company representative to the Joint Industry Projects of Azure Floating LNG Chain, LNG FPSO Safety, and Offshore Cryogenic Loading for LNG transfer.

Dr. Chiu started his career with the Lummus Company in 1969. He has worked in many LNG designs with Air Products (1973-1980) and participated in the startup of Arun LNG plant. He has worked at Exxon Production Research on Natuna LNG until 1986. He has also worked 2 years at M. W. Kellogg. At Bechtel until 1998 he has worked on baseload LNG of Kenai, Yukon Pacific, Wapet, Nigeria, and Atlantic, and LNG receiving terminals for China, India and Thailand. He briefly worked at Fluor (1998-1999). He has worked at Texaco and Chevron on Angola, Gorgon, Venezuela and offshore LNG. He also worked on LNG receiving terminals in China, Korea, Thailand, Canada and USA.

Dr. Chiu is a Fellow of American Institute of Chemical Engineers and was a Chair of AIChE Fuels and Petrochemicals Division. He has many publications and patents on LNG. He is the organizer and LNG Session chair of the AIChE Topical Conference on Natural Gas Utilization. Dr. Chiu graduated from National Taiwan University with a B.S. in 1962 and earned his Master of Engineering in 1964 and Ph.D. in 1970 from the University of Oklahoma, all in Chemical Engineering. He is a Registered Professional Engineer in Texas and Pennsylvania.

Definition of Hydraulic Efficiency of Two-Phase Expanders driven by Boiling Liquid

Christopher D. Finley
Ebara International Corporation
Sparks NV

AIChE Spring Meeting, April 2006
6th Topical Conference on Natural Gas Utilization
Orlando, Florida, April 23-27, 2006

INTRODUCTION

The reduction of liquid pressure from high pressure to low pressure is called expansion and is an integral part of cryogenic gas liquefaction processes. One present technique of expansion is the Joule-Thompson method. Another, more efficient way to expand liquids is by using a hydraulic turbine generator, which recovers the lost fluid energy as electrical power. This improves the total productivity of the process by recovering energy previously transferred to heat by the Joule-Thompson method [1]. To further increase the efficiency of the expansion process, two-phase expansion machines are being developed. This development allows the full reduction in pressure to occur across the machine, eliminating the need for a Joule-Thompson valve. Two-phase expander design concepts essentially follow existing single-phase turbine and expander technology. The hydraulic energy of the pressurized fluid is converted to electrical energy by first transforming it into kinetic energy, then into mechanical shaft power and finally to electrical energy through the use of an electrical power generator [2].

To accurately measure the benefits of expansion through an expander, single-phase or two-phase, expansion process efficiency must be defined. Currently, two methods of calculating efficiency are utilized. The first method relates the electrical power output to the hydraulic power input. This method is useful in describing the efficiency of a single-phase expander where the fluid density and quality are relatively constant through the machine. The second method uses the change in the power output of the generator, or internal energy, or enthalpy, of the fluid [3]. Although this method accurately describes the efficiency of a single-phase machine, it does not take into account the enthalpy of vaporization (or latent heat of vaporization) required during the phase-change process observed in two-phase expanders. The following paper describes and defines the method of calculating the efficiency of a two-phase expander utilizing the change in internal energy of the operating fluid while taking into account the enthalpy of vaporization. This definition of efficiency is exclusively applicable to turbines or expanders driven by boiling liquid, when boiling occurs inside the engine.

Figure 1 Two-Phase Expander

CALCULATIONS

In the analysis of certain types of process, particularly in power generation and refrigeration, the combination of properties u+Pv are encountered, where u is the specific internal energy of the liquid, P is the pressure and v is the specific volume. For the sake of simplicity, this combination is defined as enthalpy with the unit of kJ/kg. One current method of calculating single-phase expander efficiency is defined by

$$\eta_{one\_phase} = \frac{W_{generator\_out}}{\dot{m}(h_1 - h_2)}$$

Where W is the electrical output of the generator, m is the mass flow rate, h1 is the enthalpy of the liquid entering the expander and h2 is the enthalpy of the liquid exiting the expander. Although this is adequate for a single phase machine where the density and quality of the liquid are relatively constant, it does not take into account several fluid properties observed during two-phase expansion.

During the vaporization process, a substance exits as part liquid and part vapor. That is a mixture of saturated liquid and saturated vapor. To analyze this mixture properly, we need to know the proportions of liquid and vapor in the mixture. This is done by defining a property called quality [x] as the ratio of the mass of vapor to the mass of the total mass of the mixture:

$$x = \frac{m_{vapor}}{m_{total}}$$

where $m_{total} = m_{liquid} + m_{vapor}$

A saturated mixture can be treated as a combination of two systems, saturated liquid and saturated vapor. Usually, the mass of each system is not known so it is useful to consider this mixture homogenous, finding the overall fluid properties by finding the average of the individual system properties. The enthalpy of a mixture of a saturated liquid/vapor mixture can now be defined as

$$h_{av} = h_f + xh_{fg}$$

Where h_f is the enthalpy of the saturated liquid at a given temperature and pressure, x is the quality of the saturated mixture and h_{fg} is the difference between the enthalpy of the saturate liquid and the saturated vapor [3].

With the quality and total enthalpy of the fluid mixture defined, another property must be introduced. The enthalpy of vaporization h_{vap} (or latent heat of vaporization) is the amount of energy required to vaporize a unit of mass of saturated liquid at a given temperature and pressure. During the vaporization process, the enthalpy of vaporization applies no work to the system. This is due to the fact that at the moment of vaporization, the volume of the liquid remains constant. Because no work is done to the fluid during the vaporization process, it is necessary to remove this energy term from the efficiency calculation. The energy used in the vaporization process is defined as

$$h_{vap}(x_{out} - x_{in})$$

where x_{out} and x_{in} are the qualities of the saturated mixtures at the exit and the inlet of the expander respectively. Subtracting this term from the total enthalpy change across the machine yields the total useful enthalpy change across the machine that can be applied to the expander shaft as work. The useful change in enthalpy then becomes

$$\Delta h = h_{av\_in} - \left[h_{av\_out} - h_{vap}(x_{out} - x_{in})\right]$$

where h_{av_in} and h_{av_out} are the total enthalpies of the saturated mixture at the inlet and the exit of the expander respectively. This total useful enthalpy change is equal to the maximum possible work output in an ideal, 100% efficient two-phase expander. The maximum possible work output from an ideal two-phase expander is then

$$W_{\max} = h_{av\_in} - \left[h_{av\_out} - h_{vap}(x_{out} - x_{in})\right]$$

By combining the total useful change in enthalpy with the previously described method of calculating one phase expander efficiency, we can define the efficiency of a two-phase expander.

$$\eta_{two\_phase} = \frac{W_{generator\_out}}{\dot{m}\left[h_{av\_in} - \left(h_{av\_out} - h_{vap}(x_{out} - x_{in})\right)\right]}$$

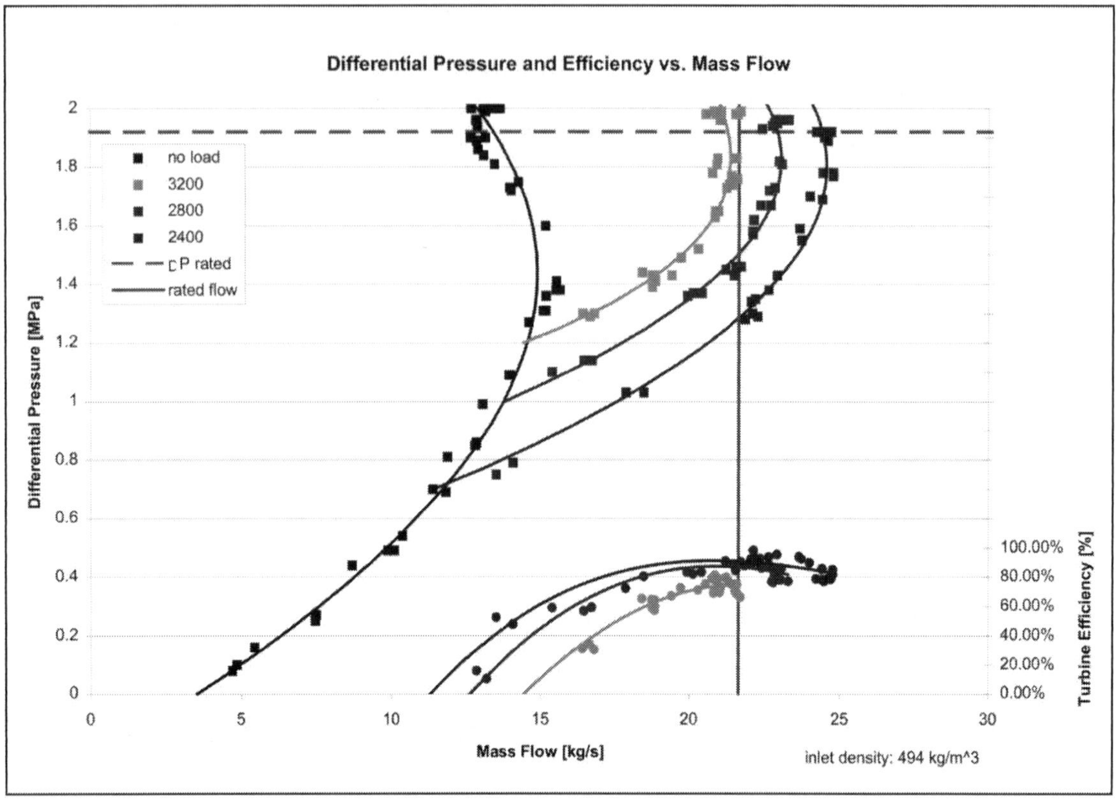

Figure 2 Hydraulic Performance of Two-Phase Expander

CONCLUSION

To accurately measure the benefits of expansion through an expander, single-phase or two-phase, expansion process efficiency must be defined. This paper describes and defines the method of calculating the efficiency of a two-phase expander utilizing the change in internal energy of the operating fluid while taking into account the enthalpy of vaporization. The above defined efficiency calculation is universal and applicable to turbines driven by boiling liquids as well as non-boiling liquids, like single-phase liquids or single-phase gases. For non-boiling liquids and gases the term (xout-xin) is equal to zero and disappears. Further measurements and tests should be performed to verify the validity and usefulness of this efficiency calculation.

REFERENCES

1. Cengal, Yunaus A. and Boles, Michael A., 2001, *Thermodynamics: an Engineering Approach*. Mcgraw-Hill College; 4th edition.
2. Finley, Christopher D. (2004), "Continuously Transient Operation of Two-Phase LNG Expanders," AIChE Spring National Meeting, New Orleans, Louisiana.

3. Habets, Gilbert (1999), "Economics of Cryogenic Turbine Expanders," *The International Journal of Hydrocarbon Engineering,* January.
4. Kimmel, Hans E. (1996), "Variable Speed Turbine Generators in LNG Liquefaction Plants," The 17th International LNG/LPG/Natural Gas Conference & Exhibition, Austria Center, Vienna, Austria.

Biography

Christopher D. Finley is the Test Engineering Manager of Ebara International Cryodynamics Division.

Mr. Finley earned a B.S. degree in Mechanical Engineering from the University of Nevada, Reno.

LNG-Fired Peak-Shaving Power Plants in China

Xuesheng Lu, Wensheng Lin, Anzhong Gu
Shanghai Jiaotong University
Shanghai, China

AIChE Spring Meeting, April 2006
6th Topical Conference on Natural Gas Utilization
Orlando, Florida, April 23-27, 2006

Abstract

Due to the rapid increase of petroleum price in the international market, the costs for small scale peak-shaving style power plants that use diesel as fuel are increasing. This results in operation difficulties for the power plants, which even suffer from losses. In the past two years, some peak shaving power plants start to explore the possibility of using liquefied natural gas (LNG) to replace diesel. The situation of peak shaving power plants in China is briefly reviewed in this paper. The designing schemes, practices and effects of using LNG alternate fuel are described. At the same time, development of the LNG industry in China is also introduced and the future prospect of using LNG in peak shaving power plants is discussed.

Introduction

There are many power plants driven by gas turbines in the cities in south China, which are also called peak-shaving power plants. Generally, the peak-shaving power plants have relatively low capacity, so that they are easy to be started up or closed down, and are suitable for the adjustment of electricity network at variable loads. However, the operation cost is often high because of small capacities and discontinuous operations. Especially with the rising of fuel oil price, the cost of power generation rises continuously. In the worst case, the cost of fuel oil exceeds the profit of power generation.

At the end of 2003, the selling price of electricity to the network is 0.72 CNY/kWh (0.615 CNY/kWh excluding tax), and the price of cruel oil in the international market is 28 USD/barrel. Power plants showed good profits at that time. The oil price has been rising rapidly since 2004 and the highest price exceeded 60 USD/barrel, while the usual price was between 40 and 50 USD/barrel, equal to about 4000 CNY/t. If a general power plant has two combined cycle sets, one LM2500 and one FT8, fired with light diesel. The oil consumption is about 210 g/kWh, and the cost is about 0.84 CNY/kWh, which means that the power plant loses 0.12 CNY/kWh if the selling price of electricity to the network is still 0.72 CNY/kWh. In this case, the power plant couldn't afford and has to stop running. The rise of oil price in the international market has severe influences on the gas turbine power plants in China.

In order to cut down the cost of power generation, some companies begin to consider substituting diesel fuel with natural gas (gasification from LNG). The government also provides some favorable policies to encourage this practice. After thorough investigations of the source, price and devices of LNG, five power plants have successfully made some necessary modifications to their gas turbine systems, and set up LNG gasification and feeding system for the gas turbines. In May and June 2005, these LNG-fired plants started generating electricity successfully, and have been running well from then on.

As a clean fuel with high-quality, LNG's applications in the fields of power generation, transportation, chemical engineering and city fuel have developed rapidly. LNG is easy to transport, and no long-distance pipeline is needed. The density of LNG is much higher than that of natural gas, and it is cleaner and environmental friendly. At current LNG price, the cost of power generation goes down considerably.

LNG Gasification and Feed Flow

At the normal pressure, the temperature of LNG is about -163°C, and its density is about 0.514 kg/m^3. The inlet conditions of the gas turbine are required at room temperature and at the pressure of 35 bar. Therefore the LNG gasification and feeding system must include LNG unloading, storage, pressurizing, gasification, pressure stabilization, safe relief, etc.. The system process is shown in Fig.1.

The LNG gasification and feeding system mainly consists of the receiving/storage devices and gasification/feeding devices. LNG is transported to the site by trailer and then transferred by unloading devices and pipelines to the LNG tank. The number of unloading points is determined by the LNG supplying quantity. For a 40-inch standard tank container, the total unloading and auxiliary time is about 1.5h. In Fig.1, the system has four unloading points and four 100m^3 double-walled vacuum-insulating tanks. Under normal operation, LNG is discharged from the tank, then pressurized to 3.5MPa by the cryogenic piston pump and heated in the water-bath vaporizer. Thus, LNG becomes high pressure natural gas at room temperature. After going through the buffer pipeline, the pressure stabilizing device and the filter, LNG is fed to the gas turbine with stable pressure.

The capacity of the LNG gasification and feeding system is determined by the gas turbine capacity, efficiency, and daily operating time. For example, the capacity of one power plant is 63 MW, the efficiency of the combined cycle set is 40%, and the combustion heat of natural gas is about 39.5 MJ/m^3. Thus, the gas turbine needs about 13000 m^3/h natural gas, and the consumption of LNG is around 22 m^3/h.

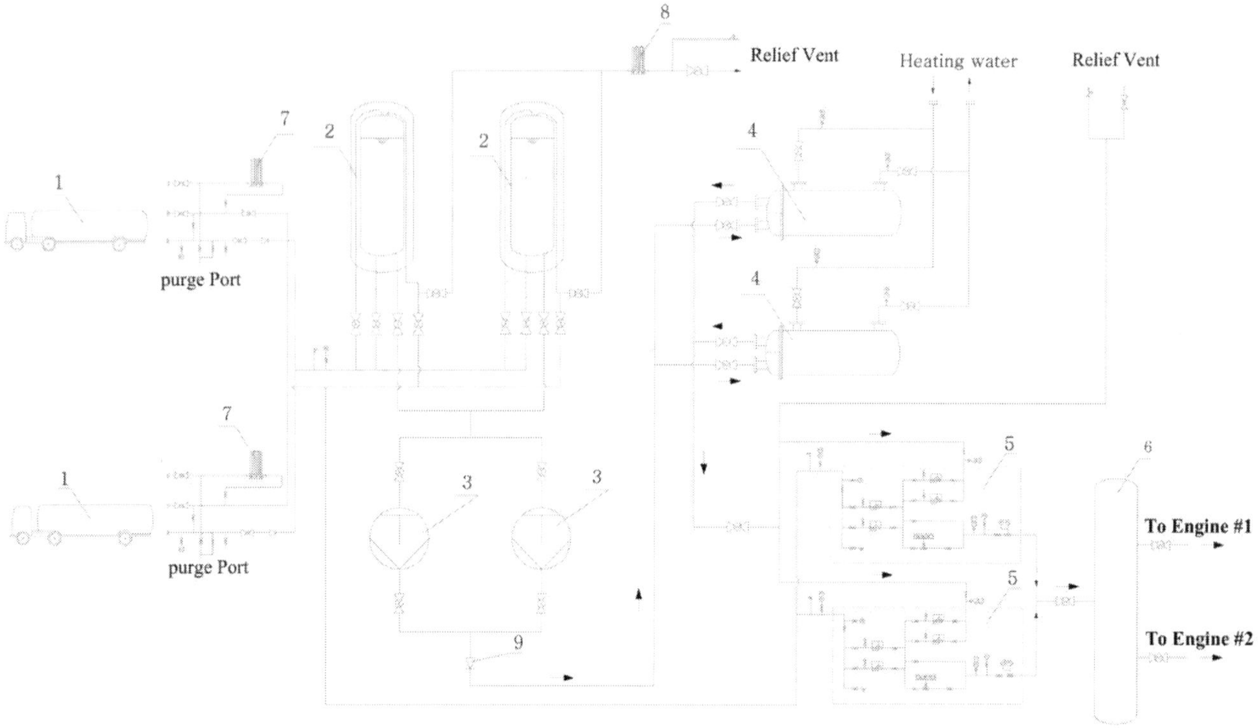

Fig. 1. Flow Process of LNG Project at Futian Power Plant, Shenzhen
1—LNG Trailer 2—LNG Storage Tank 3—LNG Pump 4—vaporator 5—Pressure Stabilizer
6—NG Buffer 7—Loading pressurizer 8—Venting Heater 9—One Way Valve

Main Devices

The LNG gasification and feeding system mainly consists of the LNG unloading system, LNG tank, LNG pump, water-bath vaporizer, buffer tank, pressure stabilizing devices and safe relief system.

LNG Tank

Less land is needed when adopting double-walled vacuum-insulating vertical tank. The tank capacity is determined by daily LNG consumption and backup storage. In order to deal with unexpected conditions in LNG transportation, such as bad weather, traffic jam and accidents at LNG supplier, which resulted in delayed LNG arrival, it is necessary to keep a reasonable amount of LNG backup storage.

If the capacity of a single tank is small, one can consider a combination of several tanks manufactured in tank firms. Because of the transportation restriction, the capacity of this kind of pre-manufactured tanks is normally less than $100m^3$. If the required storage capacity is so large that the system needs too many tanks, one should consider the possibilities of manufacturing large-sized tank(s) on site.

LNG Pump

Comparing with the compressor in gas boosting, it is more efficient and cost-saving to use cryogenic pump to pressurize LNG. The best way is to use submerged multi-stage centrifugal pump. Enough static positive pressure must be ensured to avoid cavitations in the pump.

Vaporizer

LNG must be vaporized and heated to room temperature before usage. The cooling water temperature of gas turbine is usually high, so that it is a good way to recover the waste heat from the cooling water of gas turbine to vaporizing LNG. In all the above-mentioned projects that substitutes fuel oil with LNG, the warm water tube-and-shell vaporizers are adopted. LNG flows in the tubes and warm water flows in shell, meanwhile LNG is heated and vaporizes. The structure of water-bath vaporizer is usually very compact. Two types of the water-bath vaporizers produced by THERMAX and CRYOQUIP are listed in table 1.

Table 1. Comparison of Water-Bath Vaporizers Produced by THERMAX and CRYOQUIP

Item	Unit	THERMAX	CRYOQUIP
Parameter at water side			
Flowrate of water	m^3/h	1543	302
Diameter of inlet and outlet pipes	mm	300	250
Velocity at inlet and outlet	m/s	6.0	1.64
Flow resistance of water	bar	1.50	0.50
Designed inlet temperature of water	°C	24.0	24.0
Designed outlet temperature of water	°C	13.0	12.0
Parameter at LNG side			
Flowrate of natural gas	m^3/h	13000	13000
Diameter of LNG inlet pipe	mm	75	50
Diameter of natural gas outlet pipe	mm	100	75
Inlet pressure of natural gas	bar	35	35
Outlet velocity of natural gas	m/s	14.0	27.0
Flow resistance of natural gas (pressure drop)	bar	2.5	2.5
Inlet temperature of LNG	°C	-196	-196
Outlet temperature of natural gas	°C	2.0	>=0 (tube wall)

Using the cycling water from the cooling tower to heat LNG provides a convenient and reliable heat resource for LNG vaporizing, furthermore, the temperature of cycle water is lowered by the cold energy of LNG.

Pressure Stabilizing Device

In order to maintain the steady state of natural gas at the inlet of gas turbine, it is necessary to adopt gas buffer and pressure stabilizing devices. There is certainly a delay before the LNG pump can adjust to cope with the change of load. The function of gas buffer is to supply or hold natural gas for a short time, keeping the inlet pressure of pressure stabilizing devices within designing scope. The pressure stable devices react to the minute change of inlet pressure of gas turbine, thereby to maintain the supplying pressure in a permitted range.

4 The Adjustment of System

Before filling LNG, the LNG gasification and feeding system must be pre-cooled and purged by liquid nitrogen. The system tightness, the operating characteristics of devices at cryogenic temperature, the pressure stabilizing performance of system and the reliability of safe relief devices must be confirmed. Only after all the above operations can LNG be filled in the system. After the filling-up of LNG is completed, one needs to adjust the system according to the following procedures:

1) Pre-cooling of LNG pump and liquid pipeline
2) Start warm-water filling of vaporizer
3) Start LNG pump
4) Pressure build-up in gas buffer
5) Start power generation system

It takes about 30 minutes after successful start-up, the gas turbine achieves the rated rotational speed, and electricity is sent to the network. The operation parameters of main devices are shown in table 2.

Table 2. Experimental Operation Parameters of Gas Turbine

Operation parameters	Unit	Value
LNG tank pressure	MPa	0.4
Inlet pressure of LNG pump	MPa	0.38
Outlet pressure of LNG pump	MPa	3.39
Inlet temperature of LNG pump	°C	-144
Outlet temperature of LNG pump	°C	-139
Inlet water temperature of vaporizer	°C	28.4
Outlet water temperature of vaporizer	°C	24.4
Gas supplying pressure after pressure stabilization	MPa	3.16
Inlet LNG pressure of gas turbine	MPa	3.10

The electricity generated from natural gas is about 4.0 kWh/Nm^3. Assuming that the price of natural gas is currently 2.16 CNY/Nm^3, the fuel cost of power plant is 0.54 CNY/kWh. If the selling price of electricity to the network is 0.72 CNY/KWh, the profit is about 0.04~0.05 CNY/kWh after paying the sales tax, equipments depreciation and labor cost.

5 The Application and Prospect of LNG in Power Plant

In the past ten years, power plants driven by gas turbines have increased rapidly in China. The total capacity has already exceeded 12380 MW. These gas turbine power plants are mainly located at the southeast coast of China, as shown in Fig 2. Comparatively, these areas are economically developed, consuming large amount of electricity.

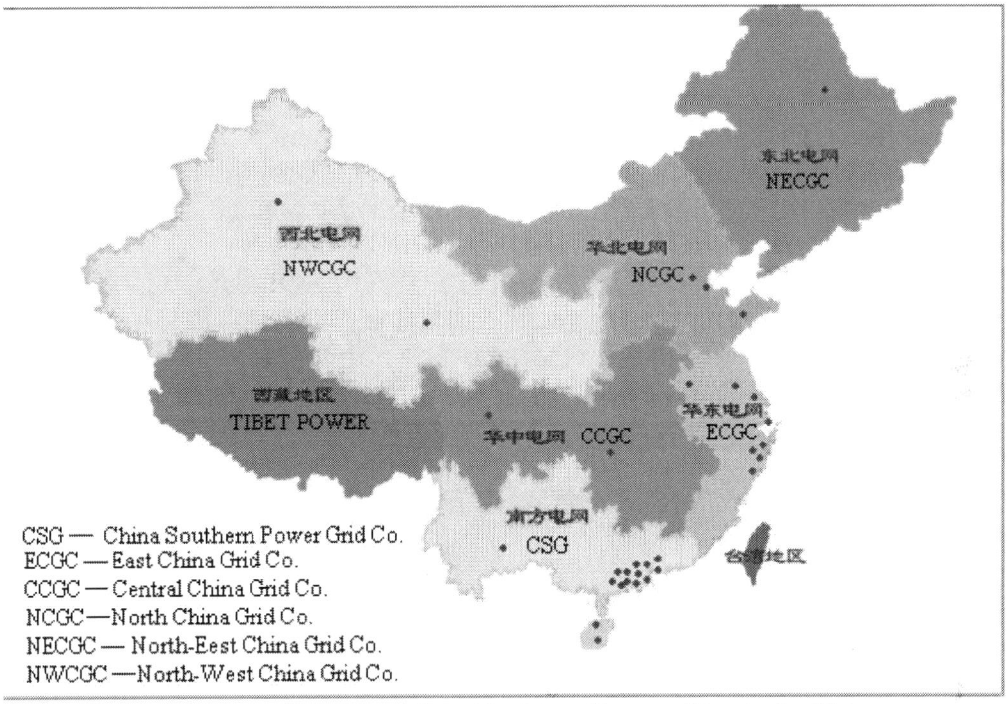

Fig. 2. Distribution of Gas Turbine Power Plants in China

According to an incomplete statistical result, there are currently more than 60 gas turbine power plants located mainly in southeast of China, with a total capacity of 12380 MW, as given in Table 3.

Table 3. Capacity of Gas Turbine Power Plants

No.	Area	Numbers of power plants	Capacity (MW)
1	Guangdong	27	7792
2	Zhejiang	8	1608
3	Jiangsu	6	392
4	Shanghai	1	600
5	Hainan	4	737
6	Others	14	1251
	Total	60	12380

Along with the continuous development of distributive cold-heat-electricity cogeneration technology, as well as the rapid development of the natural gas industry, many authorities believe that the gas turbine combined cycle power plants will increase faster in the near future. Because many power plants are not located near the natural gas pipelines, LNG will be also widely used as the fuel for gas turbine power plants.

Conclusions

From the viewpoint of fuel substitution, the program of substituting fuel oil with LNG has been successfully carried out. The gasification and feeding system runs stably, safely and reliably, and it is easy to operate. The success of this project supplies some valuable experiences for similar power plants.

In China, the main fuel of power plants is coal, which brings serious impacts to the environment. The high price of fossil oil has driven many power plants into an embarrassing position. The gradual adjustment of energy constitution is the trend of further industrial development. The efficient way to decrease the cost of energy and control environment pollution is to increase the proportion of using clean energy. Natural gas is a kind of clean energy with high quality. LNG is natural gas stored in liquid form after liquefaction, and is even cleaner than pipeline natural gas. There is wide prospect for LNG to be used in the fields of energy and transportation. Although the output of LNG is very limited at present and thus inhibits the fuel substitution in some power plants, the LNG industry is developing rapidly in China. With the increase of LNG production and importation, LNG will be more and more widely used in gas turbine power plants.

References

1. Gu AZ, Lu XS, Wang RS, et al. (2004), "*Liquefied Natural Gas Technology*", Beijing, China Machine Press.
2. Adorjan A S, (1990), "*Heat Transfer in LNG Engineering*", New York, Hemisphere Pub. Corp.
3. Hong Yuan Mi, The Present Situation and Prospect of Power Plants Driven by Gas Turbines in China and Abroad. http://energy.icxo.com/htmlnews /2003/12/29/58 100.htm accessed Feb. 03. 2006

A Flexible LNG Conditioning and NGL Recovery Process for LNG Receiving Terminals

John Y. Mak
Technical Director, Process Engineering
Dick Nielsen
Executive Director, Process Engineering
Curt Graham
Principal Engineer, Process Engineering
Energy and Chemicals Group, Fluor Enterprises, Inc., Aliso Viejo, California

AIChE Spring Meeting, April 2006
6th Topical Conference on Natural Gas Utilization
Orlando, Florida, April 23-27, 2006

Abstract

In the United States, pipeline gas is predominantly methane with some nitrogen and carbon dioxide. Heating values generally range between 1000 and 1075 Btu/scf (HHV). Many pipelines specify a maximum HHV of 1050 Btu/scf while some have a maximum heating value specification of 1075 Btu/scf. There are only a few LNG sources that have a sufficiently low heating value for direct injection to the U.S. sales gas pipelines.

This paper presents a process that can be used to control the Btu content of pipeline gas derived from imported LNG with the production of ethane and LPG for sales. The *Fluor®* *CryoGas*sm process (patent pending) utilizes the cryogenic refrigeration provided by LNG for NGL fractionation and re-liquefaction of the resulting residue gas to form a lean LNG which meets typical U.S. pipeline gas specifications. The *CryoGas* LNG conditioning unit can recover over 97% of the propane and about 70% of the ethane in the LNG and can operate in either an ethane recovery or an ethane rejection mode.

Introduction

While several North America LNG receiving terminals will soon be operational, there are increasing concerns about the composition and heating value of imported LNG. The composition and heating value of LNG vary significantly depending on the characteristics of the gas field and the level of NGL recovery at the LNG liquefaction plant. A "spot market" strategy based on acceptance of LNG of varying heating values and compositions could allow LNG to be imported more cost effectively than a long term contract of a single supply. This strategy allows acquiring supplies at a lower cost and may lower the delivered natural gas price. It is likely that this marketing strategy would

---

Fluor is a registered service mark of Fluor Corporation.
Cryo-Gas is a service mark of Fluor Corporation.

have to be supported by a LNG Btu conditioning facility at the receiving terminal that extracted C2+ hydrocarbons from the LNG to meet pipeline sales gas specifications. Extracting C2+ hydrocarbons from LNG may present an economic opportunity: ethane can be utilized for petrochemical production and propane plus components can be sold as LPG.

There are two methods for controlling the heating value of pipeline gas derived from imported LNG. The first method is by dilution. In one dilution option higher heating value gas from LNG regasification is blended with a lower heating value natural gas. This strategy requires large volumes of lower Btu natural gas. This may be the case for regasification terminals located along the U.S. Gulf Coast. Nitrogen dilution is another option; however, nitrogen dilution requires a dedicated nitrogen plant and the amount of nitrogen that can be added may be limited by the pipeline gas specification for maximum inerts content.

The second method for controlling heating value is by extracting ethane and heavier hydrocarbons from the LNG after it has been re-vaporized. Conventional turbo-expander processes can be used for recovering ethane and propane plus hydrocarbons from natural gas, but would require high power consumption. The *Fluor CryoGas* process, specifically developed for LNG receiving terminals, is a non turbo-expander based process that can achieve high NGL recovery with significantly less power consumption.

LNG Heating Values and Compositions

In most LNG liquefaction plants, NGL recovery units are usually not installed for heating value control. This is due to the fact that many current LNG importers (mainly Japan) desire LNG with a relatively high heating value. Removal of hydrocarbons (mainly C_6+) in LNG liquefaction plants is often limited to the high freezing point hydrocarbons to avoid wax formation and plugging in the cryogenic exchangers. Methane, ethane and heavier components are liquefied together and the heavier hydrocarbons raise the LNG heating value. The variations in LNG heating values for several LNG export terminals in the Atlantic and Pacific basins and in the Middle East are depicted in Figure 1. The corresponding component breakdowns for methane, ethane, propane to pentane and hexane plus components are shown in Figure 2. [1], [2]

As can be seen in Figure 1, only the Kenai, Alaska and the Trinidad LNGs can meet the typical U.S. pipeline sales gas heating value limit of 1075 Btu/scf. Other LNGs contain significantly more ethane, propane and heavier hydrocarbons, and their heating values exceed U.S. pipeline gas specifications. There are also concerns regarding LNG composition swings and heating value variations; such as lower combustion equipment thermal efficiencies, higher hydrocarbon dew points and higher emissions due to incomplete heavy hydrocarbon combustion.

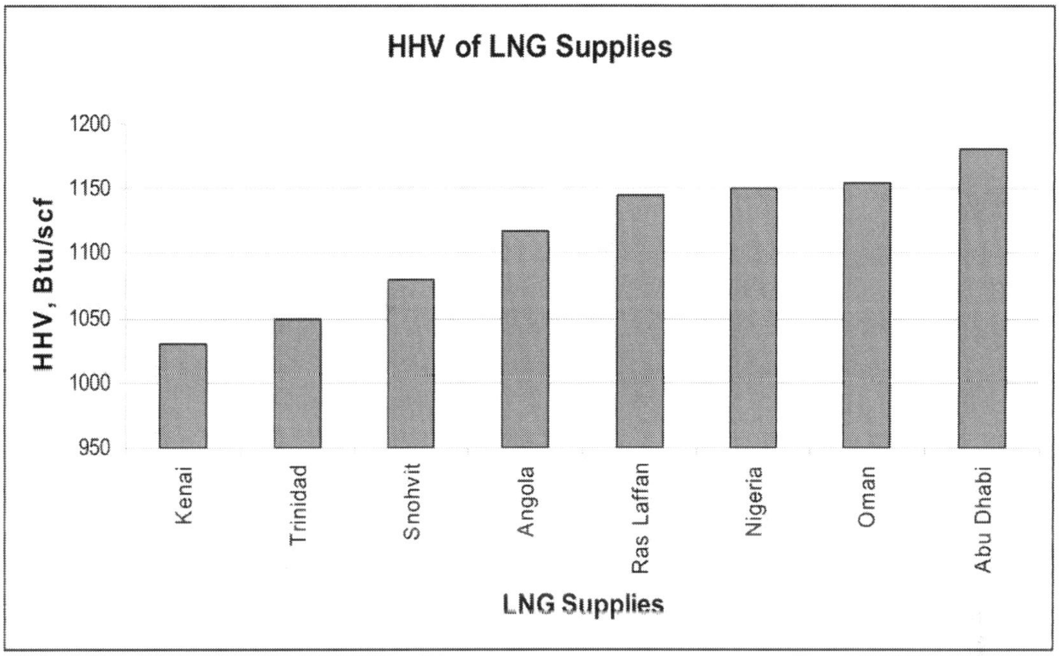

Figure 1- Heating Values of Atlantic Basin, Pacific Basin and Middle East LNGs

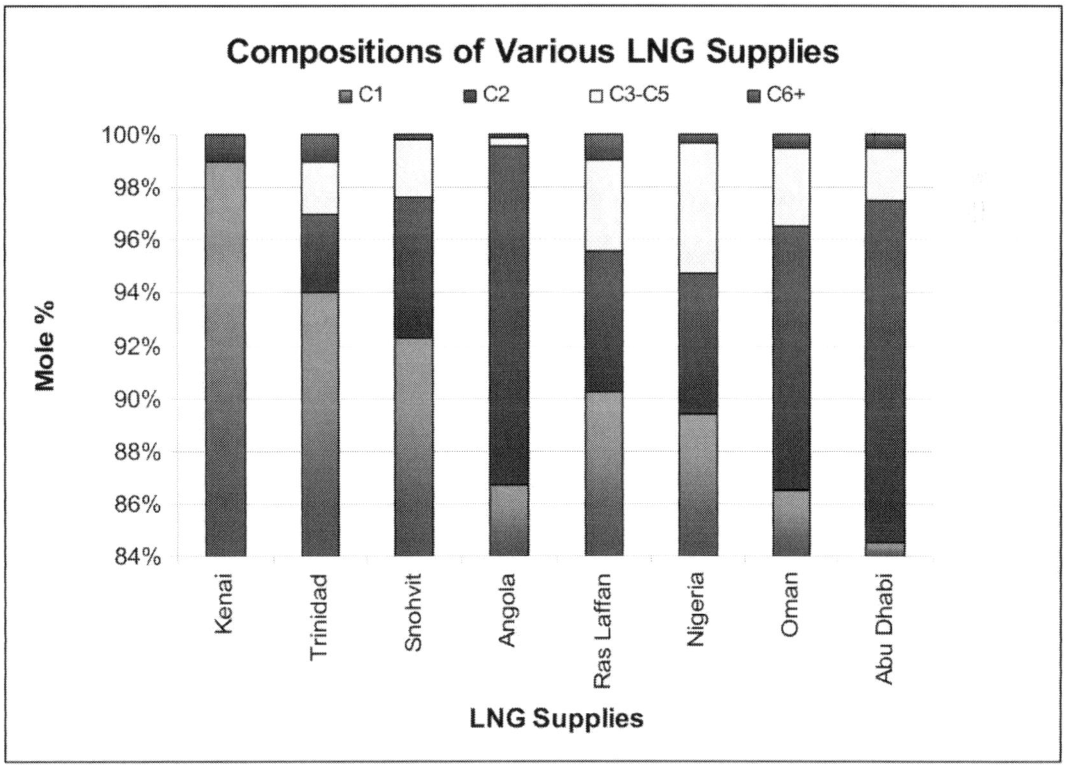

Figure 2– Compositions of Atlantic Basin, Pacific Basin and Middle East LNGs

Nitrogen Dilution

For a standalone LNG regasification terminal, a reliable source of low Btu natural gas for blending may not be available. Nitrogen dilution has been used for controlling the Btu content of the vaporized LNG in LNG import terminals. The amount of nitrogen dilution increases with the richness of the LNG. Figure 3 shows the amount of nitrogen that is required to meet a 1075 Btu/scf heating value requirement for LNG with different heating value. The nitrogen dilution requirement would increase the inerts content in pipeline gas to as high as 9 vol. % when LNG with a heating value of 1170 Btu/scf is imported. However, this amount of nitrogen would far exceed the typical US pipeline gas specification of 3 vol. % inerts. At the 3 vol. % inerts limit and a sales gas heating value specification of 1075 Btu/scf, the heating value of the imported LNG must not exceed 1100 Btu/scf when nitrogen dilution is used for Btu control.

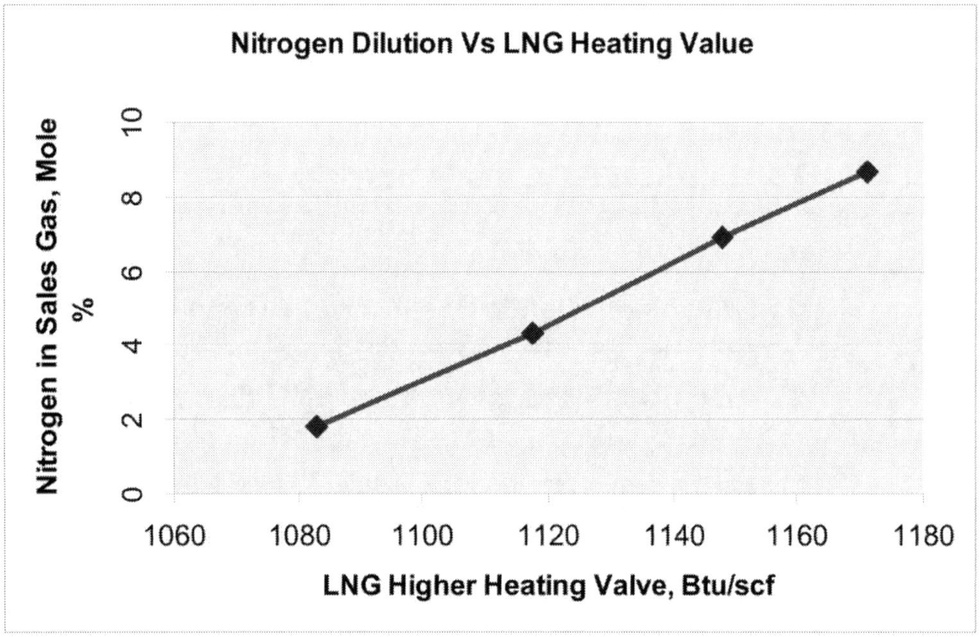

Figure 3– N_2 Dilution for Btu Control to meet a 1075 Btu/scf HHV

LNG Regasification Terminal

A LNG Btu conditioning unit can be integrated with a conventional LNG regasification terminal as depicted in Figure 4. A typical LNG receiving terminal consists of LNG ship unloading, LNG storage, LNG in-tank pumps, boil-off gas compression and a boil-off gas re-condenser. When a LNG Btu conditioning unit is integrated with a LNG regasification terminal, LNG is pumped from the boil-off gas re-condenser to an intermediate pressure at about 550 psig, the battery limit pressure of the LNG Btu conditioning unit. Only a portion of the LNG is required to be processed in the LNG Btu conditioning unit while the remaining LNG is bypassed to the LNG vaporization unit.

The LNG Btu conditioning unit produces lean LNG, LPG and ethane. The lean LNG is blended with the bypassed LNG as needed to meet the pipeline sales gas heating value specification. The blended LNG stream is further pumped by the high pressure LNG pumps, vaporized and directed to the sales gas pipeline. Conventional LNG vaporization processes, such as seawater open rack vaporizers or submerged combustion type fuel-fired vaporizers, can be used. The LNG Btu conditioning unit is designed as an add-on unit that can be totally bypassed when not required. The LNG vaporization unit is a self-sufficient unit, and is designed to operate independent of the LNG conditioning unit.

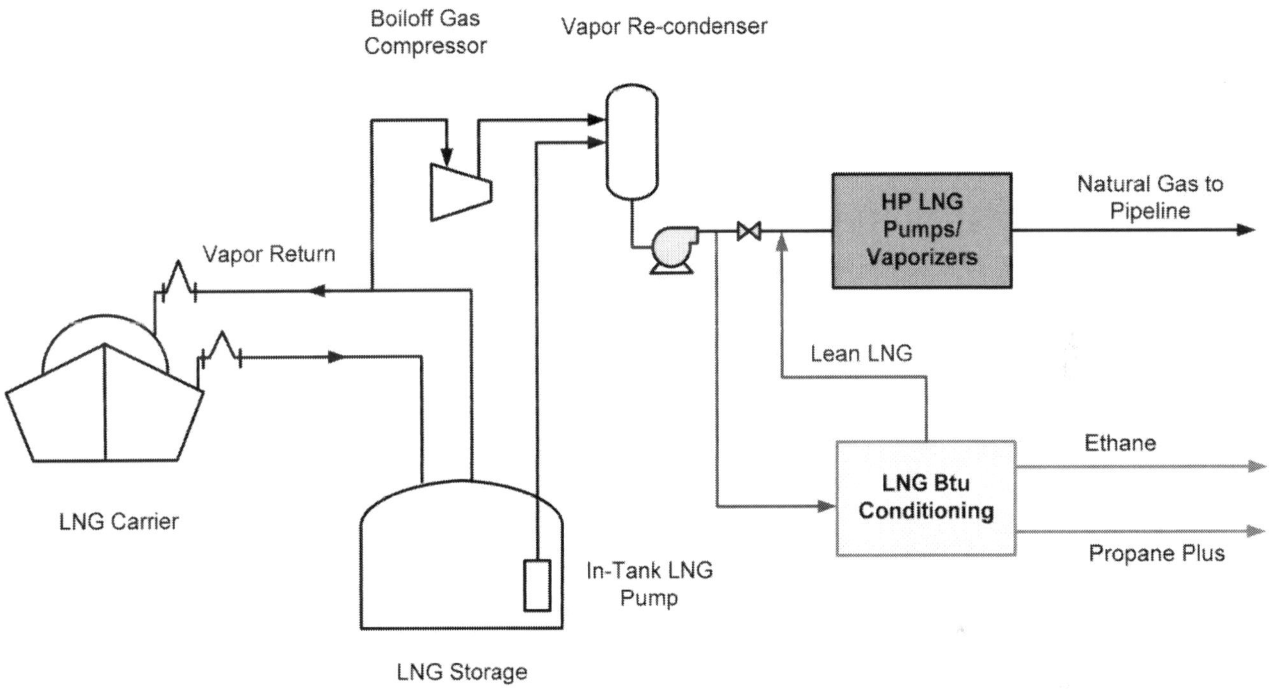

Figure 4– LNG Facility with a LNG Btu Conditioning Unit

Case Study

Table 1 shows five LNGs with heating values ranging from 1043 to 1165 Btu/scf that were used in the case study. The LNG terminal is a base load terminal with a send-out capacity of 2000 MMscfd. LNG E has a higher heating value of 1043 Btu/scf and nitrogen dilution is not required. Nitrogen dilution could be applied to LNG D, which has a heating value below 1100 Btu/scf. But for LNGs A, B and C, nitrogen dilution would produce a sales gas that would exceed the typical 3 vol. % inerts content limit.

LNG Sources	A	B	C	D	E
Composition Mol%					
Nitrogen	0.17	0.13	0.03	0.64	0.00
Methane	85.98	89.30	86.79	91.90	96.18
Ethane	8.69	5.30	12.87	5.30	3.39
Propane	3.47	3.40	0.28	1.93	0.34
i-Butane	0.85	0.75	0.01	0.10	0.04
n-Butane	0.79	1.10	0.01	0.12	0.03
i-Pentane	0.05	0.02	0.00	0.00	0.01
Total	100	100	100	100	100
HHV, Btu/scf	1,165	1,142	1,112	1,078	1,043

Table 1– Various Receiving Terminal LNG Compositions

Fluor CryoGas LNG Btu Conditioning Unit

The *Fluor CryoGas* LNG Btu conditioning unit uses the cryogenic refrigeration from LNG to supply the cooling and reflux duty for the conditioning unit's demethanizer and deethanizer. In addition, LNG refrigeration is utilized for re-liquefaction of the demethanizer overhead gas, which allows the liquid to be pumped to the pipeline pressure, eliminating the conventional gas compression system. The process can recover over 97% of the propane and 70% of the ethane from LNG, producing a lean gas with significantly lower heating value than specification. The heating value of the lean LNG produced from this unit varies from 1020 (LNG E) Btu/scf to 1035 Btu/scfd (LNG A)

The LNG Btu conditioning unit is designed to process the five different LNGs summarized in Table 1 while meeting a pipeline sales gas heating value specification of 1075 Btu/scf. The target LNG send-out for this plant is 2000 MMscfd. Note that send-out is defined as the MMscfd of LNG leaving the LNG storage tank. The LNG conditioning unit can operate in either an ethane recovery or an ethane rejection mode. The design of the LNG conditioning unit is controlled by the richest LNG composition (LNG A) which requires processing about 70% of the LNG in the conditioning unit. The remaining LNG bypasses the unit. Figure 5 is a process flow schematic of the LNG conditioning unit.

A key design feature of the *Fluor CryoGas* LNG conditioning process is the feed exchanger, E-1, which consists of two heat exchanger passes. The first heat exchanger pass uses the inlet LNG to cool and condense the lean residue gas from the demethanizer reflux drum. The second pass cools and partially condenses the demethanizer overhead stream to produce a lean, methane rich demethanizer reflux. This methane rich reflux improves ethane and propane recovery.

LNG refrigeration is also utilized to provide deethanizer reflux. With LNG as a refrigerant, the deethanizer can operate at a lower pressure (140 psig) than a conventional design (typically 350 psig). Therefore, less reboiler duty and fewer trays are required due to the higher relative volatility between ethane and propane at low pressure. The deethanizer produces an ethane liquid overhead product and a LPG bottoms product.

To reduce the demethanizer reboiler duty, LNG from the deethanizer condenser, E-2, is heated in two steps: first in heater E-3 and then in heater E-4 after the heated effluent from E-3 is flashed in separator V-2. The flashed vapor is fed to the upper section of the demethanizer and the heated liquid is fed to the mid-section of the demethanizer. Splitting the feed in this manner lowers the reboiler, E-5, duty and also lowers the vapor traffic in the lower section of the demethanizer. Lowering the vapor traffic reduces the column diameter. With this approach, a single demethanizer column with a column diameter of less than 20 ft can be used to condition sufficient LNG to produce on specification pipeline gas for 2000 MMscfd of LNG send-out for any of the LNGs in Table 1. Since most of the LNG heating occurs at cryogenic temperatures, low level heat sources such as seawater, ambient air or waste heat can be used, and it is possible to reduce or even eliminate fuel gas consumption.

The demethanizer pressure is one of the important parameters in this design. The shape of the LNG vaporization curve (temperature vs. duty) is a function of the LNG composition and vaporization pressure. The shape of the vaporization curve must match the residue gas condensation and the demethanizer reflux cooling curves and must not result in a temperature pinch in exchanger E-1. While higher pressure increases the residue gas condensation temperature, which eases the temperature pinch, high pressure operation also reduces the relative volatility between methane and ethane, making fractionation more difficult. The demethanizer pressure is also controlled by the ethane and propane content of the LNG and the required NGL recovery level. For this process, the optimum demethanizer pressure is about 450 psig. This pressure can be maintained for the different LNG compositions in Table 1 and for either ethane recovery or ethane rejection operation.

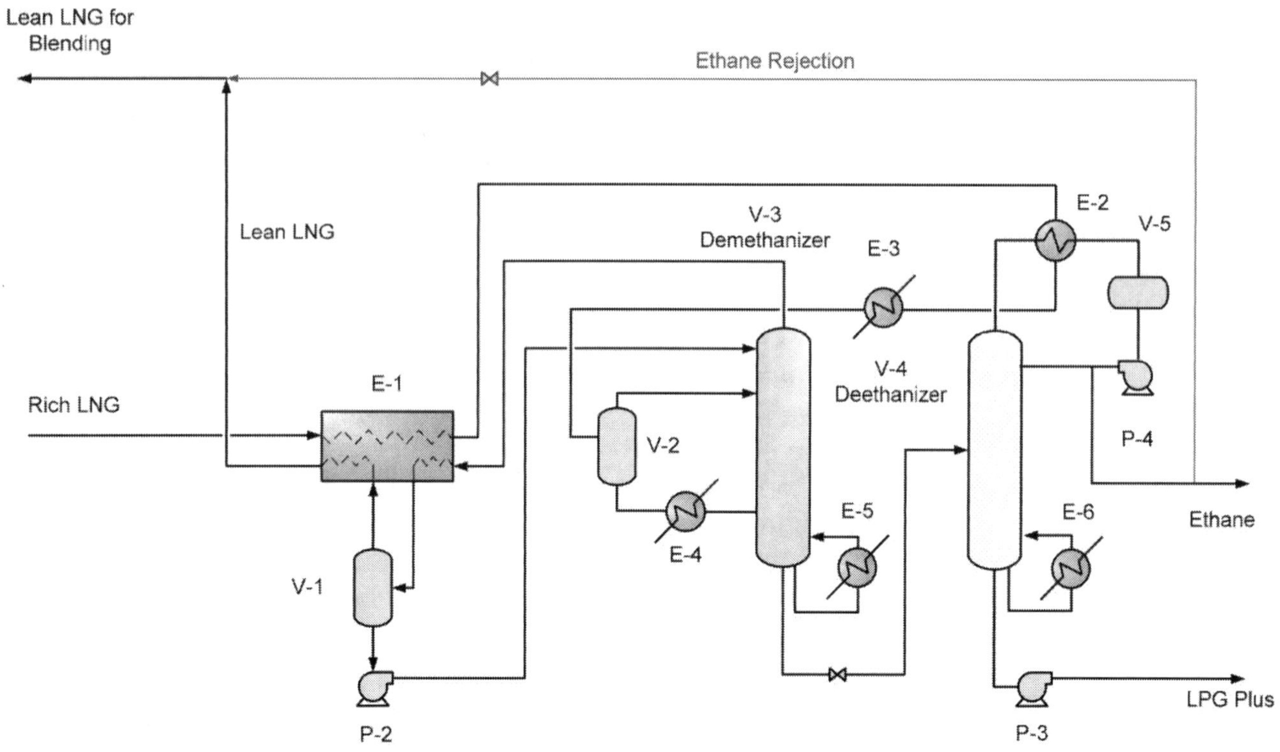

Figure 5– *Fluor CryoGas* LNG Conditioning Unit

Heat Exchange Composite Curves

The heat exchange composite curve for LNG vaporization is plotted against the residue gas condensation, demethanizer reflux and deethanizer reflux curves in Figure 6. The basis for these curves is a LNG regasification terminal with a total send-out of 2000 MMscfd. Per Figure 6, LNG is first heated from -250°F to about -117°F by the residue gas to supply the duty needed to condense the residue gas. The LNG is then heated to about -100°F by supplying the reflux condenser duties for the demethanizer and the deethanizer. As can be seen in Figure 6, most of the LNG heating duty is supplied by liquefaction of the residue gas.

Figure 7 is an overall LNG vaporization curve and includes the heat duty supplied by the seawater vaporizers. As shown in Figure 7, the LNG Btu conditioning unit contributes about 33% of the total heating duty needed for vaporizing the LNG. The remaining duty is supplied by conventional seawater open rack vaporizers.

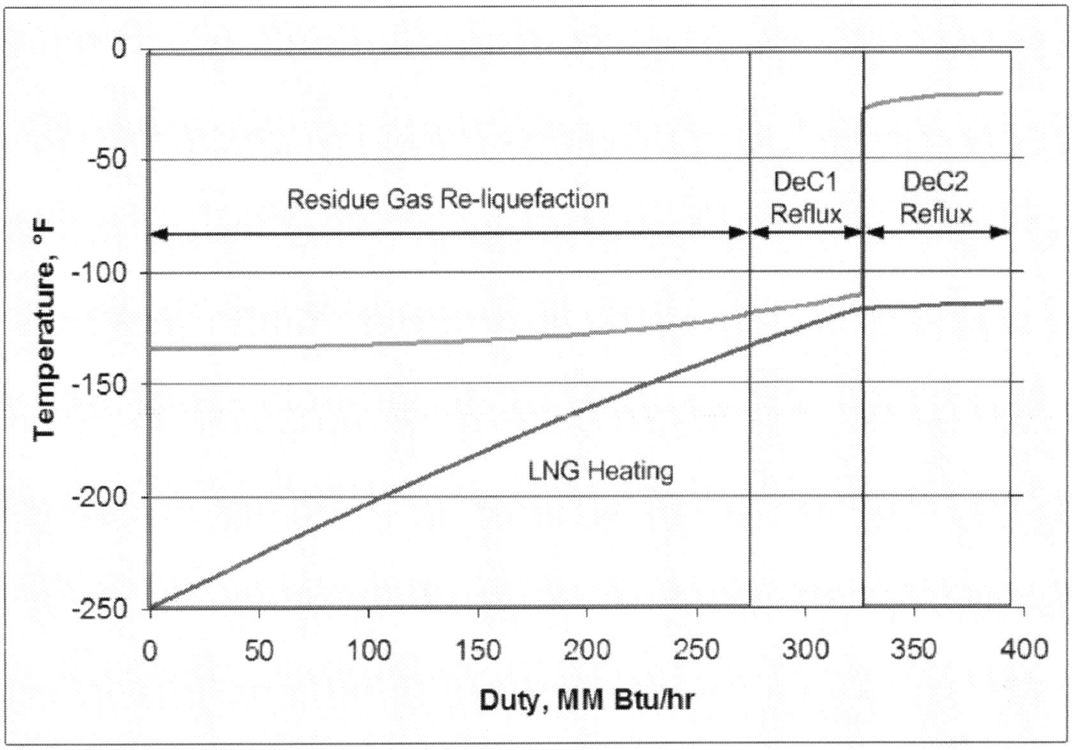

Figure 6–Composite Curves for LNG Conditioning

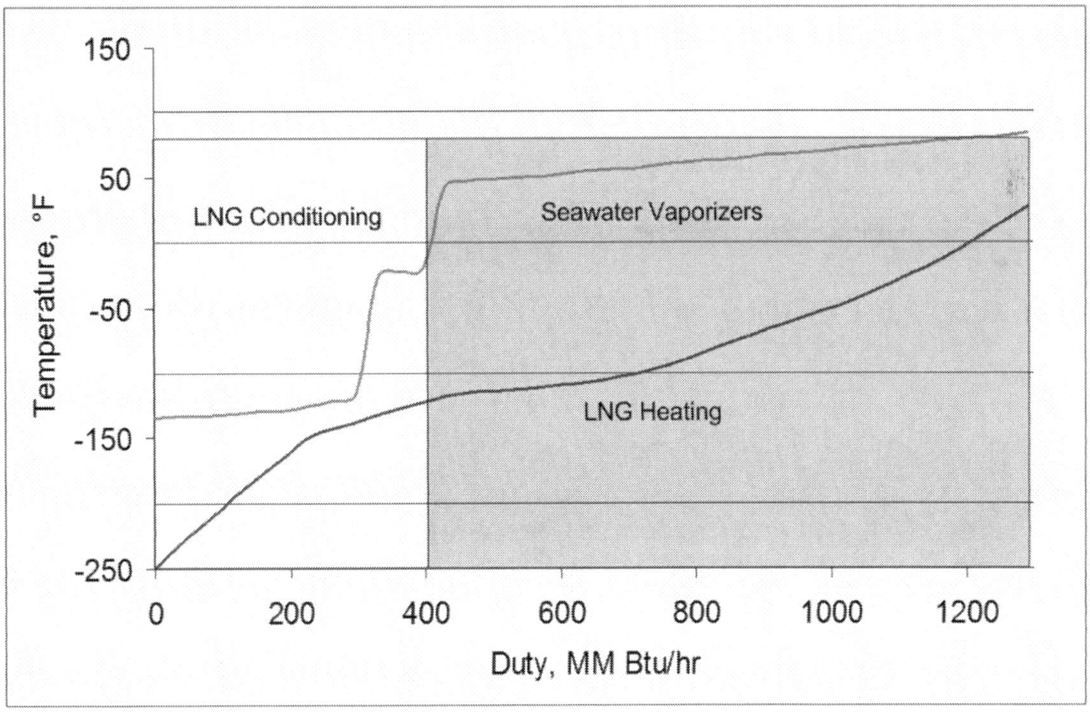

Figure 7–Overall Composite Curves for LNG Conditioning and Vaporization

Ethane Recovery Mode

Table 2 summarizes the ethane and propane production rates and recovery levels of the LNG conditioning unit for a LNG terminal with 2000 MMscfd send-out for the five LNGs of Table 1. The ethane and propane production rates are based on processing about 70% of the LNG in the LNG conditioning unit or about 1400 MMscfd of LNG. The conditioning unit produces a lean LNG with a heating value between 1020 to 1040 Btu/scf (depending on the LNG feed composition). This lean LNG is blended with the bypassed LNG to meet the sales gas heating value specification. The LNG conditioning unit can achieve 97% to 99% propane recovery and about 70 to 80% ethane recovery. It should be noted that the ethane and propane production rates could be increased by 30% if the total 2000 MMscfd send-out is processed in the LNG conditioning unit. This would require a slightly larger LNG conditioning unit.

As can be seen in Table 2, there are opportunities for ethane and propane production for all five LNG feeds. LNGs A and B yield about 45,000 BPD of propane plus components. C3+ production could be even higher if the total send-out of 2000 MMscfd is processed in the LNG conditioning unit. Ethane production is highest for LNG C at about 77,000 BPD. Since this much ethane may not be marketable, it may be necessary to reject some of this ethane to the sales gas pipeline or use some ethane for power generation.

LNG Feed	Feed Gas BTU/scf (HHV)	Ethane Productio n, BPD	Propane + Productio n, BPD	Propane Recovery %	Ethane Recovery %	Sales Gas BTU/scf (HHV)
A	1,165	49,470	44,100	99	69	1,075
B	1,142	29,760	44,670	97	69	1,063
C	1,112	77,280	2,270	99	74	1,062
D	1,078	30,380	16,760	97	70	1,037
E	1,045	21,140	2,950	99	77	1,025

Table 2–Ethane and Production Recoveries during Ethane Recovery for 2000 MMscfd of LNG Send-out

Ethane Rejection Mode

The ethane rejection mode of operation allows excess ethane that cannot be marketed to be re-injected to the sales gas pipeline. Ethane production is adjusted by blending excess ethane liquid with the lean LNG. The combined stream is then vaporized in the downstream LNG vaporizers. This method of ethane rejection avoids changing the process conditions in the LNG conditioning unit fractionation systems.

Therefore, the fractionation system parameters are the same for both ethane recovery and ethane rejection modes of operation. Another advantage, which is summarized in Table 3, is that high propane recovery is maintained in both ethane recovery modes.

Per Table 3, ethane must be removed from LNGs A, B and C to meet the 1075 Btu/scf heating value requirement. While this excess ethane is an ideal petrochemical feedstock, excess ethane could be used for blending if there is low Btu gas available in the gas distribution system or possibly used for power generation. Only LNGs D and C can meet a 1075 Btu/scf pipeline gas heating value requirement during total ethane rejection.

LNG Feed	Feed Gas BTU/scf (HHV)	Ethane Productio n, BPD	Propane + Productio n, BPD	Propane Recovery %	Ethane Recovery %	Sales Gas BTU/scf (HHV)
A	1,165	49,470	44,100	99	69	1,075
B	1,142	11,910	44,670	97	19	1,075
C	1,112	20,100	2,270	99	37	1,075
D	1,078	None	16,770	97	0	1,054
E	1,045	None	2,950	99	0	1,038

Table 3–Ethane and Propane Recoveries during Ethane Rejection for 2000 MMscfd LNG Send-out

Utility Consumption

The main utility requirements for the LNG conditioning unit are electric power and a heat source for reboiling the columns and vaporizing the LNG. Electric power is required for driving the reflux, product and heat medium pumps. The heat source can be seawater, waste heat from a power plant or a fired heater using fuel gas. The total power consumption for a LNG conditioning unit for a 2000 MMscfd regasification terminal is about 3 MW. The total heating requirement of the LNG regasification/ Btu conditioning facility is about the same with or without the LNG conditioning unit in operation.

Conclusions

LNG compositions and heating values vary. Many LNGs must be conditioned to meet U.S. pipeline gas heating value and inerts specifications. In this environment, including a LNG conditioning unit in a LNG regasification complex may provide an economic advantage. The *Fluor CryoGas* process can recover over 97% of the propane and 70% of the ethane in the LNG while producing a lean LNG for blending to meet U.S. pipeline gas heating value and inerts specifications. The *Fluor CryoGas* process can be operated in either an ethane recovery or an ethane rejection mode without changing operating parameters. The process can be used to control the pipeline gas heating value for different LNGs while adjusting ethane and propane recovery to match market conditions.

References Cited

1. Mak, J., Nielsen, R. B., Schulte, D., and Graham, C., "LNG Flexibility," Hydrocarbon Engineering, October, 2003.
2. Mak, J., "Mixed Fluid Power Rankine Cycle Using LNG as Heat Sink with LPG Extraction," AIChE 2005 Spring Meeting, LNG I – Plant & Operation, April 10–14, 2005, Atlanta, GA.

Biography of Speaker

John Y. Mak is a Technical Director and a Technical Fellow of Fluor's Energy and Chemical Group in the Aliso Viejo, California. He has been working in the Fluor's California office for over 30 years and has been involved in many oil and gas and energy projects. He leads the Fluor technical group in the development of new applications in the oil and gas and power industries. Recently, he has been focusing on LNG regasification technologies, and has developed several patented processes in the area of NGL recoveries and LNG regasification and power generation. He recently published several articles in these technical areas. He attended Oregon State University and received his BS in chemical engineering, and also received a MS degree in mechanical engineering from the University of California. He is a registered professional engineer in California.

Processes for High C2 Recovery from LNG –
Part II: Schemes based on Expander Technology

Stanley Huang, Roger Chen, Jame Yao, Douglas Elliot*
IPSI LLC
3000 Post Oak Boulevard
Houston, TX 77056
* To whom all correspondences be addressed

AIChE Spring Meeting, April 2006
6th Topical Conference on Natural Gas Utilization
Orlando, Florida, April 23-27, 2006

1.0 INTRODUCTION

Extracting ethane plus (C2+) components from LNG may serve different purposes. The most basic one is to meet pipeline gas interchangeability, for which a C2 recovery level below 50% would typically suffice [Rogers, 2003; Soto, 5]. A higher recovery level may become desirable depending on the marketability of the products. With the dwindling domestic production of natural gas, the accompanied natural gas liquid (NGL) production may also decrease [True, 2005]. Instead of importing more oil to make up the difference, it may become economically justifiable to extract the maximum amount of C2+ components from the imported rich LNG.

Schemes suitable for high C2 recovery from LNG can be classified into two approaches: refluxed DeC1 and expander-based schemes. The refluxed DeC1 relies on the refrigeration in the LNG to perform purification on the residue gas. Therefore, this approach is efficient and schemes have been discussed in Part I of this series [Huang et al., 2006].

This work will focus on the expander-based schemes. The expander technology has been widely accepted in the gas processing industry. It would be worthwhile to compare the performances of the two approaches so that rational choices can be made during process selections. Furthermore, the LNG application represents an extreme scenario that the feed stream contains plentiful refrigeration, which typical gas processing plants would be striving to generate. Studying the extreme scenario will shed light on the processes, particularly their intrinsic limitations. The conclusions will help the design works of conventional gas processing facilities.

2.0 TECHNICAL BACKGROUND

Schemes of C2 extraction from LNG can be divided into three categories: Residue Compression, Residue Compression and Condensing, and Residue Condensing [Huang et al., 2004]. Table 1 summarizes the three categories. Generally speaking, the process flexibility and the compression work go hand-in-hand. Schemes of Residue Compression are most flexible, but demand the highest compression work.

In the previous work, it was demonstrated that providing a refluxed DeC1 would enhance the performances of all the three categories. Furthermore, compressors are required to achieve a recovery level of 95%. Since Categories A or B already include compressors in their base flowsheet, this requirement does not change the designs appreciably. However, Category C does not include compressors in its base flowsheet. This requirement implies installing new compressors, in addition to reflux condensers and drums.

Table 1 also indicates that only Category A is suitable for converting to expander-based schemes. Since the expander requires finite pressure gradient to operate properly, the inlet LNG is pressure boosted by pumps to a relatively high level, sometimes close to 46 bar, the critical pressure of methane. When the inlet LNG is heated, the temperature can rise to a relatively high level without entering the two-phase domain.

After the turbo-expander, the overhead stream from DeC1 has a pressure below 35 bar. The returning residue is condensed by the inlet LNG which has a higher pressure. It is challenging to create a good temperature match so that complete condensation of the returning residue can be guaranteed. Therefore, Categories B and C are excluded in the following discussions. Only Category A will be included.

Some details of Part I are provided below because the results will be quoted later. Figure 1 depicts the conventional DeC1 designs for C2 extraction from LNG. For the scheme without reflux provisions (shown as blocked routes), the incoming LNG is boosted in pressure by the Primary Booster Pumps, goes through a pre-heater before entering the DeC1. The residue gas from DeC1 overhead is compressed to pipeline specifications and exported. A side stream from the pump discharge can be directed to the DeC1 top for rectification purposes. When the entire incoming stream is directed to the top, it represents the coldest top temperature achievable and the C2 recovery is at the maximum. The lower curve in Figure 2 shows the limit of un-refluxed DeC1 operations.

For the scheme with reflux provisions (shown as un-blocked routes in the dotted box), the refrigeration in the incoming LNG can be utilized to support the operation of the LNG pre-heater/ condenser. As a result, the C2 recovery level in the DeC1 is dramatically improved. The upper curve in Figure 2 shows the C2 recovery limit of a refluxed DeC1. The constraints to achieve even higher C2 recovery is the temperature approach in the LNG pre-heater/ condenser.

3.0 SCOPE OF WORK

Figure 3 presents a generalized scheme for an expander-based process. The common components in the expander scheme include a separator, expander, optional subcooler, and DeC1. The incoming LNG is boosted in pressure, and preheated to a two-phase region before entering the expander box. The vapor stream from the separator is directed, in full or partially, to an expander, where additional refrigeration is generated. The returning low-pressure (LP) residue stream is compressed by a compressor driven by the expander. A second compressor, driven by a motor, further boosts the pressure to pipeline specifications.

The exact arrangement inside this expander box is an art, for which hundreds of processes are protected by patents. A key to the efficiency of the box is the refrigeration integration at low temperature regions. The sub-cooler is provided for this purpose. The refrigeration recovery outside of the expander box is unimportant in this work because the incoming LNG contains plentiful refrigeration. In a typical gas processing plant, however, the refrigeration integration outside the expander box is another key to the overall efficiency of the process.

This work will discuss only the main features of selected schemes from a conceptual level. No design details will be addressed. This decision is not only from practical considerations due to the large number of patents, but also to avoid possible commercial implications. Usually, patents would address details beyond current discussions. It would be these details that dictate the overall performance of a design.

4.0 APPROACH

Table 2 lists two LNG compositions used in this work. The composition denoted as Medium LNG is consistent with the previous work [Huang et al., 2006]. The one denoted as Rich LNG [McCartney, 2003] will be used in sensitivity analysis only. The phase envelopes of both LNG compositions are shown in Figure 4. The Rich LNG has an enlarged two-phase region toward higher temperatures and pressures.

In the following discussions, a procedure will be defined which demands systematically lowering the separator temperature from a relatively high starting point. The procedure for the Medium LNG is schematically presented in Figure 4 as the Track of Testing Separator Temperature. Physically, this is achieved by controlling the heat input to the LNG pre-heater.

The lower temperature limit of the above procedure is arbitrarily set at -100 °C in this work. Since the specified compositions can only exist in liquid forms at that condition, turbo-expander will be rendered unusable. Whenever it is impossible to operate the turbo-expander, it is replaced by a JT valve and the procedure continues. It is understood that that the expander-driven compressor should be taken out of service and all compression work will be assumed by the motor-driven unit.

All simulations in this work are performed using HYSYS. Table 2 lists pertinent simulation parameters. For a selected scheme, once the DeC1 operating pressure is fixed, a series of C2 recovery level against the separator temperature can be uniquely determined. The results will be presented in the next section.

A heat pump is a mechanism capable of moving heat from low to high temperatures. Physically, a heat pump would include a compressor, a cooler for rejecting heat to a high-temperature sink, an expander or JT valve, and an exchanger to take heat from the low-temperature source. This work will adopt thermodynamic heat pumps in order to hold discussions simple and conceptual. The most famous example of a thermodynamic heat pump is the Carnot engine [Elliott and Lira, 1998].

The heat pumps in this work are assumed to have an efficiency of 50% of the Carnot engine. For example, to move 1 MW of heat from -100 °C (173 K) to -50 °C (223 K), a Carnot engine requires 0.29 (= 223/173-1) MW of work. In this work, twice of the amount, or 0.58 MW, is required due to the reduced efficiency.

5.0 RESULTS AND DEFINITION OF SIMAR

In addition to presenting the simulation results, this section will introduce a new terminology called SIMAR, which stands for System Intrinsic Maximum Recovery. The concept of SIMAR was originated from observing performances of gas processing schemes. Most expander-based gas process schemes are constrained by available refrigeration. Yet, certain schemes seem to demonstrate upper limits even when the external refrigeration supply is unlimited. Since LNG inlet stream contains plentiful refrigeration, it provides a unique opportunity to explore this intrinsic limitation.

5.1 CLASSIFICATION OF EXPANDER-BASED SCHEMES

The expander-based schemes are divided into three types in this work: XPDR 1, 2, and 3. Table 3 lists the descriptions and Figures 5, 6, and 7 show typical flows, respectively. Just briefly, the classification is based on the handling of separator top stream: whether the stream is not rectified (XPDR 1), fully rectified (XPDR 2), or self-rectified by a side-draw from itself (XPDR 3). The example of XPDR 2 in Figure 6 uses a recycled and chilled high-pressure (HP) residue as the reflux to DeC1. There are other potential sources of reflux streams, for example, side-draws from DeC1 or streams from other columns.

The examples for the three types may correspond to existing commercial processes [Lee et al, 1999]. However, this work will only highlight the characteristics of the selected schemes. It is not the intention of this work to compare performances of commercial processes.

5.1.1 TWO TYPES OF SYSTEM BEHAVIOR

Figure 8 shows two kinds of behavior as the separator temperature decreases. For XPDR 1 and 2, the C2 recovery level increases monotonically. When the separator temperature decreases, C2 recovery of XPDR 1 and 2 steadily improves. This trend continues even after the vapor phase totally disappears. The test temperature stops somewhat arbitrarily at -100 °C. Nonetheless, the trend is expected to continue until -160 °C, the LNG inlet temperature. At that condition, no pre-heater is required and the impact of JT valve is insignificant. The system performance approaches an un-refluxed DeC1, which was discussed in Part I of this series.

Yet for XPDR 3, the C2 recovery level shows a maximum at a temperature where the fluid in the separator has about 50% vapor fraction (VF). For XPDR 3, blindly increasing the refrigeration supply to chill the separator fluid would not always improve the C2 recovery.

5.1.2 SIMAR: SYSTEM INTRINSIC MAXIMUM RECOVERY

This section will introduce the procedural definition of SIMAR. It will be demonstrated that SIMAR reflects the characteristics of C2 recovery schemes. The application of SIMAR can be extended beyond LNG to the gas processing fields. The extension will be covered in PART III of this series.

The procedures to determine the SIMAR for a selected process scheme are summarized below:

1. Plot the phase envelop of the feed LNG.
2. Specify the inlet LNG pressure level. Since the inlet pressure is generated by the Primary LNG Booster Pumps, the level can be specified to facilitate expander operations. Often the pressure would be below the critical pressure of the LNG. Impact of supercritical pressure will be addressed later.
3. Select the operating pressure of DeC1.
4. Starting from a relatively high temperature, systematically decrease the separator temperature, e.g. by specifying different levels of VF. The C2 recovery level should have gradual improvements, as shown in Figure 8.
5. If the C2 recovery demonstrates a maximum, this is the SIMAR point at this column operating pressure.
6. If the C2 recovery demonstrates a monotonically increasing trend, the SIMAR point is selected at a temperature where the separator fluid has 30% VF. This selection reflects a practical consideration that an expander is unlikely to be installed if the vapor phase is below this level.
7. Change the operating pressure of DeC1 and repeat steps 4 to 6.
8. The entire SIMAR curve for the process scheme is created by plotting C2 recovery against DeC1 operating pressure.

The three lower curves in Figure 9 are the SIMAR curves for the three XPDR categories. XPDR 1 is the lowest as it has no reflux to the DeC1. XPDR 2 and 3 are comparable in C2 recovery, although each shows slight advantages over the other at different DeC1 operating pressures.

The two upper curves in Figure 9 are re-plotted from Figure 2, which represents refluxed and un-refluxed DeC1 schemes. It is obvious that direct utilization of LNG refrigeration for purifying residue can achieve higher C2 recovery at the same level of compression duty. Therefore, it is unlikely for expander technology to find applications in LNG re-gasification plants.

5.2 IMPACT OF HEAT PUMPS

In the discussions so far, all the components in the expander box are "passive", i.e. the pressure profile does not increase across the elements. Compressors represent a unique application because suitable combinations may form heat pumps. Heat pumps can alter the original system temperature distributions. As a result, the existence of heat pumps within the expander box may impact the C2 recovery levels significantly.

Figure 10 shows an example of a heat pump which moves heat from the top tray of a DeC1 to a lower one. In additional to the benefit of a lean reflux stream, the temperature at the top tray is lowered and the C2 recovery is further improved [Pitman, et al., 1998].

Figure 11 shows the impact of this type of heat pump. The heat pump has an efficiency corresponding to 50% of a Carnot engine and it moves heat from the top tray to tray No. 4. No material recycle streams were created in the HYSYS simulations.

As anticipated, the more heat moved, the more improvement in C2 recovery is observed. There is an upper limit, though. As more heat is moved to a lower tray, the temperature there will rise above its surrounding ones. The temperature spike within the column cannot go on forever without disrupting the column operations.

5.3 IMPACT OF INLET PRESSURE

Figure 12 shows the impact of varying the discharge pressure of the LNG Primary Booster Pumps. The scheme used is XPDR 3. Generally, as the LNG primary pump discharge pressure increases, so does the SIMAR temperature at the separator. The C2 recovery level suffers slightly due to the higher temperature. Furthermore, the returning residue gas is warmed to a higher temperature before entering the re-compressors. This results in a higher compression work.

When the LNG pump discharge pressure increases to 70 bars in the super-critical region, an unusual phenomenon can be observed during the determination SIMAR per HYSYS simulations. As the separator temperature was systematically lowered, the C2 recovery improved as anticipated. However, there was a minimum temperature below which the expander simply ceased to function. The reason was speculated to be that the separator fluid became too liquid-like for the expander to work properly. The actual expander performance should be verified by the vendors. In Figure 12, the SIMAR temperature is set at 2 °C above the aforementioned minimum temperature.

5.4 IMPACT OF FEED COMPOSITION

Figure 13 shows the impact of LNG compositions. The LNG Booster Pump discharge pressure for the Rich LNG is 62 bar, whereas for the Medium LNG is 50 bar. The different discharge pressure is intended to maintain the same reduced pressure of 0.85. The reduced pressure is defined as the ratio of actual pressure to the LNG critical pressure, which is also listed in Table 2. The scheme used is XPDR 3.

The shape of SIMAR curve is weakly dependent on the LNG compositions. This is anticipated because the residue compositions only deviate insignificantly from pure methane. The Rich LNG requires slightly more compression work at the same level of C2 recovery mainly due to the higher inlet pressure as explained in the previous section.

6.0 SIMAR APPLIED TO PATENTED PROCESSES

The procedures of determining SIMAR curves are applied to selected patented processes and the results are presented in Figure 14. The identification of individual processes is intentionally left out. Most of the patented processes improve the C2 recovery from the generic XPDR 3, however, there are exceptions. Users' discretion is required in the process selection.

7.0 CONCLUSIONS

The conclusions of this work are summarized as follows:

1. Comparing the two approaches of refluxed DeC1 versus expander technology, the former can achieve higher C2 recovery at the same level of compression duty. Therefore, it is unlikely for expander technology to find applications in LNG re-gasification plants.
2. The expander-based schemes can be divided into three types: XPDR 1, 2, and 3. the classification is based on the handling of separator top stream: whether it is not rectified (XPDR 1), fully rectified (XPDR 2), or self-rectified by a side-draw from itself (XPDR 3).
3. When the C2 recovery is plotted against the decreasing separator temperature, there are two types of behavior: XPDR 1 and 2 shows monotonic improvement whereas XPDR 3 shows a maximum. The behavior types may shift depending on operating parameters, such as LNG inlet pressure (LNG Booster Pump discharge pressure).

4. The procedures to determine SIMAR are defined for process schemes and examples are provided. Also, sensitivity analyses for LNG compositions and LNG inlet pressures on SIMAR are performed and presented.
5. Heat pumps used in this work are defined. Heat pumps embedded inside the expander schemes can alter temperature distributions and improve the system performances significantly.
6. The SIMAR curves for different processes reveal process characteristics. SIMAR can be a useful tool in technology selections.

8.0 REFERENCES

Elliott, J.R., Lira, C.T.; "Introductory Chemical Engineering Thermodynamics", Prentice Hall, NJ (1998).

Huang, S., Coyle, D., Cho, J., and Durr, C.; "Select the Optimum Extraction Method for LNG Regasification", Hydrocarbon Processing, July (2004).

Huang, S., Cook, D., Yao, J., and Elliot, D.; "Processes for high C2 recovery from LNG – Part I: Schemes based on Refluxed Demethanizer", paper accepted for presentation in AIChE 2006 Spring National Meeting, Orlando, FL (2006).

Lee, R., Yao, J., Elliot, D., "Flexibility, efficiency to characterize gas-processing technologies", Oil & Gas Journal, Issue of December 13 (1999).

McCartney, D.; "Gas Conditioning for Imported LNG", paper presented in 82nd GPA Annual Convention in San Antonio, TX, 2003.

Pitman, R.N., Hudson, H.M., Wilkinson, J.D., "Next generation processes for NGL/LPG recovery", paper presented at the 77th Annual GPA Convention, Dallas, TX (1998).

Rogers, D. R., "Gas Interchangeability and Its Effects on U.S. Import Plans", Pipeline & Gas Journal, September (2003).

Soto, A.K., "US Import Terminals - Interest Grows in Gas Interchangeability, Quality", Oil & Gas Journal, Issue of October 24 (2005).

True, W., "Regional shifts in worldwide capacity, production slow", Oil & Gas Journal, Issue of June 27 (2005).

9.0 List of Tables

Table 1 Three categories for schemes of C2 Extraction from LNG

Cat.	Description	Base Design		Refluxed DeC1	Expander-based technology
		Comp. work	Flexibility	Additions to achieve 95% recovery	
A	Residue Compression	High	Good	Reflux condenser and drum	Suitable for purpose
B	Residue Compression & Condensing	Medium	Fair	Reflux condenser and drum	Temperature mismatch in LNG pre-heater/ condenser for residue condensing
C	Residue Condensing	Low to none	Poor	Reflux condenser and drum + Compressor	Temperature mismatch in LNG pre-heater/ condenser for residue condensing

Table 2 Simulation parameters used in this work

LNG inlet	Medium LNG	Rich LNG (sensitivity analysis only)
Inlet temperature	-160 °C	-160 °C
Inlet pressure to C_2+ facilities	10 bar	10 bar
Primary Booster Pump discharge pressure unless specified otherwise	50 bar	50 bar
Plant capacity, inlet LNG	100 MMscfd	100 MMscfd
Send-out gas pressure	100 bar	100 bar
Send-out gas temperature	1 °C	1 °C
Components	Mole Fraction	Mole Fraction
Nitrogen	0.00127	0.00430
CO2	0.00008	0.00000
Methane	0.90692	0.84560
Ethane	0.08617	0.10930
Propane	0.00552	0.03210
i-Butane	0.00003	0.00470
n-Butane	0.00001	0.00380
i-Pentane	0.00000	0.00020
n-Pentane	0.00000	0.00000
TOTAL	1.00000	1.00000
HHV, MJ/Sm3	40.3	40.3
Btu/SCF	1,082	1,157
Critical pressure, bar	58.63	72.56
Expander polytropic efficiency	80%	80%
Compressor polytropic efficiency	80%	80%
Pressure drop through plate-fin exchanger	0.2-0.3 bar	0.2-0.3 bar
Pressure drop through air coolers	0.6 bar	0.6 bar
Minimum delta T in plate-fin exchanger	2.5 ± 0.2 °C	2.5 ± 0.2 °C
DeC1 operating pressure	17 to 42 bar	17 to 42 bar
Number of trays in DeC1 column	15	15
C1 composition in DeC1 bottom product	1%	1%

Table 3 Categories for expander-based C2 extraction schemes

Category	Figure	Description	Limiting case for C2 recovery
XPDR 1	5	Separator top not rectified	Non-refluxed DeC1
XPDR 2	6	Separator top fully rectified	Refluxed DeC1
XPDR 3	7	Separator top self-rectified	None

10.0 List of Figures

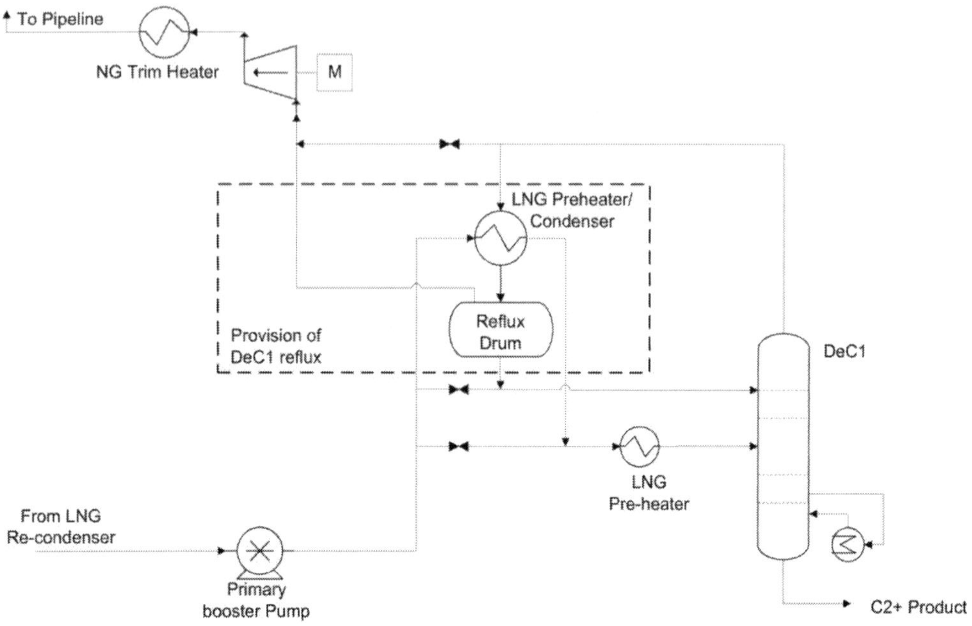

Figure 1 Residue Compression with refluxed and un-refluxed [blocked] DeC1

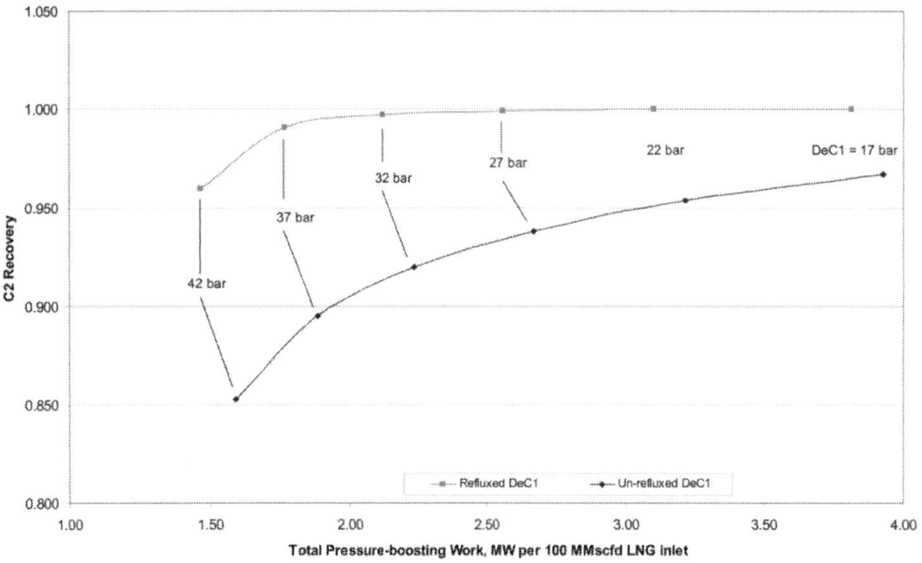

Figure 2 C2 recovery limit using conventional DeC1 designs

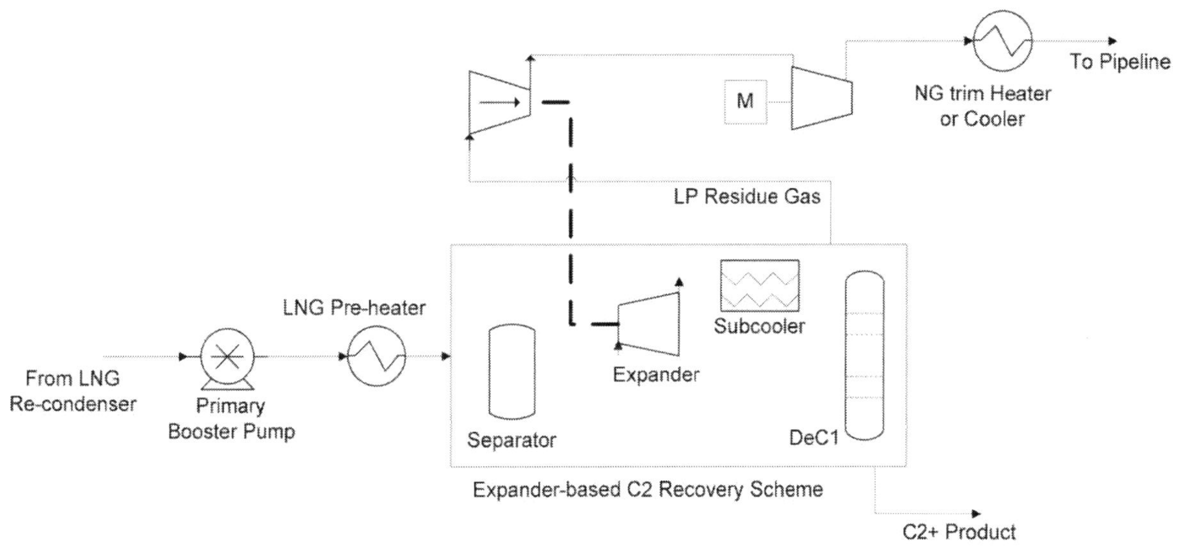

Figure 3 Residue Compression, combined with expander-based C2 recovery schemes

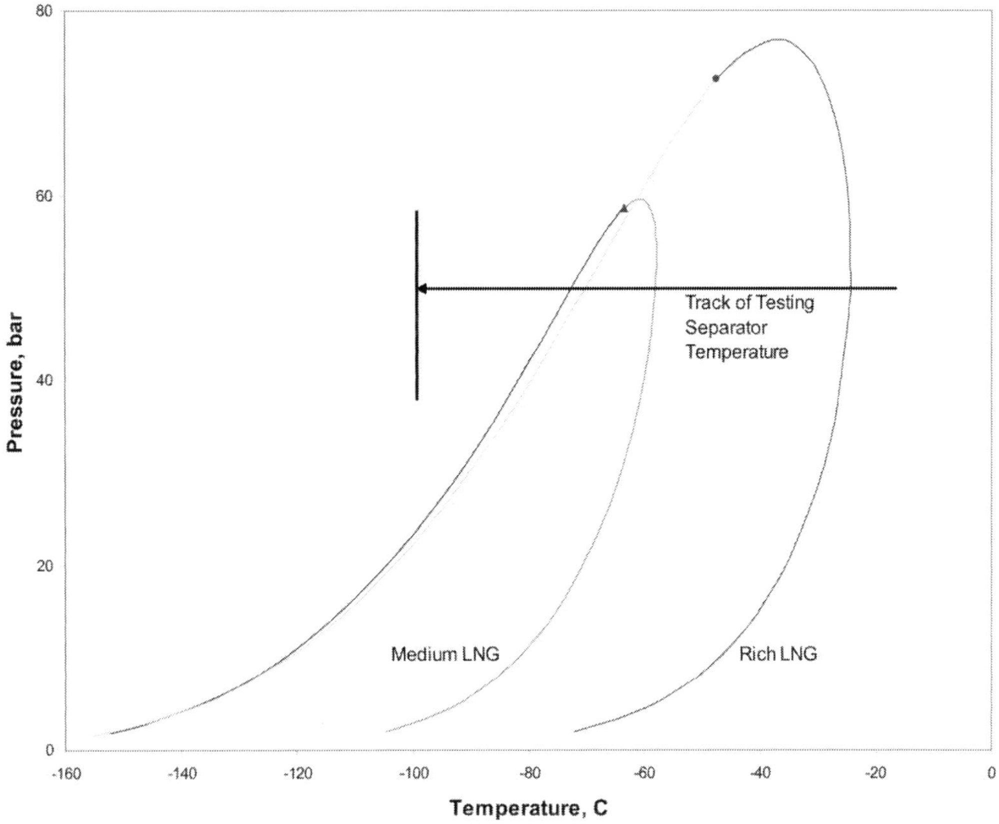

Figure 4 Phase envelope of inlet LNG

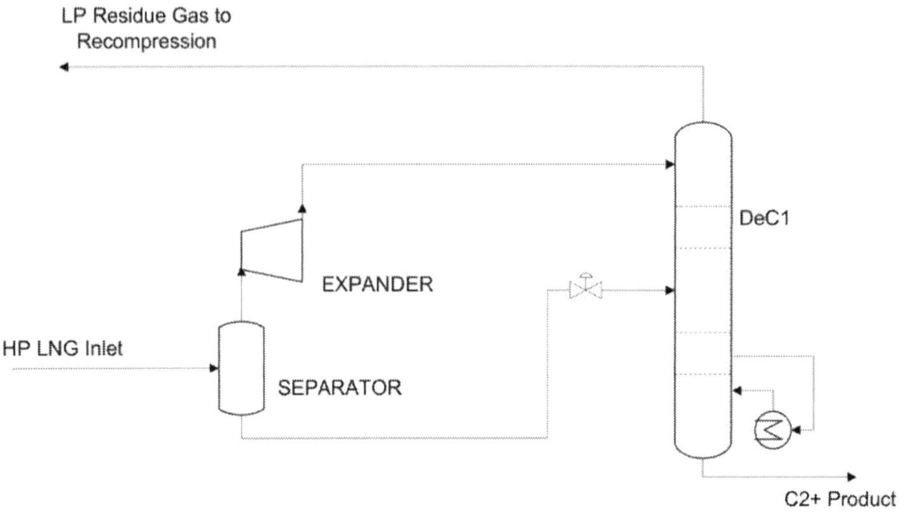

Figure 5 Category XPDR 1: separator top not rectified

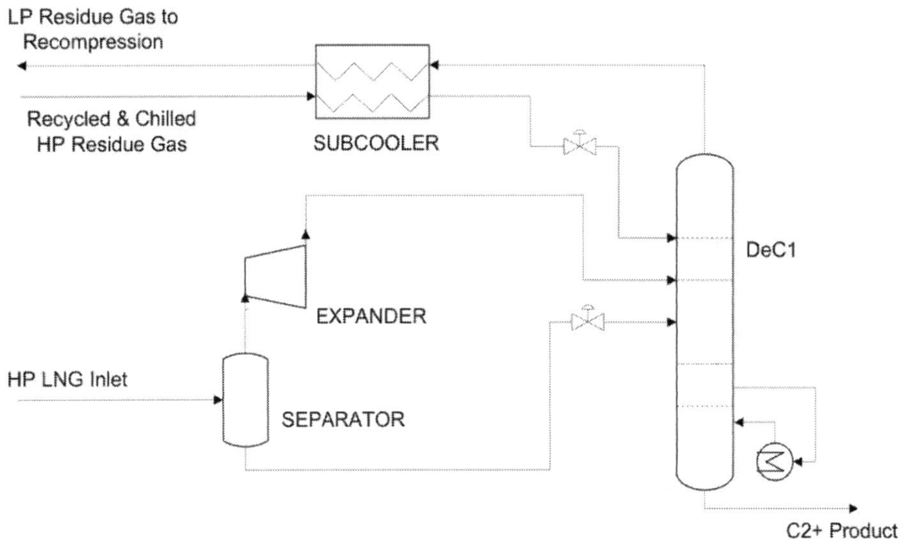

Figure 6 Category XPDR 2: separator top fully rectified

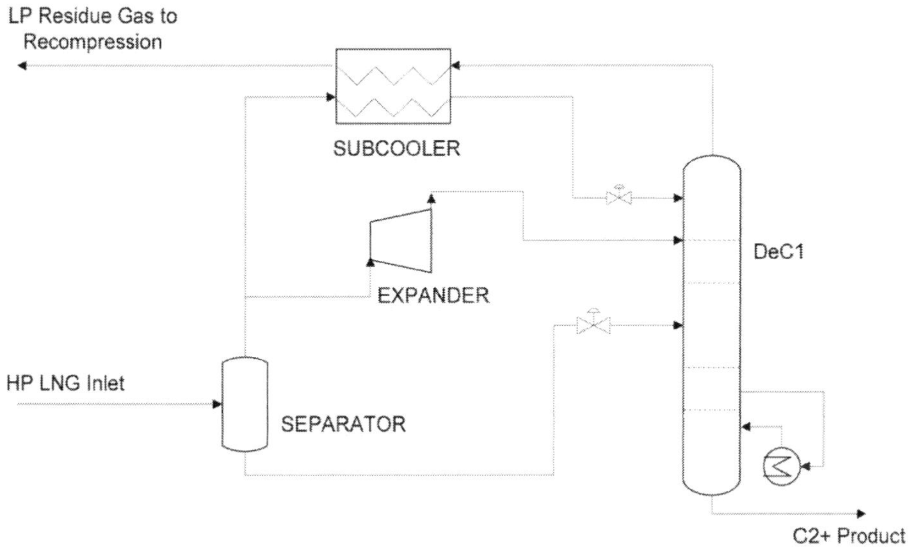

Figure 7 Category XPDR 3: separator top self-rectified

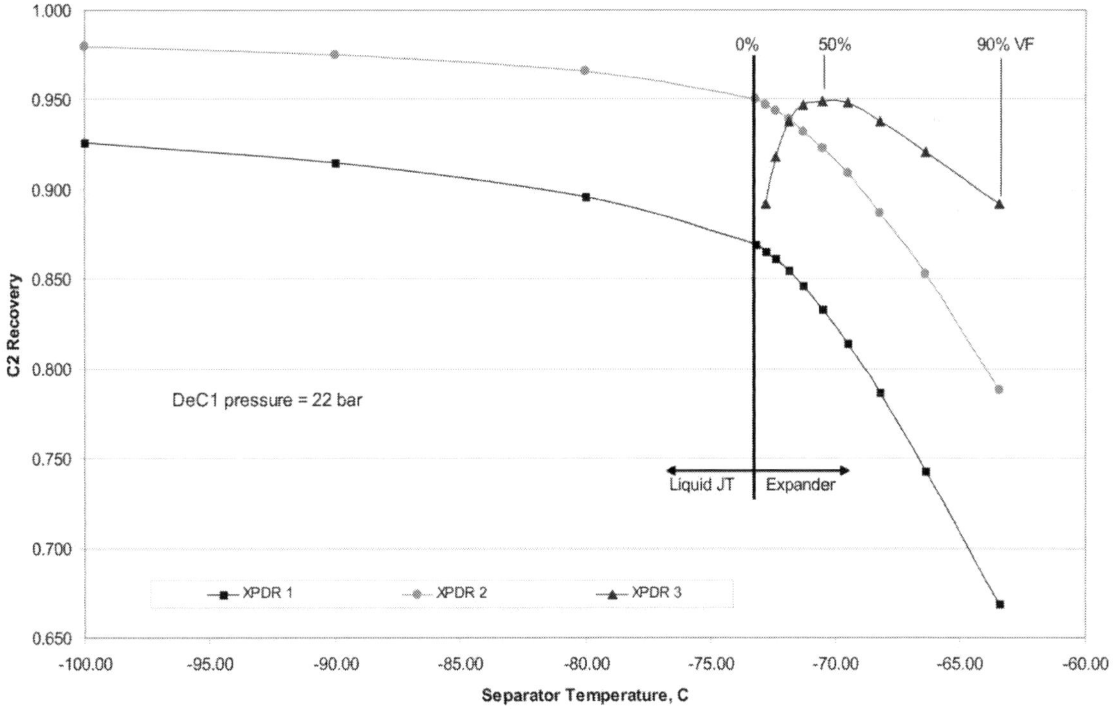

Figure 8 Two types of behavior for different categories.
VF stands for vapor fraction

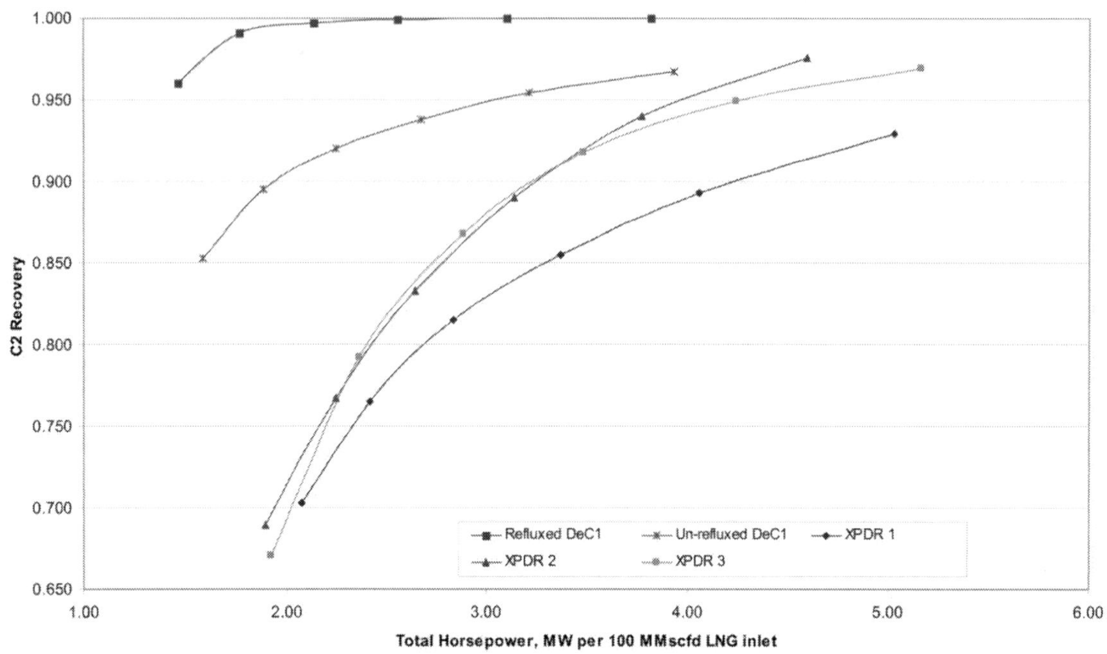

Figure 9 Comparing SIMAR curves for three XPDR categories

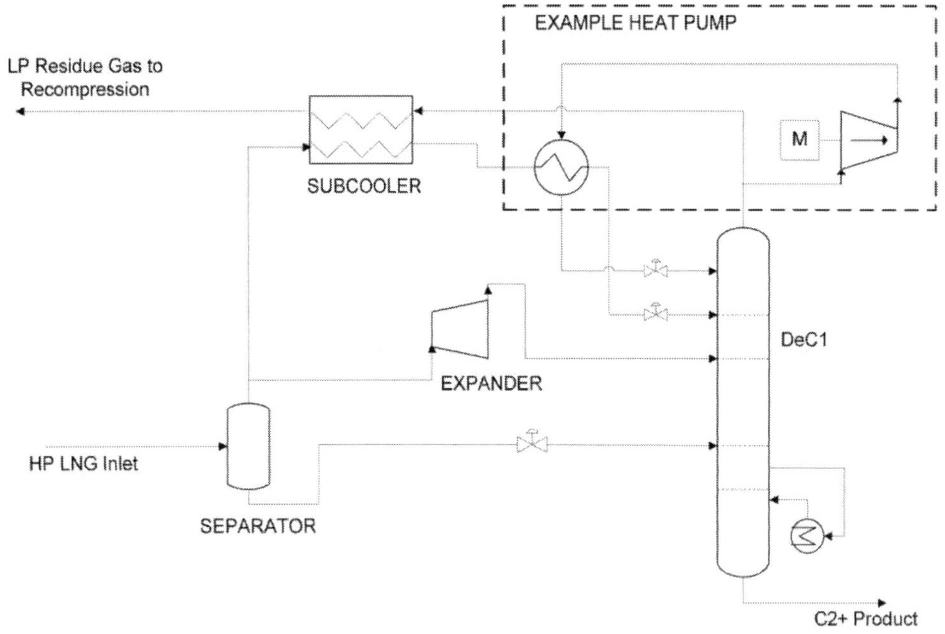

Figure 10 An example showing a heat pump embedded in Category XPDR 3

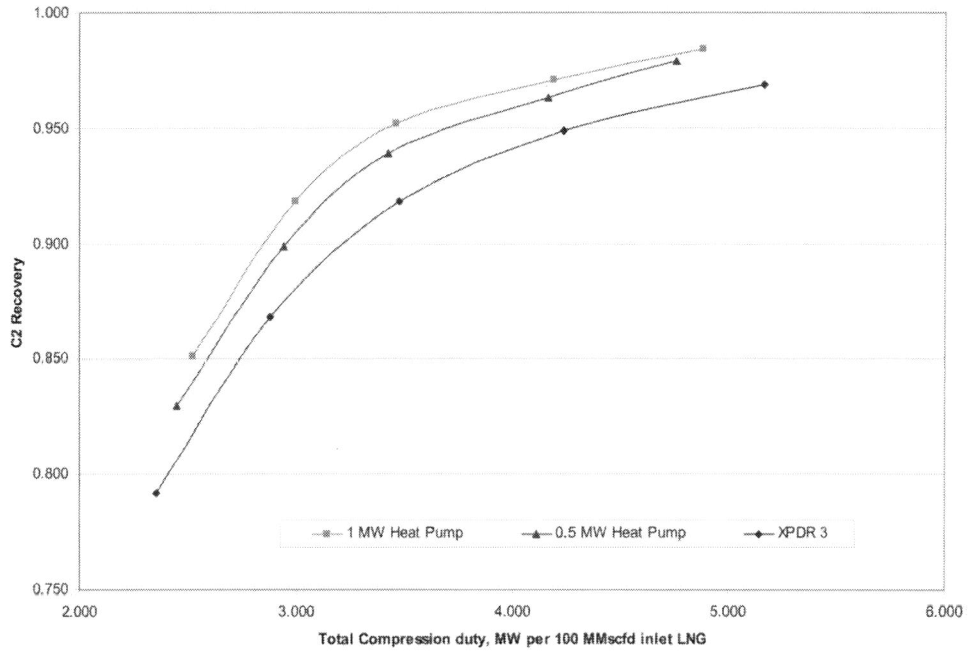

Figure 11 Impact of heat pump on C2 recovery

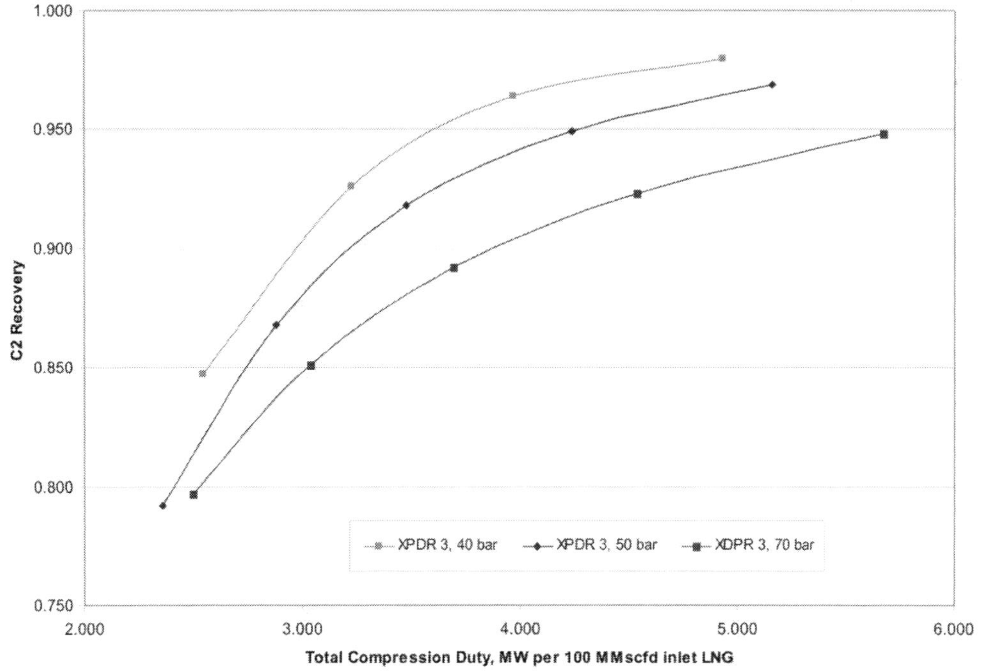

Figure 12 Impact of LNG inlet pressure on C2 recovery

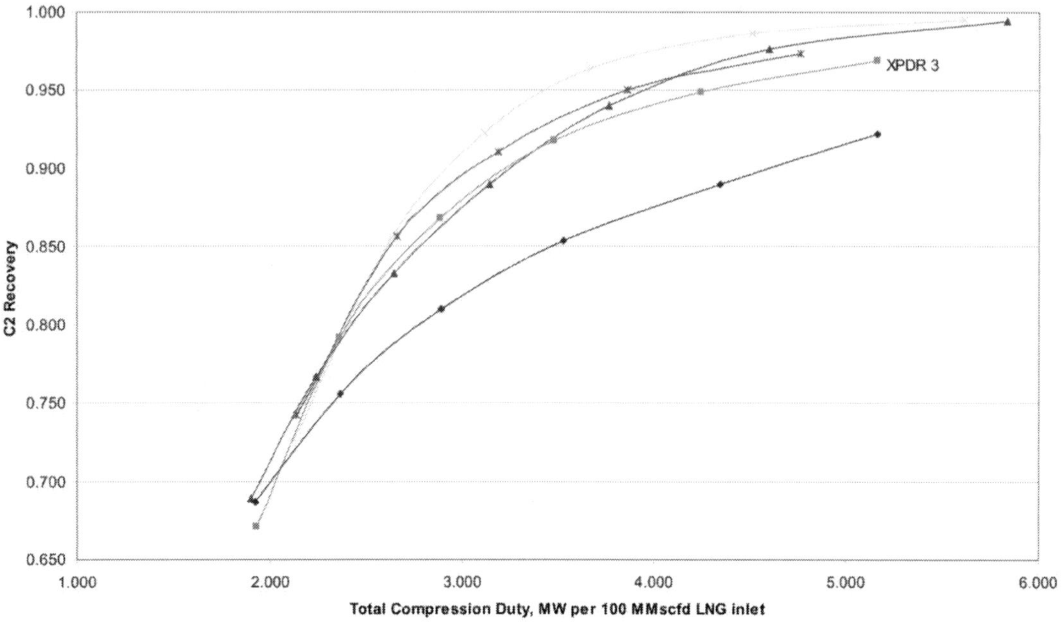

Figure 14 SIMAR applied to patented C2 recovery processes

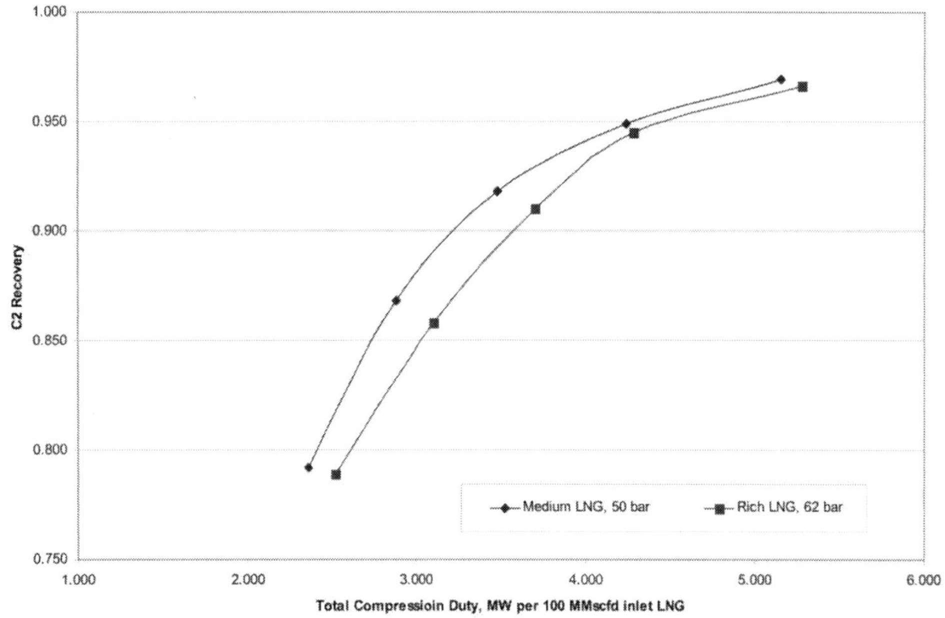

Figure 13 Impact of LNG composition on C2 recovery

Biography

Stanley Huang is a Staff LNG Process Engineer in Chevron Energy Technology Company. He is a registered professional engineer and has eighteen years of experiences in the hydrocarbon industry. Stanley graduated with a Ph. D. degree in chemical engineering from Purdue University in 1986.

Pre-cooling Concepts for Large Base Load LNG Plants

Heinz C. Bauer (speaker)
Vice President Technology, LNG and Gas Processing
Barbara Buttinger
Process Engineer, LNG and Gas Processing
Christiane Kerber
Process Engineer, LNG Technology
Hubert E. Franke
Senior Process Engineer, LNG and Gas Processing
Linde AG, Linde Engineering, 82049 Hoellriegelskreuth, Germany

Arne O. Fredheim
Senior Research Advisor, Research and Technology
Statoil ASA, 7005 Trondheim, Norway

AIChE Spring Meeting, April 2006
6th Topical Conference on Natural Gas Utilization
Orlando April 23-27, 2006

Introduction

All relevant LNG processes for large capacities are based on two or more refrigeration cycles. The pre-cooling cycle seems to be the technologically least challenging one, as the temperature level is still moderate. In reality, however, the pre-cooling cycle has to reject almost the entire excess heat from natural gas and the cold refrigeration cycle(s). This paper will discuss the main conceptual elements of pre-cooling cycles in modern LNG processes with special focus on the Mixed Fluid Cascade (MFC®) in cold climate:

- Type of heat exchanger (plate-fin, spool wound, kettle)
- Mixed refrigerant versus propane
- Load balancing between the refrigerant compressors

Heat Exchangers

The LNG industry uses many different types of heat exchangers in the pre-cooling section of the process[i]. Kettle type heat exchangers are preferred in APCI's propane pre-cooling section, ConocoPhillips uses brazed aluminum plate-fin heat exchangers and Shell Global Solutions integrates spool wound heat exchangers also in the warm section of the DMR process.

Linde AG, Linde Engineering Division (called Linde in this paper) is a new player in the base load LNG business. In the first application of the MFC® process[ii] (co-developed and co-owned by Statoil and Linde) for the Snøhvit LNG project in Hammerfest, Norway plate-fin heat exchangers have been selected for the pre-cooling cycle. This decision, however, has to be revisited on a case by case basis to identify the most convincing solution for every project's and each customer's specific needs.

Linde is not all biased as far as the selection of the optimum cryogenic heat exchanger is concerned. Linde's portfolio includes the largest vacuum brazed plate-fin heat exchangers worldwide with a core volume beyond 25 m^3, which are usually installed together with vessels, interconnecting piping and instrumentation inside of a cold box. Alternatively, the cores can be integrated in one or more pressure vessels (block-in-kettle), which again can be part of a cold box design or can be installed as loose items. Spool wound heat exchangers, the second pillar of Linde's cryogenic heat transfer equipment portfolio, presently can be manufactured and shipped with up to 40,000 m^2 surface.

In a case study for a location north of the Arctic Circle a large LNG train with about 9 mtpa LNG and 300 MW refrigeration compressor power has been considered. The question was: can spool wound heat exchangers still be manufactured as single items or is parallel equipment required?

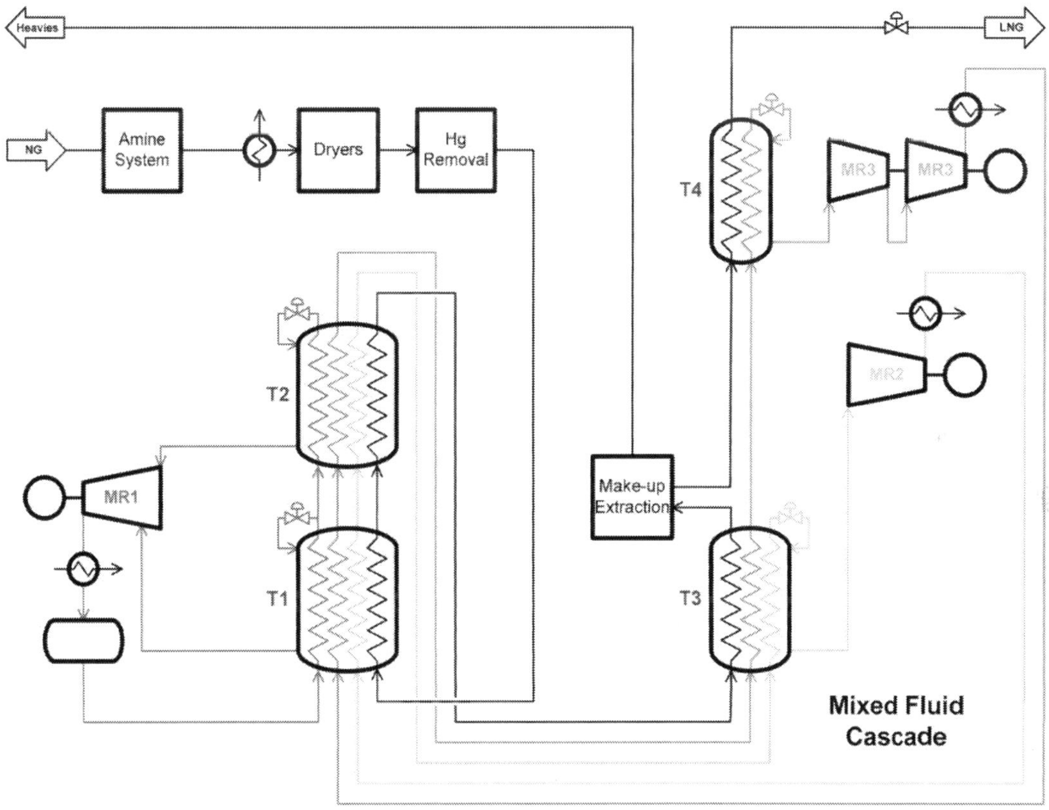

Figure 1
MFC® Process with Spool Wound Heat Exchangers for All Refrigeration Cycles

Figure 1 shows an all spool wound heat exchanger version of the MFC® process with two temperature levels for the pre-cooling cycle MR1. The details of the second and third cycle MR2 and MR3 as well as the NGL extraction scheme will not be discussed in this paper. Both pre-cooling exchangers for the temperature ranges T1 and T2 have been sized to slightly above 5 m diameter and less than 40,000 m^2 surface each. Thus, they are considered as feasible within the present limitations for workshops and transportation.

The traditional MFC® concept[iii] with plate-fin heat exchangers has been upgraded for better scalability and reduced piping complexity using two-stream heat exchangers only (see Figure 2). Eight different types of exchangers with dedicated services are used. Thus, no more than four parallel cores per service are required even for the large train capacity of 9 mtpa. This concept minimizes maldistribution between parallel heat exchangers of identical service.

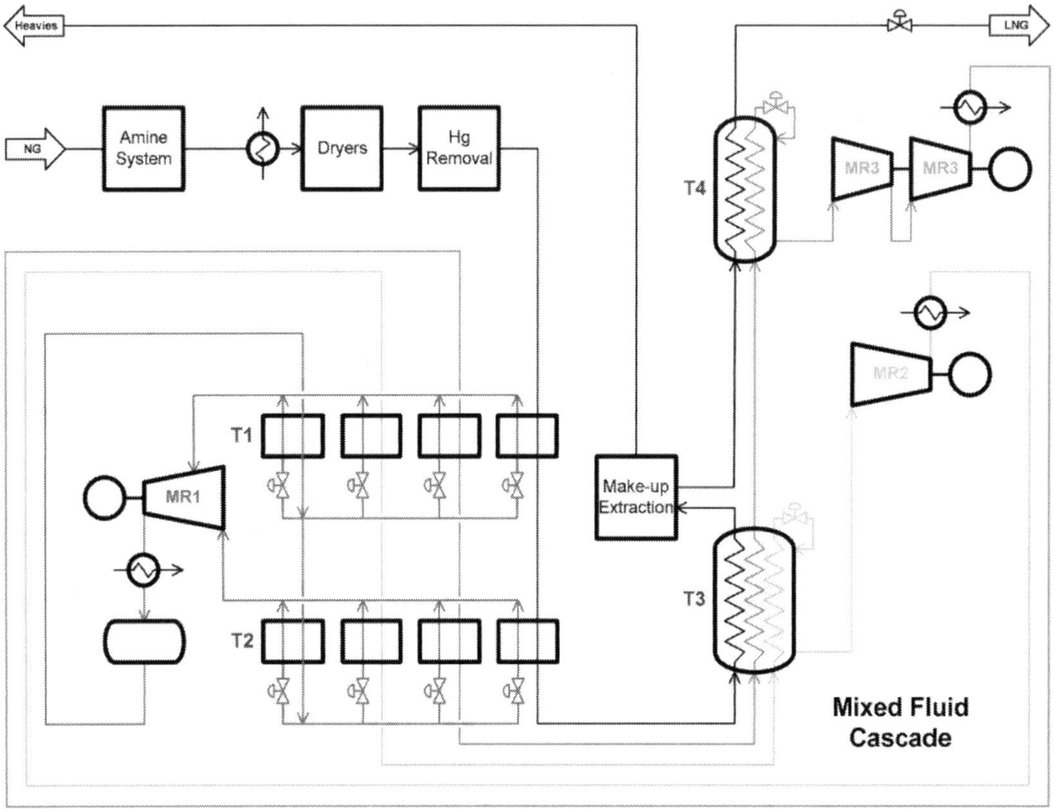

Figure 2
MFC® Process with Plate-fin Heat Exchangers in Pre-cooling

Comparing the solutions in Figure 1 and 2 the conclusion is as follows: Spool wound heat exchangers are more expensive and do not save much plot area. The arguably higher robustness, however, is offset by a reduced train capacity for a given power consumption. In more detail: the significant pressure drop (several bar) of natural gas in the tube bundles of the pre-cooling coils reduces the thermal efficiency by 3-5%. Hence, the unit cost of LNG production will be lower for the design with plate-fin heat exchangers. However, at request from the customer, Linde can design and supply spool wound heat exchangers also for the pre-cooling section of the MFC® process.

Mixed refrigerant or propane

Propane is readily available from natural gas. It is used in both APCI's and ConocoPhillips' processes for pre-cooling. However, Shell Global Solutions selected a mixed refrigerant for its Sakhalin LNG project like Linde did for the Snøhvit venture.

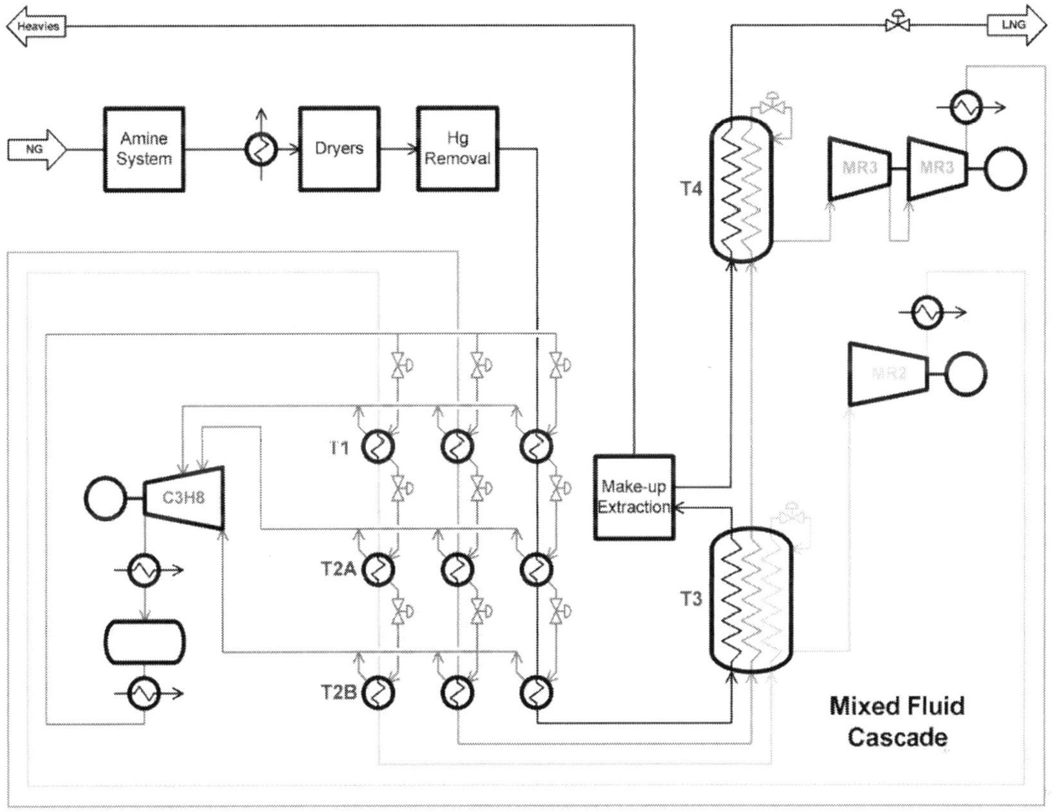

Figure 3
MFC® Process with Bock-in-Kettle Heat Exchangers in Pre-cooling

To compare a propane system with a state-of-the-art mixed refrigerant system the winner of the heat exchanger contest, i.e. the plate-fin heat exchanger concept shown in Figure 2 was benchmarked against a synthetic process (see Figure 3) with a three stage propane cycle instead of the first mixed refrigerant cycle MR1. The process sections MR2, MR3 and the NGL extraction remain unchanged. Block-in-kettle heat exchangers have been envisaged for cost effective heat transfer in the propane case.

As result of rigorous process simulations and equipment sizing the propane pre-cooling system suffers from the low ambient temperatures of the assumed arctic site location. If sea water with 5 degree C inlet temperature can be used, a major part of pre-cooling e.g. upstream of the driers does not need mechanical refrigeration. As a consequence the contribution of a propane system shrinks in comparison to MR2 and MR3 (see Figures 5 and 6) or has to be operated under unfavorably low pressure conditions. Refrigeration by means of propane at -30 degree C produces a vapor pressure of only 1.7 bar abs while the use of a 50/50 C_2H_6/C_3H_8 mixture gives a much higher (bubble point) pressure of 5.9 bar abs. Consequently, suction losses are less relevant and compressor casings stay within proven limits.

The extra investment cost due to the low operating density of propane and the additional refrigeration temperature level have not been evaluated yet, as the availability of suitable rotating equipment is questionable. Figure 4 shows a maximum volumetric flow of 500,000 m³/h to the third stage of the propane compressor. The corresponding maximum flow to the MR1 compressor is only about 300,000 m³/h.

Large duties in C3 pre-cooling concepts require special compressor configurations with parallel casings for different compression services. Solutions with several dedicated casings for the up to four C3 pressure levels have been discussed recently[iv].

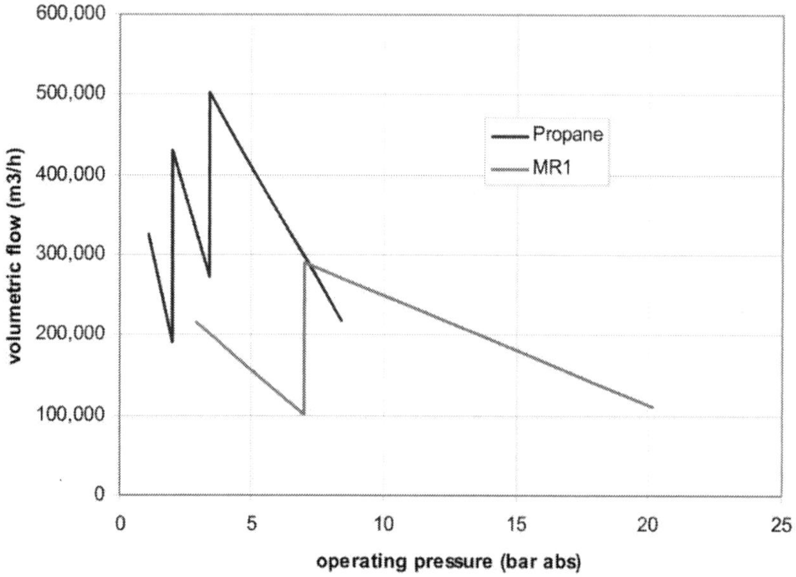

Figure 4
Operating Regime of Pre-cooling Cycle Compressors

The disadvantage of propane in arctic climate can be offset partly by using propylene instead of propane. In this case the operating density of the working fluid increases by 25% and helps to make this solution conceivable. Propylene, however, has to be imported to the LNG plant, contributes to operating cost and reduces plant availability, if supplies fall short.

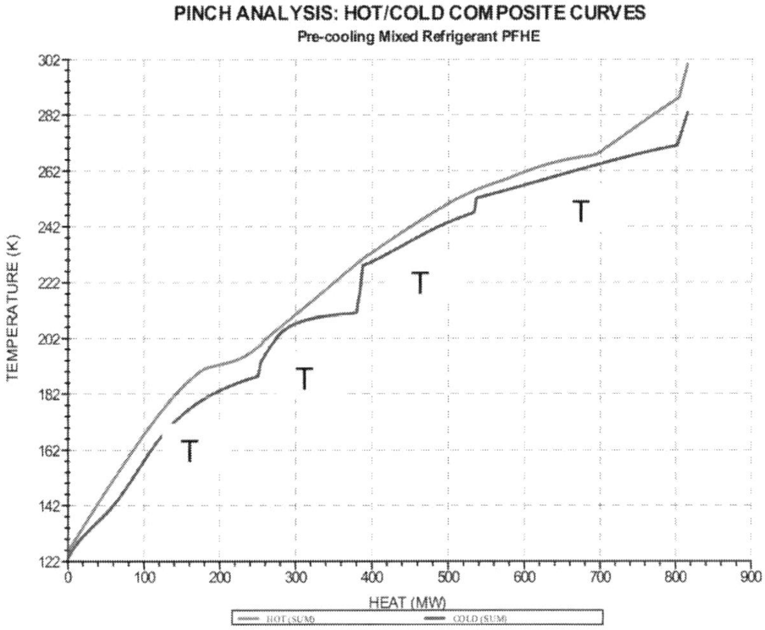

Figure 5
Typical Composite Curves for MR Pre-cooled MFC® in Cold Climate

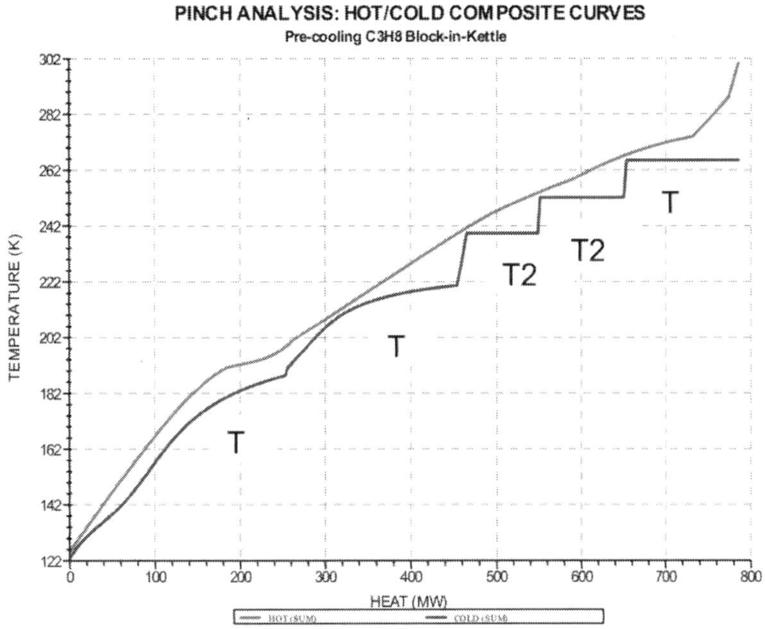

Figure 6
Typical Composite Curves for C3 Pre-cooled MFC® in Cold Climate

Load Balancing

In all previous discussions energy consumption was of the essence, if low unit cost of LNG production shall be achieved. This can be explained by the need to deploy the largest proven drivers for the refrigerant compressors. If a minimum of large and identical drivers suffices to satisfy the power demands, an optimum economy of scale can be expected.

What does this general statement mean for a 'triple flow' LNG process like the MFC® process? The number of gas turbine drivers shall be minimized in order not to jeopardize the plant availability. In case of electric motors as drivers availability is governed more by e-power production than by the number of electric motors.

A perfect solution for the gas turbine case is the selection of two GE Frame 9 turbines, which are assisted by starter/helper motors. A total power output of 300 MW is realistic including all relevant margins like fouling, aging and API. A typical compressor train configuration for this case is shown in Figure 7, where MR1 and MR2 share on common driver, whereas the two casings of MR3 are hooked up to another identical driver.

Today, electric motors with 150 MW power output are not there yet. But motors with 75 MW are a realistic evolution from the successfully string tested Hammerfest VSDS units. As the MFC® process typically requires four compressor casings a dedicated and identical driver for each casing is desirable. Actually, the forced load balancing of the overall shaft power to either two gas turbines with 2x 50% power or four motors with 4x 25% power increases the overall power only by 1-3%. This little tribute to symmetry is more than compensated by the economy of scale of large drivers. A symmetric driver concept with three identical drivers has merits in warmer climates. This solution has been discussed earlier[v].

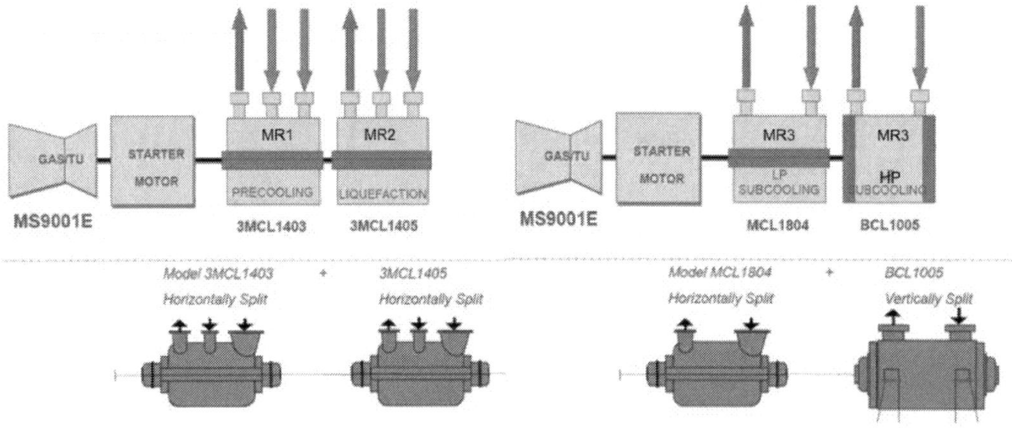

Figure 7 (Courtesy of Nuovo Pignone)
Compressor Trains for a Large LNG Base Load Plant in Cold Climate (the side stream to the MR2 compressor is required only for deep NGL recovery)

Conclusions

In a cold climate mixed refrigerant pre-cooling in an optimized plate-fin heat exchanger network results in the lowest investment and operating cost. The exclusive use of mixed refrigerants reduces the complexity of the refrigeration compressors, as the number of refrigerant pressure levels, the molecular weight of the refrigerants and the volumetric flows to and within the compressors are minimized.

In essence, larger LNG trains can be built around the same compressors, if mixed refrigerant pre-cooling is selected, as the higher operating pressure range keeps equipment dimensions and line sizes within proven limits.

Biography of Speaker

Dr. Heinz C. Bauer (heinz.bauer@linde-le.com) is Vice President Technology for LNG and Gas Processing within Linde Engineering, Hoellriegelskreuth, Germany. He has been with Linde for over 25 years and has been involved in many projects in the field of cryogenic gas processing. He graduated at the Technical University Munich in applied Physics.

References Cited

---

[i] Paradowski, H. and Hagyard, P., Comparing five LNG processes, Hydrocarbon Engineering October 2003, p. 32

[ii] Bauer, H., A novel concept, Hydrocarbon Engineering May 2002, p. 59

[iii] Foerg, W., Bach, W., Stockmann, R., Heiersted, R., Paurola, P., Fredheim, A., A new LNG base load process and the manufacturing of the main heat exchangers, LNG 12, 12th Int. LNG Conference, Perth May 4-7, 1998, Conference Papers

[iv] US Patent 6,962,060, Petrowski, J. et al

[v] Bauer, H., Franke, H., Schier, M., Owren, G., MFC®s3, the formula for high capacity LNG base load plants, Conference Proceedings, 4th Topical Conference on Natural Gas Utilization, AIChE Spring National Meeting 2004

Integrate LNG Terminal with Power Plant to Achieve BTU Control and Efficiency Enhancement

Kamal Shah
Technical Vice President
Aker Kvaerner, Houston, Texas
Nimish Dhuldhoya
Opportunity Development Specialist
Chevron Global Gas, Houston, Texas

AIChE Spring Meeting, April 2006
6th Topical Conference on Natural Gas Utilization
Orlando, Florida, April 23-27, 2006

Introduction

US Energy Information Administration (EIA) [1] estimates natural gas consumption in the United States (U.S.) at approximately 22 trillion cubic feet for 2004. EIA also forecast the U.S. consumption for natural gas to increase to 26 trillion cubic feet in 2010 and further increase to approximately 31 trillion cubic feet by 2025. At present, imported Liquefied Natural Gas (LNG) makes up part of this natural gas demand. The supply of imported LNG will make up from 0.65 trillion cubic feet in 2004 to approximately 6.4 trillion cubic feet by 2025. At present there are five LNG import terminals operating in the U.S and new terminals will be built to meet increasing demand for the import of LNG.

One of the reasons for increased natural gas demand is gas fired power plant installations in various parts of the country. Natural gas is a fuel of choice for meeting energy need as well as achieving national economic and environmental goals. Natural gas fired power plant produces the lowest levels of greenhouse gases of all hydrocarbon fuels. For example, power generation from natural gas produces 40% less CO_2 than coal, and 25% less CO_2 than oil for the same energy output. In addition, Memorandum of Understanding (MOU) [2] on the Regional Greenhouse Gas Initiative (RGGI) by the governors of seven Northeast states (called the "Signatory States") will require reduction in carbon dioxide (CO_2) emissions to meet their future goals. Each state will have a policy to conserve, improve and protect its resources and environment while producing safe and reliable electrical power supply.

The strong natural gas demand in the Northeast region as well as RGGI initiatives by the "Signatory States" supports installation of LNG terminals in this region and in US. However, one of the issues surrounding LNG import is the quality of imported LNG due to variation in composition. The heating value (BTU) of the imported LNG varies considerably and in many cases exceeds pipeline standards. According to U.S. Department of Energy [3], most of the natural gas flowing to the U.S. pipelines, at present, is processed to strip out ethane and heavier components leaving mostly methane in the natural gas. The existing infrastructure for gas processing and petrochemical industries in the Southern region supports this type of processing, but similar Natural Gas Liquids (NGL) pipeline and petrochemical infrastructure does not exist in the East coast and Northeast regions.

Due to the recent tight natural gas supply, pressure has mounted to allow natural gas with higher heating values (ethane and heavier components in the imported LNG). This "richer" gas with higher heating value may not be desirable due to a flame that may be too large or too hot to burn for certain applications, making it incompatible with U.S. appliances and industrial processes currently utilizing lower heat content pipeline gas.

Quality of Typical Imported LNG

At present there are five LNG import terminals and the supply of LNG comes from various countries [4]. The composition of imported LNG, its Gross Heating Value (GHV), and the country of origin are listed below in Table 1.

TABLE 1

TYPICAL COMPOSITION OF IMPORTED LNG BY COUNTRY

Imported LNG Origin	Methane C_1 %	Ethane C_2 %	Propane C_3 %	Butane$^+$ $C_4{}^+$ %	Nitrogen N_2 %	GHV Btu/scf
Algeria	87.6	9.0	2.2	0.6	0.6	1119
Australia	89.3	7.1	2.5	1.0	0.1	1123
Malaysia	89.8	5.2	3.3	1.4	0.3	1128
Nigeria	91.6	4.6	2.4	1.3	0.1	1109
Oman	87.7	7.5	3.0	1.6	0.2	1146
Qatar	89.9	6.0	2.2	1.5	0.4	1118
Trinidad and Tobago	96.9	2.7	0.3	0.1	0.0	1037

The GHV value varies between 1037 Btu/scf and 1146 Btu/scf, with most imported LNG having a GHV above 1100 Btu/scf, exceeding most of the U.S. pipeline standards. In the Northeast, the pipeline companies like to limit their gas to below 1075 Btu/scf and would prefer to have it below 1050 Btu/scf.

Pipeline Gas Quality Control at Existing terminals

As mentioned above, most of the natural gas flowing to the U.S. pipelines, at present, is processed to strip out ethane and heavier components leaving mostly methane in the natural gas. At Lake Charles terminal in Louisiana, the pipeline quality is maintained successfully by mixing higher BTU natural gas from the terminal with relatively low BTU gas from the regions extensive gas processing infrastructure.

At the Everett, Massachusetts facility, Distrigas uses in-tank blending of leaner pipeline gas with imported LNG to meet the pipeline BTU standards. Some East coast LNG terminals such as Cove Point facility utilize nitrogen injection into the pipeline gas to reduce BTU levels. The nitrogen separation and compression increases facility as well as operating cost to produce and compress nitrogen. Also most of the pipelines have maximum nitrogen content limitations (usually between 2-3%).

Aker Kvaerner [5] and others have developed NGL extraction technology to strip out ethane and heavier components from the imported LNG. Figure 1 below describes a typical LNG import terminal configuration. The NGL extraction process could be integrated within the LNG terminal to strip out ethane and heavier components from the richer LNG. The liquid NGL product containing propane, butanes and most of the ethane has high vapor pressure and is transported to petrochemical complex in the NGL pipeline operating at high pressure. This method is very effective in controlling pipeline quality gas from high BTU content imported LNG.

FIGURE 1

TYPICAL LNG TERMINAL WITH NGL EXTRACTION

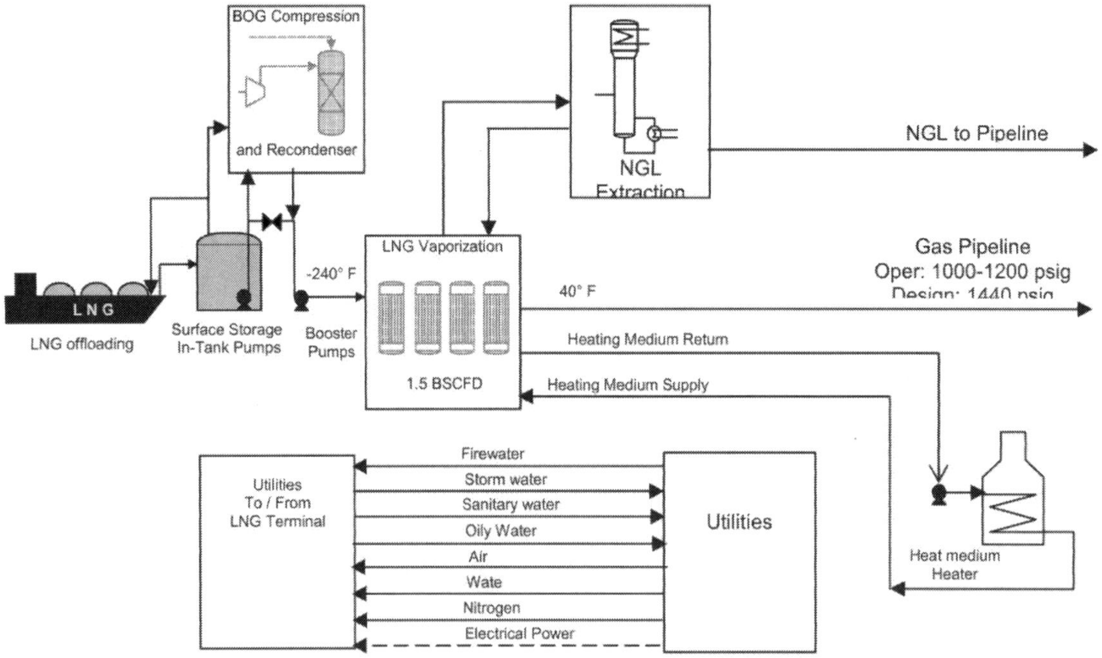

The extensive pipeline and petrochemical infrastructure supports this solution in Gulf Coast. However, this solution is not practical for the East Coast or Northeastern regions where similar infrastructure does not exist for transporting NGL containing high ethane to the petrochemical industries.

A market for LPG (propane and butanes) exists in East Coast and the Northeastern regions. Although LPG can be separated from the NGL and easily transported locally or to other regions by truck or barges due to its relatively lower vapor pressure, the issue is marketability of ethane. In much imported LNG, the ethane content is higher than the LPG content and will require removal to meet pipeline quality standards.

Combination of LPG extraction and nitrogen injection (up to maximum nitrogen allowed by the pipeline) could be considered as a potential solution; however, higher inert content is not desirable by pipeline companies, especially where peak shaving plants are employed for meeting seasonal demand. Also the capital and utility cost will be considerably higher for injecting nitrogen to control pipeline BTU.

The following discussion provides a solution to control the natural gas quality from the imported LNG, and also enhances the overall energy efficiencies and emission reductions where NGL pipeline and petrochemical infrastructure does not exist.

Power Generation and LNG Terminal Integration

The recent RGGI agreement by the "Signatory States" for reduction in carbon dioxide (CO_2) emissions, plus the requirement to increase the energy efficiency of region's power plants, will result in consideration of many gas fired power generation units. East coast and Northeast region will also benefit from having additional LNG terminals to meet their future natural gas demand. Figure 2 below provides the concept for integrating gas fired power plant with LNG terminal.

This concept utilizes a combination of proven processes. The NGL extraction methods from LNG developed by Aker Kvaerner and others, mentioned in the previous section, can be extended further to separate ethane from the NGL by using a deethanizer column, which separates ethane from LPG liquids. The LNG terminal has adequate refrigeration available from cold LNG to provide necessary cooling required for the column reflux system. The ethane separated from the NGL can be utilized to fuel the power generators. The removal of ethane from the rich imported LNG further reduces the BTU of the pipeline gas.

New power plant facilities are usually combined cycle with multiple power generation gas turbines with heat recovery steam generation (HRSG) and steam turbine to enhance the overall energy efficiency; see Figure 3.

FIGURE 2

INTEGRATED LNG TERMINAL WITH POWER PLANT

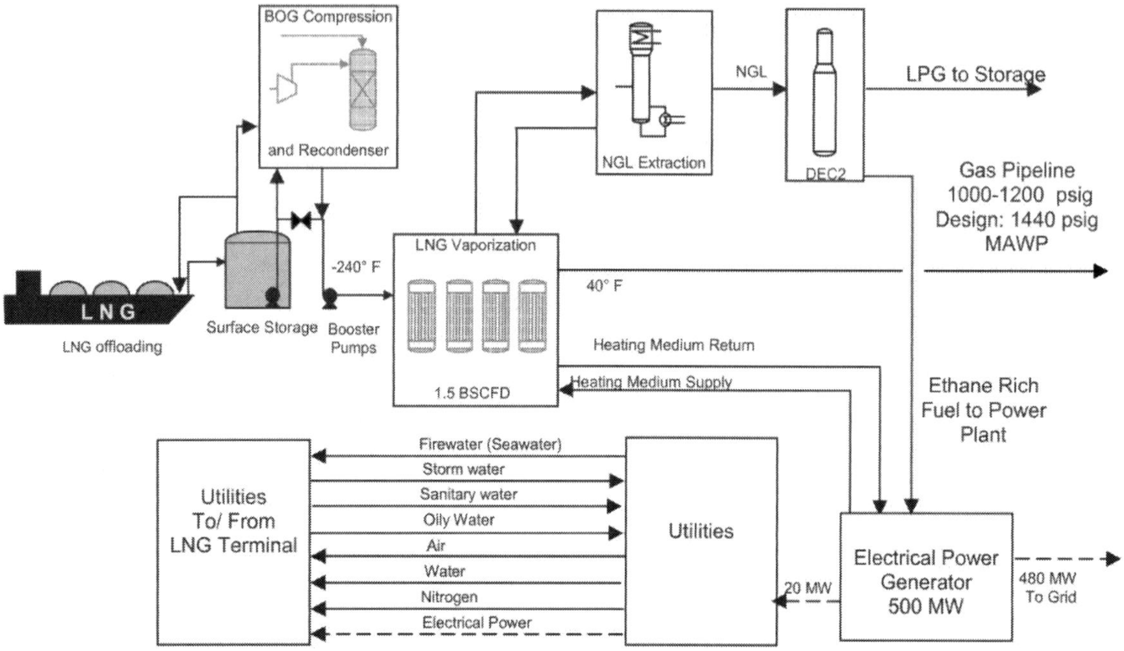

FIGURE 3

COMBINED CYCLE POWER PLANT WITH HEATING MEDIUM SYSTEM

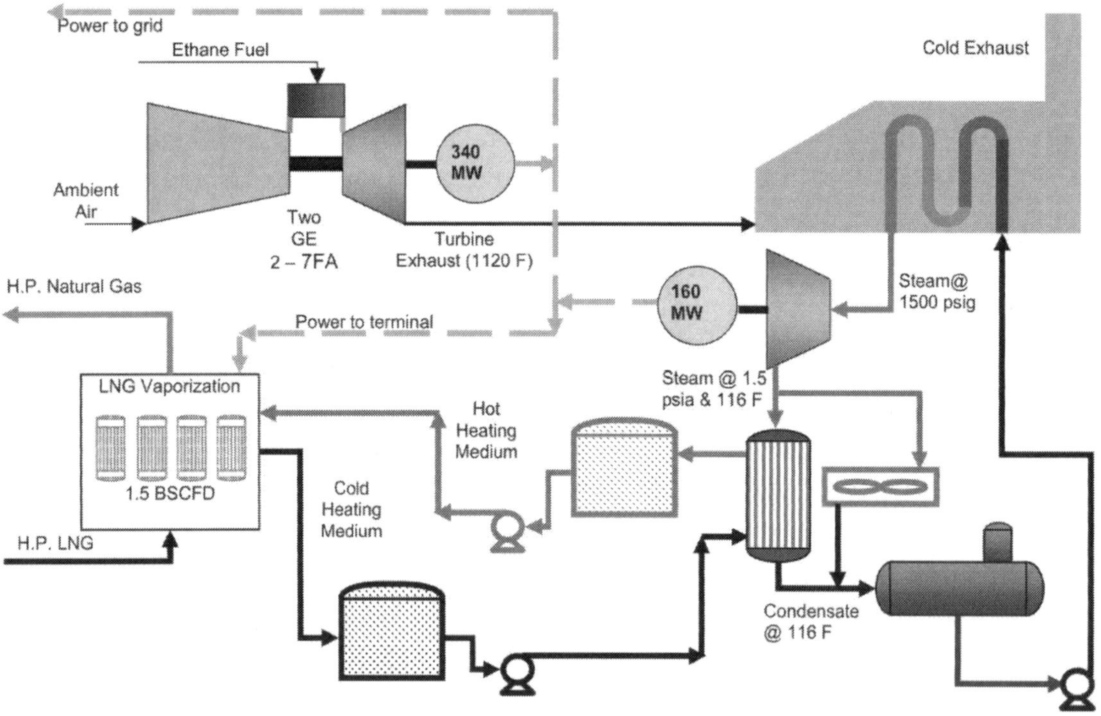

The waste heat from the steam turbine normally utilizes ambient air cooling or a cooling tower system to condense steam. The integration of power plant with LNG re-gasification, will allow recovery of the heat, normally rejected. The heat is recovered by a heating medium which supplies the required heat for LNG re-gasification in the terminal. The cold heating medium, returned from the LNG terminal vaporizers, is utilized for condensing steam reducing the energy required to run the ambient air cooler fans or cooling tower fans.

The method described above is illustrated using a typical 1500 Million Standard Cubic Feet per day (MMSCFD) re-gasification terminal and a nominal 500 MW combined cycle power plant. Table 2 shows the BTU reduction that can be achieved with LNG imported from different parts of the world, having composition described in the Table 1.

TABLE 2

REDUCTION IN GHV OF IMPORTED LNG WITH POWER PLANT INTEGRATION

Imported LNG Origin	Import LNG GHV Btu/scf	Export Gas GHV Btu/scf	Ethane To Power Plant MMSCFD[3]	Total LPG Production[4] Barrels /Day	Export Gas Volume MMSCFD[1]
Algeria	1119	1052	47.62	27571	1412
Australia	1123	1043	47.62	35004	1401
Malaysia	1128	1026	47.62	47218	1384
Nigeria	1109	1023	47.62	37470	1398
Oman	1146	1046	47.62	46630	1385
Qatar	1118	1031	47.62	37838	1398
Trinidad and Tobago	1037	1012[6]	47.62	429	1452

Notes: 1. Total equivalent inlet LNG is 1500 MMSCFD. Export gas volume after shrinkage & fuel.
2. Power plant: Total 500 MW net output using 2-GE 7FA machines and 1- steam turbine.
3. Ethane rich fuel (LHV: 1618.5 Btu/scf, GHV: 1769.4 Btu/scf) utilized for generators.
4. Estimated LPG production (propane and butanes) from 1.5 BSCFD inlet
5. Expected send out gas quality to pipeline.
6. The inlet LNG meets the pipeline quality. The data shown are if LNG is processed.

Table 2 also shows estimated quantities of LPG production after ethane rich fuel is separated from the NGL and utilized by the power plant.

Generally the fuel utilized by the LNG re-gasification terminal is approximately 1.5% of the inlet LNG when fuel gas based vaporizers are utilized. The integration of the power plant will generate fuel savings as well as reduce greenhouse gas emissions. The overall savings in fuel, and reduction in emissions for integrated and non integrated facility, is described in Table 3.

TABLE 3

FUEL AND EMISSION (GREENHOUSE GAS) COMPARISON

Type of Unit	Fuel Gas Consumption MM Btu/hr		Annual CO_2 Emissions 1000 TPA	
	Non Integrated	Integrated	Non Integrated	Integrated
Power Plant	3212	3212	1835	2017
LNG Terminal	931	0	527	0
TOTAL	4119	3212	2362	2017

The comparison in the table indicates approximately a 28% reduction in overall fuel gas consumption and a 17% reduction in green house gas (CO_2) for integrated facility compared to non-integrated facility.

Conclusions

The integration of an LNG terminal with a power plant described above provides a solution to the pipeline gas quality issue associated with imported LNG high heating values. Also, the solution presented here enhances the overall energy efficiency and supports the RGGI goals for future reduction in CO_2 emissions by the Northeast "Signatory States". This is achieved with proven processes.

However, the additional capital cost associated with the LPG and ethane extraction must be evaluated on a project by project basis. The space requirements for additional equipment in the terminal, LPG storage, loading and transportation to the local market or outside the region must also be considered. The project needs to determine commercial as well as technical feasibility before considering this option. The additional savings realized by reduction in LNG transport and terminal throughput capacity due to the fuel savings may have added benefits with this concept.

The solution described above provides the LNG re-gasification terminal operators with the much needed flexibility to process richer LNG to meet the North American pipeline standards, especially the East coast and Northeast regions.

References Cited

1 Country analysis briefs: United States; source of information is US EIA website at http://www.eia.doe.gov/emeu/cabs/usa.html

2 Memorandum of understanding on the Regional Greenhouse Gas Initiative (RGGI) by Governors of the States of Connecticut, Delaware, Maine, New Hampshire, New Jersey, New York and Vermont signed on 20th day of December, 2005.

3 U.S. Department of energy, Office of Fossil Energy, document DOE/FE-0489, Ensuring Consistent Quality for End Use, Source of information website at http://www.fossil.energy.gov
4 U.S. Department of energy, Office of Fossil Energy, document DOE/FE-0489, Typical Composition of LNG Imports by Country, Source of information website at http://www.fossil.energy.gov
5 Kulish, Shah and Wanvik, "Terminal considerations", LNG Industry, A supplement to Hydrocarbon Engineering, Autumn2005.

Biography of Speaker

Kamal Shah is a Technical Vice President, Aker Kvaerner Inc., Part of the Aker Group, Houston, Texas. He has been working with Aker Kvaerner, Inc., Houston office for 19 years and has been involved in the process design, management and technology development in many oil and gas and energy projects. Prior to joining Aker Kvaerner, he was chief process engineer at Key Engineering, Inc. in Houston, Texas. He attended Michigan State University and received his BS and MS degree in chemical engineering. He is a registered professional engineer in Texas.

E-mail: kamal.shah@akerkvaerner.com

Snøhvit LNG – Leading LNG Business into a Promising New Region

Peter A. Hortig
Project Manager, Hammerfest LNG Project
Stefan Buerkle
Executive Vice President, LNG and Gas Processing
Linde AG, Linde Engineering
82049 Hoellriegelskreuth, Germany

AIChE Spring National Meeting
6th Topical Conference on Natural Gas Utilization
Orlando, Florida, April 23-27, 2006

Introduction

The frontier of the LNG World will be pushed further north as Linde - main engineering contractor together with their client Statoil - are building a first of its kind LNG plant in Norway's Finnmark area.

This paper outlines how the specific requirements of a remote and arctic plant location have been accounted for in the engineering. The key features of the Snøhvit LNG plant are briefly discussed and the main components of the Mixed Fluid Cascade (MFC®) process are presented.

Key features of Snøhvit LNG

During the early eighties large natural gas reservoirs were discovered in the Barents Sea approximately 140 km northwest of Norway's Finnmark coast. The fields under the seabed in about 250 – 340 m water depth were named after Norwegian mythical characters Snøhvit, Askeladd and Albatross. Until then Norway's Oil and Gas activities were concentrated in sectors west of the southern part of the country. Besides Snøhvit, even bigger reserves have been located in the North, especially in the Russian Sector.

The challenges related to exploiting these northern reserves are manifold. Taking Norway's high standard in environmental protection into consideration it was clear from the beginning that all necessary efforts had to be made to protect the sensitive environment of the Barents Sea. This requisition was aggravated by the fact that the Barents Sea is an important asset for the

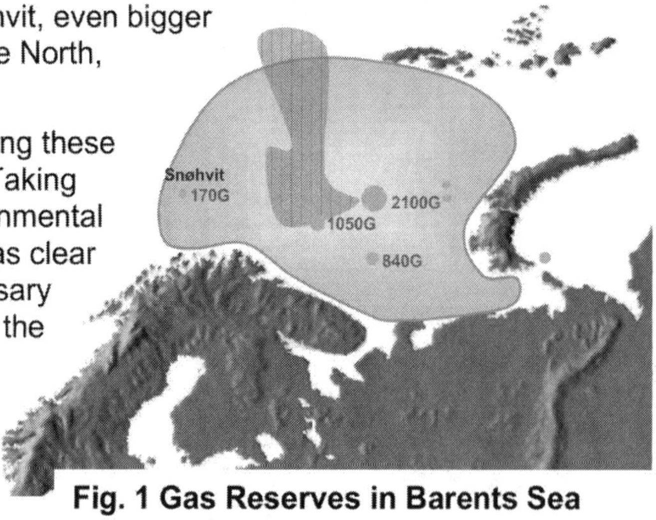

Fig. 1 Gas Reserves in Barents Sea and Russian Sector

country's fishing industry. Thus a special environmentally friendly sub sea technology was applied for the offshore part of the project by Statoil.

For the onshore plant, here referred to as the Hammerfest LNG plant, the focus was given to least possible energy consumption of the applied process and lowest emissions, especially CO_2 and NO_X. Consequently application of "Best Available Technology" (BAT) was a must for the design of this 4.3 MMTA LNG plant.

The Mixed Fluid Cascade (MFC®) process with its inherently low power consumption per ton LNG was selected and combined with an all electric drive concept for the overall plant. The concept includes on site power generation using five highly efficient aero derivative gas turbines.

Arctic Engineering

In addition to the before mentioned requirements the remote location far beyond the arctic circle called for approaches commonly addressed as Arctic Engineering.

The Hammerfest LNG Project afforded the great opportunity to expand Linde's expertise in this field beyond experience from earlier projects in arctic regions. It was mandatory to thoroughly analyze the site specific requirements together with specialized partners in an early project phase.

Existing regional weather data alone soon turned out to be not good enough for the job. A more detailed model of the weather situation on the chosen island of Melkoya needed to be created. A combination of monitoring newly installed weather stations on Melkoya and performing numerical analyses using Computational Fluid Dynamics (CFD) helped to gain a clear picture of the situation.

Having both, actual measurements on reference points as well as results from a numerical model, provided us with calibrated calculations that reflected the real situation in very much detail. Additionally, having such calibrated numerical models allowed to predict the situation in the completed plant situated on a widely reshaped island.

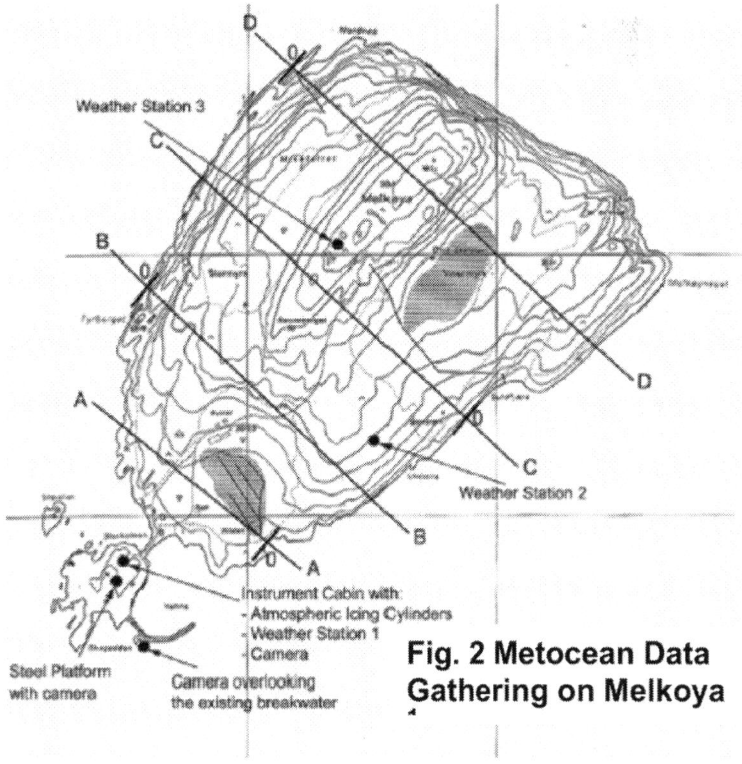

Fig. 2 Metocean Data Gathering on Melkoya

The main additional information was necessary for predicting wind velocities, icing and snowdrift as well as for calculating the Wind Chill Index (WCI) relevant for exposure of individuals to the climate on site. The latter was an important input in order to develop construction and operation concepts for the plant.

It was a challenging task to find an optimum between shielding personnel and equipment from the elements as much as possible and to ensure that explosion loads remain acceptable by keeping the area open as much as possible at the same time.

The numerical results together with measured climatic data were complemented by monitoring temporarily installed full scale model structures on site and documenting their condition via webcam during the winter period 2001/2002.

Fig. 3 Full scale model structure on Melkoya and webcam documentation [1]

With the information gathered by the various studies, simulations and measurements described before, a comprehensive set of project specifications and procedures was developed and used during the detailed engineering phase of the project.

Special focus was given to safety, operation and plant regularity as well as maintenance.

The practical outcome reflects a site specific engineering that considers:

- Structural Integrity (buildings and structures to withstand wind, snow and ice loads)
- Equipment Protection (heat tracing, precipitation shielding etc)
- Working Environment and Personnel Protection
- Plant Design Solutions (redundancy, equipment and material selection, definition of requirements towards suppliers)
- Appropriate Operational Procedures

Main components of the Hammerfest Mixed Fluid Cascade (MFC®) process

This paper focuses on the key equipment in the refrigeration cycles and the power generation units. Details of MFC® process technology are described in the papers cited at the end of this text [2,3].

The following schematic shows the three cascaded refrigeration cycles (Pre-Cooling, Liquefaction and Subcooling Cycle) with their respective compressors and electric drivers. The individually optimized refrigerant compositions and pressure levels of each refrigerant cycle are the main contributors to the high process efficiency. The specific design of the Plate Fin Heat Exchangers (PFHE) and the Spool Wound Heat Exchangers (SWHE).and the way they are integrated in the overall refrigeration cycles is the key to a reliable plant performance. After thorough examination of their respective pros and cons it was concluded to use Plate Fin Heat Exchangers in the Pre Cooling Cycle whereas the Liquefaction and Subcooling Cycle was equipped with Spool Wound Heat Exchangers.

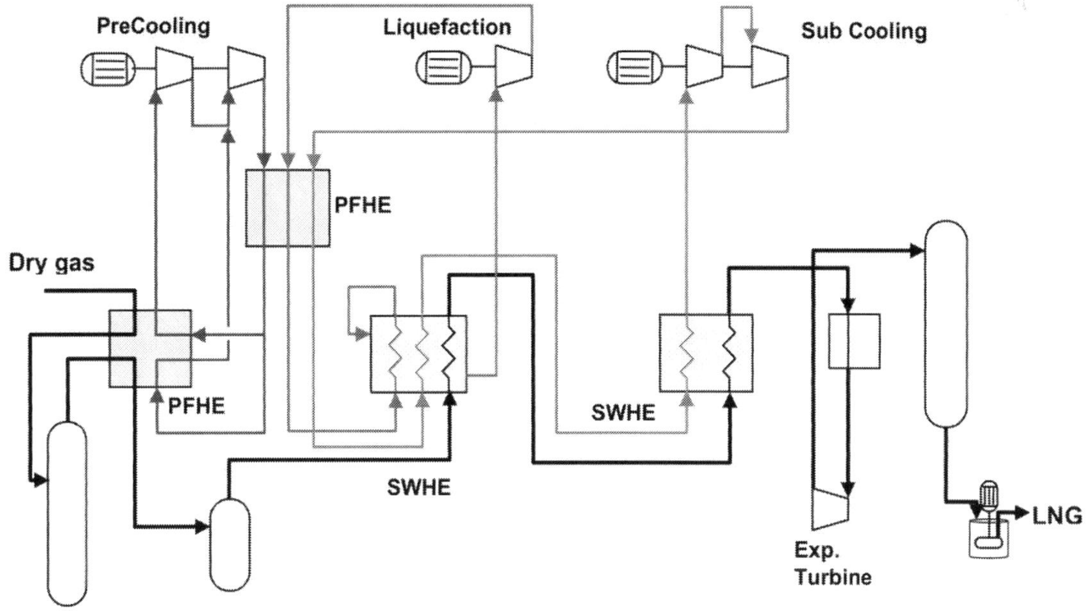

Fig. 4 Simplified Schematic of Hammerfest LNG MFC® Process

Plate Fin Heat Exchangers

This type of heat exchanger is well established and offers a variety of advantages – the possibility of complex stream arrangement and relative low cost per unit area to name two. At the same time it is important to ensure a suitable arrangement of the plate fin heat exchangers and operational safeguards as they are susceptible to temperature shocks and maldistribution. Usually these heat exchangers are arranged in parallel units combined in batteries.

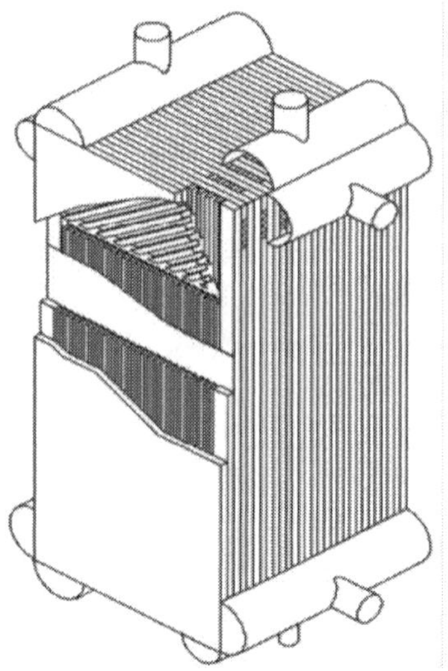

For the Hammerfest LNG plant brazed aluminum plate fin heat exchangers are used that were manufactured in Linde's Schalchen works in Bavaria.

Fig. 5 Production of PFHE batteries at Linde's Schalchen facilities

Spool Wound Heat Exchangers

They are a preferred choice when clean media are used and especially suitable for large heat loads at small temperature differences. A quality that predestines them for cryogenic service in the low temperature section of the process is their robustness regarding thermal stresses.

More than 22 km of tube had been coiled up when the Hammerfest SWHEs left Linde's production facility in Schalchen after 9 months of fabrication.

Fig. 6 Spool Wound Heat Exchanger before installation in pressure vessel

A specific feature of Linde's design for the Hammerfest LNG plant is to install the heat exchangers, vessels, interconnecting piping and instrumentation inside a cold box.

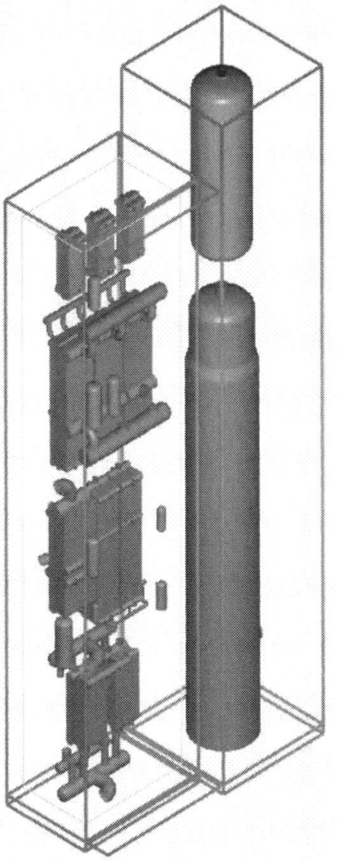

Fig. 7 Cold box – Heat exchanger arrangement and completed cold box in Antwerp on its way to the transport barge

The compact cold box provides a common insulation by filling it with perlite. Segmentation within the box improves maintainability in the unlikely case that access to a component within the box is required. Designing a cold box requires a lot of experience. Especially even temperature and mass flow distribution need to be achieved.

The cold box for the Hammerfest LNG plant was shipped after module assembly in Antwerp as a preassembled unit weighing 3,000 tons and having a height of 60 m.

All Electric Drive Concept

As Snøhvit LNG is the first base load LNG plant featuring an all electric drive approach using on site power generation by five aero-derivative gas turbines, special focus was given to the refrigeration cycle compressors with their drivers and the gas turbine modules.

Each machine had to undergo various testing steps separately before the entire chain from power generation to VSD motor driven compressor was tested in GE's facilities in Massa, Italy, in a full scale full load String Test program. Two out of the five original LM 6000 PD packages were installed and operated in parallel to generate up to 75 MW electrical power for the tests. During these tests the original control system was installed and used. The three main refrigeration cycle compressors were tested sequentially on two parallel test bed installations.

Fig. 8 Full Load String Test in GE's Massa facilities – from power generation to VSDS involving original HV switchgear, transformers and control system

The complex electrical system and its interaction with rotating equipment was analyzed in detailed engineering studies and verified during the string test program. During the start up phase a similar verification of the overall electrical system in the original plant is foreseen.

Process Barge and PAR/PAU Concept

Another key feature of the Hammerfest LNG plant is the process barge concept. The core process facilities were installed on a floatable barge during construction phase in Spain and then transported as a single 35,000 ton module to its final destination north of the Arctic Circle.

Fig. 9 Construction in Cadiz

Fig. 10 Voyage to Hammerfest

This success proves that this novel construction concept is an attractive and feasible approach for LNG plant locations with harsh and adverse weather conditions and limited regional infrastructure and local resources available.

Together with the modularized concept with Pre Assembled Racks (PAR) and Pre Assembled Units (PAU) this allowed for maximum possible prefabrication combined with the benefit derived from working in parallel at multiple construction yards and distributing the load between a number of contractors.

Fig. 11
Pre Assembled Unit – one of nearly 100 modules from The Netherlands, Belgium, Italy and Germany

Fig. 12 (below) Transformation of Melkoya 2002 - 2005

Conclusions

Many dedicated their work to turn the idea of the world's northernmost LNG plant into reality and a lot has been achieved since Statoil/Linde and their partners embarked in this challenging project.

The final part of the journey lies ahead and will be mastered.

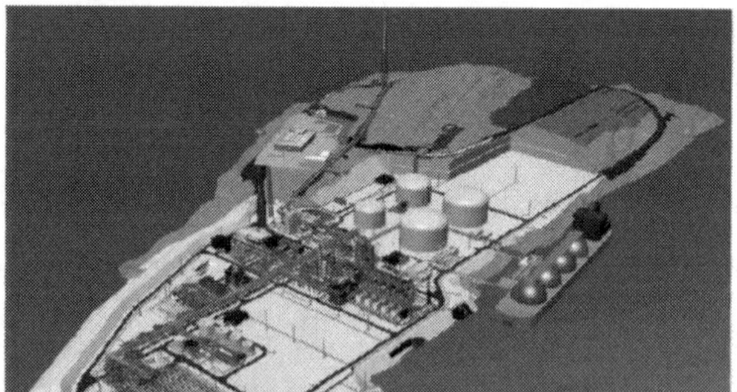

Biography of Speaker

Dr. Peter A. Hortig is Project Manager of the Hammerfest LNG Project within Linde Engineering Division, Hoellriegelskreuth, Germany. He has been with Linde for over 6 years after previous engagements in the Process and Energy Industry. He graduated at Technical University Darmstadt, Germany, in Mechanical Engineering.

References Cited

[1] Internal Report, Statoil/Linde/ Multiconsult Barlindhaug

[2] Bauer, H., Buttinger, B., Kerber, C., Franke, H., Fredheim, A., Pre-cooling Concepts for Large Base Load LNG Plants, Conference Proceedings, 6th Topical Conference on Natural Gas Utilization, AIChE Spring National Meeting 2006

[3] Bauer, H., Franke, H., Schier, T., Owren, G., MFCs3 the Formula for High Capacity LNG Base Load Plants, Conference Proceedings, 4th Topical Conference on Natural Gas Utilization, AIChE Spring National Meeting 2004

Biography of Speaker

Dr. Peter A. Hortig is Project Manager of the Hammerfest LNG Project within Linde Engineering, Hoellriegelskreuth, Germany. He has been with Linde for over 6 years after previous engagements in the Process and Energy Industry. He graduated at Technical University Darmstadt, Germany in Mechanical Engineering.

Dr.-Ing. Peter A. Hortig
Linde Engineering Division

LNG Applications of Diffusions Bonded Heat Exchangers

Tony Bowdery
Heatric division of Meggitt (UK) Ltd
Houston, Texas

AIChE Spring Meeting, April 2006
6th Topical Conference on Natural Gas Utilization
Orlando, Florida, April 23-27, 2006

Abstract

In industrial processes energy is transferred between process streams by a variety of methods, including conduction and conduction-convection in exchangers. Often exchangers operate under steady-state conditions, but in certain processes they operate cyclically, as in regenerative furnaces and some cryogenic processes.

Brazed aluminum plate fin heat exchangers have been widely used in the LNG industry where their compact size and cost competitiveness are beneficial. However, they are not suitable for services with large temperature differences between cold and hot fluids.

Diffusion bonded heat exchangers are compact, high integrity exchangers that have been widely used on offshore platforms for many years. They are typically just 20% of the size of equivalent shell & tube heat exchangers, but unlike other compact exchangers, they can be designed for pressures up to 9000psi (620bar).

A diffusion bonded stainless steel construction allows far greater temperature differences than permissible with brazed aluminum plate-fin technology, and it's also not subject to the stringent warm up / cool down rate limits associated with brazed construction.

This paper discusses the technical design features of diffusion bonded heat exchangers and their application to the LNG industry, and it also addresses some of the associated technical/operational aspects of these exchangers.

Introduction

Heatric has been involved in the commercial design and manufacture of diffusion bonded heat exchangers for over 20 years and there are now thousands of tons of such exchangers in operation on hundreds of different projects around the world. Many of them are installed on offshore oil & gas platforms where the size and weight advantages of such compact exchangers are of obvious benefit. Typical applications include gas compression cooling at pressures up to 9000psi (620bar) and gas/gas exchangers in dewpoint control systems.

More recent work in the LNG industry has highlighted a number of areas where such diffusion bonded exchangers may prove beneficial and commercially attractive and we shall investigate these further in this paper.

Diffusion Bonding Technology

Diffusion bonding is a high temperature, solid state joining process that promotes grain growth across metal boundaries, resulting in a join exhibiting parent metal strength and ductility.

This joint is significantly stronger than a brazed joint in an aluminum plate fin heat exchanger, and hence stainless steel, diffusion bonded exchangers can be designed for much higher pressures. In addition, they are far less susceptible to thermal shock and thermal fatigue than a brazed aluminum plate fin heat exchanger.

Heat Exchanger Core Construction

The high specific surface core of a compact, diffusion bonded exchanger is constructed using one of the following methods:

a) Printed Circuit style

Flow channels are created along flat metal plates using a photo-chemical milling technique analogous to that used for the manufacture of electronic printed circuit boards, giving rise to the name "Printed Circuit Heat Exchanger" or PCHE.

The plates are commonly 1.6 or 2mm thick (0.063 or 0.079"), and the chemical milled flow channels are typically semi-circular in cross-section, having a diameter of about 2mm (0.079").

Different flow geometries are designed for each stream in order to optimize heat transfer and pressure drop performance. The resultant plates are then stacked ready for diffusion bonding.

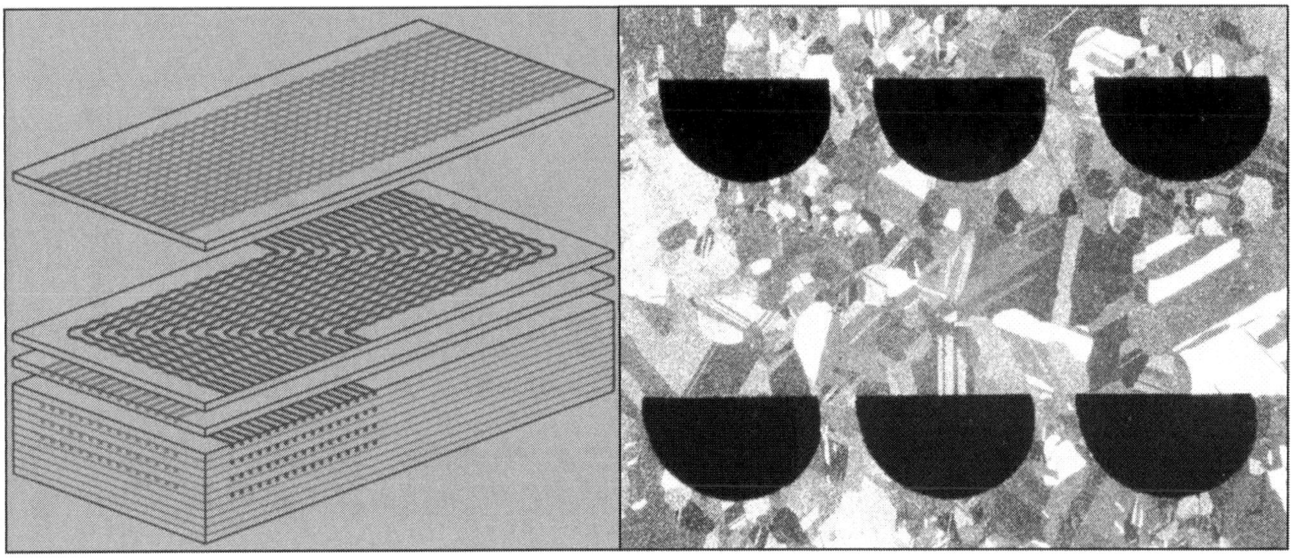

b) Plate Fin style

This form of construction is very similar to that of a brazed plate fin heat exchanger having alternating layers of corrugated fins separated by flat metal plates.

However, in this case there is no braze alloy involved and instead the peaks and troughs of each corrugated fin are diffusion bonded to the adjacent separating plates.

Many different corrugations are available with various fin densities, thicknesses and corrugation styles (plain, perforated, herringbone).

Different flow geometries are designed for each stream in order to optimize heat transfer and pressure drop performance. The resultant layers are then stacked ready for diffusion bonding.

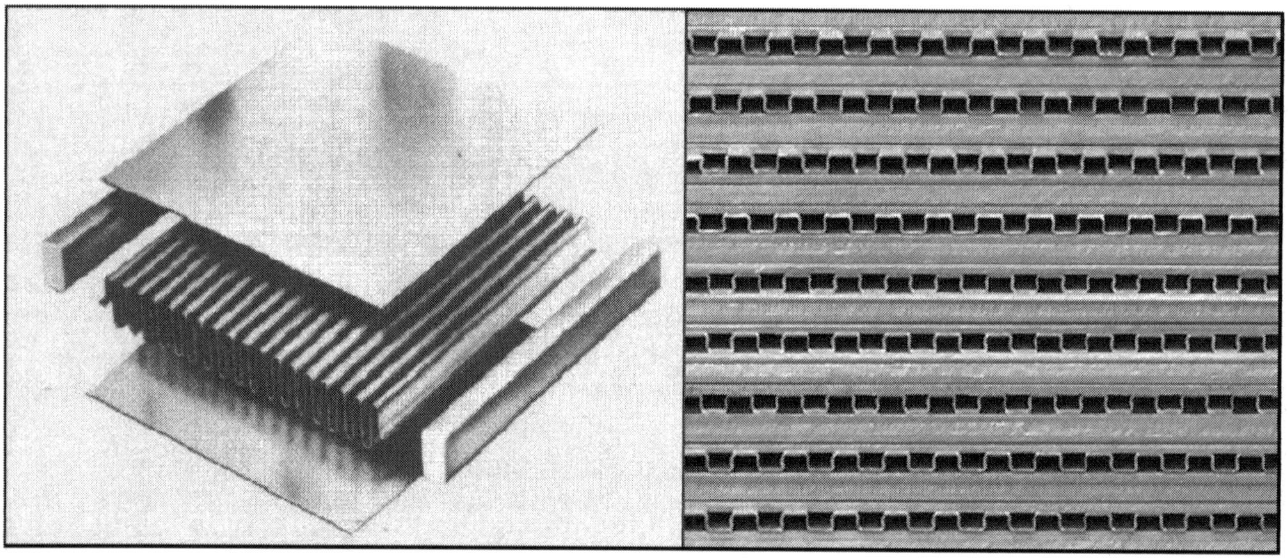

Irrespective of the chosen core construction method, the result after diffusion bonding is a solid block of metal with engineered flow channels passing through it.

Plate Fin style cores are generally more suited to lower pressure services up to around 3000psi (207bar), and Printed Circuit style cores are used for higher design pressures.

Completion of Manufacture

Multiple diffusion bonded exchanger blocks may be welded together to form larger cores. Then headers, nozzles and flanges are welded on to complete the exchanger.

Single complete exchangers can weigh up to 80,000 lb (36,000 kg) each. This may not sound like a large exchanger, but consider that Heatric exchangers are typically just 20% of the size and weight of shell & tube type exchangers. Hence this is equivalent to 400,000 lb (181,000 kg) of shell & tube in a single compact unit.

Limitations of Brazed Aluminum Plate Fin Heat Exchangers

Firstly it is worth noting that aluminum is both a cheaper material and a more thermally conductive material than the stainless steel most commonly used in diffusion bonded exchangers. Hence if brazed aluminum is considered fit for purpose then it will usually provide a cheaper exchanger solution than stainless steel.

So let us first consider the limitations of aluminum exchangers:

a) Design Temperature
In theory heat exchangers manufactured in aluminum can be designed for temperatures as high as 400F (204C). In practice however, except for very low-pressure applications, this must be limited to 150 F (65C) which is the maximum ASME design temperature for the preferred high strength alloy "5083".

b) Design Pressure
The low mechanical strength of aluminum limits the maximum design pressure to about 1750psi (120bar) according to manufacturers' current literature (although this is significantly reduced if design temperatures exceed the 150F noted above). Fin choices are severely limited at this pressure and multiple nozzles are often required per header.

c) Corrosion
Mercury is a common contaminant found in hydrocarbon feed gases, and in it's liquid form it attacks most aluminum alloys, particularly the high strength Al / Mg alloys used for headers or nozzles. Manufacturers talk of mercury tolerant construction, but in reality this only reduces the rate of destruction by limiting the potential sites likely to be attacked. Aluminum exchangers are most susceptible to mercury corrosion during plant shut downs when mercury liquid beads collect in the non-drainable dead areas.

d) Fatigue
Due to the weak nature of aluminum and particularly the brazed joints, it is very important to limit the cyclic or frequently repeated temperature fluctuations of any stream to ±1C (±1.8F) per minute, otherwise leakage can occur.

e) Temperature difference between streams
The ALPEMA guide states that for a typical geometry of brazed plate-fin exchanger under steady state conditions, the maximum permissible temperature difference between streams is approximately 50C (90F), although some manufacturers suggest even lower figures. In more severe cases such as two-phase flows, transient and/or cyclic conditions, this temperature difference should be lower, typically 20 - 30C (36 – 54F).

Capabilities of Diffusion Bonded Exchangers

In comparison, a diffusion bonded structure has far greater integrity than a brazed construction. Use of stainless steel allows design temperatures up to 1500F (815C), and design pressures up to 9000psi (620 bar) are achievable.

Much larger temperature differences are possible with diffusion bonded exchangers, including the 400F (222C) often required for LNG vaporizers.

Stainless steel is not susceptible to mercury attack, and it is much more resistant to thermal fatigue than aluminum due to its lower coefficient of thermal expansion. Hence, Heatric diffusion bonded exchangers do not have tight constraints on rates of temperature change.

Whilst frequent large temperature transients are best avoided with any compact exchanger, Heatric exchangers can accommodate several controlled start-ups and shut-downs every day over a 30 year life in the example of an LNG regasification service.

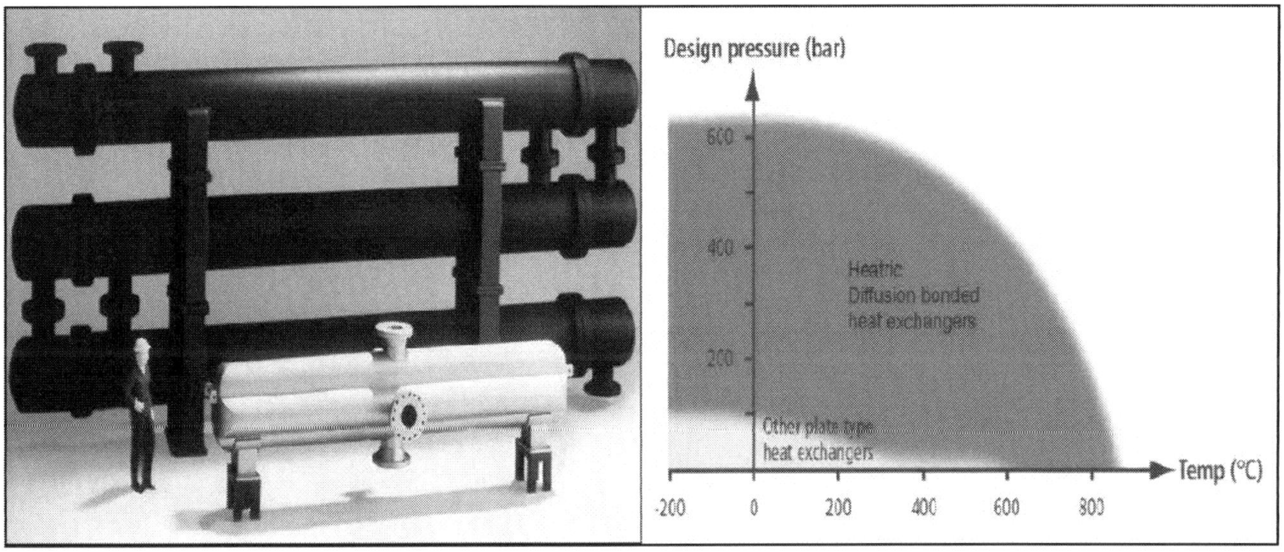

Applications for Diffusion Bonded Exchangers in the LNG Industry

Diffusion bonded exchangers offer many of the same advantages as brazed plate fin exchangers but with far greater mechanical integrity. Hence they often prove attractive where operating conditions fall outside brazed exchanger capabilities, but where the process would benefit from close temperature approaches or multiple process streams in a single exchanger.

a) LNG Regasification

Regasification of LNG can be achieved using a variety of different heat media, and diffusion bonded exchangers are best suited to intermediate fluid regas systems.

Propane and other such refrigerants result in the smallest and cheapest exchangers as the majority of the heat supplied is latent heat released as the refrigerant condenses. As a result significantly lower flow rates are required than with single phase heat media which reduces pumping costs and keeps exchanger size to a minimum.

However, propane has not been historically favored in LNG terminals to date due to its greater flammability than LNG. But consider plants where NGL extraction is being considered in addition to vaporization. Here a propane rich stream is typically one of the resulting products and the facility would have to be designed to handle the additional risk anyway.

Due to the high thermal efficiency of compact exchangers, more care is required where the heat medium can freeze, and in such cases the cold end of the regasification duty is often designed in co-current flow. However, compact exchanger designs are not always appropriate where freezing can occur and careful analysis of wall temperatures within the exchanger is required in each case. As a general observation, compact exchanger design becomes easier with increased heat medium inlet temperature.

Water/glycol mixtures are commonly used in the large shell & tube type vaporizers, but it's worth noting that the glycol has a negative impact on the heat transfer coefficient resulting in very large exchangers. Potassium formate based solutions offer greater protection from freezing with coefficients approximately twice that of glycol solutions.

The freezing mechanics of both potassium formate and glycol based solutions are unlikely to damage a diffusion bonded exchanger in the event of a power outage and sudden stoppage in heat medium flow.

(NB Refrigerant suppliers caution against using potassium formate solutions in installations containing aluminum, zinc or galvanized steel.)

b) LNG Superheating

Irrespective of the chosen method of regasification, diffusion bonded exchangers are well suited to subsequent natural gas superheating / trim heating duties.

This is especially true for some of the floating LNG receiving terminals being considered with size and weight restrictions. Here it's common for a closed loop intermediate fluid to be used such as a water/glycol heat medium. The duty is post vaporization and gas side temperatures are not usually cold enough to cause freezing of the heat medium.

In addition, the closer temperature approaches possible with compact exchangers allow greater heat recovery and a corresponding reduction in required heat medium flow rates compared to traditional shell & tube technology.

c) NGL Extraction

Certain LNG liquefaction plants (for example in North Africa) produce LNG streams which are heavier in NGLs than other plants. These "heavies" must be removed at the LNG receiving terminal in order to meet pipeline gas specification.

In some areas these NGLs are a nuisance, but here in the United States and in Canada they can be a valuable feedstock for the petrochemical industry, and as North American stocks of NGLs are depleted, demand from LNG is expected to grow.

Sizeable exchangers are required for such NGL extraction, and diffusion bonded construction is well suited to the thermal duties involved.

In one particular duty the LNG feed stream (rich in NGL) exchanges heat with the lean return gas (post NGL extraction) and brazed aluminum cannot be selected due to the temperature differences involved and two phase nature of the streams.

Spiral wound exchangers could be used, but evaluation work done by a well known engineering contractor suggests that diffusion bonded exchangers are a much more cost effective solution.

Other NGL extraction processes utilize intermediate fluids such as the potassium formate solutions mentioned above. And in some cases the extraction is done at elevated pressure (2160 psig design) so that the final product streams of methane, ethane and propane+ are at sufficient pressure to avoid the need for further compression. In these cases diffusion bonded exchangers prove attractive as pressures exceed the capabilities of brazed technology.

d) LNG Liquefaction

Heatric has supplied a number of diffusion bonded exchangers for End Flash Gas / Light Mixed Refrigerant duties where higher integrity construction was required.

Separately, a number of companies are considering offshore LNG liquefaction and diffusion bonded exchangers look attractive for nitrogen and methane compression cooling services especially from a size and weight perspective.

In closed loop refrigerant circuits, diffusion bonded exchangers may allow a greater settle-out pressure to be accommodated than possible with brazed aluminum. The resulting reduction in low pressure suction drum volume would be particularly attractive offshore where the size and weight of equipment is so important.

Conclusion

Diffusion bonded heat exchangers have been widely used offshore for many years, but they are still relatively unknown in the LNG industry.

They offer many of the same advantages as brazed plate fin heat exchangers but with far greater mechanical integrity. Hence there appears to be significant opportunity for process engineers to exploit this technology where operating conditions fall outside brazed exchanger capabilities, and where the process would benefit from close temperature approaches or multiple process streams in a single exchanger.

References Cited

1. Burn, J., Johnston, A.M., Johnston, N.M., "Experience With Printed Circuit Heat Exchangers", 1999 GPA General Meeting.
2. The Standards of the Brazed Aluminum Plate-Fin Heat Exchanger Manufacturers' Association (ALPEMA) Second Edition 2000 Revision 1, 25 February 2003, www.alpema.org.
3. Pua, L.M., Rumbold, S.O., "Industrial Microchannel Devices – Where Are We Today?", First International Conference on Microchannels and Minichannels, April 2003, Rochester, New York.

Biography of Speaker

Tony Bowdery is responsible for Heatric's technical sales activity in North & South America, and he has over 10 years experience working with compact heat exchangers. He graduated in 1995 from Loughborough University in the UK with a Bachelor's degree in Chemical Engineering. Prior to joining Heatric he was an exchanger design engineer and worked on many cryogenic & industrial gas projects. Tony lives and works in Houston, Texas and he is happily married with a son aged 3.

Tony Bowdery
Regional Sales Manager - Americas
Heatric division of Meggitt UK Ltd

11490 Westheimer Rd, Suite 850
Houston, TX 77077
Tel 713-978-7236
Tony.Bowdery@Heatric.com

Compressed Natural Gas Engines Exhaust Gas Treatment

Debora Fino[1], Nunzio Russo[2], Guido Saracco[2] and Vito Specchia[2]
(1)Materials Science and Chemical Enginering, Politecnico di Torino, Corso Duca delgi Abruzzi, 24, Torino, 10129, Italy
(2)Materials Science and Chemical Engineering, Politecnico di Torino, Corso Duca delgi Abruzzi, 24, Torino, 10129, Italy

AIChE Spring Meeting, April 2006
6th Topical Conference on Natural Gas Utilization
Orlando, Florida, April 23-27, 2006

Compressed-natural-gas (CNG) is recognized as the cleanest hydrocarbon fuel for spark-ignition engines. It represents the best compromise towards availability and well-known technology, reliability and low cost. The low emission levels of particulate matter (PM) and CO_2 reduce the dangerousness of the exhaust stream [1]. The unburned hydrocarbon emissions (93% methane) are the main negative aspect of this type of engine: even if methane toxicity is not comparable with that of the unburned hydrocarbons emitted by gasoline spark-ignition and diesel engines, it is a strong green-house gas. The global heating power is 35 higher than that of CO_2. A research line of ours, carried out in cooperation with the FIAT Research Centre (Orbassano, Italy), is aimed at developing nanostructured Pd-spinel-type-oxide catalysts employing an overall noble metal load smaller than that used in conventional converters.

Methane is indeed a rather hard-to-ignite hydrocarbon, which entails the use of three-fold higher amount of noble metals (about 200 g/ft^3) compared to the three-way catalysts for gasoline-fired engines aftertreatment. In the present contribution, innovative Pd-spinel-type-oxide catalysts are introduced and deposited by "in situ combustion synthesis" [2] over ceramic honeycombs to produce reliable catalytic converters, tested on a specific test rig and compared to conventional ones. After a large screening of methane combustion oxide catalysts obtained by combustion synthesis [2], in line with the other authors [3], the best catalyst was found to be $Co_{0.8}Cr_2O_4$(see table 1).

The catalytic converters (Chauger cylindrical honeycombs; cell density: 200 cpsi; length: 25 mm; diameter: 34 mm) were prepared by firstly depositing a layer of g-alumina (10 wt% referred to the monolith weight) and then 15 wt% (referred to the galumina weight) of $Co_{0.8}Cr_2O_4$, the most active spinel. On some monoliths, Pd was then deposited by wet impregnation with an aqueous solution of $Pd(NO_3)_2$ so as to obtain 1wt% of Pd referred to the deposited catalyst (about 80 g/ft^3 of catalytic converter). The monoliths were then kept at 500°C in order to promote the Pd stabilisation and the formation of finely dispersed Pd clusters. A calcination step at 700°C for 2h was finally performed. Adhesion tests and a full chemical-physical characterization (XRD, SEM, TEM, TPD/R/O, AAS) were carried out and will be detailed in the full presentation. Catalytic combustion experiments were performed in a stainless steel reactor heated in a horizontal split tube furnace with a heating length of 60 cm.

The catalyzed monolith was sandwiched between two mullite foams for optimizing temperature control flow distribution. A thermocouple, inserted along one of the central monolith channel, was used to measure the inlet temperature. Lean inlet conditions (0.4% CH_4, 10% O_2, N_2 balance) were ensured by mass flow controllers. The reactor temperature gradient measured in the axial direction was not significant. The GHSV was set equal to 10,000 h^{-10}, whereas the composition of the off-gases was monitored by NDIR analysers (ABB). The methane conversion results show that all gAl_2O_3-supported catalysts shift the combustion temperature range towards values significantly lower (decrease of more than 350°C) than those typical of non-catalytic combustion. Starting from the pure gAl_2O_3 catalyzed monolith, only a slight improvement of the activity is noticeable compared to the blank monolith.

Undoubtedly, the dispersion of active spinel-type-oxide phase ($Co_{0.8}Cr_2O_4$) on alumina is successful to enhance the activity towards methane oxidation. The half conversion temperature is lowered to 403°C, from the 760°C achieved for the monolith uncatalyzed converter. Further improvements (about 15°C) lowering are induced by the presence of Pd, thereby approaching the performance of the high-Pd containing catalysts. The role of Pd in methane catalytic combustion is well known and still addressed by a number of investigation [4, 5].

References

1. Kato T., Saeki K.,. Nishide H, Yamada T. , JSAE Review 22, 365 (2001).
2. Civera A., Pavese M., Saracco G., Specchia V., Catal. Today, 83, 199 (2003).
3. Cimino S., Pirone R., Lisi L., App. Catal. B: Environ., 35, 243 (2002).
4. Choudhary T.V., Banerjee S., Choudhary V.R., App. Catal., A: Gen. 234, 1 (2002).
5. Janbey A., Clark W., Noordally E., Grimes S., Thair S.,Chemosphere 52, 1041 (2003).

Versatility of Vacuum Jacketed Pipe in LNG Facilities

John W. Bonn
VJ Systems, LLC
Hilliard, Ohio

AIChE Spring National Meeting 2006
6th Topical Conference on Natural Gas Utilization.
Orlando, Florida, April 23-27, 2006

1.0 Introduction

Vacuum jacketed pipe is one of the options available today for applications in LNG terminals. This option has some advantages over other methods of insulating this cryogenic fluid. Vacuum insulation for cryogenic pipe was developed over 60 ago at Ohio State University by Herrick L. Johnson PhD. for insulating liquid hydrogen pipe as part of the rocket research development. From the development of static vacuum insulated pipe lead to applying this method of pipe insulation to most other cryogenic fluids including LNG. In some cases static vacuum insulation is the only known method to insulate such critical fluids as liquid helium and liquid hydrogen based on there extremely cold temperatures in the liquid state. Liquid helium has a very low latent heat of vaporization as well and only a small amount of heat through the piping insulation will vaporize this fluid.

Our discussion here is to bring this technology to the growing demand for natural gas and the expanding LNG terminals being built throughout the world. It is difficult to keep up with the amount of these new terminals being proposed, however there as many as 60 new receiving terminals being evaluated for North America alone. At present there are only five LNG terminals operating in North America. This information is available from the Federal Energy Regulatory Commission (FERC) and it changes quite often.

Vacuum jacketed pipe was first applied by CVI in Columbus, Ohio for use in LNG terminals. The first application occurred in 1996 at Atlantic LNG (ALNG) sending terminal. The vacuum jacketed pipe was installed at ALNG for evaluation purposes for this method of insulation on pipe. This was a 4" recirculation line and there was great skepticism about the overall performance, reliability and safety of this method of insulation for LNG lines. This insulation system proved to be very successful and the performance of this line has been witnessed by most all LNG terminal owners and after 10 years of operation there has been no maintenance or service required on this vacuum insulated LNG piping. However there has been considerable maintenance required on the standard mechanically insulated pipe that runs parallel to this vacuum insulated pipe.

2.0 Common Types of Mechanical Insulation Systems

There are several methods of insulation systems available for insulating cryogenic fluids. The insulation systems that are commonly used in LNG terminals will be addressed in the following paragraphs.

2.1 Mechanical Insulation Systems

There are three basic types of mechanically insulated pipe systems used in LNG terminals and four basic methods of insulating LNG pipe.

- Polyurethane (PUR) foam insulation
- Polyisocyanurate (PIR) foam insulation
- Foamglas Insulation
- Combination PIR and Foamglas

The different types of mechanical insulation also have a different insulation value per inch of thickness and may be from 7 inches thick to 15 inches thick. The insulation thickness of mechanical insulation takes up a considerable amount of space and adds to the cost of the pipe support system.

The R-values for the different mechanically insulation is established per ASTM A518 testing method which uses a flat slab and heat flow meter to define the heat leak per inch of thickness. ASTM C177 testing method is used on Foamglas insulation and this method also uses a flat slab guarded hot plate apparatus.

Pipe is round and the R-value test on a flat surface and some suppliers have adjusted the equivalent thickness to actual thickness taking advantage of the increase in volume as the diameter increase as the insulation is added to pipe. For example if the requested heat input into a 30" pipe line is equivalent to 15" of insulation on a flat surface that would be defined as actual thickness in the following equation.

$t = (d_1 - d)\ 0.5 + e$ *inches actual thickness*

d = *pipe diameter in inches*

$d_1 = \sqrt{(\pi\, dt_1 + (\pi/4)\, d2)/(\pi/4)}$ *inches*

t_1 = *equivalent thickness based on heat leak value required in inches*

e = *manufacturers efficiency- approximately (0.5 to 1) inches*
(e) is not included in curve calculations

t = *10.98 inches (for 30" pipe w/15" equivalent insulation)*

The following curve shows a range of pipe sizes and actual thickness of insulation on these LNG pipes.

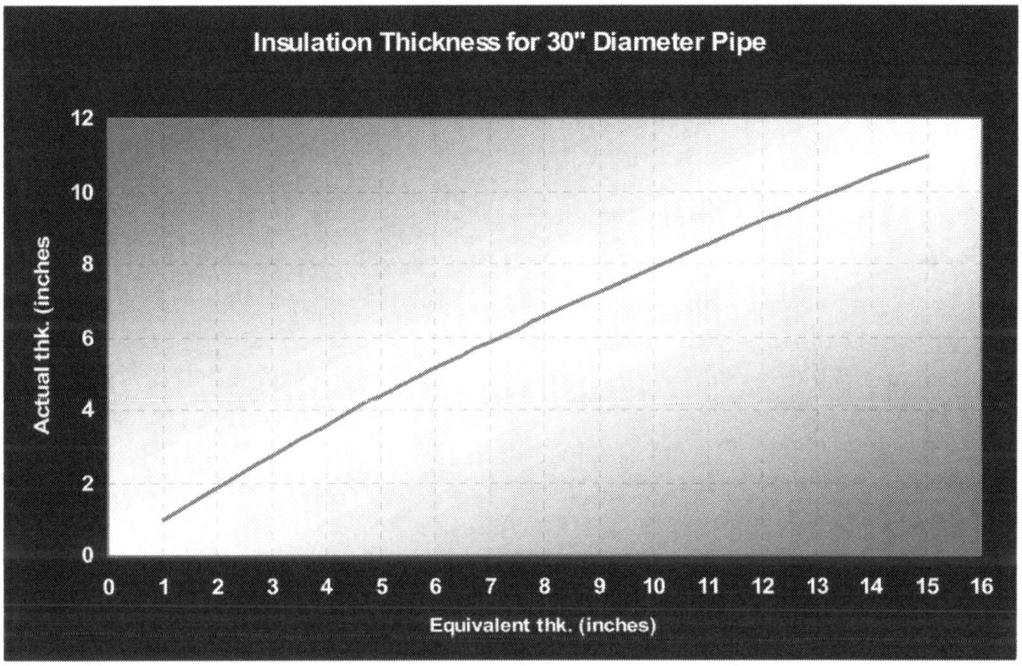

Figure 1 Mechanical Insulation Equivalent vs. Actual Thickness on Pipe

The actual thicknesses shown here may vary based on manufacturers other requirements including vapor barrier and other considerations.

The result in using mechanically insulated pipe is that it is fragile, all field fabricated, insulated and sealed with some type of barrier to reduce water vapor, low thermal performance and degrading heat leak as water permeates the vapor barrier. This will result in eventually replacing the mechanical insulation several times over the life of the LNG terminal. Vacuum jacketed pipe will last the life of the LNG terminal with very little maintenance or degradation in heat leak performance.

2.2 Corrosion Under Insulation (CUI)

This has been a problem for many years when using mechanical insulation and the cost to industry is in the billions annually. This problem is addressed in the November 2005 of "Insulation Outlook". CUI becomes a critical issue when the fluid is cold or cryogenic fluid. A cryogenic fluid will shrink the air between the pipe and insulation barrier. This will reduce the pressure in this insulated space and water vapor contaminated air will be drawn into and condense on the pipe surface through and small opening in the vapor barrier. When this happens it is not readily visible and considerable damage can occur to the pipe before the ice damage to the insulation is apparent. This is a continuing maintenance problem for all mechanically insulated pipe systems.

2.3 Pipe Supports for Mechanically Insulated LNG Pipe

The mechanical insulation on the large LNG lines cannot support the line; therefore rigid structures are welded to the pipe and then isolated from the support with a low thermal conductivity material to reduce the frost on the support and adds heat leak into the pipe. Mechanical insulation is then wrapped around this support. The support locations also require more rigid structures to meet the requirements of water hammer in the pipeline if a valve is closed rapidly. This is a very difficult engineering challenge in these mechanically insulated lines.

2.4 Field Installing Mechanical Insulation

In most cases the environment where this pipe is being installed and insulated is exposed to all the weather conditions that are not the best for LNG piping. Most of the labor is local and the quality may vary from site to site. Insulation quality cannot be established until the plant has started production. There are some methods of repairing these areas but it is difficult to determine how effective they are.

2.5 Scheduling and Installation Time

Mechanical insulation does not start until the pipe is installed and all testing is completed. The time to install the insulation is a series operation due to the critical nature of qualifying the LNG pipeline. Based on the length of pipe to be insulated and the time to complete the installation of the insulation could be on the critical path for the project. Vacuum jacketed pipe comes to the field already insulated and eliminates the time required to install mechanical insulation. This has a positive impact on the schedule that would allow the production to start a few months earlier.

3.0 Vacuum Jacketed Piping Systems

The robustness of the Pipe within a Pipe system has allowed owners to look at more cost saving and safer methods of transferring LNG to and from ships. CVI and VJ Systems in Columbus, Ohio have developed LNG transfer pipe that will allow the LNG piping to be installed aboveground, underground and underwater. This is an advantage that is not available with mechanically insulated pipe or cost effective if a method is established.

LNG terminal owners and operators are now looking at these advantages to provide a safer more reliable LNG piping system at these terminals. There are LNG terminals today that have vacuum jacketed pipe as part of their overall facility operation and it is proven to be very reliable with about 10 years of operation at some sites. As of today the maintenance cost is a non-issue. This is encouraging as this new product for LNG terminals is being evaluated for many LNG terminals proposed worldwide in the next few years.

Vacuum jacketed pipe can be achieved by two separate processes that vary in performance and maintenance requirements.

3.1 Static Vacuum Jacketed Pipe

The factory fabricated pipe is designed in long spool sections with a pipe within a pipe and sealed at the ends. These sections are between 25 to 26 meters long and each section is evacuated and sealed with a permanent high vacuum in the annular space prior to shipping. This long length and pre-insulated pipe has many benefits to installing contractor and owner.

The contractor's installation time for the LNG piping is reduced based on a reduction in weld time to less than half required for all other pipe and insulation systems.

The owner receives a high performance pipeline with very low heat input and a rugged reliable system at the same cost or less of other LNG pipe and insulation systems. This pipe can be shop fabricated in parallel with work being done at the site and shipped to the site when all field support activity is completed. This parallel operation allows the LNG terminal to get on line sooner to start the revenue stream from the terminal.

3.2 Dynamic Vacuum Jacketed Pipe

Dynamic vacuum jacketed pipe is all field fabricated and requires construction on site and the quality of the annular space for the vacuum is difficult to maintain based on normal field conditions. This operation requires one time qualification of the skilled labors and welders to fabricate this product. Other items such as internal constraints in the annular space for securing the pipe for water hammer potential further reduces the conductance for dynamic vacuum pumping. Vacuum level in the pipe is most likely in the high viscous range and heat transfer through the pipe can be very high. This heat transfer based on conductance can be greater than that of mechanically insulated pipe and requires an intermediate mechanical insulation around the LNG pipe to reduce this heat transfer.

Maintenance of the vacuum pumps and related equipment would require a redundant vacuum system if the LNG is in continuous operation.

An internal line failure will shutdown the entire line and the vacuum equipment cannot operate under these conditions. To find the leak and repair it can prove to be most difficult. Static vacuum jacketed pipe is in isolatable spool sections and if a an internal failure occurred it would only affect one spool and most likely would not shutdown the operation and repair or replacement of the single spools can easily be achieved at a convenient time.

4.0 Vacuum Jacketed Pipe Insulation Performance

Vacuum jacketed pipe has a static vacuum insulation system that surpasses most other insulation systems available today. This insulation system and technology has been used since the early 1950's with reliable performance results. The unique feature of a static vacuum system is that the performance over time does not change as with all other cryogenic insulation systems. This has proven to be especially helpful at liquefication LNG terminals when every watt is important to performance of the plant. Re-condensers can be considerably smaller and do not have to worry about eventually over powering them with degrading insulation.

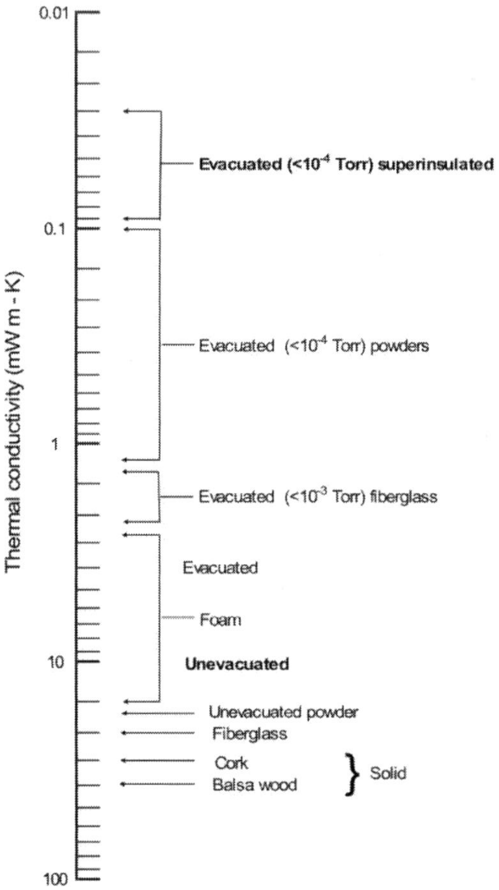

Mean effective thermal conductivity of typical insulation.
Boundary temperatures 300K (27°C) and 80K (-193°C).
(This information is courtesy of "Advanced Cryogenics" Plenum Press 1971 C. A. Bailey

Figure 2 Thermal Conductivity of Various Insulations

5.0 Cryogenic Applications for Vacuum Jacketed Pipe

The vacuum jacketed pipe is used in a variety of applications today and is a critical component in several operations at National Laboratories for research and development where cryogenic fluids such as helium, hydrogen, nitrogen and other cryogenic fluid are used extensively for testing. The industrial gas supplier use vacuum jacketed pipe for transfer of these same fluids and other air gases from the liquefier to storage tanks and from the storage tank to trucks for transporting these products over the road. The Aerospace Industry and NASA use vacuum jacketed pipe in satellite test facilities and for fueling the space shuttle and other rockets for launching satellites. Facilities for storage and testing of biological samples and medicines use vacuum jacketed pipe to transfer liquid nitrogen from the main storage tank to the specimen storage containers. Many frozen food suppliers use vacuum jacketed pipe to supply liquid nitrogen to the food freezing tunnels. Vacuum jacketed pipe is used at LNG fueling stations to fuel vehicles. The use of vacuum jacketed pipe at LNG terminals is certainly a natural progression of this product for transferring large amounts of LNG great distances. In addition to using vacuum jacketed pipe in LNG terminals it also allows for installation for LNG pipe to be installed underground and underwater as well.

5.1 Underground Vacuum Jacketed Pipe

Pipe in Pipe construction allows this pipe to be installed underground at LNG terminals for improved safety at the site and if the LNG is crossing over other unsecured property underground pipe is preferred. This does require additional monitoring of the pipeline since visibility is not possible with buried pipe. It is impractical to monitor the vacuum in each 24 to 25 meter long spool with instrumentation as this vacuum instrument is not designed for constant monitoring and is also considered a laboratory type instrument. These vacuum gauges are best left sealed to prevent potential failure.

One of the unique items using vacuum jacketed pipe for underground installation is that expansion loops are eliminated and internal bellows are installed in the pipe to compensate for the material shrinkage when cooled to LNG temperature. Otherwise underground cryogenic pipe would not be considered for underground installation. A bellows is installed in each spool of the underground pipe and a minimum of 500 to 5000 full thermal cycles based on EJMA calculations that will determine the useable life of each bellows. The bellows used in this design have the line pressure on the exterior surface of the bellows to allow for greater movement than if the pressure was interior of the bellows. The bellows is installed as shown in Figure 3 in such a manner that squirm is eliminated, torque and lateral offset are not possible and with vacuum on the inside and LNG on the outside surface corrosion is eliminated. This design has been used for cryogenic vacuum jacketed piping for over twenty years and more recently in LNG terminals for over five years successfully. LNG piping have less severe cyclic service than most other cryogenic fluid applications and when the LNG pipe is cooled down it is rarely warmed back to ambient temperature and the thermal cycles on the bellows are normally very low.

Some of the advantages to the arrangement of the bellows as shown in Figure 3 include the fact that the bellows is compressed when the inner line is cooled that will bring the convolutes together, adding some additional strength to a bellows with external pressure.

There are some reluctant and skepticism on part of some people to use bellows in LNG transfer lines and are continuing to seek other solutions to accommodate the pipe shrinkage without bellows when cooling the piping to LNG temperature.

One of the potential alternatives is the use of Invar material that has a very low shrink rate of about 12% of austenitic stainless steel. Based on the length of the overall pipe run and if bellows are not incorporated in the design the pipe will be in tension and anchoring the pipe at each end will be required. If the pipe run is long enough the consideration for a bellows may be required. The cost of Invar is 3 to 4 times that of stainless steel based on 36% nickel content and availability of Invar is limited. The present day price for nickel is volatile and it continues to increase which makes it very difficult to justify using this material. There is very little difference in the allowable tensile strength between Invar and stainless steel and this does not allow you to decrease the thickness of the material to reduce the amount or pounds of material being used.

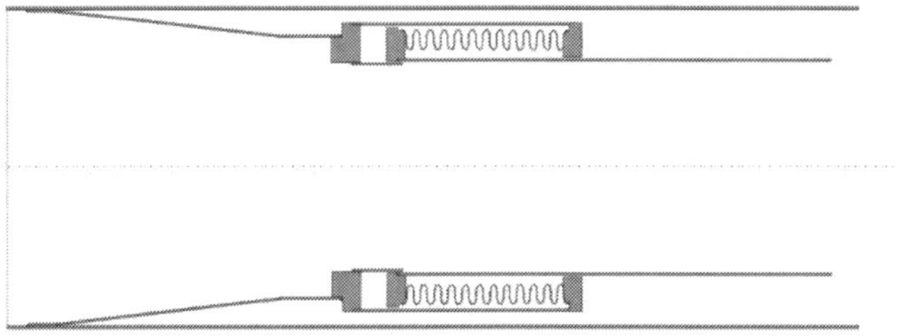

Figure 3 - Internal Bellows Assembly with Female Bayonet End Connection

5.2 Underwater Vacuum Jacketed Pipe

The construction of LNG vacuum jacketed pipe for underwater service is similar to the underground piping except the consideration for potential physical damage of a ship anchor or other potential hazards normally considered for this environment. A heavier jacket wall and installing in a trench is recommended for this pipe. In most cases this seawater depth is not expected to be five to ten meters below the maximum draft of LNG carriers. The assumption made here is the ship draft is about 18 to 20 meters and the lowest pipe burial depth would be 30 meters below sea level at low tide. Also consideration for maintaining the pipe secured in the water with concrete or other methods is required. The length of pipe and size will also determine the jacket material and method of installing this pipe. S-Lay using an offshore vessel and “bottom pull” are the most likely methods that would be evaluated for installing the LNG vacuum jacketed pipe.

Due to the unpredictable potential of a sled from a fisherman’s net and / or an anchor being lowered on the pipe, consideration should be given to installing the underwater piping below the natural terrain of the seabed. This would require trenching the area where the pipe is going to be installed and after installation allowing the silt to cover the pipe. The top of the pipe should be at least one meter below the surface of the seabed. The jacket material for underwater LNG vacuum jacketed piping should be 9% nickel for the adding tensile strength and the compatibility with LNG temperatures. This would be a secondary containment for this piping and using the self contained vent device described in Paragraph 5.4.3 there would not be any release of LNG into the water if there is a line failure.

5.3 Underground/Underwater (DTS) System

There is no visible access to monitoring underground or underwater pipe and a Distributed Temperature Sensor (DTS) is a system that can be installed on the entire length of the jacket to monitor the temperature of each spool section. If the LNG pipeline is 2 to 8 kilometers the DTS system can detect a temperature variance within one meter the entire length of the pipe run.

By having isolated vacuum spaces for each pipe spool this makes this method of monitoring and location of the leak very easy to find. If a leak occurred in a dynamically pumped system with a common vacuum the entire length of the pipe run would be almost impossible to isolate the leak since the entire length of the pipe is affected.

5.4 Leak Containment

LNG leak containment is probably one of the major concerns when the pipe is buried or underwater. This can cause a real hazard in both of these pipe installations that has to be evaluated at each site.

5.4.1 Underground open vent

There are buried LNG pipe installations in California that are about 32" underground and access openings aboveground are provided for venting the gas to stack if a LNG leak occurred. Also access is provided to these vent devices for maintenance if required.

5.4.2 Underground/Underwater with manifold vent header pipe

Another method of controlling any release from underground or underwater piping is to provide a vent pipe that is parallel to the LNG pipe and each vent check valve that is part of each static vacuum pipe spools is manifolded to this parallel vent pipe. This vent pipe may have a nitrogen purge in it with very low flowrate to keep moisture out of it and to reduce the aging of the elastomer seals in the vent valve. The end of the vent line would be located in a safe area and a methane detector would monitor the gas exiting the vent line for a potential leak. If a leak would occur in this system there is a potential that it could affect adjoining pipe spools as the cold gas passing by these spool could cause the elastomer o ring seals to fail since they are not designed for these cryogenic temperatures. Also the vent pipe would shrink from the cold gas passing through it from the failed spool to the end of the pipe. Shrinkage of the vent line must be compensated for to assure the stresses at each connection to the LNG pipe will not fail. This will require expansion loops in the vent pipe or a bellows to compensate for the differential movement between the LNG pipe and vent pipe.

5.4.3 Self contained vent device

Each pipe spool has its own isolated vacuum space in the 24 to 25 meter long spools. Therefore each one of these pipe sections or spools will have an evacuation valve and vent valve. The vent portion of this combination valve will open if the carrier pipe containing the LNG is breached. When this occurs the vent valve will open as the LNG is expanding in this vacuum space. It must be vented to assure there is no overpressure in the vacuum space. VJ Systems, LLC has recently received a US Patent for secondary containment device that will vent all leaks back into the LNG carrier pipe. This device eliminates the release of and liquid or vapor outside of the LNG vacuum jacketed piping system. There is no requirement for a parallel vent pipe or the open venting of this LNG or vapor to the atmosphere for underground or underwater LNG vacuum jacketed pipes. This also reduces the cost of the piping system while adding a considerable margin of safety to the installation and operation of LNG vacuum jacketed piping. There is no additional field work required for this system as it is all contained in each spool section in the factory further reducing the installation cost of this unique venting system.

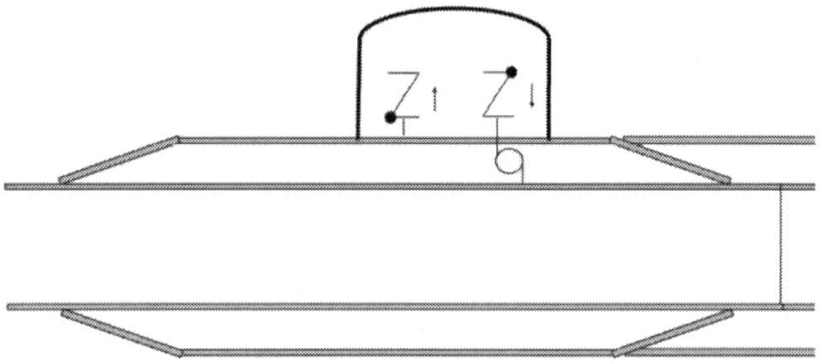

Figure 4 – Self Contained Vent Device

6.0 Applicable Codes and Standards

Vacuum Jacketed Pipe is unlike other insulation systems since the insulation system is constructed of a pipe within a pipe and is assembled in a factory environment. The facility and personnel insulating the pipe must comply too codes and standards of the pipe manufacturer as well as commercially accepted vacuum industry standards. This further separates the technology of mechanically insulated pipe systems that conform to a variety of material quality standards of testing the product to ASTM standards including thermal conductivity to C 177 & C 518, and other standards for composition, density, linear coefficient of expansion, absorption of moisture, water vapor permeability, combustibility, flame and smoke and other standards for materials.

Vacuum jacketed pipe meets all ASTM material standards for the selected material but also meet the manufacturing and testing standards as defined below and/or special customer test and material specifications.

The following are the most common codes and standards required for LNG terminal piping:

- ASME B31.3 Process Piping
- ASME Section VIII Boiler & Pressure Vessel Code, Division 1
- NFPA 59A Standards for the Production, Storage and Handling of Liquefied Natural Gas (LNG) 2006
- 33 CFR Part 127 Waterfront Facilities Handling LNG and Liquefied Hazardous Gas
- 49 CFR Part 193 for all LNG Piping
- EJMA Expansion Joint Manufacturers Association

The Federal Energy Regulatory Commission (FERC) and U.S. Coast Guard have some additional standards for the entire site and may have an impact on the codes and standards for LNG terminals located in North America.

7.0 Materials of Construction

Vacuum jacketed pipe for the most part is constructed using all metal construction with exception of some materials in the annular space that include internal spacer to center the pipe within the pipe and radiation shielding that reduces this heat input considerably. It is a multilayer insulation that is wrapped around the inner pipe before it is inserted in the jacket pipe.

7.1 Pipe Material

All LNG wetted material and adjoining materials are normally constructed using ASME B31.3 approved materials such as ASTM A312 304/304L material and ASTM A553 9% nickel material. The jacket material may vary from ASTM A312 304/304L material to carbon steel ASTM A53 or other materials such as 9% nickel ASTM A553 depending on the service specification for this material. If this is an aboveground installation then in most cases carbon steel is acceptable. If this is an underground or underwater installation you may consider the austenitic 304/304L or 9% nickel material. Jacket material can also provide a secondary containment for LNG if the jacket material meets the operating cryogenic temperature requirements. Another recommendation for 9% nickel pipe is that it a higher tensile strength than 304/304L stainless steel and is better suited for long bottom pull LNG pipe runs and it adds some additional strength against physical damage.

7.2 Pipe Elbows and Tees

The pressure boundary elbows and tees are butt weld fittings and comply with ASTM A403.

7.3 Flanges

All flanges are per ASTM A182 type 304/304L and are rated a minimum of 150 # class.

7.4 End Connections

The end connections may vary in configuration based on pipe size, pressure rating and other design requirements. The material for these can be from sheet material ASTM A240 304/304L or pipe ASTM A312 304/304L materials. The end connections will transition in temperature from -162°C to 27°C along the length of the material.

7.4.1 Bayonet End Connections

The bayonet end connection is considered the standard termination for static vacuum jacketed pipe spools. This connection is designed to reduce the field installation time and labor. Figure 8 is a typical example of 550 meter runs of pipe that was completely installed and ready for LNG in three days. The bayonet joint is a precision end connection that is male and female connection that has a reduced heat leak design and requires only one weld per joint. The bayonet joint standard sizes range from ¾" to 16" and larger sizes may be considered based on the application. The larger sizes as well as the smaller sizes are design pressure limited based on the male bayonet that will be exposed to external operating pressure.

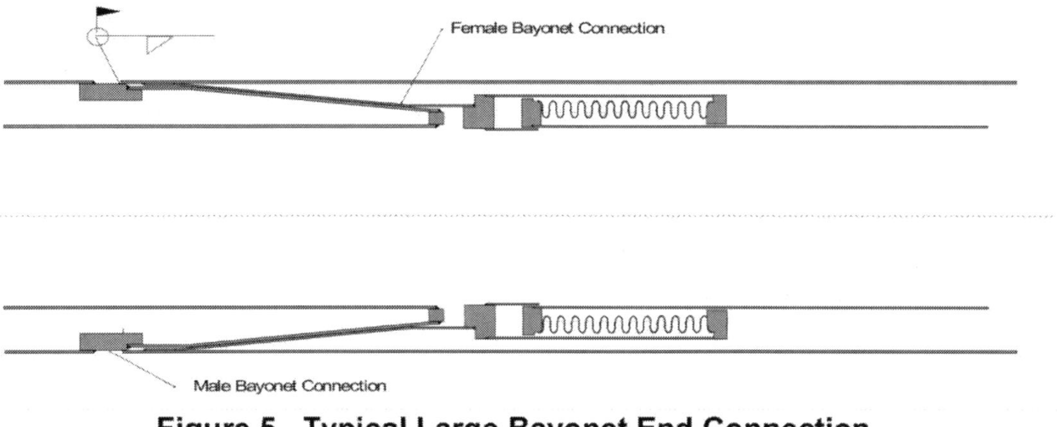

Figure 5 - Typical Large Bayonet End Connection

$$Q_1 = \left(\frac{A_1}{L_1} + \frac{A_2}{L_2} \right) \int_{200^\circ R}^{540^\circ R} \lambda \Delta T Btu/hr$$

Q_1 = *Btu/hr*

A_1 = *sq.ft end closure spool one or female bayonet*

A_2 = *sq.ft end closure spool two or male bayonet*

L_1 = *lg. of closure one*

L_2 = *lg. of closure two*

$$\int_{200^\circ R}^{540^\circ R} \lambda \Delta T = 2623 Btu/hr - ft \text{ for LNG}$$

Determine of number of joints in the length of pipe x Q_1 will give you the total conductive heat leak.

There are other conductive heat leaks that also need to be considered through intermittent spacer between the inner pipe and jacket pipe normally made of low heat leak materials such as glass filled epoxy resin. These items may vary based on the design of the spacer.

The resultant heat leak for static vacuum jacketed pipe is most always considerably lower that all other insulation methods and can be as low as 10% of other methods. The length of pipe spools, expansion loops and other pipe bends and constraints have an impact on the overall heat leak. Vacuum jacketed pipe is best used in long straight runs. The longer the pipe runs the better and the more efficient the LNG pipe is obtained using vacuum jacketed piping.

15.0 LNG Terminal Locations with Vacuum Jacketed Pipe

As mentioned above that vacuum jacketed pipe was first installed and in operation in 1996 at Atlantic LNG (ALNG) in Trinidad. This first installation proved successful and several other LNG terminals now have this static vacuum insulation system on some or most all of the LNG piping and vapor piping. The following is a list of LNG terminals and the size pipe and date of installation.

4" line at Atlantic LNG Train 1 (about 950 meters) – 1996

12" line at Distrigas Upgrade (about 250 meters) – 2000

8" line at Egyptian LNG (about 5,700 meters) - 2004

6" line at Atlantic LNG Train 4 (about 3,600 meters) -2004

30", 24", 12", 8" line for Darwin LNG train 1 (about 1,900 meters each) -2004

There is several other projects LNG reviewing vacuum jacketed pipe for the new terminals that are in various stages of design or approval.

16.0 Conclusion

Introduction of this product for the transfer of LNG is going to continue to grow as more companies become aware of the benefits and saving it has to them and overall energy costs and added safety when this product is employed. It has a

Figure 6 – Typical Bayonet Installation

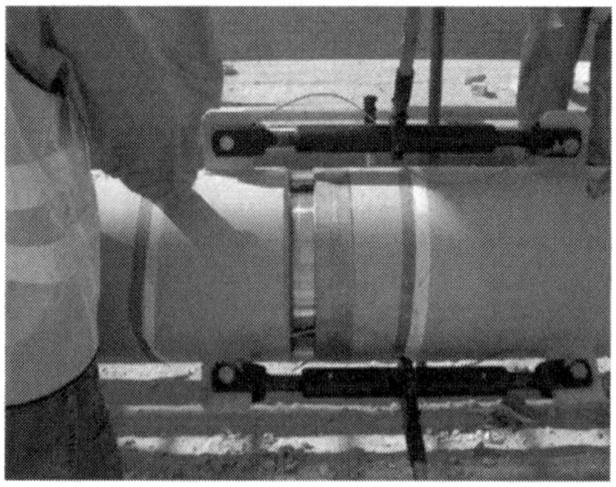

Figure 7 – bayonet Assembly Secured with Ratchet Jacks before Welding

Figure 8 – Bayonet Connected Vacuum Jacketed Pipe, approximately 550 meters installed in 3 days, ready for LNG.

7.4.2 Mechanically Insulated Field Joint

The mechanically insulated field joint requires more space and each end connection will require mechanical insulation to transition from the LNG temperature to ambient temperature. This butt weld connection is made on the inner line and does require x-ray or ultrasonic testing. This joint is used on large pipes where bayonets are not available or the system pressure rating cannot be met with the bayonet. The butt weld end connection is no different than if the pipe was all field fabricated. The initial cost of this joint on larger pipe sizes may be less than a bayonet joint but will need additional maintenance and it does take up more space than a bayonet joint. Figure 9 shows a typical field joint with cone closures and mechanical insulation installed after the joint is tested.

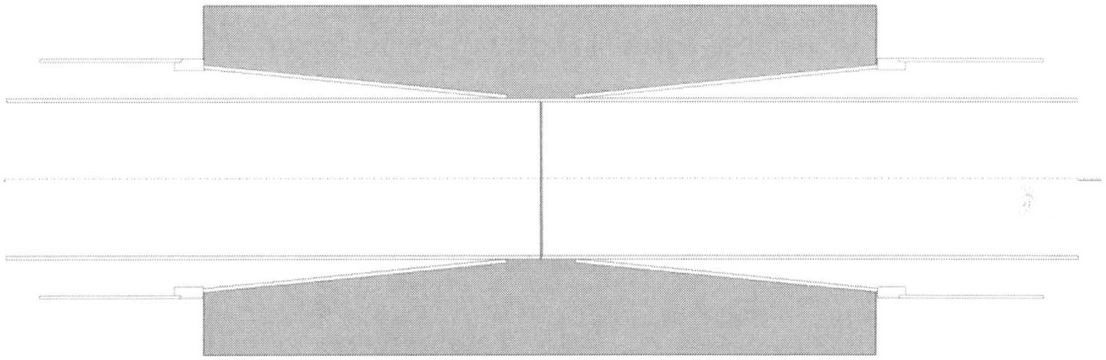

Figure 9 – Mechanically Insulated Field Joint

7.4.3 Vacuum Insulated Field Joint

In some cases where bayonets cannot be used and mechanical insulation is not acceptable for the environment where the piping is installed vacuum insulated field joint may be used. These joints have proven to be the most expensive of all other field joint connection but do work well in underground and underwater applications as well as the bayonet joint. This joint requires additional equipment in the field to evacuate and leak test the completed joint. Figure 10 shows the basic elements of this evacuated field joint. The evacuation is completed using a small portable vacuum pump and is not as high a vacuum as the shop fabricated pipe spools. The vacuum range may be $1x10^{-2}$ mmHg or lower. This is acceptable for this joint since it is a very small area compared to the line length.

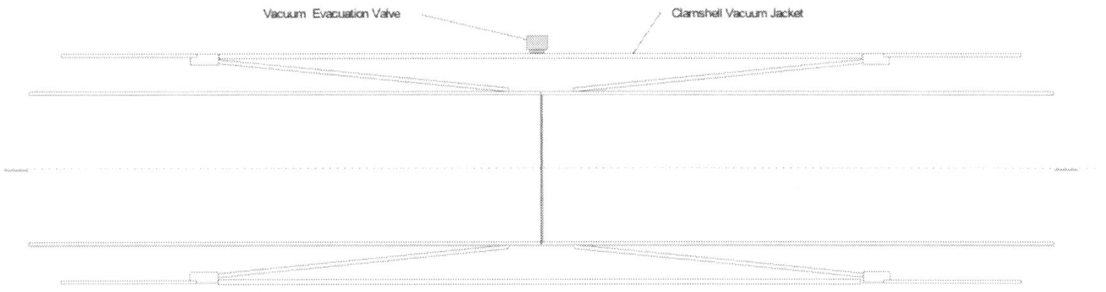

Figure 10 – Vacuum Insulated Field Joint

7.5 Internal and/or External Bellows

There is a requirement for bellows in pipe within pipe construction for static vacuum pipe spools. These bellows may be located in the inner carrier pipe or in the jacket pipe to compensate for the carrier pipe shrinkage when cooled to LNG temperature. The bellows are formed from sheet material rolled and welded, then convolutes are added to make the bellows. The material for these bellows commonly used for this application is ASTM A240 321 or 347 materials and have 304/304L butt weld ends.

7.6 Vacuum Seal-off Valve

Every vacuum jacketed pipe spool is supplied with a vacuum seal-off valve that also serves as a vent valve if there is an over pressure in the vacuum space. This valve is constructed from 304 stainless steel material and is supplied with elastomer o rings to seal the vacuum in the annular space.

8.0 Vacuum Jacketed Pipe Fabrication

Vacuum jacketed pipe requires unique fabrication requirements in order to assure the static vacuum integrity of the insulation system that will need very little to no maintenance for the normal expected operation of the terminal. All fabrication will comply with the requirements as listed above in paragraph 6.0

8.1 Welding

All welders for fabricating the carrier pipe pressure boundary welds must be qualified to ASME Section IX Welder Qualifications, Article III of the Boiler and Pressure vessel code.

8.2 Welding Methods

GTAW, GMAW and orbital welding is the recommended standard for fabricating vacuum jacketed pipe. Inert gas shielding and qualified filler metal to be part of the weld process.

8.3 Cleaning

It is critical to clean the pipe surfaces in the vacuum space to remove all hydrocarbons that will continue to outgas in a static vacuum process. Also other dust, grit and residue that may contaminate the elastomer seal to be removed from these surfaces. Cleaning the pipe is to be completed prior to wrapping the inner pipe with multi-layer insulation and before inserting into the jacket pipe.

9.0 Examination and Testing

The manufacturer of the vacuum jacketed pipe to use qualified personnel reporting to the quality assurance manager and to examine and test all pipe spools to a qualified procedure in a quality assurance manual. All examination and tests to be properly recorded and maintained that will include material and welder certifications.

9.1 Mass Spectrometer Leak Testing

All vacuum Jacketed Pipe is subject to a helium mass spectrometer leak test due to the leak rate requirements for vacuum static piping. This examination sensitivity is far greater than or test methods such as bubble testing with air or nitrogen gas and has a proven history for maintaining a static vacuum in the annular space for 25 to 30 years with no leak through a pinhole opening that may not show up in other test methods. This test is achieved by evacuating the annular space through the pump-out weld welded to the jacket pipe and attaching the helium mass spectrometer calibrated to sensitivity of 1×10^{-9} std. cc/sec. No leak indication will be allowed. Additional tests pressurizing the annular space with nitrogen and about ten percent helium gas will also be conducted while the mass spectrometer is connected to the pump-out valve. After successful completion of the mass spectrometer testing and pressure testing the pipe spool can then move to the next step.

10.0 Evacuation and Sealing

This is the most time consuming process for obtaining vacuum jacketed pipe. It requires heating each pipe spool to 100°C to 110°C to assure all water is vaporized and then evacuation of the annular space can begin. The heat to remain on during the evacuation process until the pressure is reduced to below .01 mm/Hg (10 microns). Continue pumping until the pressure is less than .001 mm/Hg. Allow the metal to degas at this pressure depending on pipe size and length of pipe spools. Close the valve and monitor the pressure for at least seven days for any indication of rate of rise that would indicate a leak.

11.0 Painting

Jacket pipe is not always required to be painted but if it is carbon steel and/or is stainless steel that is exposed to the seawater environment it is recommended that a protective coating be painted on the jacket pipe. This coating should never be installed prior to all leak testing as the paint may temporarily fill a potential leak.

12.0 Packaging and Shipping

The inner pipe is sealed from the environment it will be exposed to during shipping. Caplug's will be installed in the ends to keep out foreign material and corrosion air. Based on the size of the pipe spools being shipped they will be located in a skid assembly and secured for shipping and stacking if required.

13.0 Site Storage

Pipe spools may arrive in the field ahead of schedule and a designed protected laydown area is important for these spools. Upon arrival the vacuum level should be monitored and visual inspection completed to assure the spools were not damaged or lost vacuum during shipping.

14.0 Heat Leak Analysis for VJP

Evaluating the heat leak is based on considerable testing early in the development of vacuum jacketed pipe and analysis that substantiated these tests. Some of the early work in this field was done by Sir James Dewar in 1923 for a liquid hydrogen container that was the first vacuum jacketed container. Today we still refer to cryogenic containers as "Dewar's". The affect of these containers is that the pressure is very low and the molecular movement is almost non-existent. Radiation and conductive heat transfer must be dealt with in a different manner. Today we reduce the radiation heat transfer to the cryogenic fluid by wrapping it in super insulation that consists of multi-layers of aluminum foil and a spacer material that reduces the conduction between each radiation shield layer. Other materials used to reduce this radiation heat input is aluminized Mylar that is aluminized on both sides and each layer can be considered two radiation shields. Of course the factory fabricated vacuum jacketed spools are sealed at each end with a stainless steel closure from the inner pipe to the jacket pipe. In most cases this closure is the larger source of heat input the cryogenic fluid. Particular design is required to reduce this heat input by extending the length of this closure and reducing the thickness to the minimum required for a specific pipe size without jeopardizing the mechanical strength of this connection.

14.1 Radiation Heat Leak Analysis

Thermal radiation is still being transfer even in a high vacuum in the annular space. The emitters are assumed to be gray bodies that are not dependent on wave length but are usually assumed to be so in order to simplify the calculation. Without defining a particular example here the basic elements of this radiation heat transfer analysis is defined by the basic heat exchange between two surfaces of different temperatures and is written

$$q_r = \sigma F_e F_a A (T_1^4 - T_2^4) \text{ Btu/hr}$$

where;

q_r = *Btu/hr, rate of heat flow by radiation*
$\sigma = 0.1714 \times 10^{-8}$ *Btu/hr sq. ft °R, Stefan-Boltzman constant*
F_e = *emissivity factor*
F_a = *configuration factor*
A = *sq. ft. area*
T_1 = *°R, temperature of hot surface*
T_2 = *°R, temperature of cold surface*

14.2 Conductive Heat Transfer

The conductive heat path is determined based on the thermal conductively of the material and between the temperature differences from one end to the other. This assumes the ambient temperature of 540°R (27°C) and LNG temperature at 200°R (-162°C).

positive proven track record in the LNG terminal application that makes it more acceptable as a viable product for future terminal applications. It has also been proven underground and the applications for underwater installations is being evaluated as an alternative to long jetties and dredging required for these new and larger LNG ships that have a very deep draft.

References

- Insulation Outlook November 2005 article on corrosion under insulation (CUI).
- Federal Energy Regulatory Commission (FERC)
- Advanced Cryogenics – Plenum Press 1971 C. A. Bailey
- ASME Boiler & Pressure Vessel Code Section VIII, Division 1
- ASME B31.3 Process Piping
- NFPA 59A Standards for the Production, Storage and Handling of Liquefied Natural Gas (LNG) 2006
- 33 CFR Part 127 Waterfront Facilities Handling LNG and Liquefied Hazardous Gas
- 49 CFR Part 193 for all LNG Piping
- EJMA Expansion Joint Manufacturers Association
- Russell B. Scott - Cryogenic Engineering reprint 1988 by Met-Chem Research Inc.
- David H. Holkeboer, Donald W. Jones, Frank Pagano and Donald J. Santler – Vacuum Technology and Space Simulation 1993 by American Institute of Physics
- C. M. Van Atta – Vacuum Science and Engineering, 1965 McGraw Hill, New York

Biography of Speaker

John Bonn started working in the engineering department at a cryogenic equipment manufacturing company called CVI in 1961 and has held various position in the engineering department from designer to chief engineer and vice president of product development. After 43 years of service he started his own company VJ Systems, LLC to continue engineering and manufacturing of vacuum jacketed pipe for LNG terminals.

CAREER HIGHLIGHT

Responsible for selling and designing the first vacuum jacketed piping for installation at Atlantic LNG Terminal in Trinidad.

Development of LNG fueling station with underground tank and piping for fueling truck and bus fleets.

Mirror Fusion Test Facility High Vacuum Cryopumping Design that made this achievable by improving the efficiency of the Cryopumping Array to capture stray deuterium in the high vacuum space inside the large chamber. This cryopump design received awarded in 1985 by the Ohio Society of Professional Engineers for Outstanding Engineering Achievement for this cryopump design.

First Filtering System approved by the Atomic Energy Commission to be 99.999% efficient to remove radioactive Methyl Iodine from Nuclear Fission Containment Facilities.

Mars Temperature Simulation System led to John Bonn's signature on the Viking Satellite that was the first successful landing on Mars.

Space Simulation Chamber for testing of Astronauts in Space Suits in outer space environment that received complements from Dr. Werner Van Braun for the simple, safe and efficient design.

EDUCATION
Ohio State University Mechanical Engineering School 1958 to 1961
Bucknell University 1962 to 1965
TECHNICAL PUBLICATIONS
The American Society of Mechanical Engineers – Large Scale Cryopumps for Fusion Power Systems
Space Simulation Conference – High Vacuum Facility for Hydrazine Thruster Testing
TRADE PUBLICATIONS
Multiple articles in trade publications for various cryogenic products.
PROFESSIONAL AFFILIACTIONS
Reviewing Editor: Cold Facts Publication, Cryogenic Society of America
U. S. PATENTS
Several patents relating vacuum jacketed piping, LNG fueling system, cryogenic components, deuterium cryopumping in nuclear fusion devise and a high efficiency carbon absorber for nuclear power plants to remove radioactive methyl iodine.
John W. Bonn VJ Systems, LLC, 3815 Darbyshire Drive, Hilliard, Ohio43026.
Tel: (614) 876-8231, Fax (614) 876-8241, Email: johnbonn@vjsystems-llc.com

Processes for High C2 Recovery from LNG – Part III: SIMAR Applied to Gas Processing

Stanley Huang, Roger Chen, Dennis Cook, Douglas Elliot*

IPSI LLC
3000 Post Oak Boulevard
Houston, TX 77056
* To whom all correspondences be addressed

AIChE Spring Meeting, April 2006
6th Topical Conference on Natural Gas Utilization
Orlando, Florida, April 23-27, 2006

1.0 INTRODUCTION

Expander technology has played an important role in the gas processing industry since the 1970's [Collins et al., 1984]. This is particularly true in ethane (C2) recovery applications as the expander is an efficient way of generating refrigeration by pressure letdown in low temperature regions. Furthermore, the connected booster compressor can recover a significant portion of the pressure letdown. The combination of the two factors makes this approach cost-competitive against other approaches.

There are hundreds of patents describing different processes with various claims [Lee et al., 1999]. The sheer number makes technology selection anything but a straightforward task. Quite a few articles appeared in recent years discussing the principles and guidelines for process screening [Barclay, 2005; Pennybaker et al., 2000]. These publications also provided insight of the art. This work follows the trend and shares the same intentions.

In our previous works on high C2 recovery from liquefied natural gas (LNG), a procedural definition of System Intrinsic Maximum Recovery (SIMAR) was provided [Huang et al., 2006a and b]. It has been demonstrated that without any constraints in the refrigeration supply, the maximum C2 recovery is capped by the physical configuration of a system, which includes an expander, sub-cooler, demethanizer (DeC1) column, and reflux streams.

This work extends SIMAR to gas processing where the supply of refrigeration is finite. In additional to the intrinsic constraints imposed by the physical configuration of the system, the limitation of refrigeration supply is considered. More specifically, the refrigeration to the feed chilling section may be drawn from the returning residue gas and the lower section of the DeC1 in a traditional design. If there is a deficiency, then external refrigeration must be provided. The total compression work should include both residue gas compression and external refrigeration.

The impacts on SIMAR by inlet gas pressure, composition, and the number of side-boilers of DeC1 will be discussed. Also, impacts of heat pumps on the system will be addressed separately because heat pumps change temperature distributions which affect C2 recovery directly.

2.0 TECHNICAL BACKGROUND

Figure 1 shows a generalized scheme for C2 recovery based on expander-technology. The process is designed to strip the inlet natural gas (NG) of its heavier components. The residue gas is recompressed and returned to the pipeline.

Sweet and dry natural gas goes through an inlet chilling section, where the gas is chilled to a suitable level before entering the heart of the plant: an expander-based C2 recovery scheme. The refrigeration of the inlet chilling section is mainly provided by the returning residue gas and supplemented by side-draws from DeC1 (not shown). Depending on actual requirements, external refrigeration, such as propane refrigeration or cascaded propane and ethylene systems, may be required (not shown).

The main components of the expander section include a separator, expander, and DeC1. A subcooler is usually provided for improved refrigeration integration in the low temperature regions. The addition would enhance the performance significantly. The exact design of this section remains an art and many popular schemes in industry are protected by patents.

Following our previous works using LNG as feed, the expander schemes are grouped into three types: separator top not rectified (XPDR 1), separator top fully rectified (XPDR 2), and separator top self-rectified (XPDR 3). The three types are shown in Figures 2, 3, and 4, respectively. When LNG is the feed, the pre-chilling section in Figure 1 is not required. Also, the returning residue gas is allowed to retain the refrigeration, which reduces the compression duty dramatically.

Figure 5 plots the C2 recovery level against the separator temperature for all the three types. Both XPDR 1 and XPDR 2 show monotonic trends of improvement as the temperature decreases. The trend continues even after the inlet gas is totally liquefied. At that point the expander is replaced by a Joule-Thomson (JT) valve and no expansion work is recovered. In contrast, XPDR 3 shows a different trend, though. As the separator temperature decreases, the C2 recovery goes through a maximum. In other words, simply increasing the refrigeration supply to the inlet chilling section may not always improve the C2 recovery.

A procedural definition of SIMAR at a specified DeC1 operating pressure is summarized below. For XPDR 3, the SIMAR is defined at the maximum of the curve. For XPDR 1 and XPDR 2, the SIMAR is defined at the temperature level where the separator fluid has 30% of vapor fraction (VF). The choice of 30% is based on a practical consideration that no expanders would be installed if the gas flow is below this level.

Following the aforementioned procedural definition, Figure 6 depicts typical SIMAR curves for the three types of process schemes. The characteristics of the schemes are observable: (1) XPDR 1 has the lowest C2 recovery level because it has no reflux stream to the DeC1. (2) The reflux stream makes XPDR 2 more efficient than XPDR 1 throughout the entire range, which corresponds to the operating pressures of DeC1 from 17 to 42 bar. (3) The efficiencies of XPDR 2 and XPDR 3 are comparable, while each has its advantages over a certain span.

3.0 SCOPE OF WORK

Figure 1 with NG as the feed is the focus of this work. The direct consequence of the shift from LNG to NG is that it is necessary to recover the refrigeration in the returning residue gas in order to support the operation of the inlet chiller. Accordingly, the re-compression work increases significantly. Supplementary refrigeration can be extracted from the bottom section of the DeC1 (not shown). If the refrigeration is still in deficit, then it is required to introduce external refrigeration.

This work will also investigate the impacts of inlet gas pressure, composition, and heat pumps on the SIMAR. In case the supply of external refrigeration is limited, operations below SIMAR level, called sub-SIMAR, will be presented as well.

4.0 APPROACH

Table 1 lists the four gas compositions used in this work. They represent different richness in C2+ components. The richness of a gas sample is reflected in its C2+ or C3+ components, which are indicated in gallons per thousand SCF (GPM). The phase envelopes corresponding to the four compositions are shown in Figure 7. The richer the gas, the wider its envelope becomes. Gas 1 composition is identical to the medium LNG composition in our previous works. The critical properties of the gas are listed as well. The inlet pressure of each gas is set at a level corresponding to 85% of the reduced pressure, which is defined as the ratio between actual pressure and the critical pressure of the gas.

All simulations in this work are performed using HYSYS. Table 2 lists pertinent parameters. The delivery pressure to pipeline is assumed to be 100 bar. Four temperature levels of external refrigeration are defined for conceptual discussions. The high-temperature (HT) sink represents air coolers and is the default setting in this work. The low-low temperature (LLT) level represents a cascade system including ethylene loops. Those levels in-between are different stages of propane refrigeration loops. The determination of external refrigeration requirement is described below.

When external refrigeration is required, it is assumed to be provided by conventional cascade loops using propane and ethylene as refrigerants. Although provision of refrigeration at several temperature levels is possible in an application, this work assumes only one level of refrigeration at an optimal temperature (not limited to the listed levels in Table 2). The accounting of the external refrigeration is through a corresponding compression work to remove the heat from the specified temperature. The conversion from the heat removed to the compression work follows an assumption of 50% Carnot efficiency. For example, to remove 1 MW of heat from -30 °C (243 K) to 40 °C (313 K), a Carnot engine requires 0.288 (= 313 /243 -1) MW. In this work, 0.576 (= 0.288 x 2) MW of compression work will be added in calculating the total compression work.

In HYSYS simulations, the external refrigeration is provided at the highest temperature level possible. Also, only the minimum amount is added to satisfy the refrigeration balances. The intention is to minimize the additional compression work and provide a fair comparison for all cases.

Refrigeration balance in the inlet chiller is mainly from the returning residue gas, and supplemented by extracting refrigeration from the lower section of the DeC1. Usually, the latter is achieved through side-boilers, which are typically thermal-siphons. The more side-boilers are used, the easier it is to align cooling curve (cold side) against the heating curve (hot side) and achieve better overall efficiency. However, increasing the number of side-boilers also increases the difficulties in control. One additional constraint resulted from multiple thermal siphons is that they may not be easily integrated into a single exchanger block, as in typical plate-and-fin exchanger designs.

For purposes of conceptual discussions, the simulations in this work will use one integrated exchanger to balance the refrigeration in the inlet chilling section. In other words, the exchanger will handle streams of inlet gas, returning residue, side-boiler(s), and external refrigeration. This simulation represents an idealized model that excess energy from one exchanger can be used in another. In actuality, when several exchangers are used, the energy flow between exchangers is impeded and there will be a reduction in plant efficiency.

Using one side-boiler reduces the aforementioned impact of inter-exchanger energy flows. However, the refrigeration in DeC1 cannot be utilized efficiently. Using two discrete side-boilers has the opposite trends. In the following discussions two side-boilers in an integrated exchanger will be assumed, unless specified otherwise. The actual efficiency of using two discrete exchangers is anticipated to be lower than the presented results, but is expected to be higher than using only one side-boiler. The difference between one and two side-boilers will be established in Section 5.1.1.

5.0 RESULTS

The simulation results are presented following a sequence from simple to complex cases. Section 5.1 assumes that the inlet chiller would supply sufficient refrigeration so that the system would operate at SIMAR levels. Any possible deficiency in refrigeration supply will be compensated by external refrigeration. Section 5.2 adopts a different approach. The inlet chiller will only chill the feed to an appropriate level short of SIMAR if the returning residue gas and side-boilers do not provide sufficient refrigeration. This approach will reduce the external refrigeration at the cost of a reduced C2 recovery. Results of this subsection are referred to as sub-SIMAR operations.

Section 5.3 will investigate the impact of introducing a heat pump to the system for enhanced system performances. Heat pumps have demonstrated their versatility in previous works. Finally, Section 5.4 will discuss special features in handling very rich gas inlet.

Gas 1 is the default gas composition and XPDR 3 is the default process configuration. Although there are no specific disadvantages using XPDR 1 or XPDR 2, it is easier to identify the SIMAR level using XPDR 3 due to the its unique maximum as shown in Figure 5.

5.1 OPERATION AT SIMAR LEVEL

SIMAR represents the maximum C2 recovery level achievable by a scheme without constraints in the refrigeration supply. The case of using LNG as feed represents this scenario and the result is plotted as the curve at far left side in Figure 8. The refrigeration in the returning residue gas from DeC1 reduces the work of recompression.

When the feed is natural gas, the refrigeration in the returning residue gas has to be recovered to support the operation of the inlet chiller. As a result, the recompression work increases. The second curve from the left in Figure 8 shows the case of natural gas feed. The case is separated from the LNG case by a dotted line. The two curves are roughly parallel to each other and the difference is about 2 MW per 100 MMscfd.

5.1.1 IMPACT OF SIDE-REBOILERS

No external refrigeration is required for LNG cases and the recompression work is the total work required. For NG case, the total compression work should also include the external refrigeration. The two thick curves on the right side of Figure 8 present the total compression work by using one or two side-re-boilers. The case of two side re-boiler requires a lower total compression work because the provided refrigeration matches better with the chilling requirements.

Moving along the curves in Figure 8 from high to low C2 recovery implies increasing DeC1 operating pressures. When the DeC1 operating pressure increases, the recompression work decreases. Therefore, the refrigeration generated by the expander also decreases. For the system to operate at SIMAR level, it demands a certain refrigeration to chill the inlet gas to a specified temperature. Hence, the demand for external refrigeration increases with the increasing DeC1 operating pressure. This explains the observed curve-to-right phenomenon in total compression at low C2 recovery levels.

It is possible to replace the external refrigeration by increasing expander contributions at high C2 recovery levels. This is demonstrated by the convergence of the curves of recompression and total compression. The final choice is an economic decision.

In the following work, only the results of two side-boilers will be presented. The relative distribution of loads on the two side-boilers is mainly determined by the requirement of the pre-chiller. The C2 recovery level is insensitive to the distribution.

5.1.2 IMPACT OF GAS INLET PRESSURE

Figure 9 shows the impact of inlet pressures. The results of three inlet pressures, 40, 50, and 70 bara, are presented. The thin curves on the left are recompression works only, and the thick ones at right are total compression works. The thin curves form a relatively narrow band because the compositions of the residue gas do not deviate significantly from pure methane.

Generally, as the inlet pressure increases the SIMAR C2 recovery decreases slightly at a specified DeC1 operating pressure. This is caused by higher SIMAR temperatures at the separator. The recompression work also decreases because higher inlet pressure translates to more recovered compression work from the expander.

The most pronounced impact is on the external refrigeration. At a specified DeC1 operating pressure, the refrigeration generation increases with the inlet pressure. Therefore, the requirement for the external refrigeration is reduced. At an inlet pressure of 70 bara, the system almost can be self-sustained.

When the inlet gas pressure was increased to 70 bar in the super-critical region, an unusual phenomenon was observed during the determination SIMAR using HYSYS. As the separator temperature was systematically lowered, the C2 recovery improved as anticipated. However, there was a minimum temperature below which the expander simply ceased to function. The reason was speculated to be that the separator fluid became too liquid-like for the expander to work properly. The actual expander performance should be verified by the vendors.

5.1.3 IMPACT OF FEED COMPOSITION

Figure 10 shows the impact of inlet gas composition by presenting the results of three gas compositions: Gas 1, Gas 2, and Gas 3. The inlet pressure is maintained at 85% of the reduced pressure in all cases. Thin curves on the left are recompression works only, and thick ones at right are total compression works.

The recompression works for three gases cluster in a relatively narrow band, indicating the insensitivity toward inlet gas compositions. This is attributable to the similar residue gas compositions after DeC1. The general trend of reduced recompression work as the inlet gas becomes richer is mainly caused by the higher inlet pressure.

As the gas richness increases the external refrigeration, indicated by the gap between recompression and total compression, required to achieve a specified C2 recovery level decreases. The reasons for this apparent contradiction to traditional experiences, i.e. richer gas demands more external refrigeration, are listed as follows:

1. It is listed in Table 2 that richer gas has higher inlet pressure in order to maintain a constant reduced pressure of 0.85. The high inlet pressures of rich gases reduce the requirement for external refrigeration.
2. As the gas richness increases, the liquid traffic in the lower section of the DeC1 also increases. This is due to the aforementioned increment in refrigeration generation by the expander and larger amounts of intermediate components in the feed stream. As a result, the refrigeration integration in the inlet chiller improves significantly and the requirement for external refrigeration is reduced.
3. The compression work for external refrigeration is determined by two factors: amount of refrigeration supply and its temperature level. This work specifies the target C2 recovery at SIMAR level, and adjusts external refrigeration accordingly. The required external refrigeration for Gas 1 may go as low as -65 °C. This low temperature level accelerates the demand for the compression work.

   In contrast, the traditional approach for determining the external refrigeration fixes the lower bound of refrigeration supply, say at -40 °C, then adjusts the refrigeration supply to match the target C2 recovery. The richer gas results in a reduced returning residue gas, while the refrigeration supply by the expander remains relatively constant if the inlet pressure is fixed. Therefore, the rich gas feed demands more external refrigeration.

For the very rich feed, such as Gas 4, complications do occur. The issues will be discussed in Section 5.3.

5.2 OPERATION BELOW SIMAR LEVELS

SIMAR operation is uniquely defined as far as the separator operation conditions are concerned. This is no ambiguity. In contrast, sub-SIMAR operations involve some judgments. The simulation results in this section are presented by feed compositions following the sequence of increasing richness: from Gas 1 to Gas 3.

5.2.1 SUB-SIMAR OPERATIONS FOR GAS 1

Figure 11 compares the SIMAR and sub-SIMAR operations for Gas 1. Two cases of sub-SIMAR operations are presented with progressively raised separator temperatures. This is to reduce the load on external refrigeration from LLT to LT levels. The thin curves on the left side are recompression work. As anticipated, to achieve a specified C2 recovery, the farther the deviation from SIMAR, the lower DeC1 operating pressure would be required. Hence, the recompression work increases.

The thick curves on the right are total compression work, which includes external refrigeration. The farther the deviation from SIMAR, the less external refrigeration would be required. Hence, the gap between recompression and total works reduces.

It is observed that total compression work remains the same regardless of deviations from SIMAR operations. In other words, once the process configuration is fixed, the total compression required to achieve a specified C2 recovery level is insensitive to pre-chilling levels. The choice of the separator operating conditions would be optimized by other factors. For example, if operating a system at SIMAR level requires a supply of external refrigeration at -50 °C. The decision may shift toward sub-SIMAR operations in order to avoid cascaded refrigeration systems including ethylene. Ethylene refrigeration systems are relatively expensive to provide.

Therefore, given the conditions of an inlet gas, SIMAR can serve as an expedient tool to establish the relations between C2 recovery and total compression work. Then, the exact pre-chilling level will be determined by other factors, such as economics [Barcley, 2005].

5.2.2 SUB-SIMAR OPERATIONS FOR GASES 2 AND 3

Similar to the previous section, Figure 12 compares the SIMAR and sub-SIMAR operations for Gas 2. Only one case of sub-SIMAR operation is presented because the external refrigeration requirement is relatively minor. The motivation to go deep into sub-SIMAR operations is very much reduced. Nonetheless, it is observed that total compression work is insensitive to sub-SIMAR operations.

It has been demonstrated in Figure 10 that as inlet gas becomes richer, the external refrigeration requirement decreases due to the improved refrigeration integration. The external refrigeration requirement for Gas 3 becomes insignificant that there is no incentive to pursue sub-SIMAR operations.

5.3 HANDLING VERY RICH FEED GAS

Figure 7 shows that when the inlet gas is very rich, the two-phase region expands and shifts toward higher temperatures and pressures. Consequently, the separator temperature for SIMAR operations may reside outside of the two-phase region. For example, Gas 4 at the specified pressure of 108 bar (equivalent to 85% reduced pressure), the bubble temperature is higher than -5 °C. However, it would require a temperature of -80 °C or lower to achieve a reasonable C2 recovery in DeC1. This gap is simply too wide for direct applications of SIMAR definitions.

It is advisable to perform coarse separations at a relatively warm temperature, say 0 °C, and make the feed gas leaner before directing it to the cryogenic section. It is against economical principles to chill the entire feed stream, including intermediate and heavy components, to an excessively low temperature level when the heavier components should be heated later to drive out lighter components. After this coarse separation, the feed gas will be leaner and the standard SIMAR principle can be applied.

The liquid stream from the aforementioned coarse separation may be directed to the same DeC1 in the cryogenic section or be handled by separate columns. There are pros and cons for either approach. The final choice will be determined by economic considerations. Nonetheless, the following statements would be useful guidelines. Handling by the same DeC1 has the advantage of reduced equipment count. However, the DeC1 bottom temperature would be high due to the existence of heavy components. Also, the traffic in the column would increase. To handle the increased traffic and avoid excessive upward heat transfer into the cryogenic section, the DeC1 column tends to increase in the diameter and height. The increased column height may lead to an additional side-boiler for improved refrigeration integration with the inlet chiller.

The other approach of using separate columns would avoid the aforementioned disadvantages. However, it increases the equipment count and may require compressors to return the stripped light components to residue gas streams.

5.4 HEAT PUMPS

It was demonstrated in our previous works that heat pumps, with the help of compressors, can move heat from one point to another within a system. Therefore, the temperature distribution can be altered significantly inside the system. The C2 recovery can be different from typical systems consisting of "passive" components. This section is dedicated to heat pump applications.

Figure 13 shows two examples of heat pumps embedded in an XPDR 3 schemes. Example Heat Pump 1 [Pitman et al., 1998] takes a side-draw from DeC1 overhead, which is lean in C2. A cryogenic compressor compresses the side-draw and the discharge is chilled by the incoming cold stream from the sub-cooler. The chilled or condensed stream is letdown in pressure and directed to the top tray as reflux. This heat pump effectively moves heat from the top tray to a lower one in the DeC1. The improvement in C2 recovery can be attributed to both lean C2 concentration and the added refrigeration at the top tray.

Example Heat Pump 2 [Yao et al., 1999; Nasir et al, 2003] takes the liquid as a side-draw from a suitable point in the lower (stripping) section of the column. The side-draw is letdown in pressure to generate refrigeration, which is used in the inlet chiller. The warmed side draw from the inlet chiller is then separated into vapor and liquid streams. The vapor stream is compressed and the discharge is cooled by an air-cooler before retuning to the column. The liquid stream is boosted in pressure by a pump. The discharge can be sent out as product or can be returned to the column. This heat pump effectively moves heat from the inlet stream to the bottom of the column or rejects the heat to the environment. In some cases, the reboiler of DeC1 can be totally eliminated or used as spare only.

Figure 14 shows the impact of using Example Heat Pump 1 to move 0.5 MW of heat from the top tray to a lower one. The simulation is performed based on energy transfer only. No material flows shown in Figure 13 are incorporated. The energy transfer decreases the temperature at the top tray and the C2 recovery is improved. In Figure 14, the shift from the lower curves to the higher ones is almost vertical because the compression work remains virtually unchanged.

The impact of Example Heat Pump 2 is not shown in Figure 14. The C2 recovery levels remain unchanged, however the total compression work will be reduced. The HT heat sink in this work has a temperature of 40 °C, which is higher than the column bottom temperatures for all cases. Therefore, moving energy from the inlet chiller to the lower section of DeC1 takes less work than rejecting it to the heat sink. The total compression curve is anticipated to move horizontally to the left (not shown).

6.0 CONCLUSIONS

The conclusions of this work are summarized as follows:

1. SIMAR is applied to gas processing, where refrigeration supply to the inlet chiller is limited. The refrigeration is mainly recovered from the returning residue gas and supplemented by extraction from the lower section of DeC1. Any deficiency should be supplied by the external refrigeration. The total work includes the recompression of residue gas and external refrigeration.
2. The recompression work for cases using NG as feed is significantly higher than LNG cases. This is due to the recovery of the refrigeration in the residue to fulfill inlet chiller demands.
3. Using two side-boilers reduces the total compression work from cases using one, because the refrigeration can match the chilling requirements better.
4. It is possible to eliminate the external refrigeration by increasing expander contributions. The choice is an economic decision.
5. Increasing the inlet pressure decreases the recompression work because the recovered work from the expander increases. The most pronounced impact is on the external refrigeration. Since the refrigeration generation increases with the inlet pressure, the requirement for the external refrigeration can be reduced. At a sufficiently high inlet pressure, the system can be self-sustaining.
6. As the gas richness increases the total compression work required to achieve a specified C2 recovery level decreases. The decrease is related to the increased inlet pressure as assumed in this work and explained in the study basis.
7. The total compression work is fairly independent of deviations from SIMAR operations. Therefore, given the conditions of an inlet gas, SIMAR can serve as an expedient tool to establish the relations between C2 recovery and total compression work. Then, the exact pre-chilling level will be set by other factors, such as economics.
8. Heat pumps have direct impact on the system performances. A heat pump at the DeC1 top portion improves the C2 recovery without impacting the compression work significantly. On the other hand, a heat pump at the DeC1 bottom section

reduces the requirement of external refrigeration, hence total compression. These applications are case-specific and licensor's involvement would be recommended.

7.0 REFERENCES

Barclay, M., "Study Identifies Optimal C2 Recovery for NGL Plants", Oil & Gas Journal, Issue of August 22 (2005)

Collins, C., McKee, D., Chen, R.J.J., Elliot, D.G., "Trends in NGL Recovery from Natural and Associated Gases", paper in Gas Tech 1984 Proceedings, Amsterdam (1984).

Huang, S., Coyle, D., Cho, J., and Durr, C.; "Select the Optimum Extraction Method for LNG Regasification", Hydrocarbon Processing, July (2004).

Huang, S., Cook, D., Yao, J., Elliot, D.G., "Processes for High C2 Recovery From LNG – Part I: Schemes Based On Refluxed Demethanizer", Paper to be presented in AIChE Spring Meeting, Orlando, Florida (2006a).

Huang, S., Chen, R.J.J., Yao, J., Elliot, D.G., "Processes for High C2 Recovery From LNG – Part II: Schemes Based On Expander Technology", Paper to be presented in AIChE Spring Meeting, Orlando, Florida (2006b).

Lee, R., Yao, J., Elliot, D., "Flexibility, efficiency to characterize gas-processing technologies", Oil & Gas Journal, Issue of December 13 (1999).

Nasir, P., Sweet, W.C., Elliot, D.G., Chen, R.J.J., Lee, R.J., "Enhanced NGL Recovery Process Selected for Neptune Gas Plant Expansion", Oil & Gas Journal, Issue of July 21 (2003)

Pennybaker, K.A., Wolverton, S.E., Chafin, S.W., Ruddy, T.R., Pritchard, C.W., "Comparative Study of Ethane Recovery Processes", paper presented at the 79th Annual GPA Convention, Atlanta, GA (2000).

Pitman, R.N., Hudson, H.M., Wilkinson, J.D., "Next generation processes for NGL/LPG recovery", paper presented at the 77th Annual GPA Convention, Dallas, TX (1998).

Yao, J., Chen, J.J., Elliot, D.G., "Enhanced NGL Recovery Process", U.S. Patent 5,992,175 (1999).

8. List of Tables

Table 1 Gas compositions used in this work

Natural Gas (NG)	Gas 1	Gas 2	Gas 3	Gas 4
Inlet temperature, C	40	40	40	40
Inlet pressure, bar	50	65	79	108
Critical T, C	-63.45	-48.40	-26.36	32.76
Critical P, bar	58.63	75.69	92.22	126.9
Components	Mole Fraction	Mole Fraction	Mole Fraction	Mole Fraction
Nitrogen	0.00127	0.0000	0.0000	0.0124
Methane	0.90700	0.8695	0.7787	0.6163
Ethane	0.08617	0.0792	0.1326	0.1476
Propane	0.00552	0.0332	0.0549	0.1076
i-Butane	0.00003	0.0074	0.0128	0.0333
n-Butane	0.00001	0.0055	0.0100	0.0292
i-Pentane	0.00000	0.0037	0.0045	0.0172
n-Pentane	0.00000	0.0015	0.0020	0.0111
n-Hexane	0.00000	0.0000	0.0045	0.0118
n-Heptane	0.00000	0.0000	0.0000	0.0135
n-Octane	0.00000	0.0000	0.0000	0.0000
TOTAL	1.00000	1.0000	1.0000	1.0000
Btu/SCF	1,082	1,165	1,281	1,602
C2+ content, GPM (Note 1)	2.45	3.63	6.20	11.05
C3+ content, GPM	0.15	1.52	2.66	7.10

Notes:

1. GPM stands for Gallons Per Thousand SCF.

Table 2 Simulation parameters used in this work

Natural Gas inlet	Unit	Value
Plant capacity, dry & sweet NG at inlet	MMscfd	100
Send-out residue gas pressure	Bar	100
Send-out residue gas temperature	°C	40
General		
Expander polytropic efficiency		80%
Compressor polytropic efficiency		80%
Pressure drop through plate-fin exchanger	Bar	0.2-0.3
Pressure drop through air coolers	Bar	0.6
Air cooler discharge temperature	°C	40
Minimum delta T in plate-fin exchanger	°C	2.5 ± 0.2
DeC1 operating pressure	Bar	17 to 42
Number of trays in DeC1 column		15
C1 composition in DeC1 bottom product		1%
Minimum delta T in plate-fin exchanger	°C	2.5 ± 0.2
LMTD	°C	5.5 ± 1.0
Max impact on C2 recovery		-0.005
High-temperature (HT) sink	°C	40
Medium-temperature (MT) sink	°C	-10
Low-temperature (LT) sink	°C	-35
Low-Low-temperature (LLT) sink	°C	-45

9. List of Figures

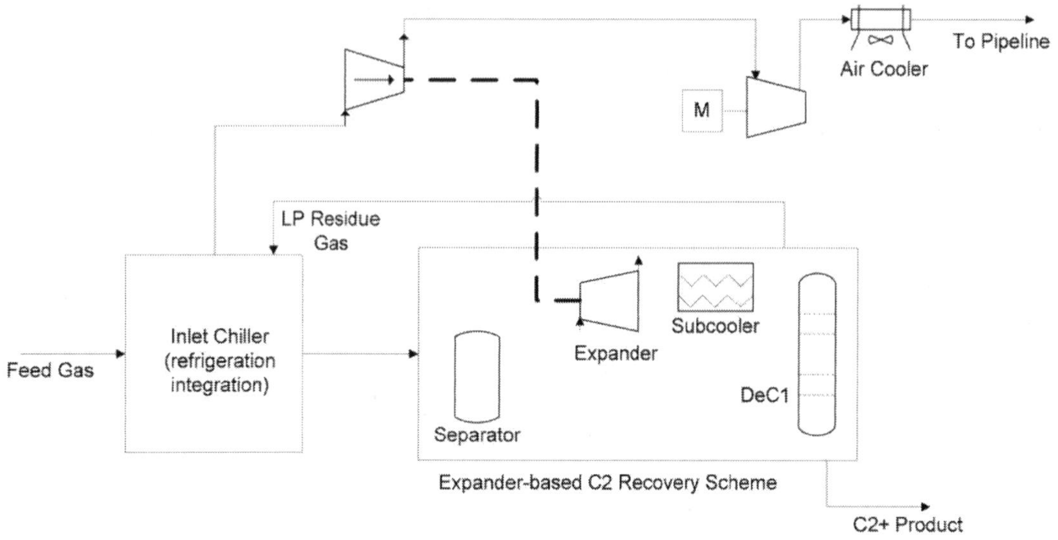

Figure 1 Generalized gas processing scheme for C2 recovery

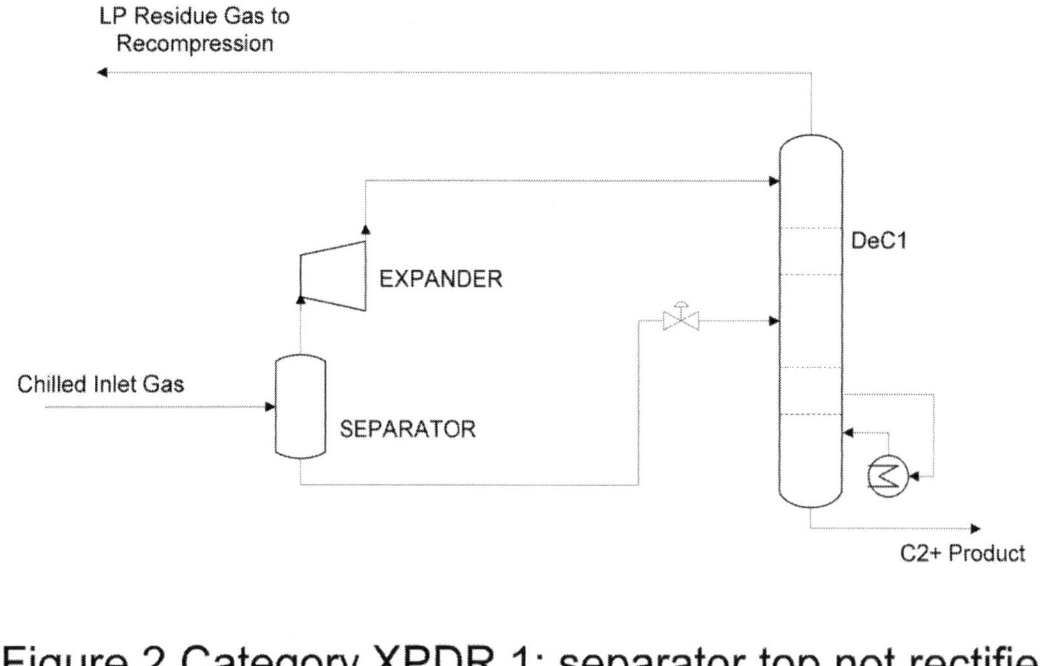

Figure 2 Category XPDR 1: separator top not rectified

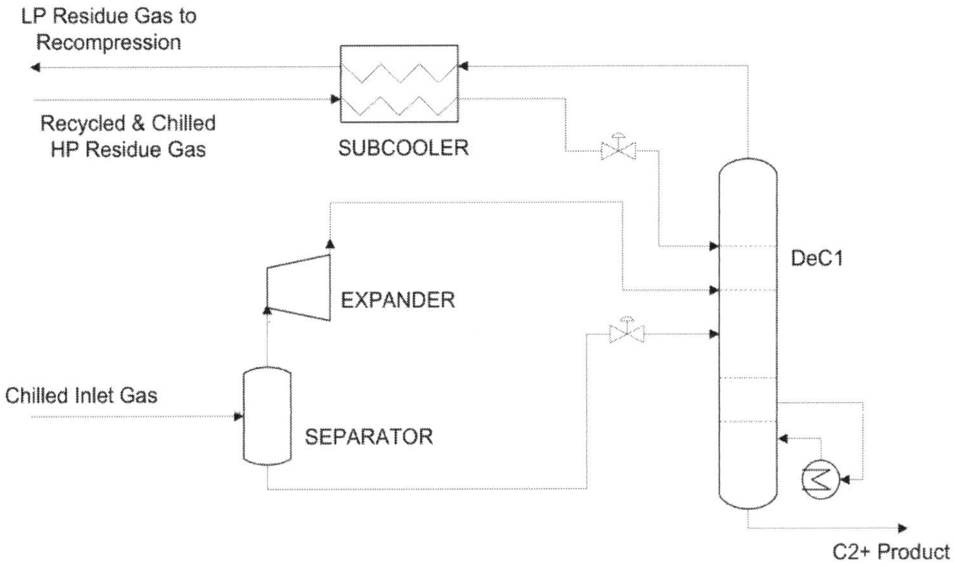

Figure 3 Category XPDR 2: separator top fully rectified

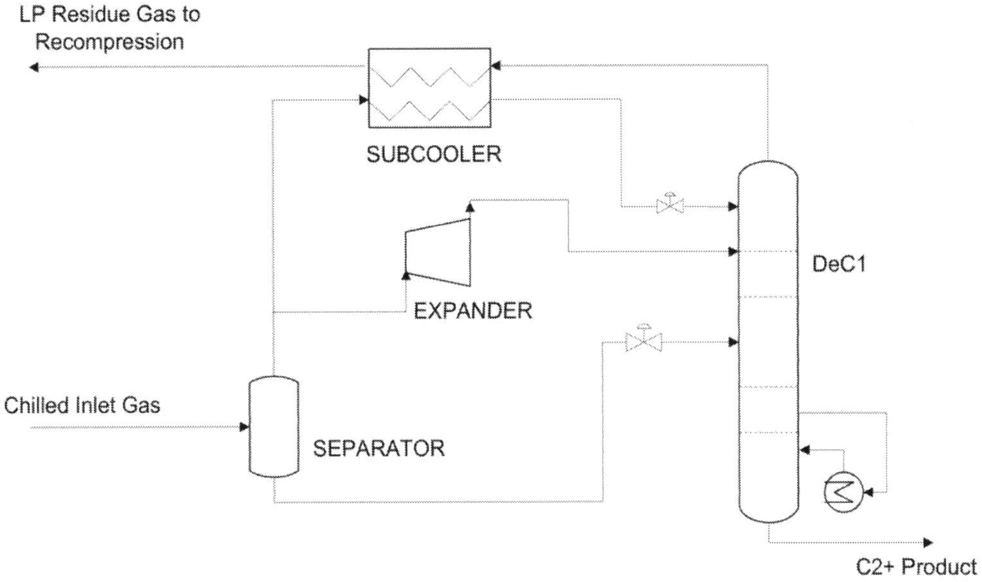

Figure 4 Category XPDR 3: separator top self-rectified

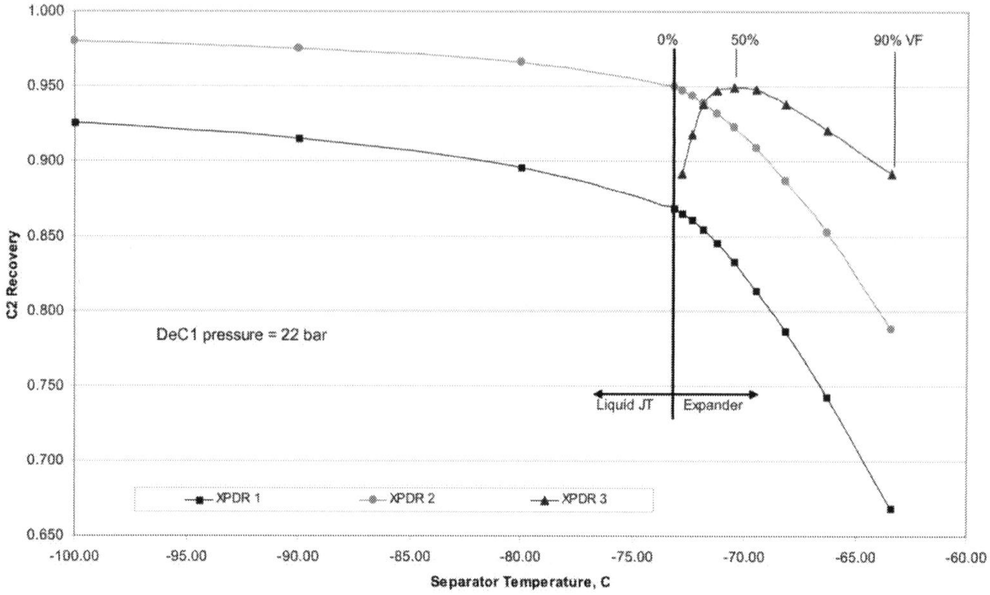

Figure 5 Two types of behavior for different categories. VF stands for vapor fraction.

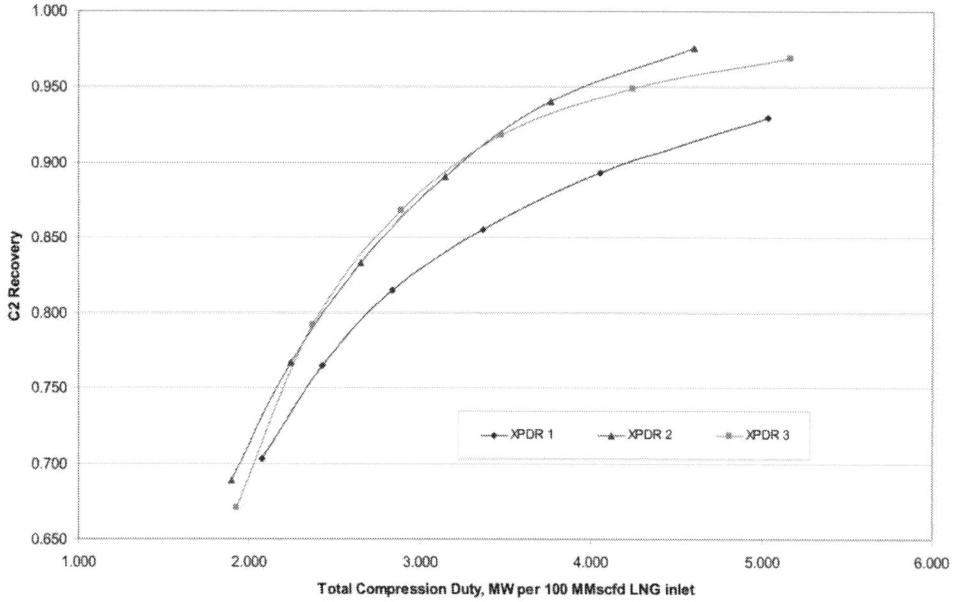

Figure 6 Comparing SIMAR curves for three XPDR categories, LNG feed

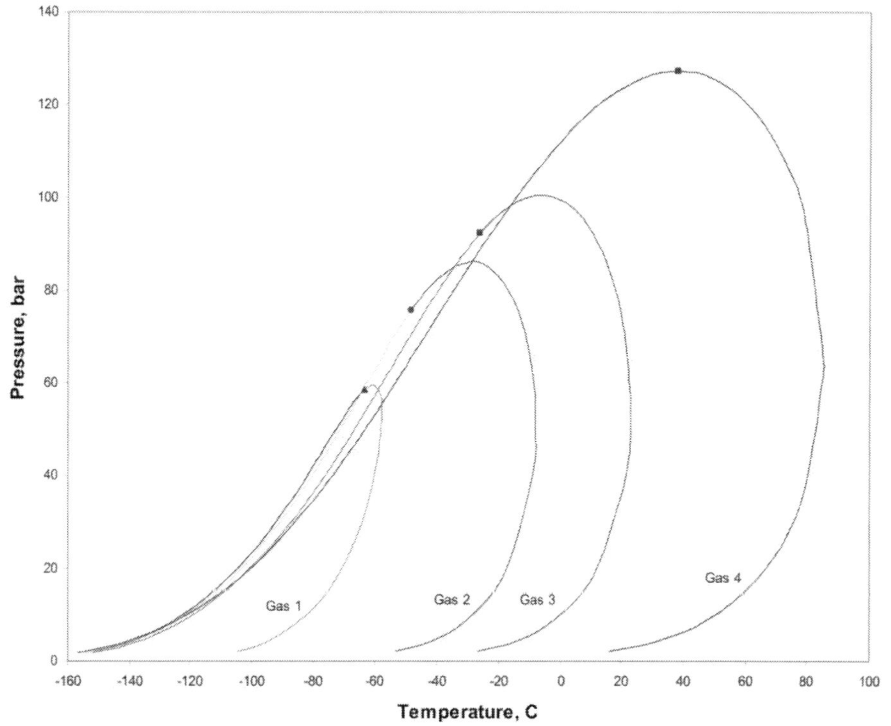

Figure 7 Phase envelope of inlet gas

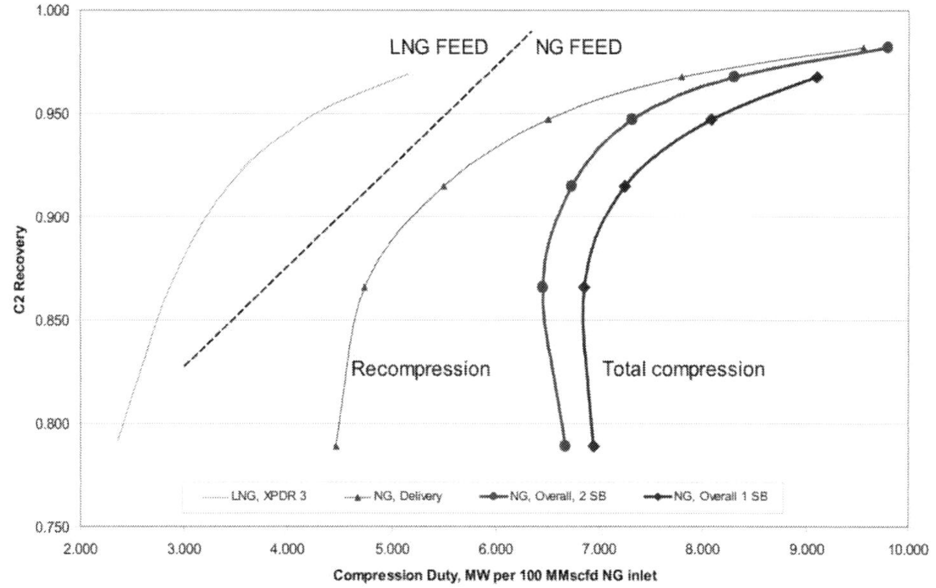

Figure 8 Comparing compression duties and Impact of side-boilers (SB)

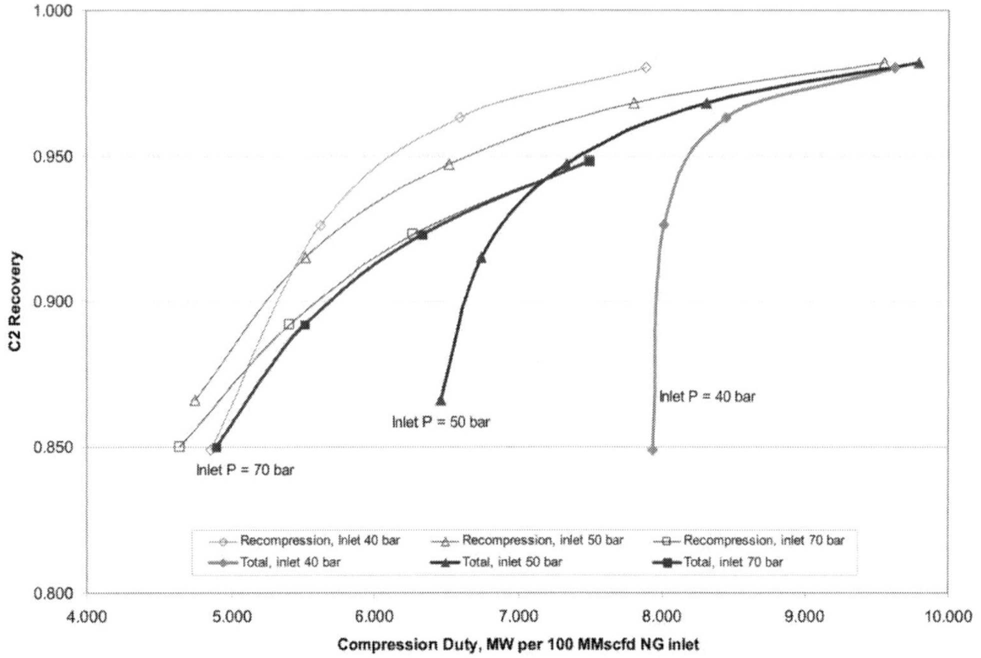

Figure 9 Impact of inlet pressure

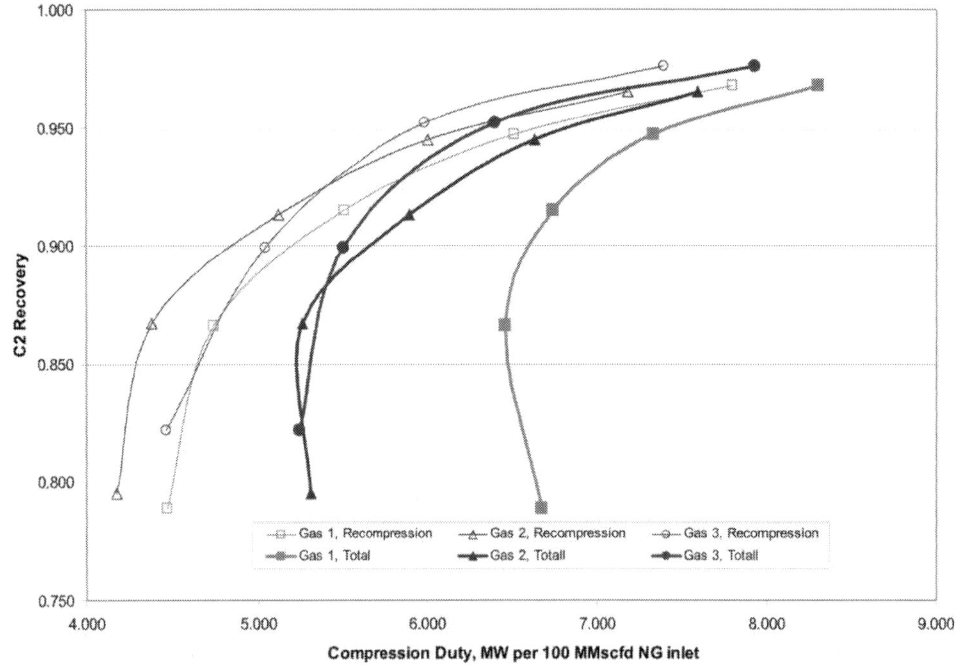

Figure 10 Impact of inlet gas composition

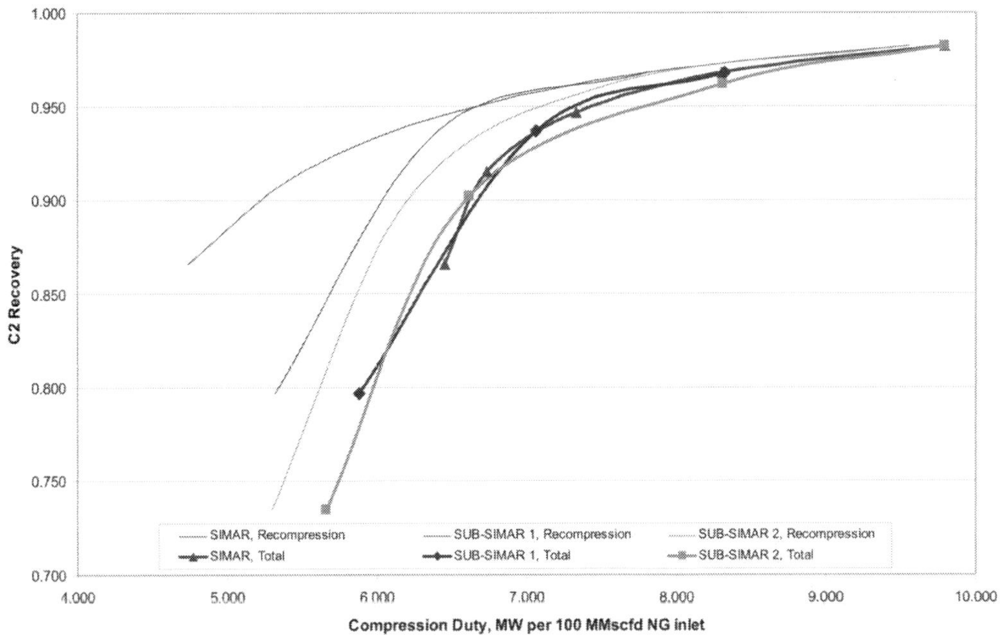

Figure 11 Sub-SIMAR operations for Gas 1

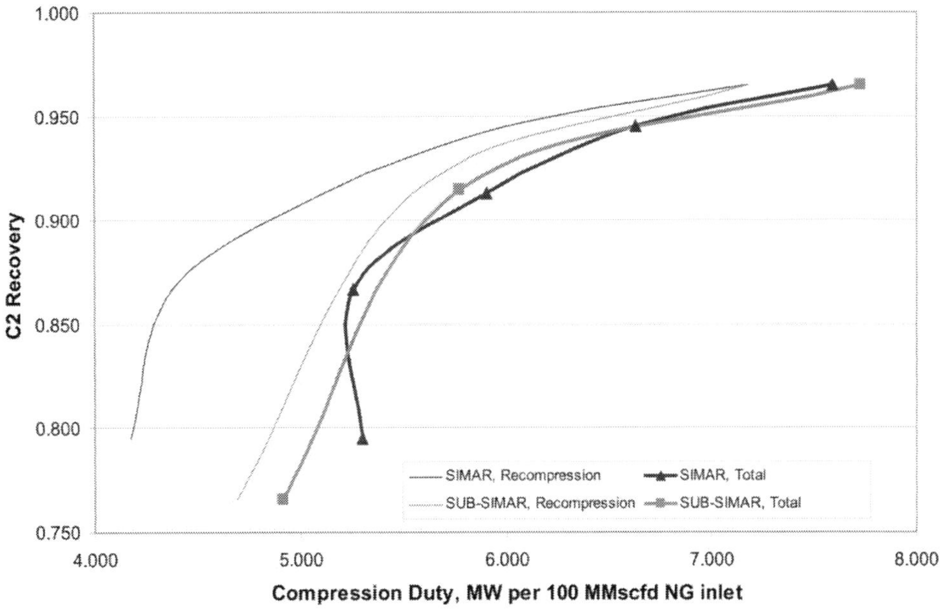

Figure 12 Sub-SIMAR operations for Gas 2

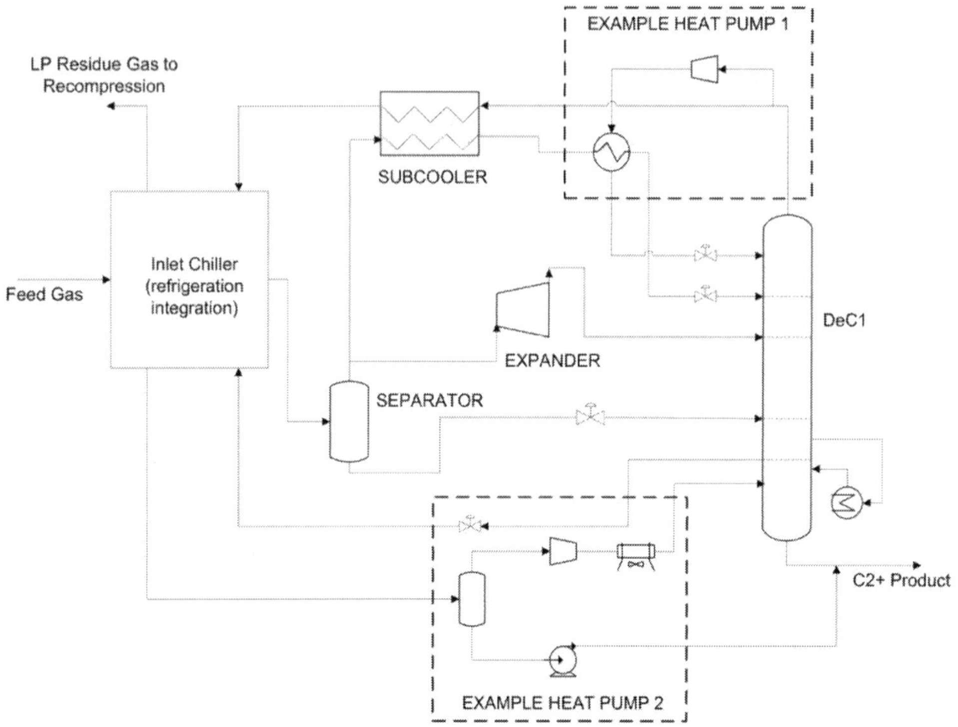

Figure 13 Example heat pumps embedded in XPDR 3

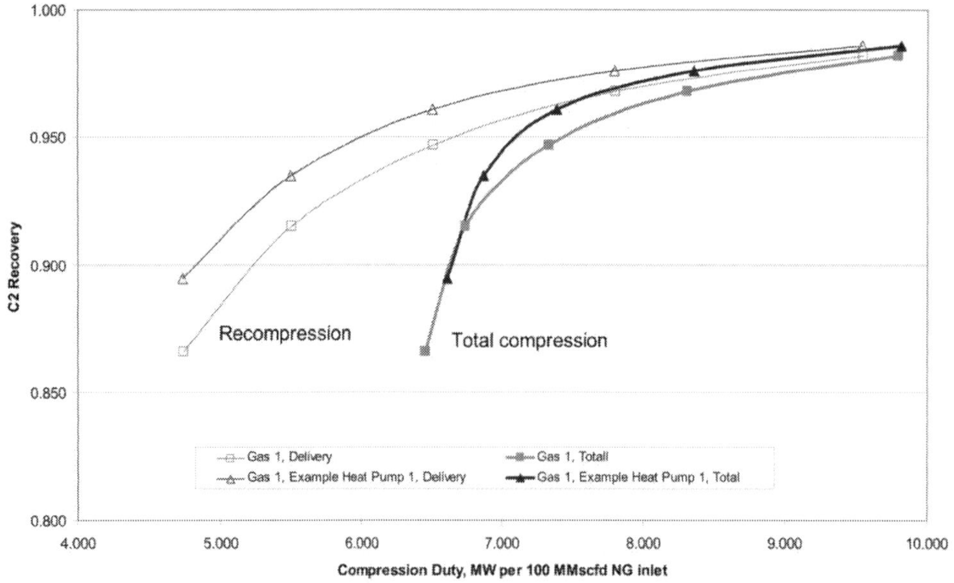

Figure 14 Impact of Example Heat Pump 1 on C2 recovery

Biography

Stanley Huang is a Staff LNG Process Engineer in Chevron Energy Technology Company. He is a registered professional engineer and has eighteen years of experiences in the hydrocarbon industry. Stanley graduated with a Ph. D. degree in chemical engineering from Purdue University in 1986.

NGL Recovery Process Synergies with the LNG Value Chain

Michael Barclay
Principal Process Design Engineer
Foster Wheeler Energy Limited
Shinfield Park, Reading, Berkshire, RG2 9FW, UK
C.C. Yang
Process and Technology Director, LNG
Foster Wheeler North America Corporation
2020 Dairy Ashford, Houston, Texas, 77077, US

AIChE Spring National Meeting
6th Topical Conference on Natural Gas Utilization
Orlando, Florida, April 23-27, 2006

ABSTRACT

Successful integration of NGL recovery into the LNG value chain can have a significant effect on both process design and overall project profitability. Several decisions must be taken fairly early in project development to define the process. Firstly, should NGL be recovered in the project? This is primarily a function of feed gas composition and required product specifications in the destination LNG market. Assuming that a project supports NGL recovery; the next question is where should the NGL be extracted? The choice is either at the point of production or the receiving terminal. Integration into the chilling and liquefaction process may take advantage of synergies and result in a decrease in equipment and CAPEX and OPEX. Alternatively, NGL products can be extracted at the receiving terminal particularly where lower BTU gas is required, as in the US where there is an end-user market.

This paper discusses insights Foster Wheeler has gained designing NGL recovery facilities upfront of liquefaction facilities and at receiving terminals during the past two years. Potential integration of NGL recovery into the liquefaction process is also discussed from a fundamental perspective. A process for C2 and C3 recovery at the receiving terminal is described with an emphasis on C3.

INTRODUCTION

Increasing global demand for natural gas is supporting the rapid growth of worldwide LNG production capacity. The demand for natural gas in North America and other energy-hungry and rapidly growing countries is predicted to continue (DOE/EIA, 2004). Part of this rising demand will be satisfied by importing liquefied natural gas (LNG). Most LNG trade consists of long-term contracts where an LNG facility provides defined LNG cargoes, to a limited number of customers, with a dedicated shipping fleet. This has led to the concept of an LNG value chain where value is added in five stages from stranded gas in a gas-rich region through to high value sales gas delivered to customers as seen in Figure 1.

The LNG value chain may be broken down into five distinct components: upstream production and transport, liquefaction, shipping, regasification, and distribution to customers. This chain presents two locations well-suited for NGL recovery. The liquefaction facility is the traditional location for separation of some of the heavier components from natural gas. This has often consisted of a scrub column that coarsely separates a condensate product stream from the natural gas.

An alternative location for NGL recovery is the receiving terminal. NGL extraction at receiving terminals can take advantage of the LNG's cold energy to cost-effectively recover a second product. Receiving terminal recovery may have several advantages including allowing stringent sendout gas specifications to be met from cargoes that have varying compositions, recovery of a higher value product, and effective use of energy.

Following a discussion of the LNG value chain and NGL recovery, this paper will discuss the potential synergies associated with NGL recovery at both the liquefaction facility and the receiving terminal.

The LNG Value Chain

The LNG value chain is schematically shown in Figure 1 and is described in more detail in the following sections. Each link in the value chain has an associated cost that depends on project-specific details such as shipping distance and field development costs. The sum of these ultimately determines the cost at which the LNG may be delivered.

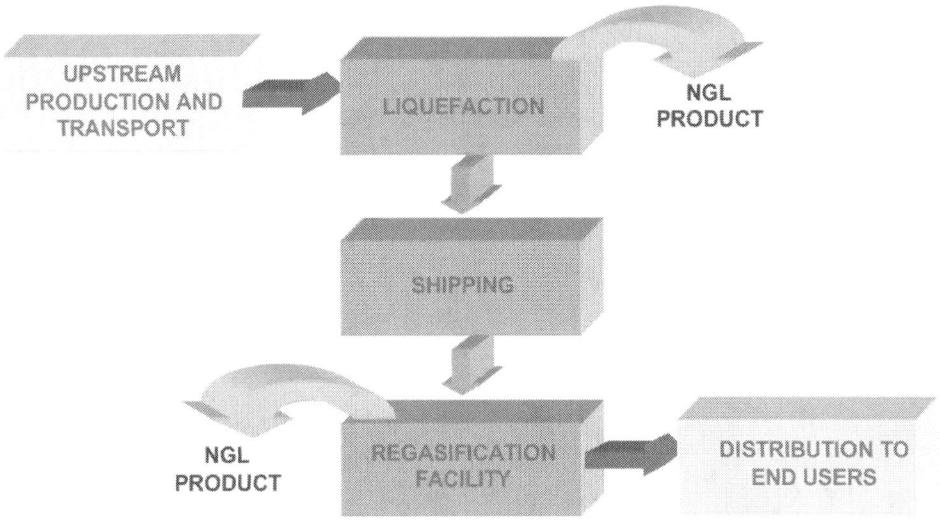

Figure 1. The LNG Value Chain showing NGL recovery opportunities

Upstream Production and Transport

Worldwide proven reserves of natural gas were approximately 5,500 trillion cubic feet in 2003 (DOE/EIA 2004). Most of this natural gas is stranded, far from large markets. The largest reserves of natural gas are in Russia, Iran, and Qatar who together have more than half of the proven natural gas reserves (Oil & Gas Journal,

2002). Some significant LNG exporters include Qatar, Indonesia, Australia, Nigeria, Trinidad & Tobago, Egypt, Oman, and Algeria.

Feed gas for an LNG facility can be associated or non-associated and can be located onshore or offshore. Field development costs are highly variable and can have a significant effect on the gas feedstock price and the resultant cost of LNG. Recent LNG frontier projects have some of the highest field development costs because they require extensive pipe networks and design for extreme climatic and subsea conditions. Regardless of the field location, the facilities must be developed in order to transport the natural gas to the liquefaction facility and will include everything from production wells to inlet facility battery limits.

The Liquefaction Facility

The liquefaction facility includes all of the process equipment required to treat, liquefy, store, and export the LNG and other liquid products. Incoming well fluids typically enter the facility through a slug catcher, often followed by a pressure control station. H_2S is a hazard and CO_2 will freeze in LNG process equipment. These components must be removed, typically in an amine-based acid gas removal unit. Following sweetening, water and mercury are removed to prevent freezing or damage to process equipment. At a minimum, benzene and condensate are usually removed in a scrub column in the liquefaction facility. Benzene freezes at relatively low concentration and other heavy components are likely to freeze or cause wax deposits in cold process equipment.

Several liquefaction processes are available for baseload liquefaction including the propane precooled mixed refrigerant (C3MR), AP-X™, and optimized cascade (CoP by Conoco Phillips) processes that are selected for the majority of projects (Shukri, 2004). These processes use single and mixed refrigerants to reach the very low temperatures required to condense the natural gas. The liquefaction plant often consists of multiple trains.

Process developments have enabled significant liquefaction train capacity increases in the last decade, which have helped to reduce the cost per tonne of LNG produced. Approximately six years ago Foster Wheeler and Chiyoda completed the engineering, procurement and construction (EPC) contract for Oman LNG. At the time of start-up (Feb. 2000 for Train 2), this plant had the largest trains in operation at 3.3 million tonnes per annum (Mtpa) and set a benchmark for process efficiency with a reported average specific power of 10.15 kW per tonne per day of LNG (McLachlan et. al., 2002). Five years later, operating train capacities are over 5 Mtpa and Foster Wheeler is involved in projects with capacities in excess of 7.5 Mtpa.

Larger trains make NGL recovery more economic because the increased product volumes and revenues support the complexity and costs associated with NGL recovery, storage, and export facilities. This has resulted in deeper NGL recovery being considered and specified in liquefaction facility development. NGL recovery processes often include deep C3+ NGL recovery, possibly with a provision for C2 recovery in the future. Condensate and LPG are usually more valuable than LNG so recovery is justified on an economic basis. These additional product streams must be exported by tanker and require their own storage and port facilities.

Expander-based processes are those most suited for deep NGL recovery at the liquefaction facility. This is because they are able to efficiently reach deep NGL recoveries and share many synergies with the liquefaction process since they are also cryogenic processes. These processes often have requirements for the same pretreatment as LNG process equipment, supplemental refrigeration, and a heating media for the reboilers. Therefore, effective integration can improve the overall profitability and operation of a LNG facility.

The liquefaction facility also includes liquid product storage, offloading jetties, and tanker berths. Once liquefied, the LNG is stored in insulated tanks at atmospheric pressure.

Shipping

LNG carriers are double-hulled ships specially designed and insulated to allow safe and efficient transport of large volumes of extremely cold product. The LNG is stored in special containment systems within the inner hull where it is kept at approximately atmospheric pressure and -256 °F. Three types of cargo containment systems are used in LNG carrier fleets: the spherical (Moss) design, the membrane design, and the structural prismatic design. Shipping costs are highly dependent on shipping distance and secondary factors such as ship capacity and fuel type.

Regasification and distribution

The LNG receiving terminal includes the jetty, berths, loading arms, storage, LNG pumps, vaporizers, and other equipment required to sendout natural gas to end users via pipeline. To return LNG to a gaseous state, it is fed into a regasification plant. Upon arrival at the receiving terminal in its liquid state, LNG is pumped first to a double-walled storage tank, similar to those used in the liquefaction plant, at atmospheric pressure, then pumped at high pressure through various terminal components where it is warmed in a controlled environment.

Conventional LNG regasification processes generally require an external heat source. Heat is usually provided by using various types of vaporizers; such as open rack seawater vaporizers, submerged combustion vaporizers, intermediate fluid vaporizers using a glycol/water mixture, or ambient air vaporizers.

Seawater heating is the most commonly chosen method of vaporization as it minimizes fuel consumption and LNG shrinkage. However, seawater vaporizers, pumps, and intake/ out-fall structures are costly. LNG vaporization is an energy intensive process. Vaporizing a 145,000 m^3 LNG cargo to send-out approximately 3 Bcf of natural gas requires ~51,000 MMBtu of energy input. This is the energy equivalent to cooling 1.1 billion gallons of seawater by 5.4 °F or burning 48 MMscf of natural gas with a nominal heating value of 1050 Btu / ft^3. A large seawater flowrate is typically needed to meet environmental guidelines of the World Bank that require a temperature drop change of no more than 5.4 °F at the edge of the zone where initial mixing and dilution takes place (World Bank Group, 1998).

The vaporized gas may then be exported to a pipeline network or to local large customers. In many markets such as Western Europe and the United State large pipeline networks transport the gas to customers on the "grid". Other markets feature dedicated customers relatively close to the receiving and regasification facilities.

The receiving terminal is the second place in the LNG value chain that supports the recovery of NGL. LNG is free of impurities and possesses "cold energy" or, in terms of thermodynamics, has a high availability. Therefore, there is an opportunity to recover NGL components with minimal additional equipment or power costs, and commensurately low capital cost. This investment decision must ultimately be supported by the heating value of the LNG to be imported relative to the sendout contract requirements, the local market and value for the NGL components, and other factors.

NGL RECOVERY AT THE LIQUEFACTION FACILITY

The liquefaction facility usually features at least some condensate recovery and fractionation. At a minimum, a scrub column will be used at an intermediate temperature to recover condensate. This meets three requirements:

1. the column removes the heavy ends such as benzene and longer chain hydrocarbons that may cause freezing and wax deposition in the cold process equipment,

2. the bottoms product of the column (hydrocarbon liquid) can be fractionated to provide the refrigerants required by the liquefaction process. The refrigerant often includes ethane, methane, and propane,

3. the condensate can be exported as a higher-value product to improve overall project economics.

While it has been used extensively in the past, the scrub column's performance is fundamentally limited. The column is normally not refluxed with a cold lean stream and not subject to a large flow of an absorbent hydrocarbon stream, so NGL recovery is limited. A scrub column normally operates at the liquefaction pressures downstream of the first set of precooling heat exchangers so residue gas compression is not feasible. Operation at high pressure, close to the critical pressure, means that the relative volatility between the separation key components is small. A final problem with operation at elevated pressure is that the reboiler must operate at higher temperature. In some cases, the reboiler operating temperature is limited by the fluid critical temperature resulting in a bottom product with light component carryover. The net result is low recovery and a volatile bottoms product that contains significant methane.

While the performance of the scrub column is fundamentally limited by critical conditions and feed pressure requirements, the degree of integration has generally been high. The column's cold source is the main refrigeration system. The columns liquid products are used as refrigerant and the column ensures that waxes and solids do not form in the cold process equipment.

Deeper NGL recovery poses challenges to successful process integration because it will generally require lower pressure operation in an absorber column, usually coupled with propane refrigeration. Cool, lower pressure feed gas from an absorber, while suitable for NGL recovery, would result in less efficient natural gas liquefaction. Liquefaction occurs at higher temperatures when at higher pressure and thus has a higher Carnot efficiency and requires less energy input. Liquefaction at a higher pressure than the critical pressure also generates smoother cooling curves that can more readily be matched by mixed refrigerant process fluids. Fundamentally, the liquefaction process favors high pressure feed while the NGL recovery benefits from lower pressure operation of the columns. This creates a challenging integration problem but some synergies are available.

Process Synergies

All NGL recovery shares some inherent process synergies with the liquefaction process. First, because expander-based NGL extraction is a cold process, much of the pre-treatment required by the liquefaction facility is also required by the NGL recovery process. NGL extraction usually requires dehydration, mercury removal if aluminum plate-fin heat exchangers are used, and some CO_2 removal in the case of many C2+ recovery processes.

PROCESS CHALLENGES

The first major decision regarding integration usually relates to the refrigeration system. LNG facilities have multiple propane refrigeration requirements that can be served by common or separate refrigeration systems. The NGL recovery system is another potential consumer of refrigeration power along with refrigerated product storage, and the liquefaction process itself. However, should these systems be integrated or separate?

Propane refrigeration system sharing between the liquefier and the NGL recovery process can result in capital and operating cost savings. However, this approach suffers from potentially high NGL recovery refrigeration duties further loading the liquefaction propane system that is often approaching the feasible capacity limits. Too much integration will result in an extremely complicated refrigeration system. A major challenge is integrating the cooling into the process so that the refrigeration is not lost when the residue gas is recompressed in the case of lower pressure absorber column operation.

Another integration opportunity is to use lean flash gas to rectify the absorber column. The availability of a lean, cold reflux stream to the column has the potential to dramatically improve NGL recovery and is the basis of most deep C2+ recovery processes with some form of residue compression and recycle.

LNG processes have boil-off gas and flash gas compression that are often compressed and returned to the fuel gas system. After compression, these lean streams could be partially condensed and flashed into the absorber column of an NGL recovery process as an improved alternative to recycling the liquefaction module's feed gas. Many alternatives are available.

A significant integration challenge is to support existing facilities and contractual commitments while recovering additional NGL from new trains. Solutions must be developed on a case-by-case basis. In many cases NGL recovery can be incorporated into a major expansion that includes additional product storage and port facilities in addition to new trains. In these cases the two product grades can be kept separate. In other cases more flexible operation or other solutions may be considered.

Greenfield liquefaction facilities such as those in Egypt, Norway, and Yemen do not need to support legacy LNG contracts and facilities. Such projects do not need consider producing, storing, and exporting two LNG products but do have a different set of problems. New facilities that start with a single train are dependent on a functioning NGL recovery process to export on-spec product LNG. How can a facility manage the loss of the NGL recovery process? The scrub column is simple and robust and has a high availability relative to the liquefaction process. Expander-based NGL facilities have additional rotating equipment and the potential to provide various heating value products to the downstream processes. In many gas plants, the consequence of a failed expander is operation in JT mode with a decreased production rate. For the LNG facility, the consequences can be severe, depending on downstream equipment design and contractual arrangements.

What happens to the downstream equipment and customers? What is the effect on the liquefaction module? Can the off-specification LNG be stored and shipped via existing facilities? Can it be exported to downstream customers with a higher heating value? Well considered answers to these questions as well as others are critical to successful integration.

This challenge is not limited to sales contracts, but also to the fuel gas system. Gas turbines (GTs) are sensitive to the heating value of their fuel. A range of operating cases should be considered to ensure that the GTs are protected during all modes of operation.

Setting Recovery Targets and Selecting the Right Process

Selecting the appropriate process and correctly specifying the NGL recovery target is equally important as sound integration (Barclay, 2005). Foster Wheeler is well positioned to identify the optimal process because it has a technical understanding of the available processes, no vested interest in particular NGL recovery processes, and good working relationships with major NGL process licensors.

Identification of potential "sweet spots" for NGL recovery where the project economics are dramatically improved by a step change in CAPEX or OPEX is also important. For NGL recovery at the liquefaction facility, step changes may include the recovery process itself, requirement for supplemental refrigeration, the requirement for residue gas compression.

Construction and other Synergies

For the combined facility, other possible synergies exist in addition to the process integration that can offer the potential to decrease CAPEX and OPEX. These are primarily due to the fact the NGL recovery unit has the potential to be a relatively small portion of a large, integrated project. This is true during engineering, construction, and through to operations.

During construction incremental effort required to construct NGL recovery process facilities is small. Because the NGL recovery absorber will generally process a very large gas flow, possibly up to approximately 1,000 MMSCFD, at high pressure, it will be a large, heavy vessel. On-site heavy-lift cranes and transport capabilities required to move the large main cryogenic heat exchanger, and the amine contactor can also be used for the NGL recovery heavy lifts.

Integration into the liquefaction facility also offers several operational benefits. An NGL recovery plant both in gas plants and at receiving terminals are usually the most operationally challenging process blocks due to rotating equipment, column operation, pretreatment requirements, and low temperatures. An NGL recovery plant operated as part of an overall LNG facility represents is typically well within the experience of the site.

The Potential for Ethane Recovery

In some cases, it may be possible to export lean LNG to major markets, such as the US, with some ethane recovered. Deeper C2 recovery generally comes at increased capital and operating costs. Therefore, the local value of ethane relative to LNG is the economic basis for its recovery.

Unlike the crude, condensate, and LPG products, ethane value is highly dependent on local market conditions. Regions that have well-established petrochemical industries have a C2 value that fluctuates with ethylene demand are generally not LNG exporters. In areas with developing petrochemical industries, C2 value may change dramatically when an ethylene cracker is developed because they usually drive the market value for ethane. This creates a "chicken and egg" type dilemma in which the petrochemical industry depends on the availability of ethane and the economic viability of C2 recovery is based on a C2 market, which does not yet exist. In emergent petrochemical industries a C2 recovery facility may initially be justified on the basis of political and future value for the C2.

Projects that can market and export a lean LNG may consider ethane recovery if it has a current or future higher value than the LNG. Ideally, this will be considered during the conceptual or early FEED project stages to ensure that an optimized process design is developed. Dedicated ethane recovery processes often comprise single-column processes featuring some form of residue recycle for rectification. Alternatively, if the provision for ethane recovery is considered early, a two column deep C3 recovery process (including the Single Column Overhead Recycle process) can be a starting point with the provision for future ethane recovery and rejection modes, as needed.

Many of the integration opportunities described above for C3+ NGL are also relevant. Other process synergies that should be considered in the early engineering phases include integration with the propane precooling circuit in the liquefaction module, column reflux with LNG or LNG boil-off gas and/or flash gas, and residue gas compressor and driver integration into the liquefaction module compressor strings.

EXTRACTION AT THE RECEIVING TERMINAL

C2+ or C3+ separation facilities at an LNG receiving terminal allow it to receive LNG cargoes of varying compositions and either meet stringent calorific value send-out gas requirements or produce a second, more valuable, product (Yang, et. al., 2003).

Recovery Options

As described above, an optimal conventional regasification process is one that most cost-effectively destroys the cold energy (availability) of the LNG that was imparted at the power-intensive liquefaction plant. NGL recovery options at the receiving terminal offer the potential to use the LNG's cold energy to affect the separation of the heavier components.

Foster Wheeler has recently developed C2+ and C3+ NGL recovery receiving terminal processes (patent applied for) that are feasible, effective, and economic. The designs use low-temperature LNG as a cooling medium for the column overhead to achieve an economic separation. The Foster Wheeler design does not need supplemental gas compression because the LNG is used as a refrigerant in a direct-contact condenser, which recondenses the C1 vapor from the column overhead to produce a lean LNG for liquid pumps, vaporization, and send-out. The vaporizers act as the heat source for the separation column. This process features low cost, no additional gas compression, uses the equipment already present at a receiving terminal, and may be retrofit to existing facilities.

Process Description

A schematic of the process for C3+ recovery is shown in Figure 2. This figure shows the traditional LNG sendout process where offloaded LNG is pumped to pipeline pressure and then vaporized to produce a high pressure natural gas stream suitable for export. The C3+ recovery process is shown in the shaded area.

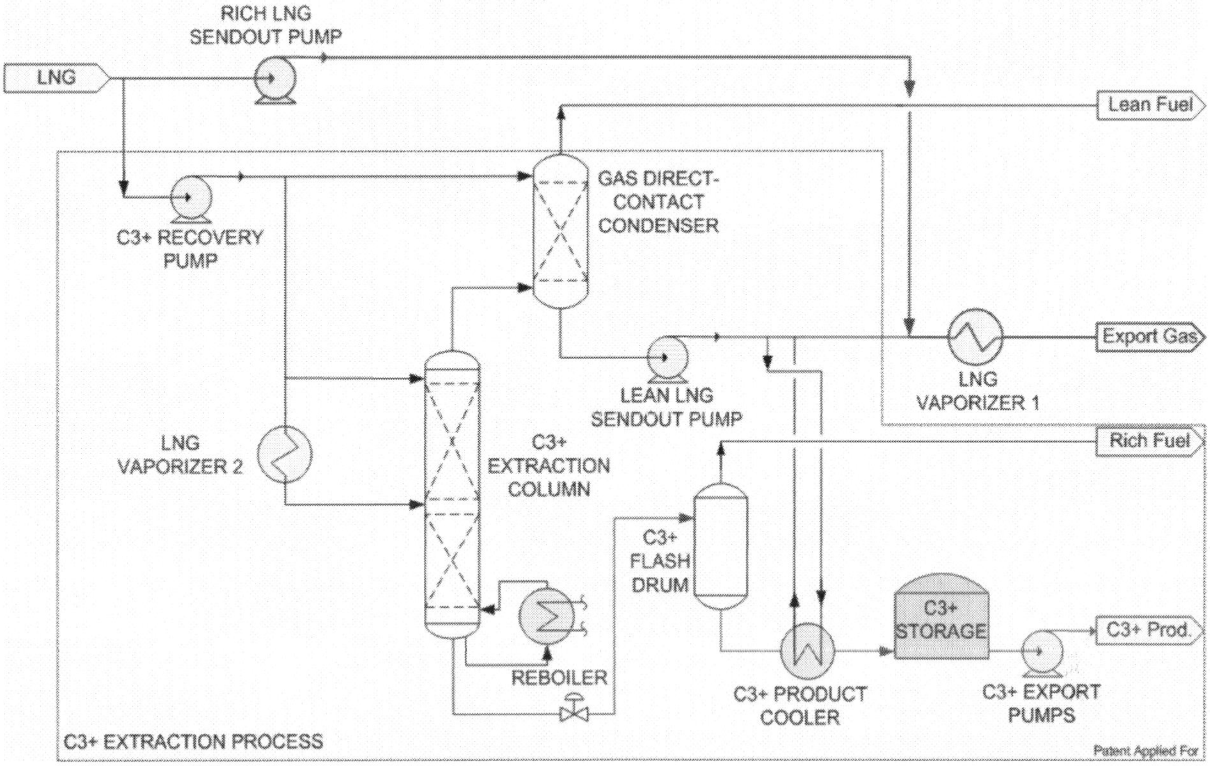

Figure 2. Receiving terminal C3+ extraction process (patent applied for) showing possible product streams

The LNG cargo being offloaded is split into two streams. A portion of the rich LNG is pumped to just above the sendout pressure, approximately 85 bar, and flows to LNG Vaporizer 1. The second stream, that will be subject to C3+ extraction, is pumped to an intermediate pressure of approximately 40 bar in the C3+ recovery pump. The pumped fluid is split into three streams that flow to the gas direct-contact condenser, the C3+ extraction column, and LNG Vaporizer 2. The optimal relative flow rates are dependent on C3+ recovery target, the composition of the LNG, local fuel gas consumers, and other variables. One portion is sent as a subcooled stream to the gas direct-contact condenser, where it recondenses and subcools the lean gas stream from the C3+ extraction column overheads.

A second stream, typically the smallest flow of the three, flows directly to the C3+ extraction column as an absorbent. This liquid absorbs NGL components from the vapor in the upper packed section improving the liquid C3+ recovery.

The third stream is heated in LNG Vaporizer 2 prior to entering the C3+ extraction column between the two packed sections. The bottoms product is a C3+ NGL stream that requires additional processing (shown as a flash) and cooling to C3+ storage. Depending on the composition of the LNG, the C3+ product may undergo further processing and fractionation into condensate and LPG streams, or may be saleable directly as an LPG product. This may be stored as a refrigerated or pressurized fluid and transported off the site via rail, road tanker, or pipeline depending on local infrastructure and demand.

The C3+ extraction column's overhead product is a lean gas that is re-condensed and subcooled by the subcooled LNG stream, as previously described. The lean LNG leaving the gas direct-contact condenser is pumped to approximately 85 bar, joins the rich LNG being offloaded, and is vaporized before being sent-out to the gas pipeline.

The relative splits between rich and lean LNG send-out and the direct-contact condenser and extraction column are dependent on the value of LPG, the existence of local customers for either the lean or rich fuel gases, and other variables. The process can be tuned to flexibly meet site-specific requirements and conditions.

The C2+ process developed by Foster Wheeler shares many of the same features as the C3+ process such as the gas direct-contacting condenser, NGL extraction column, and intermediate pressure option. This process, while highly relevant for Californian import terminals, has been previously described (Yang et. al., 2003) along with a case study and will not be discussed further in this paper.

CONCLUSION

Effective integration of an NGL recovery process into the LNG value chain can significantly increase overall project profitability. The most appropriate link in the LNG value chain for NGL recovery integration is dependent on multiple project-specific variables including the existing site facilities, the local ethane demand, and target LNG customer heating value requirements. Well concieved conceptual and early-phase scoping studies can assess the potential for NGL recovery at the liquefaction facility and receiving terminal. These study results can support some of the key project decisions discussed in this paper to define the process, products, and ulitmately profitability

REFERENCES

DOE/EIA, The Global Liquefied Natural Gas Market: Status and Outlook, Energy Information Administration, U.S. Dept. of Energy, DOE/EIA-0637, Dec. 2003.

DOE/EIA, Annual Energy Outlook – 2004, Energy Information Administration, U.S. Dept. of Energy, DOE/EIA-0383(2004), Jan. 2004.

Oil & Gas Journal, Dec. 23, 2002.

T. Shukri, *LNG Technology Selection*, Hydrocarbon Engineering, Feb. 2004.

G. McLachlan, C. Ayres, K. Vink, M. Al Mukhainy, *Efficient Operation of LNG from the Oman LNG Project*, Gastech 2002, Doha, Qatar, Oct. 2002.

World Bank Group, "Pollution Prevention and Abatement Handbook", Effective July 1998.

M.A. Barclay, *Study Identified Optimal C2 Recovery Level for NGL Plants*, Oil & Gas Journal, Vol. 103, Issue 32, Aug. 22, 2005.

C.C. Yang, A. Kaplan, Z. Huang, *Cost-effective design reduces C2 and C3 at LNG receiving terminals,* Oil & Gas Journal, May 2003.

A. Smaal, *Liquefaction Plants: Development of Technology and Innovation*, Working Committee Contribution, 22nd World Gas Conference, Tokyo, 2003.

Biography

CC Yang, Process and Technology Director, LNG

C.C Yang is Process and Technology Director, LNG, for Foster Wheeler North America Corporation. Dr. Yang has thirty (30) years experience in the engineering and construction industry, engaged in domestic and international engineering projects. In the last 17 years, he has been dedicated in both LNG baseload plants and receiving terminal design. He holds a BS from National Cheng-Kung University, Taiwan and an MS and PhD from the University of Rochester, NY, all in Chemical Engineering. He is a registered Professional Engineer in Texas and a member of AIChE.

Advanced Design of Submerged Combustion Vaporizer for Low Emission Operation

Yoshimichi Ikeda
System Engineer, Heat Exchangers Engineering
Nobuya Himoto
General Manager
Go Kato
Marketing Supervisor
Sumitomo Precision Products, Co. Ltd.

AIChE Spring Meeting, April 2006
6th Topical Conference on Natural Gas Utilization
Orlando, Florida, April 23-27, 2006

ABSTRACT

As the demand for natural gas in the United States continues to grow, there is a renewed interest in the import of LNG as an energy source. Imported LNG is stored, pressurized, and vaporized before it is injected into the national pipeline grid. The LNG vaporizer, which regasifies the LNG from liquid to gaseous form, is one of the main components of an LNG import terminal. Various LNG vaporizer technologies have been developed and utilized. Most of the existing LNG terminals in the US use Submerged Combustion Vaporizer (SCV) technology as a baseload regasifier. The SCV's have high thermal efficiency, easy start-ups, and are capable of long-term operation.

However, due to the fuel combustion process, the conventional SCV generates air emissions such as nitrogen oxides (NOx), carbon monoxide (CO), and other emissions, which is a major environmental concern for the energy industry today. For the sustainable development of this sector, the LNG regasification process requires more stringent air emissions control. This paper discusses advanced designs of the SCV that have been newly developed to reduce NOx and CO emissions in the exhaust gases from the SCV. Conventional SCV designs, including the process and major factors impacting emissions, are also included in this paper.

INTRODUCTION

Natural gas volume can be reduced by about 1/600 by liquefaction, which allows long distance transportation using LNG carriers. Upon arrival at an LNG receiving terminal, LNG is stored in insulated atmospheric LNG storage tanks and regasified to meet natural gas demand. LNG vaporizers are the most important pieces of equipment to ensure a stable and cost-effective regasification process.

The base load LNG terminals have primarily used two types of vaporizers: Open Rack Vaporizer (ORV) and Submerged Combustion Vaporizer (SCV). In addition to these, several other types of vaporizers have also been used. They include the Shell and Tube type Vaporizers (STV) with or without intermediate fluid [1].

Most of the existing LNG terminals in the US use Submerged Combustion Vaporizer (SCV) technology as a baseload regasifier. The SCV's have high thermal efficiency, easy start-up, and are capable of long-term continuous operation. However, the conventional SCV generates air emissions due to fuel combustion, which then produces nitrogen oxides (NOx), carbon monoxide (CO), and other emissions. Recently, the concern on the environmental impact by the energy industry has been dramatically increased. This trend advocates and aims for more stringent emission controls on the LNG regasification process.

The available technology of a conventional SCV can reduce the NOx and CO emissions below 40 ppmvd for each component. Since these emission levels do not satisfy the emission requirement in environmentally sensitive areas, low emission SCV's have been developed. These include Water Injection, Flue Gas Recirculation (FGR), Rapid Premix Combustion, and a combination of each technology.

This paper also discusses the concept of Regenerative Thermal Oxidation (RTO) technology for achieving ultra low NOx and CO emissions in the exhaust gases from the SCV.

SUBMERGED COMBUSTION VAPORIZER (SCV)

The SCV vaporizes LNG which flows through stainless steel tubes in a submerged water bath. The water bath is heated and maintained at a certain temperature with the help of hot flue gases. A large single combustion burner, mounted at one end of the water bath, uses fuel gas as a heating source. The hot flue gases are injected into the water below the heat exchange tube bundle by a special flue gas distribution system (sparge system). LNG flowing through the tube bundle is vaporized by heat transferred from the water bath. Figure 1 illustrates the main features of a SCV.

Injection of the combustion products forms a frothing two-phase mixture containing many small gas bubbles. Exhaust from the combustion products in this manner imparts a high turbulent motion and enhances bubble formation. The reduced density of the two-phase mixture creates a lift action which causes the mixture to rise. This rising flow is confined within a weir assembly which surrounds the tube bundle. The froth flows up through the tube bundle and over the weir and water

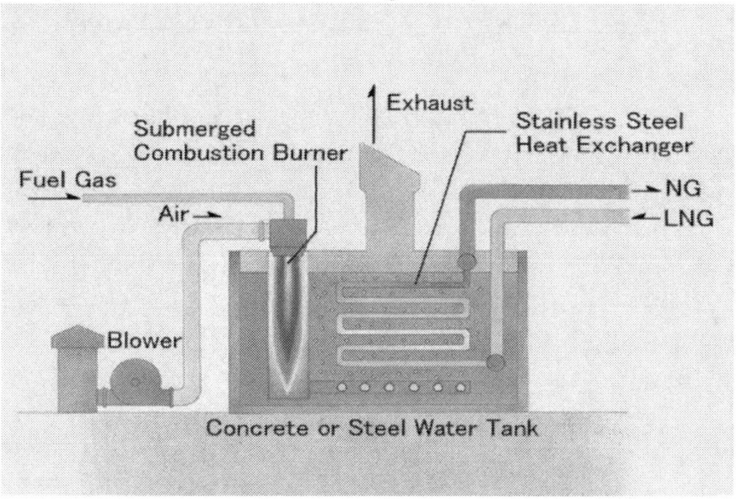

Fig. 1 – SCV Schematic Section Side View

falls back over the weir. Water recirculates back into the bath. It is necessary to maintain the frothing action, even at low heat duties, to maintain uniformity and prevent icing in the water bath.

The heat exchange tube bundle is immersed in the water bath, within the weir space and above the burner distributor system, so that its tube surface is efficiently scrubbed by the high velocity motion of the exhaust gas/water mixture. The gas/water mixture creates a large heat transfer area and a high recirculation rate, creating unusually high thermal efficiencies. The burner combustion products, which disengage from the gas/water mixture after it has passed over the tube bundle, are discharged to the atmosphere via a carbon steel stack. Adequate disengagement space and low exhaust stack discharge velocities assure minimal carryover.

The water bath temperature ranges between 15°C and 40°C. At this water bath temperature, most of the water products of combustion condense, resulting in a blowdown. The bath water becomes acidic as the combustion products are absorbed in it. Alkaline chemicals (e.g. dilute caustic, sodium carbonate and sodium bicarbonate) may be added to the bath water to control pH, and the excess combustion water will be neutralized before discharge. The excess water generation is a function of water bath temperature, burner firing rate, relative humidity, and operational time.

For the SCV operation, since the thermal capacity of the water bath is high, it is possible to maintain stable operations even during sudden start-ups/shutdowns and rapid load fluctuations. Therefore, there is great flexibility for quick start-up after shutdowns, and the ability to quickly respond to changing demand requirements.

The burner starts automatically by pushbutton from a freestanding local control panel or from a customer remote DCS control panel. The DCS panel provides remote start-up and shut-down control, operating status, first out alarm annunciation, and alarm annunciation. Vaporized LNG temperature is used to adjust the firing rate. Start-up, shut-down, and normal operation are handled automatically or semi-automatically through the use of completely automatic safety interlocks.

The SCV's are designed to utilize low pressure fuel derived from the boil-off gases of the facility and/or the let-down gas from send-out. In the baseload terminal SCV, a large single burner is used rather than multiple smaller burners because it is more economical and achieves low NOx and CO levels. Due to the larger amount of flue gas, there is a concern for increased emissions. However, a low NOx emission of less than 40 ppmvd is attainable by the Low NOx burner using a water injection system.

ADVANCED DESIGN OF THE SCV

From combustion of natural gas as a vaporization fuel, emissions from SCV fuel combustion include nitrogen oxides (NOx), carbon monoxide (CO), carbon dioxide (CO_2), volatile organic compounds (VOCs), and particulate matter (PM). Since LNG contains traceable sulfur, sulfur dioxide is normally not a concern associated with SCV emissions.

Nitrogen oxides (NOx) are a combination of nitrogen oxide (NO) and nitrogen dioxide (NO_2). Government environmental agencies have promulgated regulations requiring stricter control of NOx emissions from stationary point sources. The emission control requirements of the existing, as well as future, regulations require an understanding of NOx-emitting sources, pollution prevention opportunities, emissions reduction technologies, costs of controls, and impacts on existing operations and plant safety.

The following are descriptions of applicable technologies for low emission SCV operations:

Water Injection

Natural gas combustion produces three types of NOx: "thermal NOx", "prompt NOx" and "fuel NOx" [2]. Most of the NOx are thermal NOx formed through thermal dissociation and subsequent reaction of nitrogen (N_2) and oxygen (O_2) molecules in the combustion air. The formation of thermal NOx is affected by three furnace-zone factors: (1) excess air, (2) combustion temperature, and (3) time of exposure at peak combustion temperature. NOx is primarily a function of flame temperature, but it is also affected by oxygen concentration, and the time of those conditions [3]. Figure 2 shows the relationship between adiabatic flame temperature and thermal NOx formation.

Prompt NOx is formed as N_2 transforms into NO at the edge of the zone of rich hydrocarbon flames. Prompt NOx reactions occur within the flame and quantities are usually negligible when compared to the amount of thermal NOx [2]. Fuel NOx results from the evolution and reaction of fuel-bound nitrogen compounds with oxygen. Because of the low fuel nitrogen content in LNG, NOx formation through the fuel NOx mechanism is insignificant.

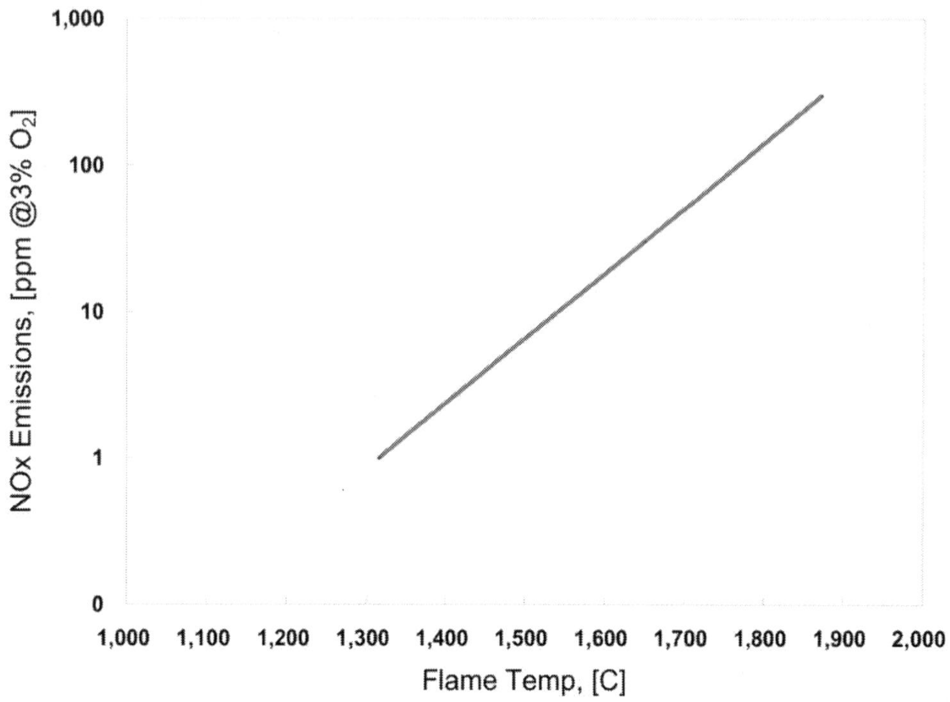

Fig. 2 – Calculated Thermal NOx vs. Adiabatic Flame Temperature [3]

With conventional burners and water injection, water is injected through the back of the burner and into the flame to lower the flame temperature. Total NOx emission from SCV combustion is normally in the range of 35-45 ppmv. The rate of CO emissions depends on the efficiency of natural gas combustion. CO emission increases when the burner is improperly tuned and/or operating at off-design levels. With conventional burners, total CO emission from SCV is normally in the range of 50-100 ppm. The relationship between water injection rate and NOx emission is shown in Fig. 3.

The conventional burner already achieves low NOx emissions. However, a small amount of water injected directly onto a burner flame decreases NOx emissions further. Based on the accumulated factory test database, the adjusted water injection rate achieves 50% lower emissions than a conventional burner.

When combined with other NOx reduction techniques, such as mixing of flue gases with the combustion air (FGR), fuel staging combustion and fuel mixing can be utilized as part of the total NOx reduction system. Water injection has a minor impact on the efficiency of the system. Water absorbs heat as part of the vaporization process and carries that heat out of the combustion chamber. However, this vaporized moisture will be saturated and heat will be recovered.

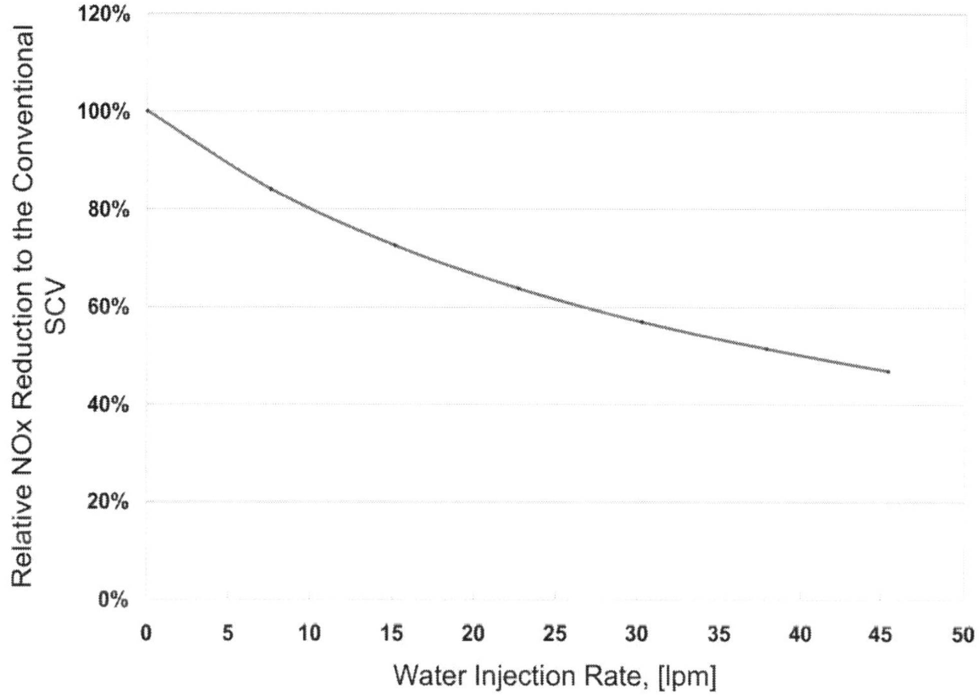

Fig. 3 – NOx Reduction with Water Injection

Flue Gas Recirculation (FGR)

In the drive to reduce NOx emissions into the lower 30 ppm range, FGR technology was considered to achieve low emission SCV. This technology has been originally developed for and applied to the incineration industry [1]. The basic concept is that a portion of the cold flue gas is recycled from the stack and combined with combustion air prior to being fed to the burner. The installation of FGR requires a motive force to recycle the flue gas. This has typically been accomplished by a forced draft blower.

Since the flue gas is water saturated and at a low temperature, a portion of flue gas is recycled to the blower suction which helps lower the combustion temperature. With 60% flue gas recirculation, the flame remains stable. A 20 – 30% flue gas recirculation rate does not begin to approach the burner's performance limits. FGR also reduces NOx emission by lowering the oxygen concentration in the primary flame zone. Recycling of flue gas reduces combustion capacity by about 5-7%. This can be compensated by an increase in the blower power. However, FGR does not suffer from an additional heat release and results in almost no impact to SCV efficiency.

The FGR can simply be applied to the conventional SCV with an increase in the blower power. However, a burner with a short flame length can achieve the maximum effect of lowering emission. The burner maintains stable combustion at the lower flame temperatures, which is one of the key factors when adding this FGR to an existing system. It also eliminates the potential flame impingement, which is one of the most common problems experienced with conventional low NOx burner retrofits. Figures 4 illustrates a schematic diagram of the flue gas recirculation burner technology.

The following points were addressed during the study of the FGR system:

- Lowering the NOx and CO emissions
- Flame stability at the lean combustion
- Easy flame temperature control
- Minimize the increase in blower power requirements

Additionally, the ability of the combustion system to handle the impact resulting from the amount of diluents required also was addressed.

This FGR system requires that control systems be designed to allow much tighter control of air-fuel ratios. This can lead to limited burner turndown, slower ramp rates, and less tolerance to air-fuel upset conditions. However, the benefit of this approach is its ability to achieve the required NOx emissions levels with less impact to system operational costs and efficiency, as well as improved burner operational characteristics.

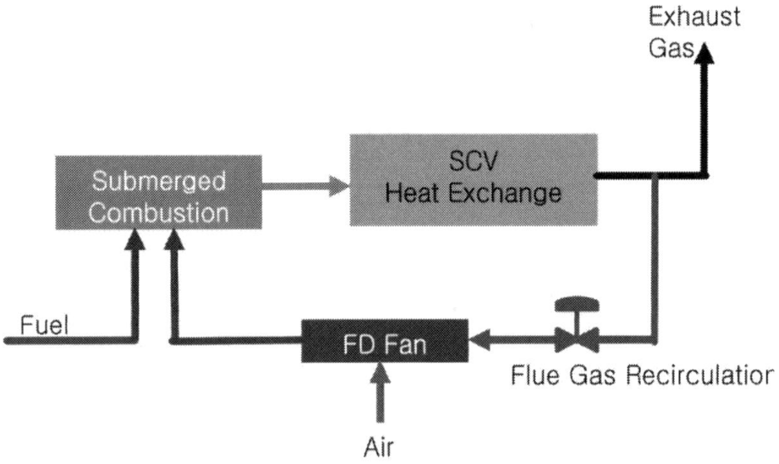

Fig. 4 – Schematic Diagram of Flue Gas Recirculation

In order to keep flammability, the flue gas recirculation valve is shut off at the starting operation of SCV. As soon as the SCV reaches normal operation conditions, the flue gas control valve gradually opens and feeds the flue gas to the suction of the forced draft blower. The valve position will be adjusted to maintain low emissions.

Rapid Premix Submerged Combustion Burner

To achieve NOx levels in the 20 ppm range, the relationship between temperature, stoichiometry, and NOx formation was re-evaluated as the basis for the design of the Rapid Premix Submerged Combustion Burner (RPSCB). It was determined that the most direct method of achieving very low NOx emissions from a natural gas flame is to avoid fuel-rich regions, which increases the potential for prompt NOx generation and lower the flame temperature to reduce thermal NOx to the desired level. To accomplish this, a burner design that avoids fuel-rich regions by rapidly mixing gaseous fuel and air near the burner exit was studied.

In RPSCB, the fuel and the oxidizer are completely mixed before combustion begins. Premix is a term applied to burners that inspirate part or all of the total air required for combustion [4]. The operating envelope of a premix burner is defined primarily by the limits of flammability of the air-gas mixture and by the velocity of the mixture passing through the ports of the burner nozzle. This burner provides for intimate mixing of the fuel and combustion O_2 prior to the ignition zone.

The mixing section is designed to provide a uniform concentration of mixed oxidant and fuel at a uniform velocity at the head end of the section. By Rapid Premixing, fuel and oxidant are mixed near the point of combustion, enhancing safety and permitting the construction of a larger burner. It is also designed to avoid the potential for flashback into the mixer and further into the chamber. The desired result at the outlet mixing chamber is a uniform fuel oxidant mixture having a lean flammability limit.

The premix burner system consists of two key components, the burner head or nozzle, and the gas-air mixing device that feeds it. The mixer includes a mixer inlet with a fuel input and an oxidant input for respectively admitting fuel and oxidant into the mixer. The mixer also includes a mixer outlet for discharging the mixed fuel and oxidant. Combustion occurs at or near the mixer outlet.

The studied RPSC has a shorter flame length compared to nozzle mix burners, the air/gas ratio is easier to control, and the burner is generally easier to set up. Special flow regulators can be fitted to simplify the air flow adjustment to raise and lower the temperature for accurate temperature control.

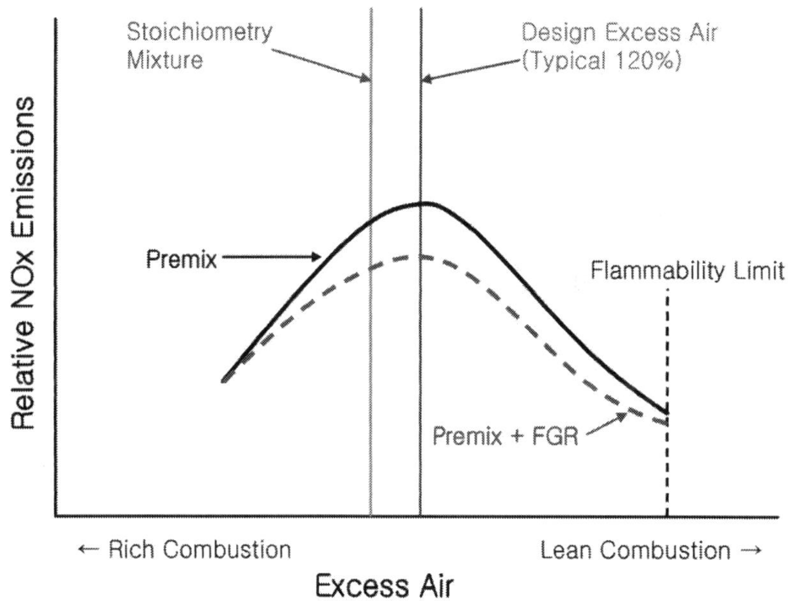

Fig. 5 – Comparison of NOx Emissions

Figure 5 shows the variation in NOx with flame stoichiometry for two different types of flames. The upper solid curve characterizes the NOx emissions expected with a premixed flame. The lower dotted curve is for the same flames with the addition of FGR. The figure shows that operating with more than one combustion zone can lead to very low NOx emissions with reasonable overall excess air in the final combustion products. Firing the remaining fuel under fuel rich conditions with flue gas recirculation will produce low NOx emissions in another combustion zone.

Regenerative Thermal Oxidation

In order to compete with post NOx treatment systems, such as Selective Catalytic Reduction (SCR), Regenerative Thermal Oxidation (RTO) has been studied and is further being developed. The original RTO technology has been widely used in the incineration industry to treat exhaust gas prior to exiting the stack and dispersing into the environment [5]. Such process units are also called "afterburners," as an RTO is often the last step in an air pollution control application. Afterburning is used to destroy volatile organic compounds (VOCs) and malodorous gases that would otherwise exit the stack into the environment.

Regenerative Thermal Oxidizers offer a highly energy efficient method of burning organic emissions from fuel. But when the emissions contain high amounts of combustible vapors, the energy efficiency becomes a handicap. The heat of combustion from the vapors produces more heat than required, causing overheating and reduced equipment life. Temperatures in the heat transfer beds and the combustion section frequently exceed the design parameters of the unit. The excess temperature causes serious structural weaknesses in the unit, which results in lower reliability and higher maintenance cost. However, it can be solved by adopting the FGR process or providing more air, which increases heat transfer. This will have a better oxidation temperature control and will thus extend equipment life.

The RTO consists of an oxidation chamber with a premix fuel transfer tube. The RTO uses multiple regenerator packed beds filled with ceramic heat transfer material to achieve regenerative heat recovery by cycling hot, clean, oxidized gas, while inlet process gas flows in individual beds, alternately adding and subsequently removing heat from an individual bed. A premixed fuel flows down and enters the ceramic media through the fuel transfer tube. To further increase thermal efficiency of the SCV application, the RTO uses forced draft fans to apply a positive pressure. Under positive pressure, the exhaust gas can be sent to the conventional SCV sparge pipe. An outline of FTO is shown in Fig. 6.

Pre-mixed fuel is preheated and is partially oxidized when it enters the combustion chamber. Here it is heated by auxiliary fuel combustion to a final oxidation temperature typically between 850°C to 950°C. The combustion chamber temperature is maintained to achieve the continuous minimum emissions requirement.

Pre-mix and oxidation are accomplished and can be optimized separately. Ceramic media has high thermal mass, with radiative and convective characteristics. This media also provide nearly homogenous temperatures, which is essentially one of the most important parameters for lowering emissions in the oxidation process.

In the RTO process, since the fuel and oxidizer is premixed before it reaches the reaction zone, a stable self-sustaining reaction wave is established across the flow axis. The ceramic media absorbs and radiates heat generated exothermically by oxidation of pre-mixed fuel. This provides both the required thermal driving force to initiate the reactions and an energy storage capacity that has a stabilizing effect on the reaction.

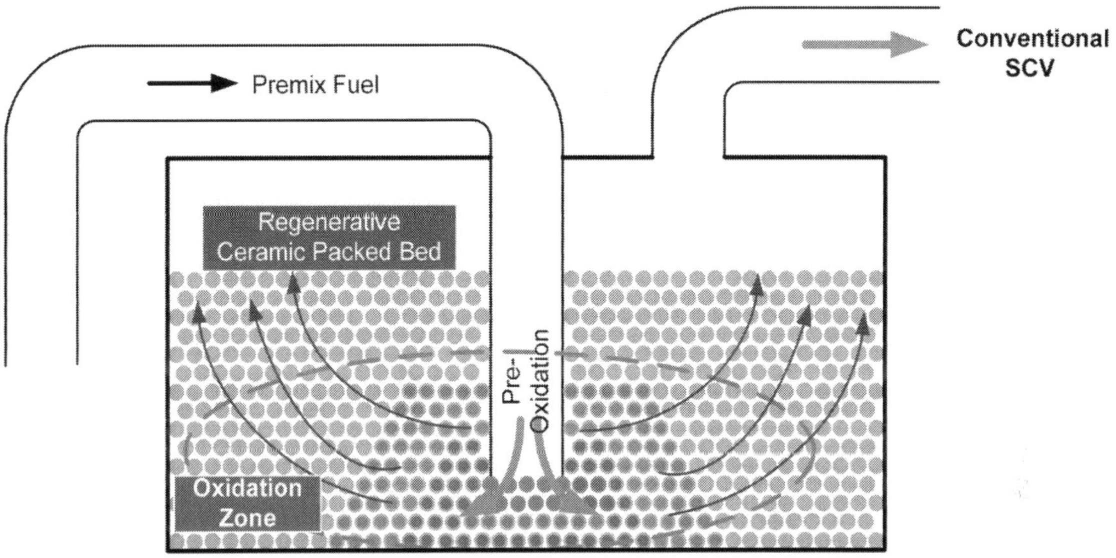

Fig. 6 – Outline of Regenerative Thermal Oxidation SCV

EMISSION AND ECONOMIC COMPARISON

Relative emissions, installed costs, and operating costs for each of these options are presented below in Table 1 for an SCV with a duty of 100 MMBtu/hr. A fuel gas high heating value (HHV) of 1,100 BTU/SCF with SCV thermal efficiency of 99% (HHV basis) and water bath temperature of 25°C are used as the basis for comparison.

Table 1 – Comparison of Relative Emissions and Costs for Various Technologies

Options	NOx Emissions (ppmvd)	Relative Capital Cost (%)	Relative Operating Cost (%)
Conventional SCV with water injection	40	100	100
Low Nox Burner / FGR	30	130	103
Premix Burner SCV	20	160	103
Regenerative Thermal Oxidizer	10	200	105

Note: Above data may vary depending on design and operation conditions.

All emissions shown above are corrected to 5% O_2 (dry basis). Operating costs include costs for power, fuel, caustic for pH control, and ammonia injection associated with SCR operation. Prices for fuel and power are assumed at $6/MMBtu and $0.08/kW-hr respectively.

CONCLUSIONS

New submerged combustion technologies successfully reduce NOx and CO emissions and achieve the emission setting goals, and can be advantageous for an environmentally sensitive area.

Water injection and flue gas recirculation can be easily combined with other technologies and implanted into the conventional SCV. The combination of water injection and flue gas recirculation can reduce the NOx emission to below 30 ppmvd and does not impact thermal efficiency.

Rapid Premix Submerged Combustion Burner (RPSC) technology reduces NOx emission to below 20 ppmvd. Newly designed burner characteristics are shortened flame length and a stable operation with lean fuel gas mixing. Rapid Premix combined with flue gas recirculation can further reduce NOx emission by 15 to 30%.

Regenerative Thermal Oxidation (RTO) can reduce NOx and CO emissions to a range of ppmvd single digits. It can be an alternative to the post treatment emissions, such as Selective Catalytic Reduction.

REFERENCE CITED

1. Cho, J.H., Mathur, G., Kotzot, H., and Durr, C.: "Applicabilities and Limitations in LNG Vaporization Processes", LNG Journal, p. 24-26, June 2005.
2. Bradford, M., Grover, R., and Paul, P.: "Controlling NOx Emissions", CEP, p. 42- 46, March 2002.
3. Webster, T.: "Burner Technology for Single Digit NOx Emissions in Boiler Applications" presented at the CIBO NOx Control XIV Conference, San Diego, CA, 13 March 2001.
4. Bennet, D.: "Burner History 103," http://www.process-heating.com/CDA/ArticleInformation/Energy_Notes_Item/0,3271,93428,00.html
5. Shah, K., Wong, J., and Reeves, G.: "Air Emission Control Technologies for Submerged Combustion Vaporizers", presented at the AIChE 2005 Spring Meeting, Atlanta, GA, 10-14 April 2005.

BIOGRAPHY of SPEAKER

Yoshimichi Ikeda graduated from College of Engineering Sciences in University of Tsukuba, Japan in 1987.

He joined Sumitomo Precision Products in 1987.
He was involved in heat exchanger design including LNG vaporizers.

He has been engaged in development and a design of several types of LNG vaporizers (SCV, ORV and the others) since then.

Email : ikeda-y@spp.co.jp

Overview of Open Rack Vaporizer and Intermediate Fluid type Vaporizer, and possible Impact of the Warming Water to Environment

Masao Endo
Manager, Process Equipment
Takasago Equipment Plant
Koichi Shinkai
Product Manager, Process Equipment
Takasago Equipment Plant
Shinji Egashira
Manager, Process Equipment,
Takasago Equipment Plant
Machinery & Engineering Company, Kobe Steel, Ltd.
Takasago, Japan

AIChE Spring National Meeting 2006
6th Topical Conference on Natural Gas Utilization
Orlando, Florida, April 23-27, 2006.

Introduction

LNG at receiving terminal is regasified to ambient temperature gas using vaporizers before leaving the terminal. Kobe Steel has been the leader in LNG vaporizers, and has developed, designed and manufactured versatile selections of vaporizers for smallest to largest vaporization needs, and has supplied them to the users around the world.

Open Rack Vaporizer and Intermediate Fluid type Vaporizer supplied by Kobe Steel, both utilize sea water as heat medium, and are used in majority of major receiving terminals in the world, notably in Europe and Asia.

This article firstly introduces safety, and maintenance features of the vaporizers in detail, with their possible environmental impacts. Other special features of the vaporizers are also introduced.

Secondly, environmental issues associated with sea water heat exchangers, and solutions to the respective points are discussed.

1. Open Rack Vaporizers (ORV)

1.1 Working Principle:

ORV is a vaporizer using sea water as heat source to regasify LNG.
See Fig.1 and Photo 1 for general assembly.

LNG in the storage tank is pressurized with LNG primary and secondary pump and supplied to ORV at LNG nozzle(s).
LNG is distributed to each panel by LNG manifold(s).

LNG is further distributed to heat transfer tubes, where it is vaporized to make NG (gaseous natural gas). NG goes out at NG nozzle(s) through NG manifold(s).

Seawater, heating source for LNG vaporization, is supplied to ORV at seawater piping connection(s). The seawater is distributed to each seawater trough through seawater manifold, butterfly valve, and seawater header pipe. Seawater overflows from the edge of troughs, and flows down along the surface of heat transfer tubes heating LNG. See Fig.2.

Photo 1 ORV appearance

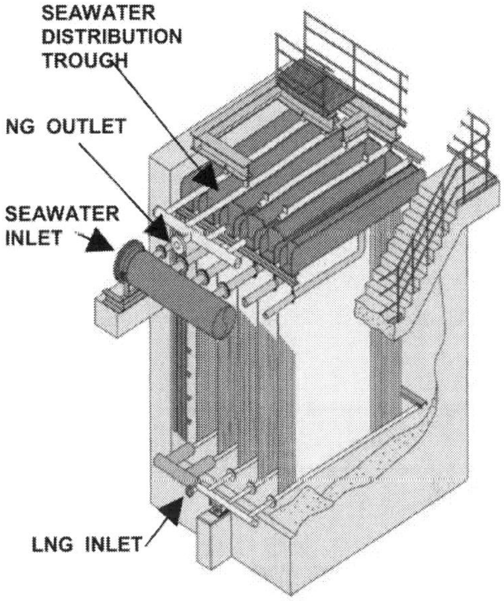

Fig. 1 ORV general assembly

1.2 Heat transfer tubes:

A number of heat transfer tubes are placed side by side to form a panel. The tube is made of aluminum alloy and metallizing is applied to protect the heat transfer tube outer surface.

1.3 Characteristics:

a) Inherent safety features and safety record:

ORV is the most preferred selection in Europe and Asia, and has been operated stably since 1978, and for the ORVs Kobe has delivered, no leak is reported from more than 100 installations in over 20 years. The excellent record has told superior safety itself. Even if any leak should occur, there is no holding area or pocket to allow gas to accumulate, making risk of explosion very remote. For the added safety monitor, gas leak detectors are placed at the top maintenance deck.

b) Easy maintenance:

The "Open Rack" configuration makes all the pressure containing parts to be visually inspected very easily.

Ice will develop at the lower side of heat transfer tube, and visual check of such icing height is also very easy. Since icing height for the heat transfer tubes are uniform across the panel for normal operation, any irregularities can be detected by visually checking the icing height departing from normal operation.

Use of sea water makes periodical cleaning to be necessary to wash off dirt, algae and mussels. They adhere on the panels or remained and/or propagated inside sea water troughs and sea water pond.

The washing can be done very easily. It is just walking inside the unit, and clean with jet water. There is just no need to disassemble any part of the unit.

Metallizing of zinc-aluminum based alloy is applied to protect base aluminum alloy. Limited areas of the metallizing may need to be recoated with the interval of several years, and all of the recoating work can be done in-situ, without the need to disengage the panels out of the place.

c) Simple operation:

Operation of the system is very simple and allows most flexibility. As long as water is flowing, LNG flow rate can be changed as fast as is desired, and during the process, variation of outlet gas temperature is very small.

d) Low operating cost:

Operating cost is quite low, since sea water heat is used to regasify LNG. The only notable operating cost is one for running sea water pump. Compared to the fuel-burning type vaporizers, which consumes about 1.5% of sendout gas, the operating cost is the one tenth or lower.

e) Minimum sea water temperature:

ORV can be designed to the lowest seawater supply temperature of 42F (5 C). Since wide range of sea water temperature is acceptable, it adds flexibility to the regas system.

1.4 Design refinement of ORV –SUPERORV®:

The performance of the ORV is dictated by amount of heat exchange that a tube can handle. Since LNG temperature is very low, the bottom part of heat transfer tube will develop icing, which hampers heat transfer.

With the introduction of patented dual tube design, icing thickness is reduced, thus heat transfer is improved. See Fig. 3. As a result, LNG flow rate per tube is increased by 50%, and footprint per unit is reduced by more than 20% typically. See Table 1. *SUPERORV®* was jointly developed with Osaka Gas Co. Ltd.

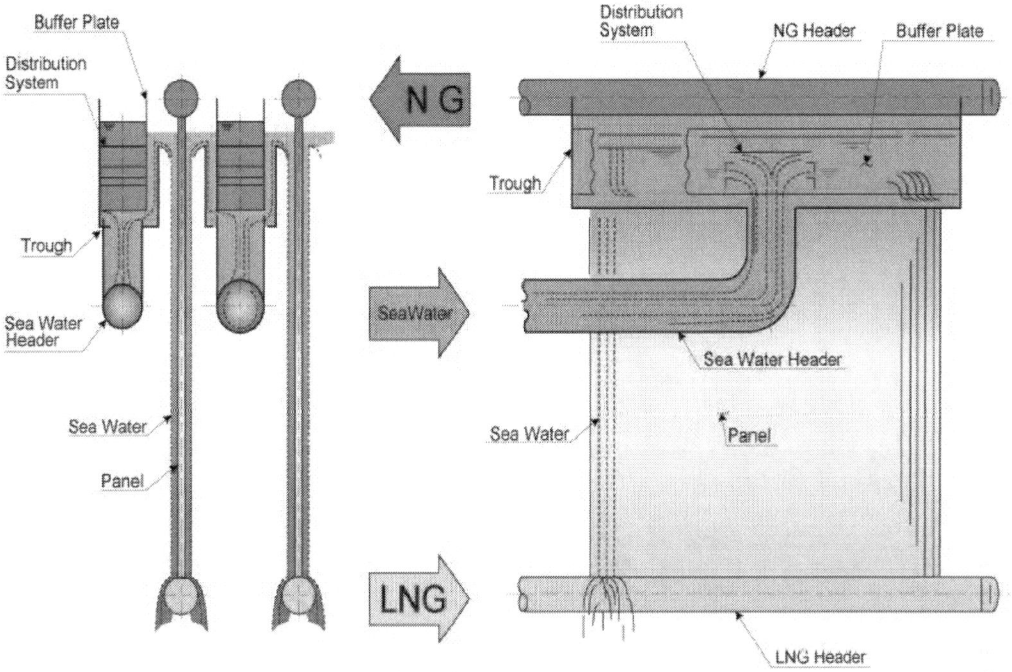

Fig. 2 ORV working principle

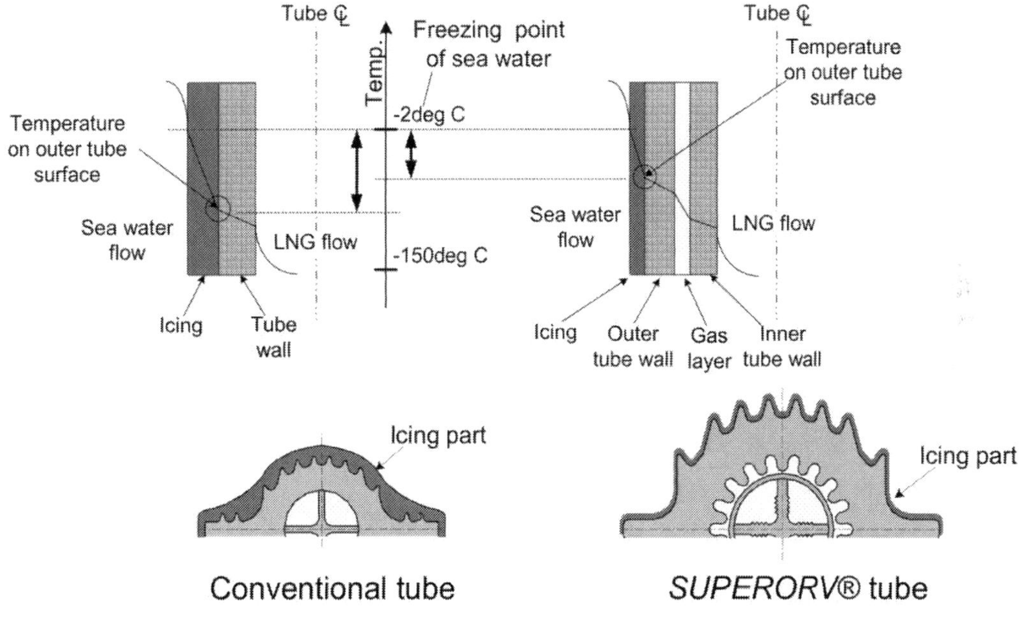

Fig. 3 Dual tube design mechanism of SUPERORV®

LNG : 0.25 Bcfd (210 ton/hr) Seawater Temp. : 55F(12.8C)		
TYPE	conventional	SUPERORV®
Tube length (m)	6	6
Q'ty of Tubes per Panel	100	86
Q'ty of Panels per Unit	18	12
Seawater flow (ton/hr)	6,000	5,000
Seawater Δt	10F (5.6C)	12F(6.7C)
SIZE (L×W×H: m) FOOTPRINT (m^2)	9.0×17.8×7.9 160	9.3×12.9×7.9 120 (25% less)
Max. D.P. (psig)	2,200	2,200
	PLAN VIEW	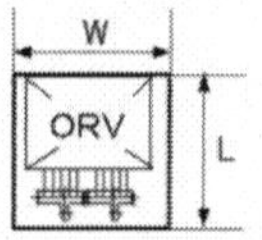

Table 1 Design improvement with SUPERORV®

1.5 Modularized ORV for offshore (GBS) application

ORV is delivered to the installation site either as an individual panel block, or as a modularized unit. The selection is generally dependent on the cost associated with assembly of sea water troughs and sea water piping.

For the offshore applications, modularized unit is preferred, as it will minimize on-site assembly and adjustments. See Photo 2 for the photo of the mockup.

1.6 Thermal Integration with waste heat of power plant

When waste heat from neighboring power plant is available, thermal integration becomes the best fit. Waste water flow rate and heat are roughly ten times larger, so a part of the power plant can be branched. Higher inlet sea water temperature makes the size of ORV smaller, also makes waste water mixture temperature lower, making the system more global friendly. See Fig. 4 for the typical configuration.

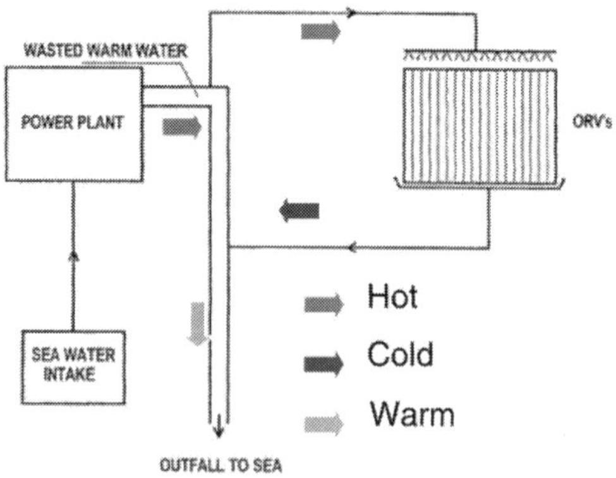

Photo 2 Mockup of modularized ORV Fig. 4Typical thermalintegration with ORV

2. Intermediate Fluid type Vaporizer

2.1 Working Principle:

IFV is a vaporizer using evaporation and condensation of intermediate fluid to regasify LNG. For the IFV, Kobe has developed with Osaka Gas Co. Ltd, sea water can be used as heat source.
See Photo 3 for general assembly. See Fig. 5 for the working principle.

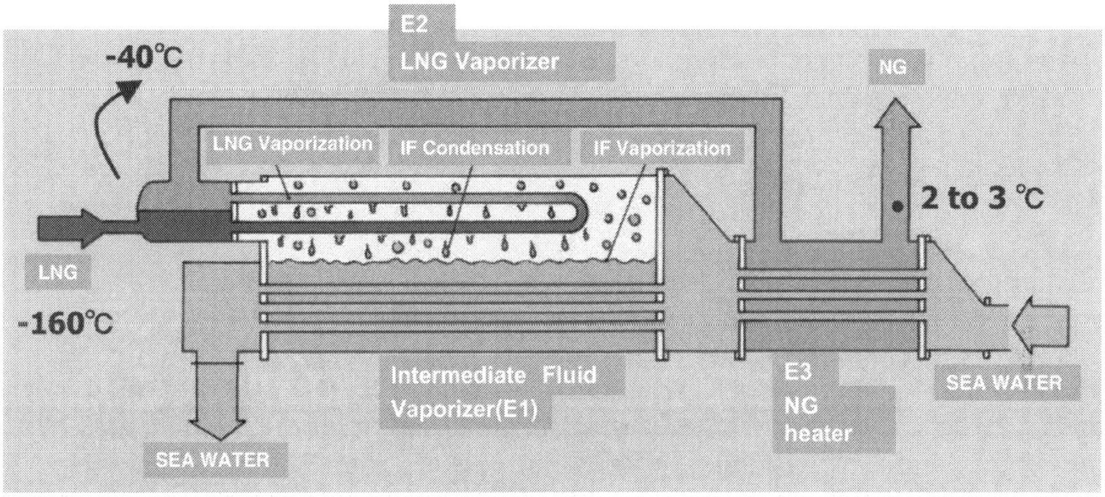

E-1: Intermediate fluid (shell side) is vaporized by seawater (tube side)
E-2: LNG (tube side) is vaporized by heat of condensation of intermediate fluid (shell side) Intermediate fluid is condensed by LNG on the surface of tubes and drops to bottom of shell
E-3: NG (shell side) is heated by seawater (tube side) up to ambient temperature

Fig. 5 IFV working principle

Photo 3 IFV appearance

2.2 Heat transfer tubes:

Heat transfer tube is made of titanium alloy to enable sea water use. Titanium is currently the only metal commercially proven to be compatible with sea water in the field.

2.3 Characteristics:

a) No restriction on the amount of suspended solids:

ORV is usually the best vaporizer selection using sea water, however, for the cases where sea water contains very high amount of suspended solid, they tend to erode metallizing faster. For such condition, the IFV is most recommended as titanium tube will not be affected by sand included in the sea water.

b) Inherent safety features and safety record:

IFV has been used stably for more than 20 years for baseload vaporizers in Japan, and no leak has been reported. For the shell and tube type vaporizers, one of the risks is associated with freezing problem of heat medium.

For the IFV, LNG tubes do not contact with sea water directly, therefore, risks of the freezing is avoided.

c) Easy maintenance:

The growth of marine plants in sea water passage is always the problem, as they will grow at any point in the sea water side where the velocity is lower. The marine plants can block water passage, leading to partial blockage of heat transfer. In order to avoid the bio-fouling, injection of anti bio-fouling chemical is useful, however excessive amount of injection will do harm to the environment, therefore, periodical washing of water passage is desired. For the IFV, washing inside of the titanium tubes can be done easily by opening manhole and with jet washing using clean water. This is one of the few maintenance items.

d) Low operating cost:

Like ORV, operating cost is low by using sea water as heat source.

e) Minimum sea water temperature:

IFV can be designed to the lowest seawater supply temperature of 42F (5 C).

3. Comparison of vaporizers

Comparison of key characteristics of ORV and IFV are summarized in Table 2.

a) Load change

Load change flexibility contains two meaning.

The first is turn down ratio, and tells if the vaporizer can handle very small flow rate. For ORV, the flow rate can be zero to one hundred percent, so the turn down ratio becomes infinity. For IFV, "keep cool" operation with minimum flow rate is necessary to avoid excessive thermal stress.

The second is load change speed, and tells if the vaporizer load can be changed fast enough responding to change in demand. For both ORV and IFV, no limit is present for the load change speed.

b) Maintenance for bio-fouling

For sea water system, some users use sodium hypochlorite or other chemicals to control bio-fouling to make periodical maintenance interval longer.

For ORV, the maintenance is easier, so other users choose not to use anti-biofouling chemicals. For IFV, addition of small amount of anti-biofouling chemical is very effective to ensure maintenance intervals. The amount of bio-fouling is greatly dependent on sea water temperature, season and location.

Vaporizer type	ORV	IFV	
Sea water quality requirement	Yes	No	IFV: Titanium, Seawater is flowing tube side
Operation cost	Very Low	Very Low	
Leak Safety	Excellent	Excellent	ORV: enclosed accumulation pocket not present IFV: Titanium
Safe operation experience	100+ units 20years No Leaks	20+units 20years No Leaks	
Maintenance For bio-fouling	HOCl- not required with periodical cleaning	HOCl- advised at 2 mg/liter	HOCl- to be non-detectable at facility exit
Load change	Most Flexible	Very Flexible	

Table 2 Comparison of key characteristics of ORV and IFV

c) CAPEX and OPEX

ORV start with relatively higher capital cost, as it requires sea water supply/discharge piping and sea water pumps, however combustion type vaporizer consumes some 1.5% of sendout gas, making operation cost very high. See fig 6 for the comparison. The price of fuel for a 1 Bcfd terminal is 70 MMUSD/year with 90% efficiency and fuel price of 12 USD/MMBtu.

Because of this, almost all the current baseload LNG receiving terminals use ORV, and reserves SCV for peakshaving or backup purposes.

Vaporizer type	ORV	Combustion type**
Capital cost	100*	40
Operation cost (per year)		
@ 7$/MMBtu	8	110
@ 12$/MMBtu	8	170

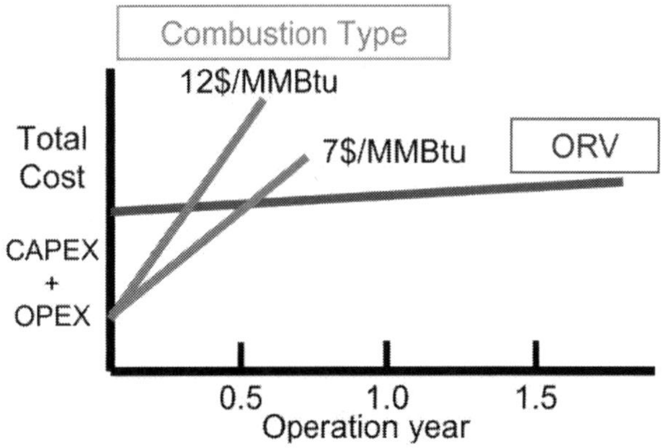

*includes seawater system

** assuming 90% efficiency Consumes 1.5% of vaporization

Cost does not include CO_2 credit

Fig. 6 Comparison of capital and operation cost

4. Environmental Impact

ORV is more global friendly than combustion type vaporizers, in the sense of NOX and CO_2 emissions. CO_2 for a 1 Bcfd terminal with combustion type vaporizer/system amounts to about 850ton/day.

With regard to the possible impact of sea water usage, no environmental influence has been reported in any of 100 plus installations.

4.1 Possible influence factors for sea water usage

Firstly, chilled water due to DT could have negative effect on Ichthyoplankton, zooplankton, fish larvae and eggs.

Secondly, High concentration sodium hypochlorite used to prevent bio-fouling will have negative effect on Ichthyoplankton, zooplankton, fish larvae and eggs

Lastly, marine habitats could be damaged by simply going through water pathways (mechanical impingement)

4.2 Chilled water

Temperature drop across ORV is generally about 8 to 12 F (4 to 6C), which will quickly deplete to 2 to 3 F by mixing at discharge point.

The12 F difference is generally well within seasonal or daily variation of sea water temperature, and found to be acceptable in Japan and worldwide.

See Fig. 7 for typical monthly average of sea water temperature.

Fig. 8 shows public survey result for a power plant in Japan, where cooling water temp. delta t = +13F (7C) , and cooling water flow rate of 200,000 ton/h, or 1268MGD. Still, the 2F (1C) temperature affected zone is limited to mostly 2 km (1.3 mile) radius.

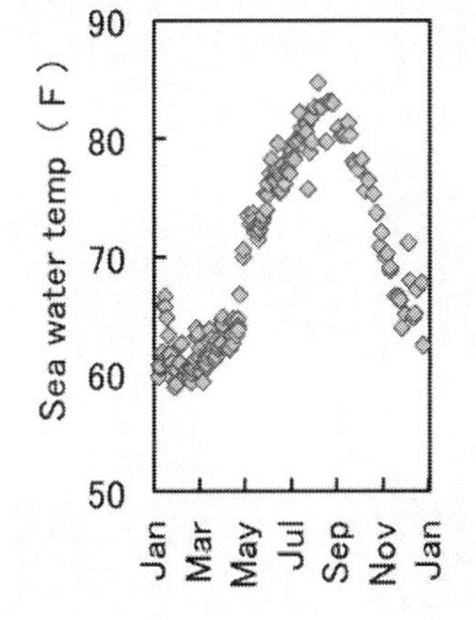

Fig. 7 Typical sea water temperature variation

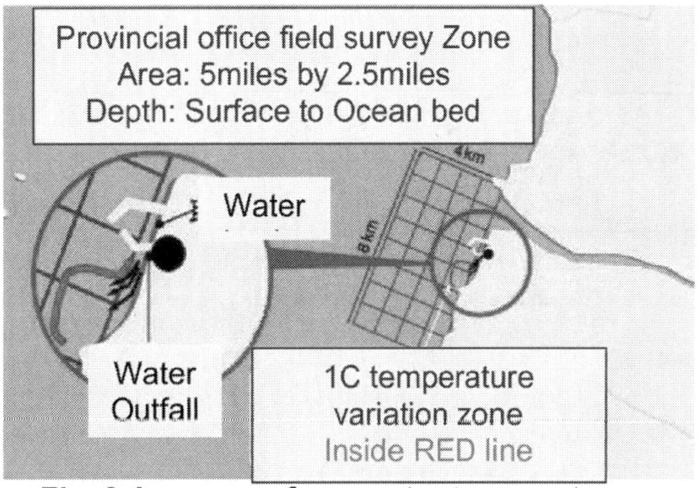

Fig. 8 A survey of sea water temperature variation in a Japanese power plant
(http://www1.kyuden.co.jp/sendai_topix_index#3)

4.3 Anti biofouling chemicals

For ORV, with periodical cleaning, injection of sodium hypochlorite is NOT a must, based on experience in Japanese receiving terminals. Conversely, excessive amount of the chlorine ion could influence life of metallizing.

An extensive study conducted in Japan from 1996 to 2003 for power plant waste water shows environmental influence smaller than anticipated. *4)

It was known that influence on ichthyoplankton, zoo plankton and fish eggs are very limited, influence on fish larvae is smaller than was believed historically.

Moderate amount of sodium hypochlorite decomposes while passing through seawater distribution system, and will become trace levels at seawater discharge See Fig. 9.

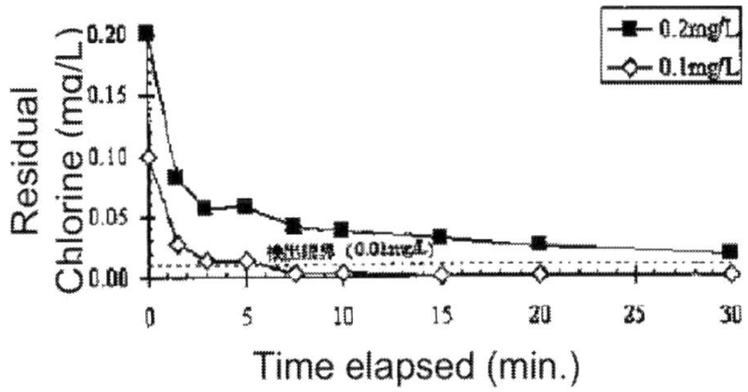

Fig. 9 Depletion curve of residual chlorine

4.4 Water Intake

Minimizing influences onto fish eggs may be necessary, considering spawning grounds. This has been accomplished by choosing the intake point carefully.

Influence on ichthyoplankton, zoo plankton and fish larvae are minimal as a result of passing through the seawater distribution system, according to the same study.

5. Conclusion

The characteristics of the ORV and the IFV, comparing other types of vaporizers, are discussed.

- Both are superior in safety and maintenance with proven field records.
- Influence of sea water discharge was studied in Japan, and judged negligible, or very limited.

References

1. Masao Endo, "Overview of ORV and IFV characteristics and operation in LNG receiving terminals in Japan and worldwide" LNG Express Workshop Latest Advancement in Regasifiers, November 30, 2005, Houston, Texas
2. Hara et al., "The predation by sessile organisms and cooling water intakes of power plants", Sessile Organisms 22(2) (2005)
3. Kyushu Electric Power Company Homepage
4. Marine Ecology Research Institute (Japan), "The fiscal 2003, Investigation report of impact by water intake and water discharge at large-scale electrical power plants (Investigation of impact on organisms in taken water) - Summary of investigation result from 1996 through 2003", March 2004

Bibliography of Speaker

Mr. Masao Endo is a manager of Takasago Equipment Plant, Kobe Steel, the leading steel, aluminum and copper company in Japan, also a major fabricator of heavy wall pressure vessel, and aluminum heat exchangers and LNG vaporizers. He has 19 years experience of energy and high pressure gas handling system engineering, including gas separation plant engineering, adsorbent process, high pressure hydrogen/oxygen facilities. His recent involvement includes ORV, SCV, AFV (direct Air Fin Vaporizer) and HWV (Hot Water bath Vaporizer) engineering and projects.

Mr. Endo earned his bachelor's degree at Tokyo University in aeronautical engineering. He lives in a suburb of Kobe city in Japan.

Low Emissions with Advanced Pipeline Gas Heater

Jung-Bin Lee
President, Wonil T&I Co., Ltd.
Jong_Ho Kim
Deputy Manager, Heat Exchanger
Wonil T&I Co., Ltd.

AIChE Spring Meeting, April 2006
6th Topical Conference on Natural Gas Utilization
Orlando, Florida, April 23-27, 2006

ABSTRACT

To comply with the Natural Gas industry's requirements for minimizing life cycle costs, many Pipeline Gas Heaters have been proposed and developed. However, the constraints in the selection of a gas heater are often site-specific environmental issues and thermal efficiency issues, especially for the planned North American pipelines. Without proper attention to these site-specific issues a project can be seriously delayed or its realization may even be jeopardized.

In recent years, the concern to minimize the environmental impact of the energy industry has dramatically increased. This trend results in tighter air emissions control and includes minimizing emissions from pipeline gas heaters. This paper discusses newly developed advanced design of Pipeline Gas Heaters that both reduces NO_x and CO emissions and increases thermal efficiency. This paper also presents the economical advantages of the advanced Pipeline Gas Heaters in comparison to conventional gas heaters.

INTRODUCTION

Natural gas is a highly desirable energy source. It burns cleanly, with less pollution than other hydrocarbon fuels. Natural gas is projected to be the fastest growing component of the world's primary energy consumption. Consumption of natural gas worldwide is forecasted to increase by an average of 2.3 percent annually from 2002 to 2025, compared with projected annual growth rates of 1.9 percent for oil consumption and 2.0 percent for coal consumption [1]. Proven reserves of natural gas were estimated at 6,040 trillion cubic feet (Tcf) as of January 2005 [2]. However, much of the world's known natural gas reserves are inconveniently located in remote and thinly populated areas, such as Western Siberia and the Persian Gulf.

Natural gas transportation from remote areas to market should be efficient, reliable and economic. Options that are often compared include pipeline transmission and LNG. Gas transmission via pipeline is common, accounting for 75% of international gas trade [3]. Although a pipeline provides the least expensive solution over short distances, disadvantages are that a pipeline runs on a fixed route and has limited capacity.

In order to enhance transport efficiency, high pipeline operating pressure is normally used since the gas transport is driven by pressure energy. However, the end user pressure requirements vary from 5 barg to 35 barg for city gas pipeline network and power plants, respectively. The pipeline pressure should be let down to a proper operating pressure before the natural gas is supplied to pipeline gas customers.

When pipeline pressure is let down from high operating pressure to the customer receiving pressure, the gas temperature drops, known as the Joule-Thomson effect. The temperature decrease is insignificant with a small pressure let down. However, when a large pressure drop is required, the temperature drop has a significant impact on the pipeline gas system. To compensate for this temperature drop, Pipeline Gas Heaters (PGH's) have been used. The heater warms up the pipeline natural gas (PNG) to maintain the pipeline gas temperature. The heating consumes a small fraction of transport gas as fuel. Consequently, the industry is concerned with the thermal efficiency of PGH's in order to lower operating costs and lower the emissions from fuel combustion.

This paper discusses the development of high thermal efficiency and lower emission of pipeline gas heaters. Several economic and technical issues related to the PGH design and operation are also discussed.

EMISSION LIMITATION

Pipeline projects face several engineering design challenges: minimization of environmental impact on pipeline construction and emissions control on pipeline operation. Pipeline operation requires energy, which is the driving force to push the natural gas inside the pipeline. Compression energy is used to boost its pressure and heat the gas to compensate for the temperature drop at the pressure governing stations.

Pipeline Gas Heaters (PGH) have widely been used to heat natural gas. Since it uses natural gas as a fuel, NO_x emission is an issue for industry because both state and federal governments regulate the emissions from industrial plants. The federal governing body that is driving state agencies to bring air quality to a designated level is the Environmental Protection Agency (EPA). Each state is given authority by the EPA to develop and implement plans to reach and maintain a minimum air quality standard. In Texas, this state body is the Texas Natural Resource Conservation Commission (TNRCC). In Louisiana, the state body is the Department of Environmental Quality (DEQ). The state is given the authority to prosecute individuals and companies that do not follow the regulations in Title 40 in the Code of Federal Regulations (40 CFR).

Two federal documents affect industry for NO_x and sulfur dioxide (SO_2) emissions, the Clean Air Act (CAA) and the New Source Performance Standards (NSPS). These regulations are forcing industry to reduce NO_x emissions for both existing and new installations of fired process heaters and boilers.

As the result of the Clean Air Act, the EPA initiated the Acid Rain Program to reduce emissions of NO_x and SO_2. The Acid Rain Program sets targeted emission limits for NO_x and SO_2 in terms of pounds per year and pounds per Btu fired. The program provides for trading of emission credits and a tracking system. Phase I began in 1995. Phase II of the program began in the year 2000, and tightens the annual emission limits on large, high emitting plants, and sets restrictions on smaller plants that use coal, oil, and gas for fuel [4]. To comply with the energy industry trend to minimize environmental impact, many different technologies have been developed.

Cost effective solutions are required to implement the environmental limitations imposed by the restrictions associated with the Best Available Control Technology (BACT) and Lowest Achievable Emissions Rate (LAER) requirements. BACT limitations often apply to existing sources and consider environmental, energy and economic impacts. LAER limitations reflect the most stringent limitation achievable or the most stringent limitation achieved in practice, whichever is more stringent.

NO_x FORMATION AND FACTORS

NO_x Formation

Natural gas combustion produces Nitrogen Oxides (NO_x) and Carbon Monoxide (CO) that is often addressed in BACT and LAER requirements. CO formation results from inefficient combustion. The principle oxides formed are nitric oxide (NO) which represents 90-95% of the NO_x formed and nitrogen dioxide (NO_2) which represents most of the remaining nitrogen oxides. NO_x are formed primarily in the high temperature zone of a combustion chamber where sufficient concentrations of nitrogen and oxygen are present. Fuel nitrogen and nitrogen contained in the combustion air both play a role in the formation of NO_x. The largest percentage of NO_x formed is a result of the high temperature fixation reaction of atmospheric nitrogen and oxygen in the primary combustion zone.

There are three types of NO_x: "thermal NO_x", "prompt NO_x" and "fuel NO_x" [5]. Most of the NO_x are thermal NO_x formed through thermal dissociation and subsequent reaction of nitrogen (N_2) and oxygen (O_2) molecules in the combustion air. The formation of thermal NO_x is affected by three furnace-zone factors: (1) excess air, (2) combustion temperature, and (3) time of exposure at peak combustion temperature. NO_x is primarily a function of flame temperature, but it is also affected by oxygen concentration, and the time of those conditions [6].

Furnace Design

The size and design of furnaces have a major effect on NO_x emissions. As furnace size and heat release rates increase, NO emissions increase. This results from a lower furnace surface-to-volume ratio which leads to a higher furnace temperature and less rapid terminal quenching of the combustion process. Furnaces generate different amounts of NO according to the type of firing. Units employing less rapid and intense burning from incomplete mixing of fuel and combustion gases generate lower levels of NO emissions. Tangentially fired units generate the least NO because they operate on low levels of excess air, and because bulk misting and burning of the fuel takes place in a large portion of the furnace [5]

Burner Design and Configuration

Burners operating under highly turbulent and intense flame conditions produce more NO. The more bulk mixing of fuel and air in the primary combustion zone, the more turbulence is created. Flame color is an index of flame turbulence. Yellow hazy flames have low turbulence, whereas, blue flames with good definition are considered highly turbulent [5].

Excess Air

A level of excess air greatly exceeding the theoretical excess air requirement is the major cause of high NO emissions in conventional boilers.

Combustion Temperature

NO formation is dependent upon peak combustion temperature, with lower temperatures producing low NO emissions.

Firing and Quenching Rates

A high heat release rate (firing rate) is associated with higher peak temperatures and increased NO emissions. A high rate of thermal quenching, (the efficient removal of the heat released in combustion) tends to lower peak temperatures and contribute to reducing NO emissions.

Mass Transportation and Mixing

The concentration of nitrogen and oxygen in the combustion zone affects NO formation. Any means of decreasing the concentration, such as dilution by exhaust gases, slow the diffusion of fuel and air; or alternately fuel-rich/fuel-lean burner operation will reduce NO_x formation. These methods are also effective in reducing peak flame temperatures.

GAS HEATER SYSTEM

Natural gas transport in a pipeline generates frictional pressure losses. Frictional pressure loss is defined as the energy lost in transporting a fluid through a pipe, due to friction between the fluid and pipe wall. A gradual pressure drop decreases the natural gas temperature. However, the natural gas temperature will be maintained by heat gain from soil to pipeline unless the pressure drop is too high.

When pipeline gas needs to let down pressure to supply natural gas to a customer, a pipeline gas heater will be installed before or after the pressure let down valve. A sudden pressure drop by a pressure let down valve causes the following concerns:

- Downstream material selection of a pressure let down valve and associated piping
- Gas hydrate formation
- Icing inside pipe due to moisture contents
- Ground freezing

Figure 1 illustrates the pipeline gas configuration. There are many debates in locating the gas heaters. Generally, high pressure trunk lines locate the PGH before a pressure let down valve. This means that the natural gas is heated before the pressure is let down. The expected advantages are as follows:

- Pressure control valve can be operated without icing on valve stem;
- Material selection is easy (not necessary to be cryogenic or cold temperature);
- Preventing hydrate formation after pressure let down;

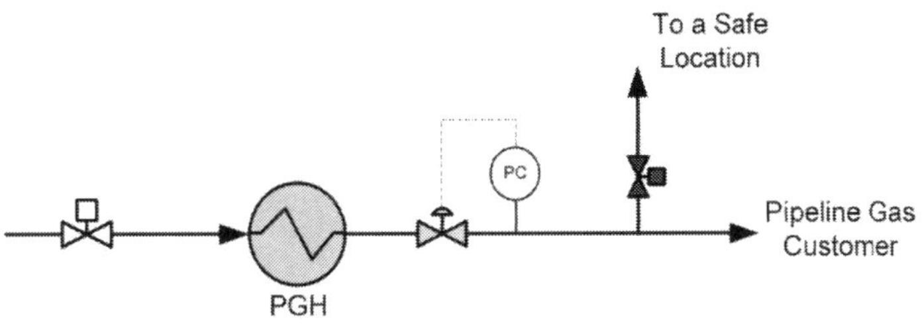

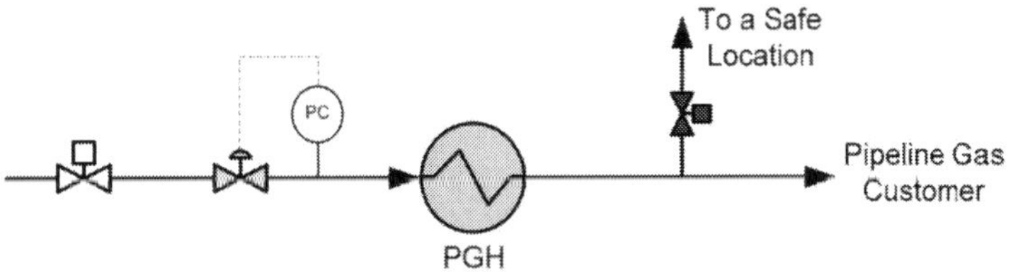

Fig. 1 – Pipeline Gas Heater Location

One of the disadvantages is that "Before" pressure let down requires higher design pressure than that of "After" pressure let down. However, cost difference is generally insignificant because the tube side pressure differences will not have much impact on the heat exchanger cost.

Heat requirement of a PGH is a function of the let down pressure, flow rate, and temperature differences. It can be expressed by Eq. 1.

$$H = f(\Omega, \Delta P, \Delta Q, \Delta T) \quad (1)$$

Where, H is heat requirement of a PGH

Ω is the fluid composition

ΔP is the let down pressure across the pressure control valve

ΔQ is the flow rate

ΔT is temperature difference between the inlet pipeline gas and outlet pipeline gas from the PGH.

Heat requirement increases as natural gas flow increases. High let down pressure requires more heat to compensate the temperature drop. The relationship between heat requirement and natural gas flow with the different let down pressure is illustrated in Fig. 2.

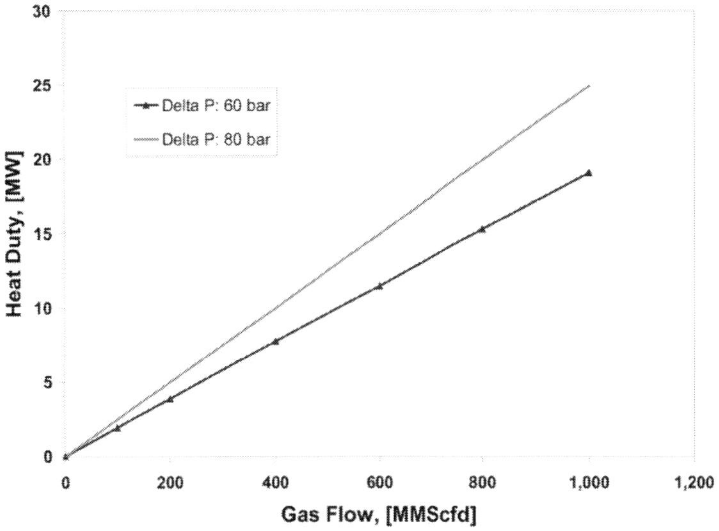

Fig. 2 – Heat Requirement for Compensation

of Pressure Letting Down

Operation of a PGH requires fuel gas, which consequently has a cost. The electricity costs are incurred by burner blower operation. Figure 3 presents the operating costs of a 15MW duty PGH with the different thermal efficiencies: 90% and 98% based on high heating value (HHV). Price of power is assumed at 8 ₵ /kW-h. Thermal efficiency of the PGH directly affects the pipeline operating costs. The annual operating cost difference will be about 0.3 million dollars for each gas heater with a fuel cost of $6/MMBtu. This trend increases with an increase in fuel costs.

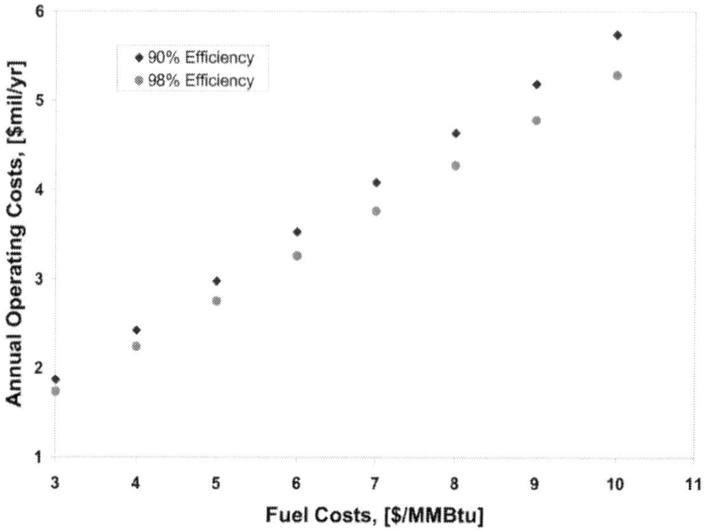

Fig. 3 – Pipeline Gas Heater Operation Costs

CONVENTIONAL GAS HEATER DESIGN

The pipeline gas heater (PGH) is skid mounted for preheating natural gas to maintain its temperature above the dew point before letting down its pressure. The main components are gas burner, combustion chamber, heat exchanger, water bath, control panel, pumps, and an expansion tank. Figure 4 illustrates the schematic diagram of a typical PGH. The PGH system also includes fuel gas pressure reduction system, glycol water circulation system, complete wire electric motor drivers and motor starters. A NEMA 4x, Z-purged, forced draft burner management panel for operation in a Class I, Division 2, Group D area classification is provided. The control panel is supplied complete with flame relay for burner.

The fuel gas is provided by letting down pressure of a slip stream from the pipeline. For the initial start-up an electric heater is used. Once it operates and bath water heats up, the fuel gas is heated by exchange with the warm water in the PGH.

A typical gross heating value (GHV) range of the fuel is somewhat limited based on emission limitations. Normal ranges of fuel gas GHV are 9,500–10,800 Kcal/Nm^3 (1,010 – 1,170 Btu/Scf). This range can be adjusted based on PNG compositions.

The main part of the heater furnace that contains the burner flames is the radiant section. The water heating tubes are located in the radiant section. The heating tubes are located close to the wall of the furnace to achieve low temperature gases exiting the radiant section. Radiation is the dominant heat transfer mechanism in the radiant section.

Convection section is located downstream of the radiant section. Flow is through the convection section and then through the stack to be vented to atmosphere. A pre-heater may be installed in the path to the stack extract for the flue gas. However, this preheat may also increase NO_x emission.

A specifically designed burner is equipped with a heater in order to achieve the stable operation and an acceptable flame pattern under the operating conditions. The specified maximum level of pollutant emissions that can be generated through the combustion process has been considered in designing the burner.

The heater operating system is designed to monitor the total flow of fuel to the heater. For a multi-burner heater, the fuel gas system is to be designed to ensure that fuel is properly distributed to all burners. The burner control system checks and ensures uniform fuel pressure to each burner for the proper operation.

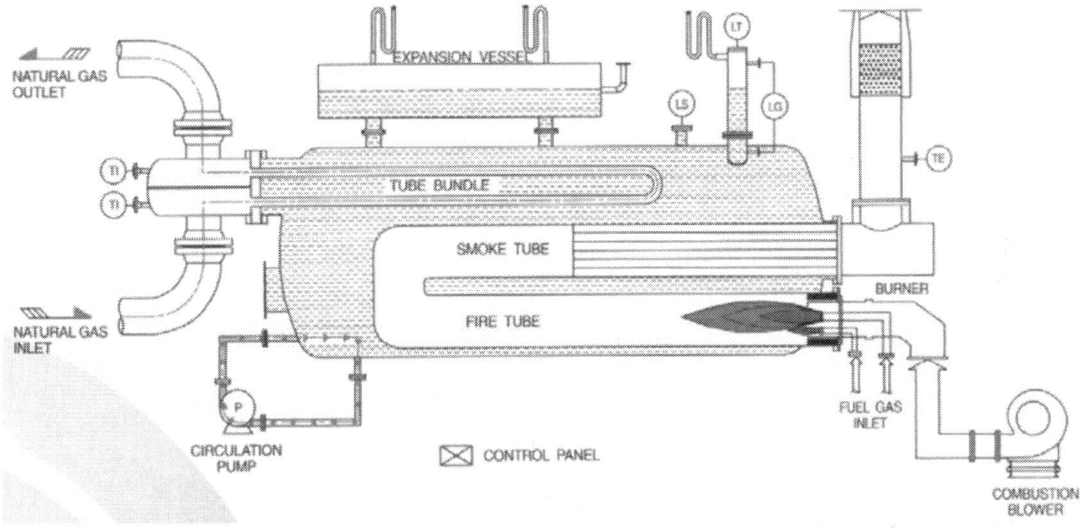

Fig. 4 – Schematic Diagram of Conventional Gas Heater

ADVANCED DESIGN OF GAS HEATERS

Prevention of NO_x formation during the combustion process is the most economical form of NO_x emission reduction. The basic approach is to lower the flame temperature and reduce excess air to prevent NO_x formation. The following technologies developed for reduction in NO_x and CO emissions are discussed.

Flue Gas Recirculation System

Low NO_x burners are the most common application of NO_x prevention in industry. Low NO_x burner technology stages the introduction of air or fuel in the burner to reduce the flame temperature. As the need for further NO_x reductions is imposed, technology for Flue Gas Recirculation (FGR) has been used to supplement low NO_x burners. The basic concept of this process recycles exhaust gas with the combustion air prior to being fed to the burner, resulting in reduction in flame temperature by lowering the excess air during combustion. The installation of FGR requires a motive force to recycle the flue gas. This has typically been accomplished by a forced draft blower. The control concepts for both designs are shown in Fig. 5.

A damper is used to control the amount of recirculation flow. The maximum NO_x removal is limited by the amount of recirculation, which is further limited by flame stability. The amount of excess oxygen or heater draft may also limit the amount of recirculation [4].

With 60% flue gas recirculation, the flame remains stable. A 20 – 30% flue gas recirculation rate does not begin to approach the burner's performance limits. FGR also reduces NO_x emission by lowering the oxygen concentration in the primary flame zone. Recycling of flue gas reduces combustion capacity slightly. This can be compensated by an increase in the blower power. However, there is no impact to PGH efficiency.

Figure 6 presents NO_x reduction with the flue gas recirculation rate. With an increase in recirculation rate, NO_x reduction rate increases logarithmically. About 40% of NO_x reduction was achieved with 20% of FGR rate. It is less expensive than Selective Catalytic Reduction system (SCR) on gas heaters.

Depending on the level of accuracy desired, control system needs to be designed to get much tighter control of air-fuel ratios. The percentage of available draft utilized for throat metering can leave little room for the design of an adequate control mechanism. Dampers and burner registers were sized for 40 to 100% of burner capacity. The benefit of this approach is its ability to achieve the required NO_x emissions levels with less impact to system operational costs and efficiency, as well as improved burner operational characteristics.

The higher the level of turbulence and shear between fuel and air streams, the more uniform the fuel and air mixture and the more rapid and complete the combustion reaction. It was observed that the higher the level of mixing, the more desired intense and complete the combustion. However, there are a number of conditions and factors that should be considered, including flame shaping and control of pollutant emissions [7].

Highly mixed fuel and combustion O_2, in the proper proportions will generate the maximum flame temperature. This will result in the formation of a large quantity of nitrogen oxides (NO_x), which is a highly regulated pollutant [7]. The conventional burner can be used for FGR. However, the advanced burner was developed to shorten the flame length and extending the reaction zone to achieve delaying of the combustion process. This newly developed burner brought low flame temperature and stable operation with the high flue gas recirculation.

In the control system, sensors such as pressure switches, valve position switches, flame scanners, and temperature switches do all the safety and sequence monitoring. The discrete control system performs safety monitoring and sequencing startup. If a monitored parameter is on when it should be off, or vice versa, the startup process is aborted and the system must be reset before another startup is permitted. The heater control system adopted the integrated control system of discrete and analog control. The analog control system usually has few tasks to perform until the system completes the startup sequence and is ready to maintain normal operation. Analog devices are part of a control loop. A simple loop consists of a sensor, a final element, and a controller. The controller reads the sensor, compares the measured value to its set point set by the operator, and then positions the final element to make the measured value equal the set point.

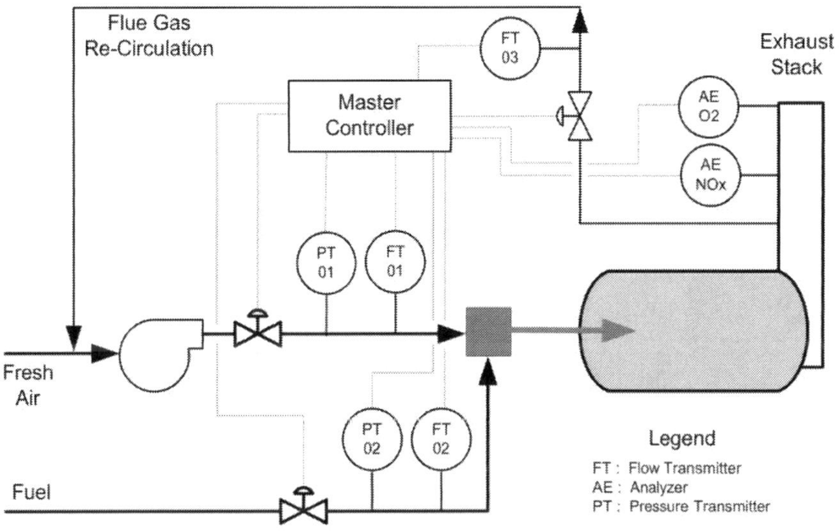

Fig. 5 – Flue Gas Re-Circulation Control

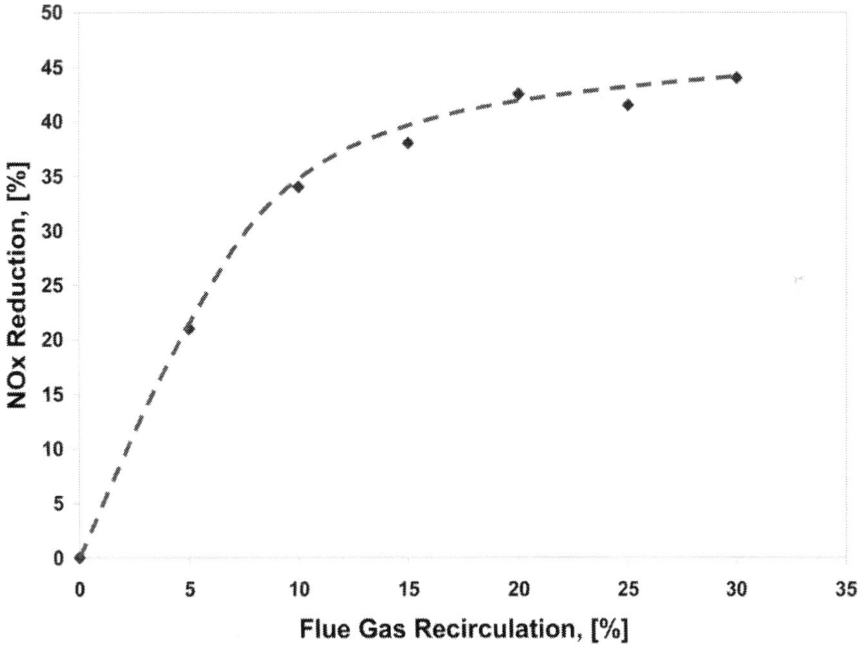

Fig. 6 – NO_x Reduction vs. Flue Gas Recirculation

Premix Burner

In order to achieve very low NO_x emissions from a natural gas flame fuel-rich regions should be avoided. This increases the potential for prompt NO_x generation and lowers the flame temperature to reduce thermal NO_x to the desired level. To accomplish this, a burner design that avoids fuel-rich regions by rapidly mixing gaseous fuel and air near the burner exit was studied.

In the Premix Burner, the fuel and the oxidizer are completely mixed before combustion begins. Premix is a term applied to burners that inspirate part or all of the total air required for combustion [4]. The operating envelope of a premix burner is defined primarily by the limits of flammability of the air-gas mixture and by the velocity of the mixture passing through the ports of the burner nozzle. This burner provides for intimate mixing of the fuel and combustion O_2 prior to the ignition zone.

The fuel gas discharging from the orifice is used to entrain air for combustion. The burner primarily utilizes entrainment as its fuel/air mixing mechanism. The fuel gas is metered through an orifice at the entrance to the mixer. The mixing section is designed to provide a uniform concentration of mixed oxidant and fuel at a uniform velocity at the head end of the section. By Premixing, fuel and oxidant are mixed near the point of combustion, enhancing safety and permitting the construction of a larger burner. It is also designed to avoid the potential for flashback into the mixer and further into the chamber. The desired result at the outlet mixing chamber is a uniform fuel oxidant mixture having a lean flammability limit.

The premix burner system consists of two key components, the burner head or nozzle, and the gas-air mixing device that feeds it. The mixer includes a mixer inlet with a fuel input and an oxidant input for respectively admitting fuel and oxidant into the mixer. The mixer also includes a mixer outlet for discharging the mixed fuel and oxidant. Combustion occurs at or near the mixer outlet.

Figure 7 shows the variation in NO_x with flame stoichiometry for two different types of flames. The upper solid curve characterizes the NO_x emissions expected with a premixed flame. The lower dotted curve is for the same flames with the addition of FGR. The figure shows that operating with more than one combustion zone can lead to very low NO_x emissions with reasonable overall excess air in the final combustion products. Firing the remaining fuel under fuel rich conditions with flue gas recirculation will produce low NO_x emissions in another combustion zone.

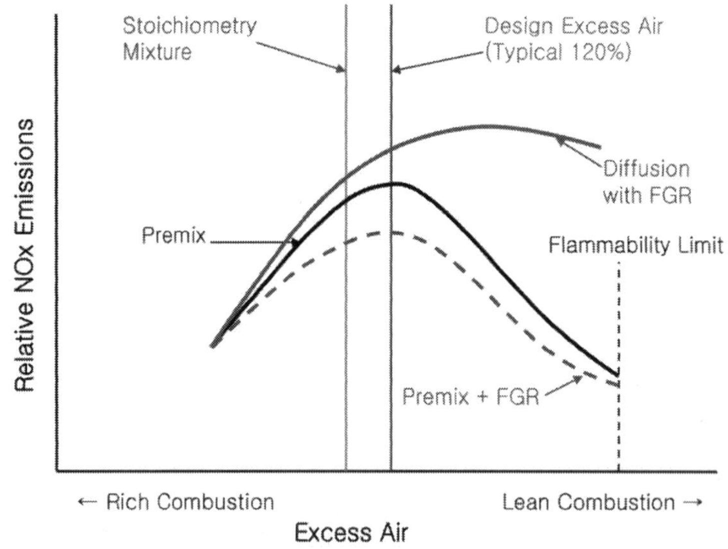

Fig. 7 – Comparison of NO_x Emissions for Premix and Premix-FGR

A fully metered control scheme for modulating fuel and air to a burner was developed with the electronically linked parallel positioning scheme. Flow meters are linear with flow rate. Meter output signal scaling provides the firing rate and air/fuel ratio required for the application. Addition of high and low signal selectors provides cross limiting of the fully metered control scheme. The low signal selector (<) compares demanded firing rate from the TIC to the actual combustion air flow rate and applies the lower of the two signals as the set point to the fuel flow controller. The high signal selector (>) compares demanded firing rate from the TIC to actual fuel flow rate and applies the higher of the two signals as the set point to the airflow controller. The oxygen signal is used to "trim," since oxygen analyzers are high maintenance and/or too slow in response to be used as a primary control in the combustion process [7]. A multiplication function (Y) in the combustion airflow transmitter signal makes the oxygen trim adjustment. Substituting a summing function for the multiplier would result in high trim gain at low flow rates and could produce a combustion air deficiency. Figure 8 illustrates the burner control scheme with O_2 trim of air flow rate.

Flue Gas Condensing

The conventional PGH has a lower thermal efficiency compared to a boiler because of the high exhaust temperature. The conventional PGH with 90% thermal efficiency (LHV basis) is designed for an exhaust temperature of 200 – 250°F. One of the reasons for designing it with a high exhaust temperature is to avoid water condensation in the flue gas. Acid gas contained in the flue gas dissolves in the condensate, resulting in acidic condensate and requiring special corrosion resistant material for the convection tubes. However, the vaporized LNG contains only a trace amount of sulfur, but CO_2 is present in substantial quantities after combustion.

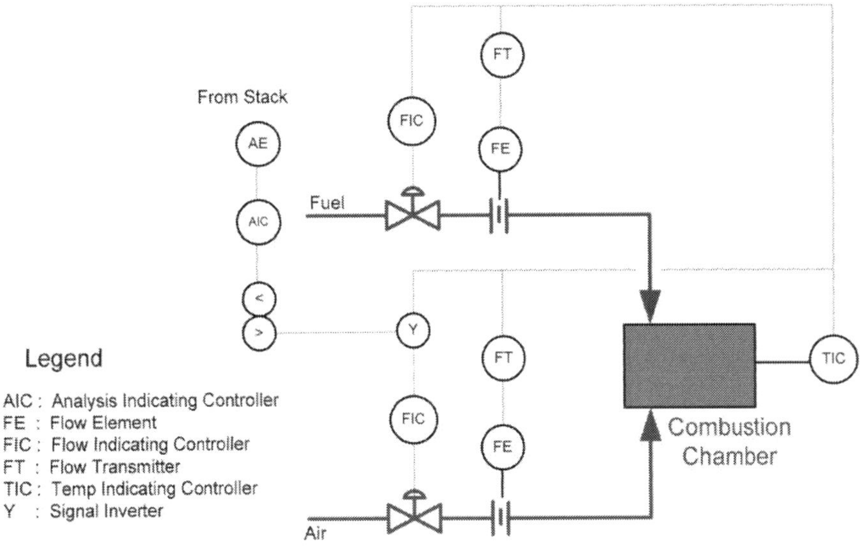

Fig 8 – Burner Control Scheme

The concept of the condensing heat exchanger (CHX) is based on removing latent heat from the flue gas. The flue gas enters the Teflon coated exchangers through a carbon steel inlet plenum at the top and flows downward across the horizontal banks of exchanger tubes. The flue gas exits the heat exchanger through a Fiberglass Reinforced Plastic (FRP) outlet plenum on the bottom of the exchanger as shown in Fig. 9. In a typical application water is heated in the Teflon CHX, flowing countercurrent to the flue gas [8].

As the flue gas temperature approaches the water vapor dew point, condensation occurs on the tubes. Droplets of condensate form and fall as constant rain over the tube bundle. This enhances the latent heat transfer and at the same time cleans the tube surface [8]. The condensate is collected and removed at the bottom of the heat exchanger.

CHX optimization studies indicate that the flue gas temperature can be reduced below the water bubble point, which improves the thermal efficiency to about 99%. Increasing the thermal efficiency of the condensing heat exchanger results in lower operating costs and improves the economics. The additional cost for the installation of the exchanger can be recovered by fuel cost savings. There will be some increase in the air blower power consumption due to the added heat transfer surface of the CHX. Depending on the fuel gas price, the payback period can be just a few years. (See Fig. 10).

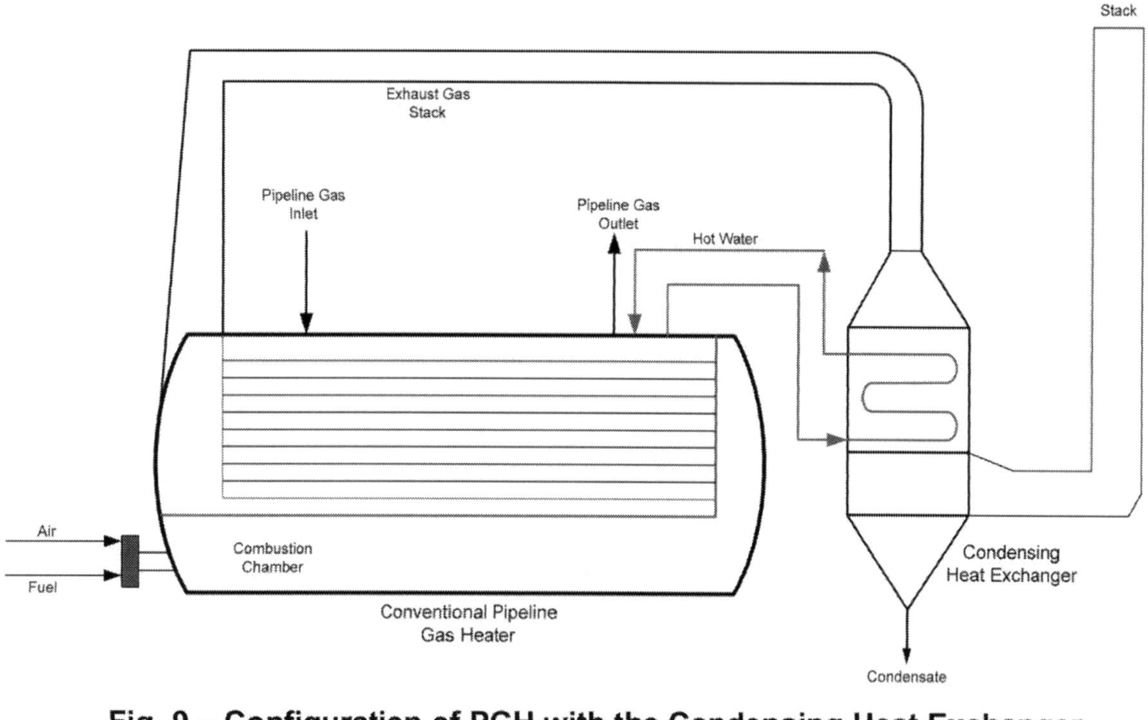

Fig. 9 – Configuration of PGH with the Condensing Heat Exchanger

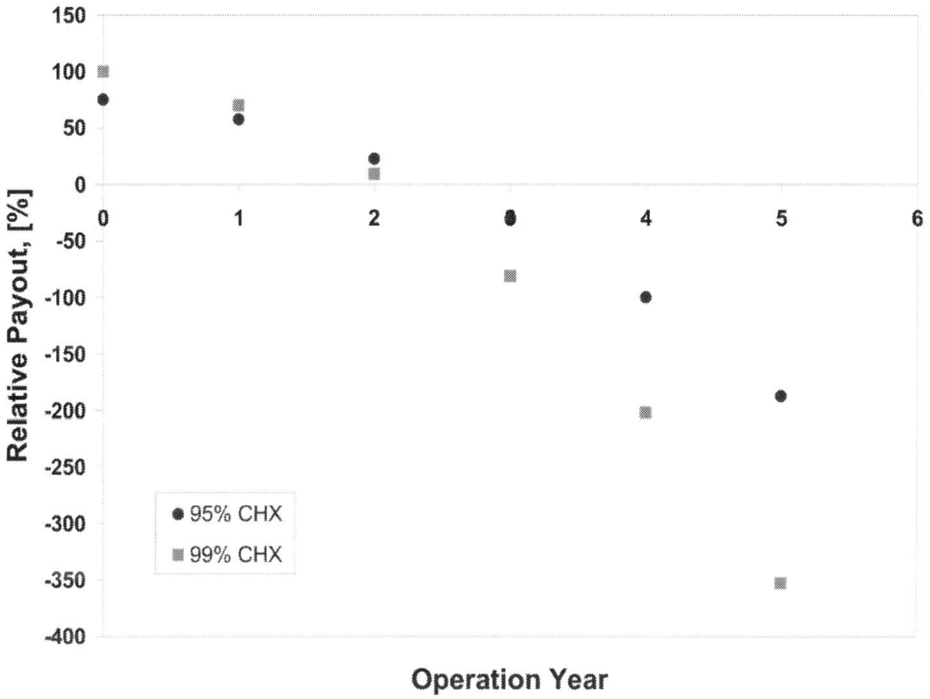

Fig. 10 – Pay-Back Period of the Condensing Heat Exchangers

CONCLUSION

The main emission sources of pipeline operation generally come from its gas compression and gas heating process. Pipeline projects require attention to minimize emissions and environmental impacts from its pipeline operation. Site-specific environmental constraints and permitting issues must be carefully reviewed in the selection of appropriate equipment for a pipeline project. Without proper attention to these site-specific issues a project can be seriously delayed or its realization may even be jeopardized.

The installation of Flue Gas Re-circulation (FGR) requires a motive force to recycle the flue gas. A 20 – 30% flue gas recirculation rate does not begin to approach the burner's performance limits. With an increase in recirculation rate, NO_x reduction rate increases logarithmically. About 40% of NO_x reduction was achieved with 20% of FGR rate. It is less expensive than Selective Catalytic Reduction system (SCR) on gas heaters.

Premix Burner technology reduces NO_x emission to below a single digit NO_x emission. Newly designed burner characteristics are shortened flame length and stable operation with lean fuel gas mixing. Premix combined with flue gas recirculation can further reduce NO_x emission by 15 to 30%.

With an increase in energy costs, the concern to minimize pipeline operation costs has dramatically increased. This trend results in integration of a newly developed low NO_x burner with the condensing heat exchanger (CHX). This increases in thermal efficiency by up to 99% based on HHV.

REFERENCES

1. "International Energy Outlook 2005", Document DOE/EIA-0484, Energy Information Administration, Washington, July 2005.
2. "Worldwide Look at Reserves and Production," *Oil & Gas Journal*, Vol. 102, No. 47, pp. 22-23, December 20, 2004.
3. Gower, S. and Howard, M: "Changing Economics of Gas Transmission," Proceeding, GPA Europe, 20th Annual Conference, Heidelberg, Germany, September 2003.
4. "NO_x Control", Plant Automation Services, Inc., http://www.pas.com/white_papers/NOxControl.pdf
5. Bradford, M., Grover, R., and Paul, P.: "Controlling NO_x Emissions", CEP, p. 42- 46, March 2002.
6. Webster, T.: "Burner Technology for Single Digit NO_x Emissions in Boiler Applications" presented at the CIBO NO_x Control XIV Conference, San Diego, CA, 13 March 2001.
7. Baukal, C.E.: "The John Zink Combustion Handbook", CRC Press, New York, p. 360-398, 2001.
8. Web page, http://www.chxheat.com/perform.html.

BIOGRAPHY of SPEAKER

Mr. Jung Been, Lee is President & CEO of Wonil T&I Co. Ltd., which designs and produces specialized equipment such as Gas Heater, Gas Filter, Liquid Hydrocarbon Vaporizer and Cryogenic pressure vessel for the gas, electricity, and LNG industry. The majority of his career has involved equipment design and management of Wonil for sixteen years. He graduated from the University of Kyungwon in Korea with a Ph.D degree in Business Administration.

E-Mail: jblee@woniltni.co.kr

The Effects of Uncertainty on Quantified Risk Assessment for LNG Facilities

R.P. Cleaver, D. Hobbs and A.R. Halford

Advantica
Houston, TX 77056
doug.hobbs@advantica.biz

ABSTRACT

Following the renewed interest in the importation of LNG into USA and Europe, there has been much discussion within the technical literature and media on the hazards associated with LNG importation. There have been a number of studies drawing attention to uncertainties in the current consequence modeling techniques and it has been shown that this can lead to variations in the predicted hazard distances associated with a range of release scenarios. The purpose of this paper is to take this analysis one step further by examining the impact that such uncertainties in predicting the consequences of releases have on the overall quantification of the risk arising from LNG import facilities.

1. INTRODUCTION

Following the renewed interest in the importation of LNG into USA and Europe, there has been much discussion within the technical literature and media on the hazards associated with LNG importation. There have been a number of studies drawing attention to uncertainties in the current modelling techniques [1] and it has been shown that this can lead to variations in the predicted hazard distances associated with a range of release scenarios. The purpose of this paper is to take this analysis one step further by examining the impact that such uncertainties in predicting the consequences of releases have on the overall quantification of the risk arising from LNG import facilities.

In order to put the discussion in context, a hypothetical LNG import terminal layout is considered, as discussed in Section 2. A number of the major accident scenarios that arise during typical studies of such facilities are listed in Section 3. Some of the more important input parameters for these scenarios are listed also in Section 3, where the source of data and range of values associated with them is discussed. A series of calculations are then reported in Section 4, using credible values for these parameters, to demonstrate the sensitivity of the results to variations in their values. This includes physical input parameters within individual calculation models, such as the heat transfer rate from the underlying surface into a spreading LNG pool and the minimum pool depth used, as well as parameters whose values are uncertain at the time of an accident, such as the prevailing wind speed and direction. The effect that these different values or assumptions on the hazard range and risk are discussed in Section 5, where the conclusions of this particular study are also given.

2. HYPOTHETICAL TERMINAL LAYOUT

The hypothetical layout for an LNG import terminal is illustrated in Figure 1. This shows an LNG jetty situated in a wide estuary to the far left of the figure. The LNG is transferred from LNG ships moored at the jetty and is pumped via an all welded pipeline into double containment concrete tanks situated on the left hand side of the site. It is assumed that LNG is exported from the storage tanks, via the ex-tank pump area to the vaporiser area in the centre of the site. Here, the LNG is converted back to natural gas ready for export through a metering area on the right of the site to the natural gas transmission pipeline system. Adjacent to the vaporiser area it is assumed that facilities exist for handling any boil-off gas from the LNG tanks, including a boil-off gas compressor enclosure.

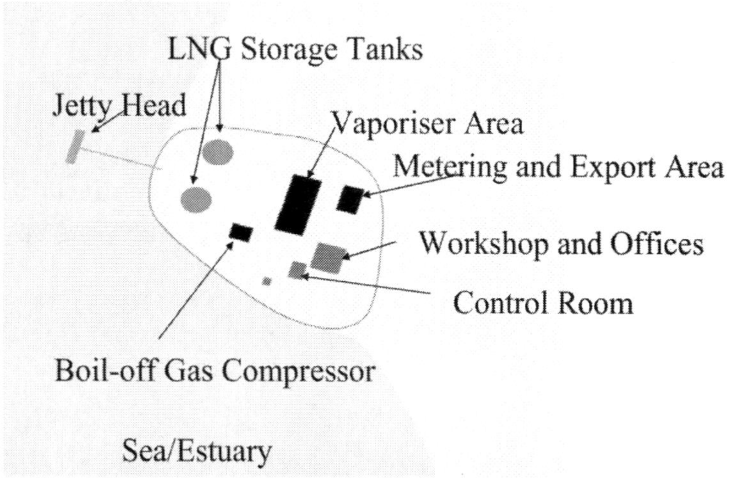

Figure 1: Hypothetical layout for LNG Import Terminal

It is assumed that no-one lives immediately adjacent to the site, but that some distance away there are a number of isolated houses, with several smaller villages and one major centre of population situated further away. This is shown schematically in Figure 2, with the main centres of population cross-hatched in grey. The remaining isolated houses are not shown on this figure.

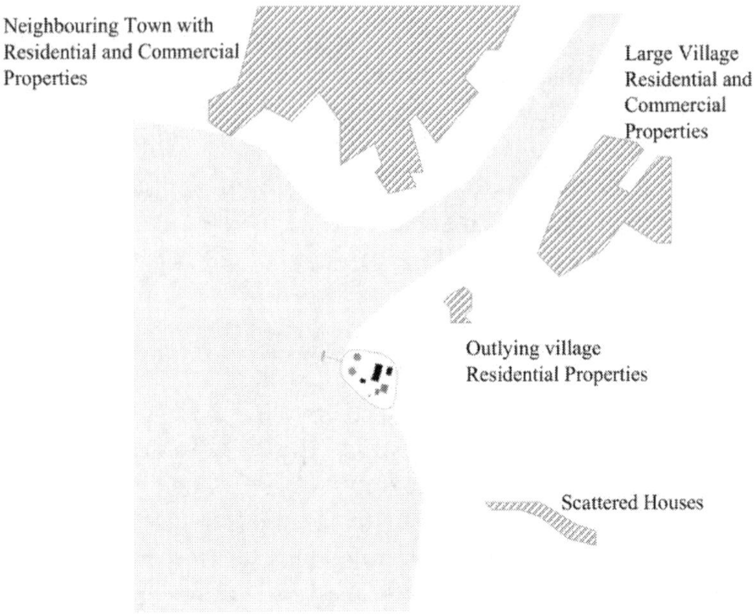

Figure 2: Hypothetical layout of population in neighbourhood of site

This situation has a number of features that are common to proposed LNG facilities in many different countries in the world. The site selection criteria have ensured that events that might be expected to occur in the lifetime of a single LNG facility have no off-site impact. However, there are a number of less frequent events that have the potential to interact with a small number of the nearby population and, at very low frequency, there are a number of high consequence scenarios that might adversely affect a greater number of people. The purpose of this example is to illustrate how the profile of risk is affected (or not) by key modeling assumptions and hence to focus attention on those factors that govern the risk.

3. RELEASE SCENARIOS

There are many different approaches that can be taken towards identifying events that might occur that could lead to a release of LNG or natural gas from the site. For the purposes of illustration here, a number of release scenarios that have been considered in previous quantified risk assessments have been selected. The scenarios have been grouped by their location on the hypothetical site, as follows.

LNG Jetty Region
Major release from ship
Failure at jetty during transfer operations, with or without successful PERC operation

Product Transfer - Import
Rupture of the LNG import line to the tanks

Storage Area
Major release from storage tank

Product Transfer – Export
Rupture of the LNG export line to the vaporisers

Gas Handling Area
Small or medium release in vaporiser region
Small or medium release in compressor area for boil-off gas handling

Gas Export
Small or medium release in gas metering area
Rupture of natural gas export line

In analysing these cases, there are a number of key parameters whose values have to be assigned. Considering the atmospheric parameters, there is a need to specify appropriate wind speeds and direction and the atmospheric stability category. Typically these are derived from a site wind rose and other meteorological information, with a number of representative wind speed and directions being used in a quantified risk assessment. Similarly the relative humidity and temperature can be determined from site records. The water content in the atmosphere is known to have an effect on the cloud density for dispersing LNG vapor clouds and also on the atmospheric transmissivity for thermal radiation for jet or pool fires. The surface roughness that is used will have an influence on the dispersion of dense gas clouds. There may be a need to consider an appropriate value for estuaries or over flat coastal plains, as a low background roughness may allow a flammable vapor cloud to travel for a longer distance than in rougher terrain.

The number of people that are present on or off-site is important when considering the individual and societal risks. These can be determined from site surveys or census data. However, there is a need to consider the different amounts of time people spend indoors or outside, as the vulnerability of people to fires or explosions may be sensitive to this. Similarly, the occupancy of nearby commercial or industrial properties may vary with the time of day, reflecting local working patters and social factors. It is important to include places of public assembly that may be occupied by large numbers of people for a short time each week, as taking average occupancies may not give a true picture of the risk.

There will also be a large number of engineering inputs for each potential release. It may be important to consider the variation in pressure or inventory. The gas composition will be determined by the project, although normally the LNG has a high methane content. However, it is important to consider this, especially as it may change the lower flammable limit concentration and hence dispersion distances.

The parameters that are required within the different mathematical models will vary depending on the model type. In modeling LNG pool spread and dispersion, it has been suggested in [1] that it is important to consider the minimum pool depth in a spreading liquid pool as this has an influence on the source size and subsequent dense gas dispersion. For spills on land, this may need to be related to site features (for example, half the height of any retaining curbs or bunding around process areas), depending on the size of the spill. For spills on the sea, questions have been raised over whether there is a need to consider the impact of waves, in either limiting the spread of the pool or enhancing the surface area for heat transfer. The precise value to be used for the heat transfer coefficient for spills on sea is also the subject of considerable uncertainty. If the spreading pool is ignited, there are then questions as to the effective mass burning rate and effective emissivity of large pool fires.

There is also a need to consider the frequency with which any of the scenarios may arise. The frequency has a direct, linear impact on risk values. Typically, the techniques that could be employed to determine the release frequency include, but are not limited to, fault tree analysis, the use of industry databases for the range of releases, generic historical data and industry specific databases. Parts counts may be used as a way of accounting for the many different sources of small or medium sized leaks.

Once these input values are determined there is then a need to consider how a scenario may evolve. This brings in factors such as the ignition probability given a release has occurred. This may be determined generically, from past incidents for example, or by modeling based on ignition source densities for example. Past experience has shown that such factors may have an unexpected influence on the risk, with a tendency for early or late ignition to have different effects in different locations, depending on the location of people relative to the event. Related to this, it is also important to consider the influence of isolation of various plant items if the release is detected. Isolation times will vary depending on whether it is automatic or manually activated and the reliability or availability of such systems should also be considered.

In the worked examples presented in Section 4, the physical consequences (gas dispersion, thermal radiation levels, explosion overpressure levels) of the accident scenarios have been analyzed using tools and methods developed over many years by Advantica (formerly British Gas Research and Development). The models have been calibrated or validated by comparison with experimental data, often obtained at full-scale (see [2] to [6], for example).

During a quantified risk assessment of a 'real' site, it would be normal to carry out some more detailed assessment to derive appropriate failure frequencies. However, for the purposes of illustration here representative failure frequencies have been assigned to the scenarios, based on engineering judgement but also taking into account a number of factors, such as a postulated frequency of ship visits and estimated length of pipelines, for example.

The frequency of each event and the consequences it has are combined to determine the risk at an array of locations where people might be present. The calculations are performed for a range of wind speeds (typically 5), wind directions (typically 12) and release directions (typically 12) and vessel inventories, as appropriate. The analysis presented here does take into account the varying population distribution throughout the day at these locations and also whether people are situated outdoors or indoors. In the latter case, the protection afforded by different types of building to fire and explosion is considered. In this way, the individual and societal risk to both the on-site and the off-site populations are determined.

4. RESULTS

It is usual in a quantified risk assessment to consider location specific risk (i.e. the risk to a hypothetical person who is at a certain location for 100% of the time), the individual risk to representatives of particular groups of people (e.g. a process worker who spends so many hours in different areas on the site) and also the societal risks. Often, the societal risk is reported by means of an F-N curve, expressing the cumulative frequency, F, with which N or more fatalities are produced by the release. Different companies and regulatory bodies have proposed acceptability criteria based on individual risk and societal risk. As an example, the UK Health and Safety Executive has proposed different criteria for individual risk to workers and to members of the public [7]. Quite often the societal risk criteria will have some form of aversion factor to weigh against producing a large number of casualties. This is usually in the form of proposing an acceptability curve on an F-N plot with logarithmic scales that has a slope of –1.5 or –2, rather than a linear weighting with a slope of –1. A single measure of societal risk that is sometimes used is to derive the potential loss of life per year (annual PLL) associated with each scenario. This is obtained as the product of the event frequency and the average number of casualties that the event produces given that it takes place. In order to illustrate the sensitivity of the results to these changes, the effect on both the hazard distances and also the different risk measures outlined above have been considered.

The sensitivity of the predicted risks to the release duration has been examined for the failures of the LNG import pipework. In the analysis, it was assumed that the release continued at the transfer rate for 60s, then decayed linearly to zero over a further 30 s supplemented thereafter by drainage from the open ends of the pipes. A sensitivity study has been carried out for steady releases of 30 seconds and 90 seconds before a linear decay. This has the effect of increasing the size of the LNG pool created following a rupture of the import pipework, and hence the size of the cloud that is formed, if the release is not ignited immediately.

The larger releases create a larger hazard range, as shown in Table 1. The hazard ranges increase typically by some 25% to 35% on going from the 30 second shut-down to the 90 second shut-down.

Table 1: Typical hazard distances associated with the LNG Import line rupture, assuming different isolation and shut-down times

Assumed Shut-down Time (in seconds)	Typical Hazard Distances (in m)			
	Thermal Radiation from Pool Fire		Dispersion Distances from Unignited Spill	
	12 kW/m2	Piloted Ignition of Buildings	To the Lower Flammable Limit	To 0.5 of the Lower Flammable Limit
30	285	205	411	663
60	323	229	501	801
90	351	250	564	901

The predicted annual loss of life as a result of the incident increases from 1.07 x 10^{-5} per year for the 30 seconds shut-down time to 1.36 x 10^{-5} per year for the 60 seconds shut-down time and 1.57 x 10^{-5} per year for a 90 second shut-down time.

In modelling failures at the jetty head, the possibility that a PERC system is fitted to the unloading arms is usually taken into account within a risk assessment. Such a system operates on detection of unacceptable amounts of arm movement, for example, and would therefore prevent any major pipe work failure or release of LNG occurring. However, even if the unloading pipework itself were to be damaged, automatic operation of the PERCs and associated safety shut-down systems on the ship would effectively isolate a release from an arm within a time period of the order of 5 seconds. As with any system, there is a finite probability that it will either fail to operate or the release will take place in such a way that it is ineffective. Hence, in the assessment, a longer duration release of 90 seconds is also considered representing the case when the release is terminated from on the ship or in the jetty control room by manual intervention. This case would normally be assigned a much lower frequency. The different typical hazard ranges predicted for the case of a 5 second spill and a 90 second spill are given in Table 2.

Table 2: Typical hazard distances associated with releases at the jetty

Case	Typical Hazard Distances (in m)			
	Thermal Radiation from a Pool Fire		Dispersion Distances from an Unignited Spill	
	12 kW/m2	Piloted Ignition of Buildings	To the Lower Flammable Limit	To 0.5 of the Lower Flammable Limit
5 second spill	90	63	130	195
90m second spill	226	156	345	511

The F-N curves expressing the societal impact associated with these different cases are shown in Figure 3. Also, shown in this figure is an F-N curve for a release at the full-bore transfer rate from the ship, such as might arise from the piping manifold of the transfer lines at the jetty. In order to plot the F-N curves, the estimated frequency of a major release at the jetty head during transfer has been apportioned in the ratio of 90:9:1 for the 5 second spill, the 90 second spill and the piping manifold release, respectively.

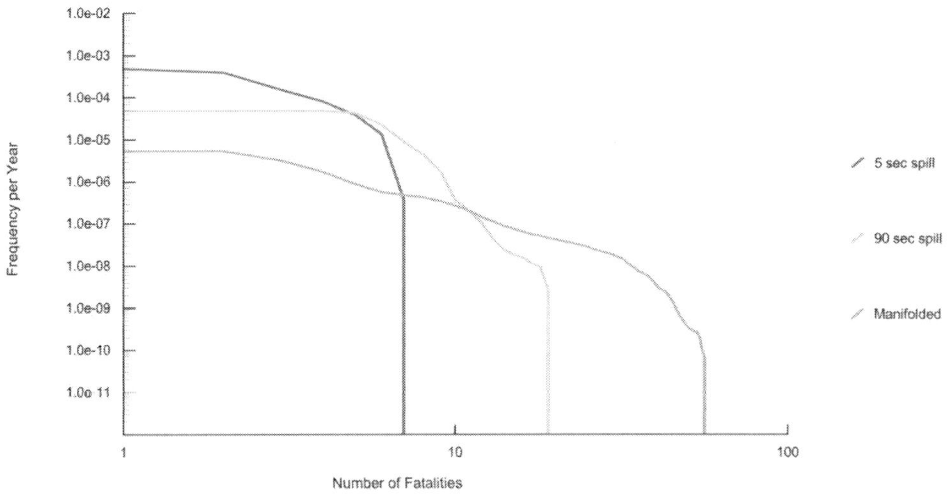

Figure 3: F-N curves associated with different cases of a release at the jetty

This is a typical result within a risk assessment, with a low frequency event having the potential to affect a larger number of people, but the much more likely release (with PERCs operating to halt the spill in 5 seconds in this case) affecting a smaller number of people in the immediate vicinity of the release.

It is interesting to note however that there is not always a similar level of benefit from isolation and shut-down. For example, considering the natural gas export line from such this hypothetical site, it was found that, in this particular example, approximately 75% of the fatalities arise from the first 30 seconds worth of a release from the export line. This result has been found in other studies of natural gas transmission pipelines, as the release rate decays rapidly as the pipeline depressurises and the first highly transient phase determines the maximum hazard range and risk.

The scenario that has attracted a large amount of interest within the technical press and the media is a release from an LNG ship [1]. The ship will normally hold the largest inventory of flammable material associated with the operations and hence has the potential to give the largest hazard ranges. There is much debate over the size of hole that it is reasonable to model and also the range of parameters that it is reasonable to use in modelling the subsequent behaviour. To this end, calculations have been performed in which 50% of the contents of a single storage tank (assumed to hold 25,000m^3 when full) are lost through a 1m^2 hole or through a 6m^2 hole. For comparison, similar calculations have been performed assuming that escalation occurs and 50% of the ship contents (assumed to be 125,000m^3 when full) are lost through a 6m^2 hole.

Table 3: Typical hazard distances associated with a major ship release

Case	Typical Hazard Distances (in m)			
	Thermal Radiation from Pool Fire		Dispersion Distances from Unignited Spill	
	12 kW/m2	Piloted Ignition of Buildings	To the Lower Flammable Limit	To 0.5 of the Lower Flammable Limit
50% of Single tank – $1m^2$ hole	676	561	1320	2230
50% of Single tank – 6 m^2 hole	1160	884	3150	4400
Contents of ship (50% full) – $6m^2$ hole	1310	1060	4120	6270

Figure 4 shows the F-N curves calculated for the three different cases. For the purposes of demonstration it has been assumed that the overall frequency for this event, obtained using the approach outlined in [8] for a representative set of input parameters, has been split in the ratio of 90:9:1 between the cases of a $1m^2$ hole in single tank, a $6m^2$ hole in a single tank and loss of a whole ship when 50% full through a $6m^2$ hole.

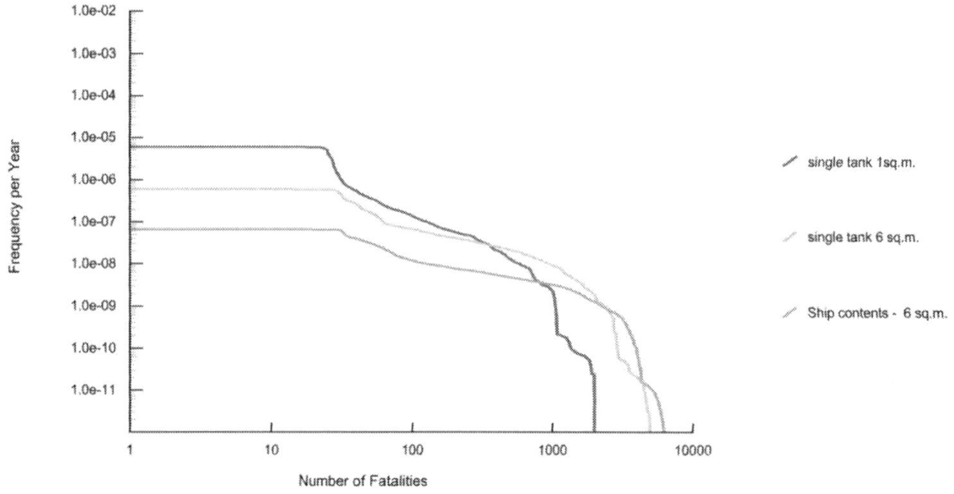

Figure 4: F-N curve for the three different cases considered for the ship spills

This figure suggests that although the hazard distances may be much worse for the larger spill, proportionately the contribution to the potential loss of life per year may be less for the larger release sizes. That is, the increase in consequence is more than outweighed by the decrease in frequency under the above assumptions. However, as noted earlier, many companies include an aversion to a large number of casualties within their societal risk criteria. This means that the acceptability criterion in the F-N curve has a slope of less then –1 (e.g. –2 for weighting with N^2). Hence, this simple comparison based on potential loss of life, which essentially gives equal weighting to component events with the same PLL, may not fully reflect such considerations.

To investigate this case further, calculations have been performed in which various modelling parameters have been varied. In the first set of results, the value of the heat flux from the underlying sea surface to a spreading LNG pool has been varied. The effect this has on the pool radius, and hence the boil-off rate, is quite marked, as shown in Figure 5.

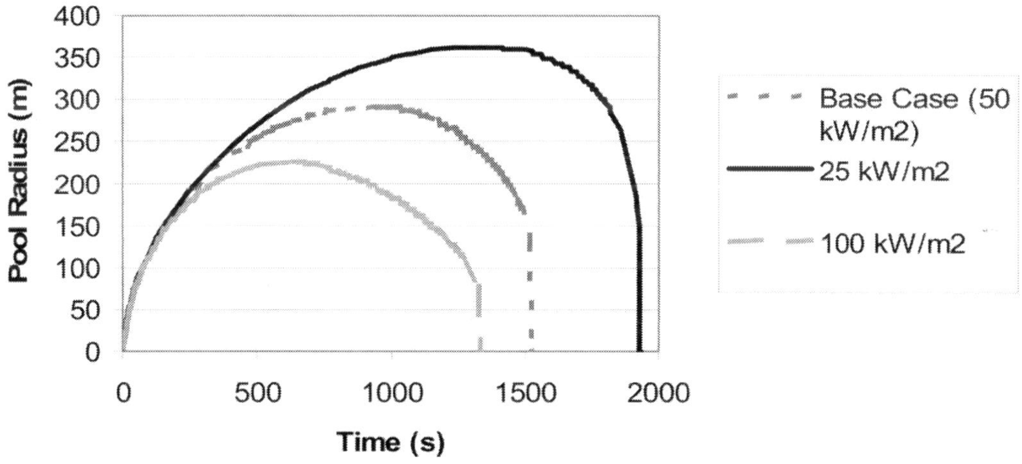

Figure 5: Pool radius for unignited, spreading LNG ship spills with different heat fluxes from the underlying sea surface

As noted elsewhere [1], the consequence calculations are sensitive to this change. However, when the effect that this has on the societal risk is considered, for this particular example, there is a less marked effect shown in Figure 6.

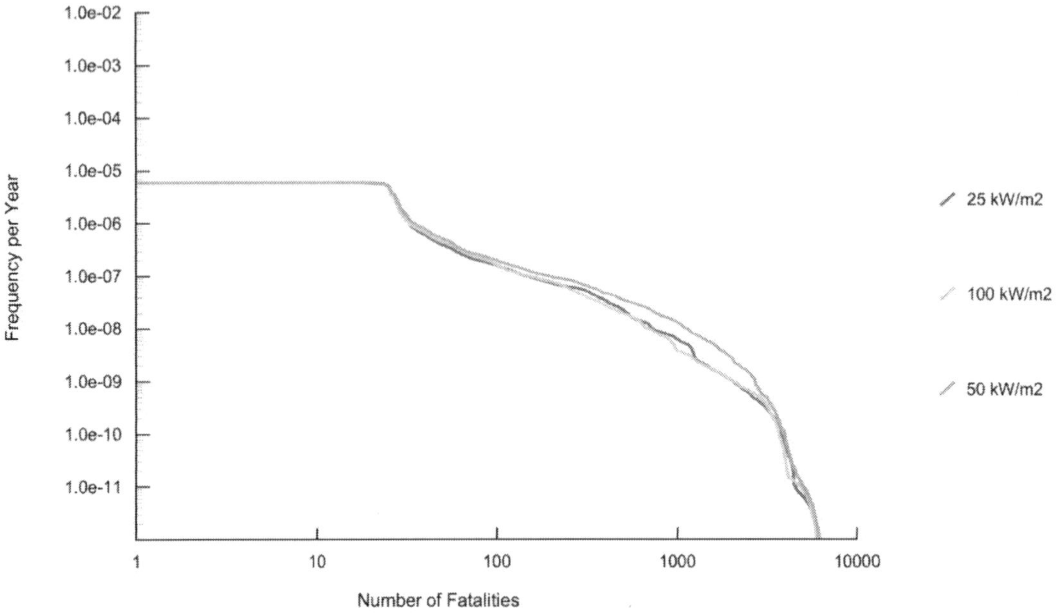

Figure 6: F-N curves for modelling LNG ship tank spill with different heat fluxes from the underlying sea surface

This relative lack of sensitivity arises through the inter-relation between the competing effects. For example, a higher heat transfer rate leads to a smaller spreading pool but a larger dispersing cloud. If the cloud is ignited at a fixed position downwind, the area covered by the cloud at this time, and hence the flash fire hazard, may be greater. However, on burning back to the LNG pool, the pool itself, and hence the pool fire hazard, will be smaller. Which of these competing effects is dominant will depend on the location of the people in relation to the release, whether they are indoors or outdoors and the availability of shelter in the neighbourhood. It does not automatically follow that increasing the 'worst case' hazard range increases the risk. Further, the people at most risk on the ship and the jetty are affected whatever the nature of the heat transfer rate. The contours of location specific risk show similar behavior, with little difference being found for the higher risk regions. However, there is an increasing degree of sensitivity at the lower risk values. Hence, provided there are not a disproportionately large number of people in this sensitive region, the net effect in an F-N curve is less than might have been expected from the large change in maximum hazard distance.

Similar results are obtained when the minimum depth to which the LNG pool is assumed to spread is varied. Again, the pool radius and hence boil-off rate are quite sensitive to this value. In terms of risk, in this particular example, the societal and individual risks are not unduly sensitive to the different values used. For example, typically there is a 25% change in location specific risk averaged over the different areas. Examination of the calculations shows that if the pool has spread to a greater extent, it is also thinner and hence is consumed in a fire in a shorter time than a thicker pool. Depending on the location, a bigger fire for less time may be more or less dangerous than a smaller fire burning for a longer time.

In a similar vein, the atmospheric conditions (wind speed and atmospheric stability) can have a strong influence on the maximum hazard distance, such as the dispersion distance for a drifting dense gas cloud, for example. In this particular case for example, the dispersion distances to the lower flammable limit in the different weather conditions span a range of a factor of approximately two for the spill from a single tank through the 1 m^2 hole. The implications of representing the range of weather conditions by different number of cases with different associated probabilities are shown in Figure 7. In this figure, the F-N curve is presented for a simulation mapping the frequency distribution from a wind rose into just two cases (2 m/s stable atmosphere and 5 ms/neutral stability atmosphere) or into 18 different cases to more fully represent the wind rose (wind speeds from 1.5 m/s to 10 m/s with different stabilities assumed for the different wind speeds).

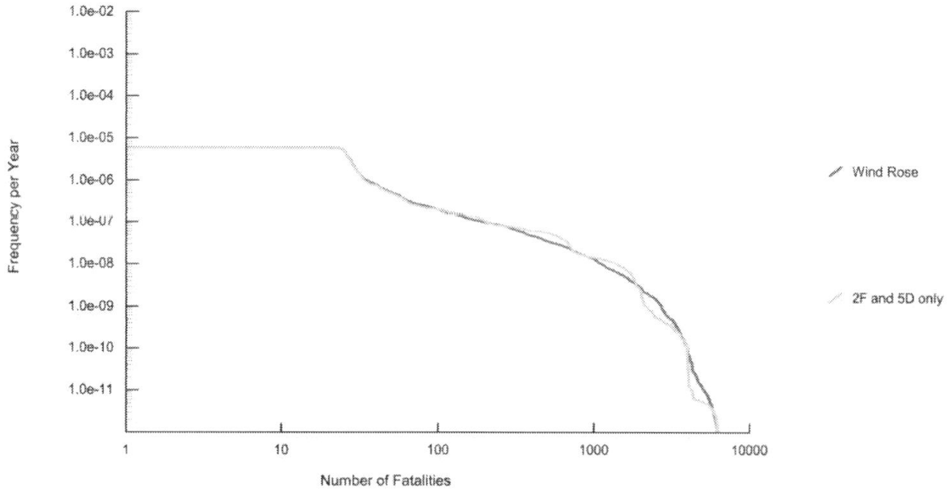

Figure 7: F-N curves illustrating effect of using 2 or 18 different weather conditions in the calculations.

This suggests that the use of more categories to represent a wind rose has the effect of 'smoothing' out the irregularities present in the coarser representation, but in this case, without altering its general properties. This may be important for consistency, when presenting the overall results.

4. DISCUSSION AND CONCLUSIONS

The results of this study show that there are some parameters that strongly influence the risks that are calculated within a quantified risk assessment. Conversely, there are others that may have a large impact on a particular component of the assessment without altering the overall level of risk. These results are based on one particular hypothetical site layout and surrounding population. Hence, it would be unwise to conclude that the results are true for every case. Nevertheless, similar trends have been obtained in practical studies for a number of different sites. That is, there are certain factors can have a large influence on a 'worst case' hazard distance without having a correspondingly large change in the societal level of risk. This may be because certain parameter changes have 'self cancelling' effects on the risk – for example, a shorter fatter cloud against a longer thinner one, or a large fire for a shorter duration against a smaller fire with a longer duration. Alternatively, it may be that those people who could be affected by a certain scenario are affected almost immediately and hence the risks are insensitive to parameter changes affecting the duration of that release. The risk assessment process offers a framework within which the importance of the uncertainties, whether in the modeling or in input values, can be put into context.

This is important from a practical point of view, as it has been shown that there are cases in which remedial action to isolate and shut down ruptured pipe work is of major importance, such as the use of PERCs at the jetty head. However, there are other cases, such as the gas export line, for which the benefit is less, unless a very short isolation time is used. This type of result is of use within a project in ensuring that expenditure to reduce risk is as cost effective as possible.

The results are also important in a more academic sense, as well. The influence of a parameter on the risk should be considered as well as the overall level of uncertainty in the parameter. For example, some parameter might only be known to within a factor of 2 but it might only have a small influence on risk, whereas another parameter may be known to 50% but have a big impact on risk. It is in this context that it is noted that the failure frequency of a scenario has a linear impact on the risk, whereas many of the modeling parameters produce less than linear changes. The examples presented here point to the importance of the failure frequency and also in determining the size of the release. The reliability and availability of safety systems, the response of the pipe work and vessels to the initiating events and how the subsequent outflow occurs are factors that can produce large changes in risk. Hence, it is concluded that these should be considered in as great a depth as the more conventional studies addressing modeling uncertainties and assumptions.

5. REFERENCES

1. Sandia National Laboratories, *Guidance on Risk Analysis and Safety Implications of a Large Liquefied Natural Gas (LNG) Spill Over Water*, Sandia Report SAND2004-6258, December 2004.
2. Linden, P., Daish, N., Dalziel, S., Halford, A., Jackson, M., Hirst, I., Perroux, J., Wiersma, S., *LSMS : A New Model For Spills Of LNG And Other Hazardous Liquids*, Proceedings of 1998 International Gas Research Conference, San Diego, November 8th - 11th.
3. Cleaver, R.P. and Edwards, P.D., *Comparison of an integral model for predicting the dispersion of a turbulent jet in a cross flow with experimental data*, J. Loss Prev. Process Ind., 3, 91-96, 1990.
4. Cleaver, R.P., Cooper, M.G. and Halford, A.R., *Further Development of a Model for Dense Gas Dispersion over Real Terrain*, J. Haz. Materials, 40, 85-108, 1995.
5. Pritchard, M.J. and Binding, T.M., *FIRE2: A New Approach for Predicting Thermal Radiation Levels from Hydrocarbon Pool Fires*. IchemE Symposium Series No 130, 1992.
6. Cleaver, R.P. and Shale, G.A., *Mathematical models for gas explosions on industrial sites*, International conference and workshop on modeling the consequences of accidental releases of hazardous materials, CCPS, San Francisco, 1999.
7. UK Health and Saftey Executive, *Reducing Risks, Protecting People*, HSE Books, HMSO Publication, ISBN 0 7176 2151 0, 2001.
8. Glynn, K., *Risk Management in LNG Import Terminals and Regasification Plants*, 22nd World Gas Conference, Tokyo, 2003.

Lessons Learned from the Application of Risk Management in the Shipment of LNG

B.R. Poblete (HSE Manager & Sr. Risk Management Specialist)
Lloyd's Register North America, Inc.
Houston, Texas
ben.poblete@lr.org

A.R. Bingham (Global LNG Business Manager)
Lloyd's Register North America, Inc.
Houston, Texas
anthony.bingham@lr.org

ABSTRACT

The objective of this paper is to first present an overview of the risks (health, safety, environment and security) during the shipment of LNG. This will involve a description of how LNG is transported internationally and the regulatory issues involved in the mitigation, reduction or elimination of the identified risks. The assessment of the impact of the risks associated with LNG transport will be discussed as well as how companies are protecting their assets from these risks. The procedures and practice that companies must attempt to undertake to protect them and be accountable will also be included during the assessment. Finally an understanding of the reality of the risks associated with LNG and the threat of a terrorist act against a LNG ship. This will include some of the myths associated with a deliberate attach on a ship and some of the more realistic scenarios.

1.0 Introduction

In order to satisfy the growing demand for natural gas as a primary source of energy, the world is beginning to rely on the shipment of Liquefied Natural Gas (LNG) over long distances to make up the shortfall in domestic gas production and increase in demand. According to the Cedigaz first estimates for 2004, LNG now accounts for 26.4% of all internationally traded gas. Although there are many advantages to shipping LNG, and the LNG shipping industry's safety record far exceeds any other shipping industry sector [1], the industry is continually challenged by those who cite the risks and doubt the credibility of experts' opinions on safety.

Do the advantages outweigh the risks? How can the industry help allay the public's concerns over the safety of LNG shipping? What are the lessons learned from previous experience in the LNG industry? In this paper, the authors will first describe how LNG is transported internationally and the regulatory issues involved in the mitigation, reduction or elimination of the identified risks. The assessment of the impact of the risks associated with LNG transport will be discussed as well as how companies are protecting their assets from these risks. The procedures and practice that companies must attempt to undertake to protect them and be accountable will also be included during the assessment. Finally an understanding of the reality of the risks associated with LNG and the threat of a terrorist act against a LNG ship. This will include some of the myths associated with a deliberate attach on a ship and some of the more realistic scenarios.

2.0 Mode of LNG Shipment[13]

When considering the transportation of liquefied natural gas by sea, the ship as a whole, can be considered as the containment system with the outer hull providing the very necessary cocoon for the safe transportation of the LNG. However for the purposes of design it is important to differentiate the system which immediately contains the cargo and provides protection to the ship hull from the extremely cold temperatures of the cryogenic cargo.

There are five basic containment system categories recognised by the Society's Rules and the IMO IGC Code and indeed there are many patented containment system designs available that meet the requirements for a specific category. However, in practice, the number of containment system designs actually 'in-service' is limited. Those containment system types currently in service with the world gas carrier fleet inc1ude:

- Membrane Containment System (primarily for LNG);
- Independent Tank Type A (for LPG);
- Independent Tank Type B (principally for LNG);
- Independent Tank Type C (for LPG);

2.1 Membrane Containment Systems:

The characteristic of the membrane type LNG tanker is the long trunk deck containing the high chamfer of the upper part of the tank. This arrangement does not restrict the view from the bridge to the same extent as the Moss tank covers and also provides a large flat deck with consequent ease of maintenance. There are three designs of such systems at a present in service in the world fleet and they are principally used for the carriage of LNG. The systems are:

- Gaz Transport,
- Technigaz MK I and MKIII, and
- CS1 system.

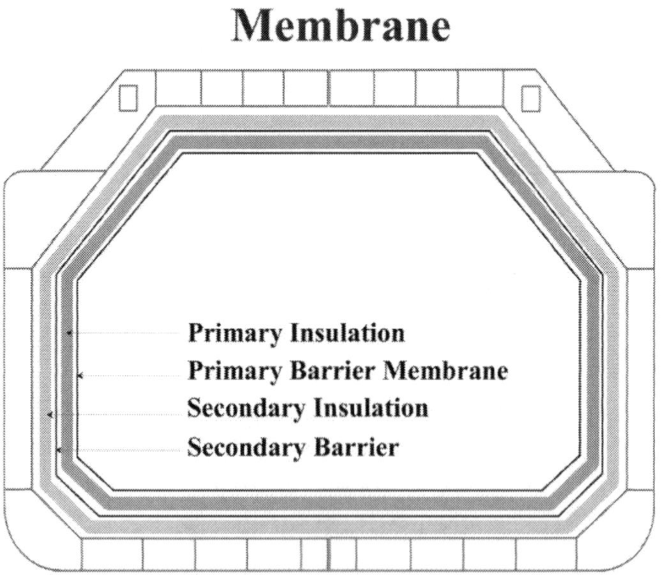

Figure 1: Example of Membrane Containment System

2.2 Independent Tank Type B:

The independent tank is self-supporting and does not form part of the ship's structure or contribute to the hull strength. Type B tanks are classified in two categories,

- Prismatic, and
- Spherical.

2.2.1 Spherical Tank

The principal design of type B system, in the world fleet, is the spherical design by Moss Rosenberg with the tanks being used for the transportation of LNG.

The Moss Rosenberg design has now been in existence since 1970 and has undergone some changes in detailed design aspects during the intervening years. The system consists of an un-stiffened spherical tank connected at its equator through a section with a special profile (called the equatorial ring) to a single cylindrical supporting skirt. The lower portion of the skirt is welded to the ship's structure.

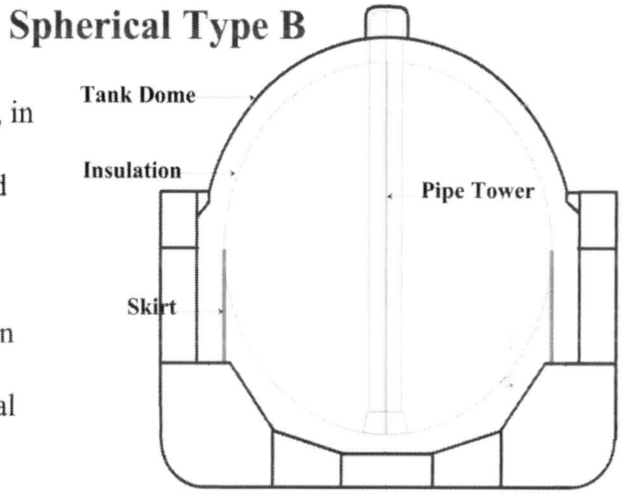

Figure 2: Example of Spherical tanks

2.2.2 Prismatic Tank

An independent prismatic tank LNG containment tank was developed by IHI in the early 1980's and is known as the IIIHI Self-Supporting Prismatic Ship IMO Type B system. In short it is known as the IHI SPB System

The prismatic nature of the tank allows it to fit below a flat upper deck. As with an independent Type A tank, this tank rests on bottom supports and is fitted with anti-rolling, anti-pitching and anti-floatation chocks.

The containment system comprises an aluminium alloy A5083-0 tank of flat plate structure with internal stiffening.

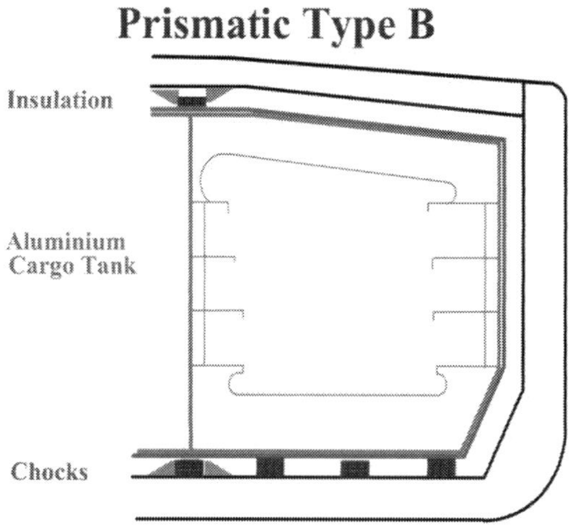

Figure 3: Example of Prismatic Tanks

2.2 Tanks Design Premise

The fundamental consideration in gas ship design is to minimise the release of cargo in the event of a casualty. In doing this it is assumed, by the designer, that these ships will be at least subjected to the abuse normally imposed during the working life of a ship. The design concept also calls for a gas cargo tank to leak before structural failure occurs allowing the leakages to be detected by the gas monitoring system and appropriate action taken to rectify the problem before significant failure occurs. Some of these issues have been readdressed with the results of project specific security and vulnerability assessment and the adequacy of this prescriptive performance standard is continually reviewed for the individual accidental scenarios.

If such a cargo tank leakage does occur in the primary barrier provision has to be made for the cargo to be contained by a second emergency barrier, which according to the code must be designed to hold the leaking cargo for a minimum of 15 days (e.g. the secondary barrier should be designed so that it is capable of containing any envisaged leakage of liquid cargo for a period of 15 days, unless different requirements apply for different voyages, taking into account the load spectrum referred to in 4.3.4.4 of the IGC code). This duration is considered the time it would take a LNG carrier to reach a suitable discharge port, unload the cargo and gas free the tank ready for repair.

3.0 Regulatory Issues

To fully comprehend the risk management issues related to the shipping of LNG, it is critical to first have a clear understanding on the regulatory issues and how they were developed. Marine safety enforcement is performed by the state in which the vessel is registered, known as the "flag state." The classification rules are in their purest sense said to be the development and worldwide implementation of a set of published rules and regulations which set and maintain standards of quality and reliability. They provide for a proven design with some freedom for innovation using the principle of "equivalence." The conventional and prescriptive hazard management process, which is the basis of the current international regulations, agreements, standards and guidelines governing the shipment of LNG, is based on the lessons learned from major and minor incidents that have occurred during the life of the shipping industry. These regulations and guidelines often represent the minimal standards for design, operations and maintenance of LNG vessels; the international economic cycle of the LNG industry will dictate the level of compliance to these standards. This reactive methodology of regulating an industry means that the risk is normally assessed by the authorities (flag state, certifying or federal authority, international regulatory body) and a prescriptive requirement or hazard treatment is developed from this evaluation (see Figure 4).

Some of the current applicable international marine codes and regulations are:

- IMO International Code for Construction and Equipment of Ships Carrying Liquefied Gases in bulk (IGC Code)
- International Load Line Convention 1966 (ILLC) and Protocol of 1988
- Regulations of International Tonnage Measurements of Ships 1969
- Safety of Life at Sea (SOLAS), 1974 and Protocol of 1978
- Convention on the International Regulations for Preventing Collisions at Sea (COLREGS) 1972
- Marine Pollution Prevention (MARPOL), 1973/78
- IMO resolutions A749(18)

- International Ship and Port Facility Security (ISPS) Code (which amended the 1974 International Convention for SOLAS)

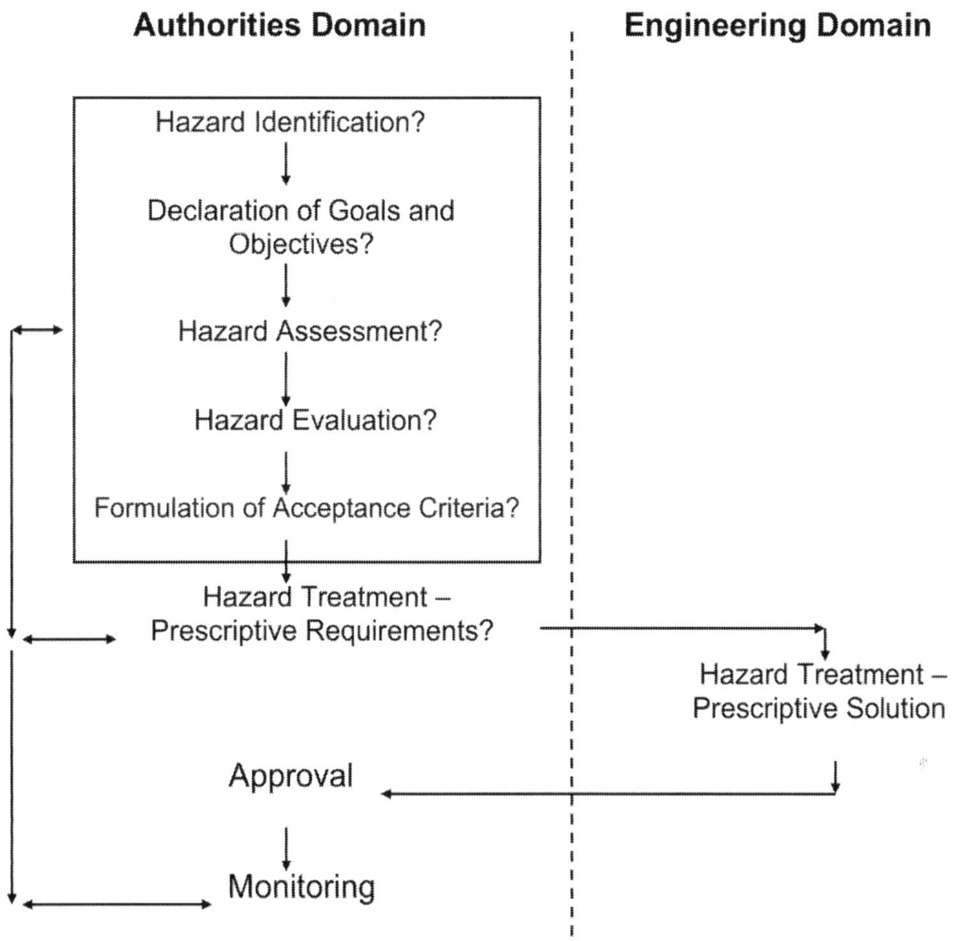

Figure 4: CONVENTIONAL HAZARD MANAGEMENT PROCESS

Examples of other national authority requirements are:

- The USCG Code of Federal Regulations for ships in US waters which include:
- USCG Rules and Regulations for foreign Vessels operating in the navigable waters of the United States;
- The U.S. Marine Transportation Security Act of 2002 (MTSA) which parallels the ISPS codes.
- USCG Rules regarding Oil Pollution, Sanitation and Navigation Safety:

    - Title 46, chapter I, subchapter O (certain bulk dangerous cargoes), part 154
    - Title 33, chapter I, subchapter O (pollution parts 155, 156, 159)
    - Subchapter P (ports and waters safety) part 164
- The Safety Case or other risk analysis concept in the statutory requirements as found in the UK or Australian waters.

The Safety Case or risk analysis concept is a proactive approach that has been adopted by countries such as Australia and the United Kingdom. This performance-based approach provides the opportunity for the LNG shipping owners to develop their own target level of safety that would be acceptable to the certifying or governing bodies (see Figure 5). With this approach, owners are responsible for the continual verification of their proposed treatment of the identified risks, giving them the opportunity to provide proactive risk management solutions to the identified hazards and risks on the LNG vessels and to have a means of focusing effort on critical areas of the vessel. This approach also has better "fit" with the requirements of the Formal Safety Assessment practiced for the shipping industry International Safety Management (ISM) code. The overall objective of both types of regulations is to ensure that all the identified hazard scenarios that will pose a major threat to not only people but also to the environment are addressed during the design, operation and maintenance of the LNG vessels. This type of approach has been accepted by most of the major LNG carriers and owners during the development and operations of these cryogenic facilities.

4.0 Risk Factors and Impacts

Risk has been defined as a measure of the exposure to, or consequences of, an adverse outcome. The LNG industry has produced many studies that assessed the likelihood of adverse outcomes. The hazards and risks during the shipping of LNG are based on the following factors:

- Momentum of the vessel during its transit and berthing;
- Relative location of the vessel with regards to marine traffic, geographical hazards and densely populated areas or environmentally sensitive locations;
- The size of the LNG inventory and type of vessel utilized during the transport.

In the mid-1970s a growing public awareness of process installation hazards resulted in the heightened concerns over LNG ports and terminals, especially with spills on water. It was determined that the LNG ship's is most exposed to risk when it approaches port, while it is in the port, during berthing and whilst performing cargo loading and unloading operations.

Risk assessments for LNG shipping are normally focused on incident scenarios, frequency estimate and population vulnerability models. These assessments help identify the root causes and safeguards required to prevent the hazardous incidents from occurring.

With the above parameters in mind the three major scenario modes of vessel failure are from fire, collision and grounding. The case histories have shown that LNG shipping incidents have been classified as grounding, leaks or spillage and lightning strike & fire[2].

The principle hazard scenarios for LNG carriers are further broken down to the following concerns:

- The large spillage of liquid onto water from grounding or collision,
- Leak during transfer operations; and
- A large fire or explosion on the ship or at the port.

4.1 Spillage

There has been significant research work on LNG spills on water and land by some of the major producers, so there are many known facts that can help to assess this risk. As LNG is released through a breached opening of a carrier onto the water, the behavior of the resultant cold, dense gas cloud is dependent on many factors. As the LNG warms a methane cloud will form which, as it warms further, will transform from a dense to a passive, neutrally buoyant gas cloud. Further warming will allow the cloud to rise and dispersion will be combined with dilution and the decrease in the relative density and concentration. An interesting fact is that if LNG is released from a height, (e.g. from the side of the LNG carrier) most of the liquid will disintegrate and evaporate before it reaches the surface. The temperature difference between the internal containment system and the external ambient temperature will determine the amount of evaporation from the outward flowing liquid.

4.2 Leaks

Methane is normally a gas that is lighter than air, but is a dense liquid if released at its boiling point of -259°F. LNG is stored and transported at atmospheric pressure and as a liquid, which reduces the volume to be transported by about 1/600th of its gaseous state. LNG carriers are double hull vessels with either air or ballast water contained in the void space between the inner and outer hull. A fully loaded carrier will have no ballast water in the ballast tanks. A leak on deck will normally result in the quick dispersion of the vaporizing gas, and if LNG comes into contact with the steel structure it will cause the carbon steel to crack. The development of a sustained fire incident is extremely unlikely due to the low potential for ignition since all electrical equipment is appropriately designed for the hazardous atmosphere on the vessel, and the quick evaporation of the gas past the flammability limits. Further design of LNG ships takes into account the possibility of hot surfaces and all such places are designed such that their maximum temperature is below the auto ignition temperature of methane which is 595° F.

If the accidental breach in the inner and outer hull was located at a certain elevation from the water line, the out-flowing LNG will first fill up the void space volume until there is enough cold liquid to spill over the outer hull opening. Evaporation of the cold liquid will commence on immediate contact with the relatively warm outer hull and inner hull insulation. The methane gas will accumulate higher up the void space between the hulls until there is enough volume of gas to flow out of the outer hull, accidental breach. Due to the volume of methane being flashed off it is extremely likely that all the air in the void space will be displaced by methane making the atmosphere rich in methane and above the upper flammable limit. There is no normal source of ignition in either the ballast spaces or the void spaces.

With the steady improvement in design, operational safety and security in the LNG shipping industry during the past 20+ years the likelihood for an LNG leak to occur during the next 30 years would be highly unlikely[9].

4.3 Explosions and Fires

The flammability range of LNG is 5% to 15% by volume. Any concentration below and above this flammability range will not support ignition due to the lack of methane gas for the lower flammability limit or the lack of oxygen for the upper flammability limit. During the slumping (evaporation and passive dispersion of LNG on water) the leading edge of the gas would more than likely pass the flammability range before it becomes too lean to burn. The inner portion of the gas cloud, with the high evaporation of natural gas would be too rich in hydrocarbons to burn. Therefore, to obtain a burning cloud an ignition source must be situated a distance from the spill source and the concentration within the gas cloud must be within the flammability limit range for LNG.

The resultant LNG dispersion and fire consequence has been heavily studied these past two years [3,4,5,9,11] and validates the appropriateness of the current utilized guidelines for exclusion zones for LNG carriers developed by the National Fire Protection Association (NFPA 59A) and the National Standard of Canada (CSA Z276-01) [6,10]. Both of these guidelines are extensively utilized for the design of LNG facilities in North America; the LNG vessels are included as part of the facility during the consequence calculations. The Society of International Gas Tanker and Terminal Operators (SIGTTO) is the major voice for the LNG shipping industry and the guidelines they have produced are extensively utilized by the LNG carriers during design and operations of the vessels.

4.4 Terrorist Attack

A deliberate terrorist attack on a vessel, while at berth or in transit, is the newest scenario that has been identified from the security and vulnerability assessments performed by LNG shipping owners, particularly since the September 11, 2001 terrorist attacks in the US. Since there has been no major incident from this mode of failure in the LNG shipping industry, the security issues and concerns are based on a proactive scenario development and credibility approach for the safety and security of the LNG vessel. Many scenarios have been discussed, developed and assessed, but the most significant from a consequence perspective is a breaching of the outer and/or inner hull of a ship with an explosive or missile attack rupturing the LNG cargo tanks. The location of the breach (on deck, above water, below water) will determine the magnitude of the dispersion and the potential harm to the general public.

Considering the current ISPS (International Ship and Port Facility Security) and local security and vulnerability code requirements, the potential for a terrorist attack on an LNG vessel having an impact to the general public is highly unlikely, especially during its transit across the oceans. The potential low economic, environmental, safety and public ramification of a single LNG vessel security accidental event along with the current barriers and safeguards developed over the LNG shipping history, makes it an unattractive and difficult target for a terrorist group. It is difficult to produce a worst case accidental release scenario with maximum impact, with all the current inherent safe and secure approaches taken by the LNG shipping industry.

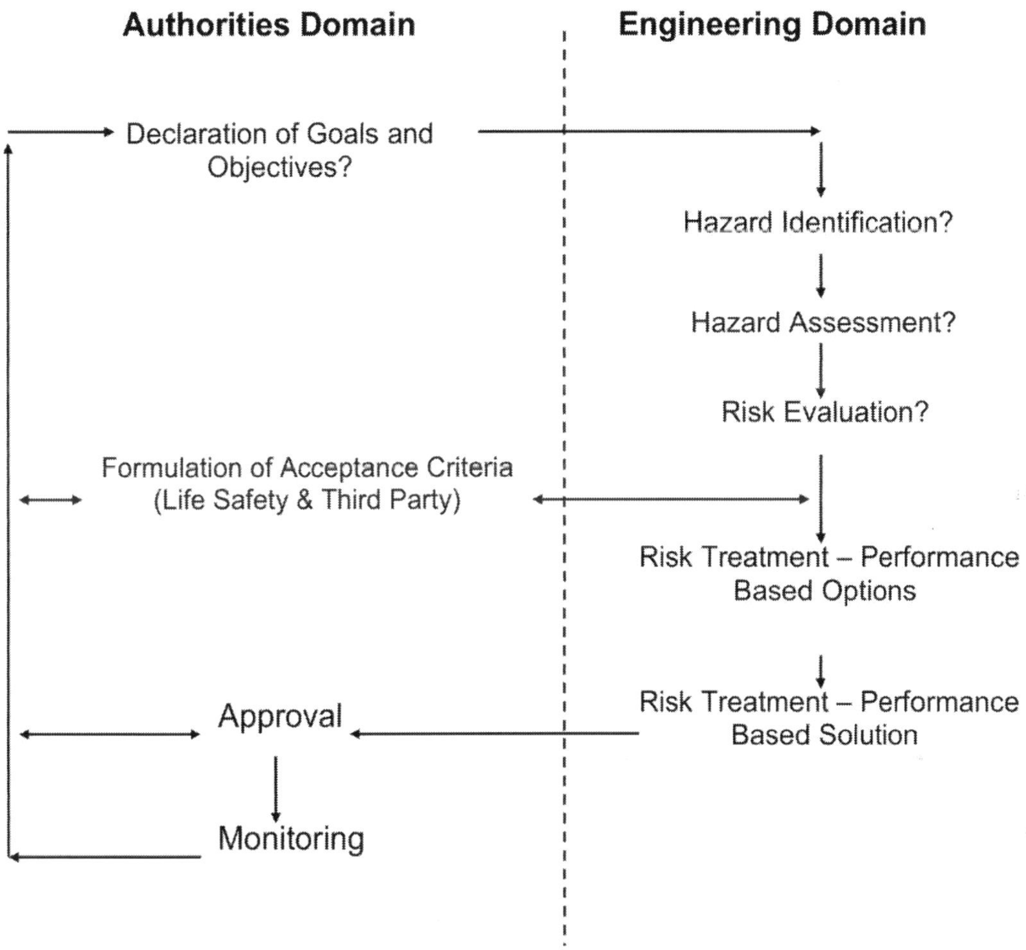

Figure 5: PERFORMANCE BASED HAZARD MANAGEMENT PROCESS

5.0 LNG Risk Management Approach

To ensure thoroughness with regard to the issues of security and vulnerability of LNG vessels all LNG carriers have currently utilized the International Ship and Port Facility Security (ISPS) Codes risk management approach as a framework in the development of a consistent method of evaluating risk and providing the governments a means of determining the appropriateness of the vulnerability risk management plans provided by the LNG shipping companies (see Figure 5).

The current practice today, especially with the public outcry in North America, is to adopt a formal risk management technique in the identification and management of health, safety, environmental and security concerns. This risk management process is supported by the ISM, ISPS and federal/country codes governing LNG shipping throughout the world. The formal process includes the:

a) Identification of the undesirable events;
b) Analysis of the mechanisms of the undesirable events;
c) Consideration of the extent of any harmful events;
d) Consideration of the likelihood of the undesirable events and specific consequences (probability/frequency);
e) Experienced judgment about the significance of the identified hazards and estimated risks;
f) Development and implementation of decisions on courses of action to manage the risk.

The success of this risk assessment process is the correct identification of the undesirable events that are to be considered for the risk assessment (step a) and the subsequent mechanisms or underlying root causes (step b) that will create an error-likely scenario for people, equipment or environment. This is the underlying reason that an initial incident occurs and barriers or safeguards that could fail to interrupt a chain of undesirable events. The quality of these first two steps of hazard identification will determine the cost-effectiveness and accuracy of the risk management solutions (step f) that will be developed for implementation. Current studies contain some uncertainties in the hazard mechanisms (how the leaks will realistically occur) and thus the resultant dispersion and fire consequence will be difficult to communicate to the general public with any confidence. The results of the assessment, from a safe business perspective, is accurate enough to make some informed design and development decisions but not effective enough to alleviate any public concerns. It is imperative that during the analysis of the mechanisms of the undesirable event, that the identification of the current barriers of protection or safeguards is included in the analysis so that the assessment and the subsequent management of the risks identified will be accurate and cost-effective. These barriers of protections have been duly included in the design guidelines and specifications offered by the LNG industry[6,10].

The development and implementation of the decision on the courses of action will entail:

- Definition of the means to reduce the risk;
- Establish priorities;
- Allocate resources;
- Communicate;
- Monitor.

The difficulty with the currently adopted risk management process is the development of acceptable criteria that satisfies both the public perception and regulatory body needs; these criteria will be the measurement of satisfaction from the public. It is the LNG shipping industry's responsibility to test the criteria with the public/government and build confidence in the industry's technical proficiency, the process and the subsequent results.

6.0 Reality

The outstanding safety record of the LNG shipping industry is due to the respect and understanding of the unchanging chemical and physical properties of the LNG fluid to be transported, the heavily regulated industry worldwide and the multiple barriers, or safeguards, used as common practice in systems and operations by the LNG shipping industry. As mentioned above the potentially low economic, environmental, safety and negative general public outcry from a single LNG vessel security or safety accidental event, especially during transit in open waters, makes the ships a very unattractive and difficult potential terrorist target. It is difficult to produce a worst-case accidental LNG release scenario with maximum impact with all the current inherently-safe barriers and secure approach taken by the LNG industry. Unfortunately, even with the inherently safe and secure design and operational efforts, the LNG shipping industry needs to improve the effective communication of the technology and its hazards to the public. The most critical aspect of this process is the communication of the risk management opportunities. With more effective communication, the public should begin to trust in the results of security and vulnerability assessments performed, concur with the technical opinions of the LNG experts who develop risk assessments, and subsequently begin to support the industry.

7.0 Conclusion:

This paper first highlighted some of the general public concerns or queries, based on their LNG experience and review of published public inquiries. This paper attempted to answer these concerns from a more methodical and some historical perspective. It is hoped that some of the lessons learned by the authors and described in the paper it will provide a different, but proactive, perspective in the management of risks by the LNG shipping industry.

The success of the LNG shipping industry is dependent on the demonstration and communication of the benefits and risks of this mode of transportation to the general public that could potentially be affected by the LNG shipping risk. It is the LNG owner's responsibility to educate the public and gain its confidence.

"A thing is provisionally categorized as safe if its risks are deemed known, and in the light of that knowledge judged to be acceptable." [11].

8.0 References:

1. Centre for LNG: "Updates on LNG Security and Safety" (Feb. 23, 2005)
2. F.P. Lees, "Loss Prevention in the Process Industries – Second Edition", Butterworth – Heinemann, London, 1996, Chapter 23 – Transport.
3. Sandia National Laboratories, "Guidance on Risk Analysis and Safety Implications of a Large Liquefied Natural Gas (LNG) Spill Over Water", Prepared for the U.S. Department of Energy (DOE), Office of Fossil Energy, under contract SAND2004-6258., December 2004
4. ABS Consulting, "Consequences Assessment Methods for Incidents Involving Releases from Liquefied Natural Gas Carriers", Prepared or the Federal Energy Regulatory Commission under contract FERC04C40196, May 13, 2004.
5. P.W. Parfomak & A.M. Flynn, "Liquefied Natural Gas (LNG) Import Terminals: Siting, Safety and Regulation, CRS Report for Congress, Order Code RL 32205.

6. National Fire Protection Association (NFPA), 2001; "NFPA 59A – Standard for the Production, Storage, and Handling of Liquefied Natural Gas (LNG)".
7. Lloyd's Register, "Classification and certification of floating offshore liquefied gas installation – Guidance notes", Revision 2, April 2004.
8. University of Houston Law Center – Institute for Energy, Law & Enterprise, "LNG Safety and Security", October 2003.
9. Pitblado et al., "Consequences of LNG Marine Accidents", CCPS Conference, Orlando, June 29 – July 1, 2004.
10. National Standard of Canada; "CAN/CSA-Z276-01 – Liquefied Natural Gas (LNG) Production, Storage, and Handling", Toronto, Canada, November 2001.
11. Aspen Environmental Group, "International and National Efforts to Address the Safety and Security Risk of Importing Liquefied Natural Gas: A Compendium", Prepared for the California Energy Commission, under contract CEC-600-2005-002, January 2005.
12. The Council for Science and Society, "The Acceptability of Risks – The logic and social dynamics of fair decisions and effective controls", Barry Rose (Publishers) Ltd., London, 1977.
13. Lloyd's Register LNG Ship Training Course 2003 - Edited by Tony Bingham

LNG Security Vulnerability Assessment

Authors

David A. Moore, PE, CSP
President and CEO
and Brad Fuller, Senior Consultant
AcuTech Consulting Group
2001 N Beauregard Street
Suite 700
Alexandria, VA 22311
dmoore@acutech-consulting.com
bfuller@acutech-consulting.com

David A. Jones, CSHO
Manager, Houston Operations
AcuTech Consulting Group
2500 CityWest Boulevard, Suite 300
Houston, Texas 77042
djones@acutech-consulting.com

Michael Hazzan, PE
Manager, Eastern Business Unit
AcuTech Consulting Group
1911 Princeton Avenue
Lawrenceville, NJ 08648
mhazzan@acutech-consulting.com

Prepared for Presentation at
American Institute of Chemical Engineers
2006 Spring National Meeting
Center for Chemical Process Safety Symposium
Orlando, Florida
April 24-26, 2006

Copyright © (AcuTech Consulting Group)
(April, 2006)

UNPUBLISHED

LNG Security Vulnerability Assessment

Authors

David A. Moore, PE, CSP
President and CEO
and Brad Fuller, Senior Consultant
AcuTech Consulting Group
2001 N Beauregard Street
Suite 700
Alexandria, VA 22311
dmoore@acutech-consulting.com
bfuller@acutech-consulting.com

David A. Jones, CSHO
Manager, Houston Operations
AcuTech Consulting Group
2500 CityWest Boulevard, Suite 300
Houston, Texas 77042
djones@acutech-consulting.com

Michael Hazzan, PE
Manager, Eastern Business Unit
AcuTech Consulting Group
1911 Princeton Avenue
Lawrenceville, NJ 08648
mhazzan@acutech-consulting.com

Abstract

Public concerns of a terrorist attack on LNG facilities and transportation is one of the many factors causing the industry to have difficulties in siting new and much needed import terminals. The LNG industry does not have a standard security vulnerability assessment methodology to help analyze and explain these risks consistently. With a defined risk assessment process, security can be more consistently assessed and risks more clearly understood. This paper examines the issues of making security risk decisions in the LNG industry and the requirements, processes and industry methods available.

For the import terminals, there is a requirement to conduct a Facility Security Assessment (FSA) in the USCG regulations under the Maritime Transportation Security Act of 2002 (33 CFR Part 105). No particular methodology is required to be used but certain issues need to be addressed as specified in the rule. Industry practice varies to achieve those goals from formal methods to expert reviews.

For transportation, which is the primary concern of most, the Department of Homeland Security (DHS) United States Coast Guard (USCG) released guidance for conducting a Waterway Suitability Assessment (WSA) on June 14, 2005. The guidance (COMDTPUB P16700.4 NVIC 05-05) is an important new development in the area of safety and homeland security for the transportation of LNG. The purpose of the guidance is to assist applicants seeking to build an onshore terminal or existing terminal operators in the analysis of safety and security of the port, terminal, and vessels and the surrounding public and infrastructure from the transportation of LNG. Since the NVIC guidance was not meant to be prescriptive in its approach, it describes a process for conducting the analysis, but does not describe a specific methodology. It is even less specific than the FSA requirements. While the WSA is a major step forward in assessing safety and security risks, in practice the process leaves the applicant with some challenging decisions that the guidance does not fully answer.

The significance of the new guidance is profound. It sets precedence for the analysis of safety and security risks of the transportation of LNG to onshore terminals at a time when the need is greatest and the concerns for safety and security are heightened. It encourages the involvement of various stakeholders for the analysis and validation of the WSA. FERC may determine the need to annually review and update the WSA. It tackles difficult issues of safety and security in port operations and intercoastal waterways by LNG tankers.

The basis of the assessment is the results of the recent Department of Energy (DOE) funded study by Sandia National Laboratories on the safety implications of intentional and accidental spills of LNG on water[1]. While this study was mostly a consequence analysis, it did recommend that a risk-based approach be used for the site specific analysis of these situations. Consequence-based zones of concern are to be overlaid on tanker routes to assess potential consequences, vulnerabilities, likelihood of attack, and mitigation measures.

This paper discusses the process and focuses on the challenges of the WSA process including the identification of vulnerabilities, the determination of consequences and likelihood, and the evaluation of risk by the USCG. This paper describes the experiences of the authors in conducting one of the first WSA studies in the United States using the published guidance.

Structuring the analysis of facility and transportation risks by using a single SVA methodology is very useful for consistent risk terminology and approach. This paper will describe the application of a SVA methodology published by API/NPRA to analyze LNG risks and provide guidance on how facility and transportation risks can be evaluated and coordinated. It will review key issues expected to be considered by USCG regulations and industry SVA methodologies.

1. Introduction

While there are increased demands for LNG and concerns for safety and security of the various existing and proposed projects, LNG terminal projects must follow a comprehensive permitting process, led by the Federal Energy Regulatory Commission (FERC) for onshore terminals and by the U.S. Coast Guard for offshore projects. The level of strict control results in a careful analysis of the risks. Owner/operators must also conduct both facility security assessments (FSA) and waterway suitability assessments (WSA) for review and approval by the USCG.

---

[1] *Guidance on Risk Analysis and Safety Implications of a Large Liquefied Natural Gas (LNG) Spill Over Water*, Sandia, 2004

For the import terminals, there is a requirement to conduct a FSA in the USCG regulations under the Maritime Transportation Security Act of 2002 (33 CFR Part 105). No particular methodology is required to be used but certain issues need to be addressed as specified in the rule (Appendix A). Industry practice varies to achieve those goals from formal methods to expert reviews. The author's experience was that a more defined process for safety and security analysis is required to effectively accomplish the FSA and WSA, and using the same methodology to accomplish both requirements is helpful in that a standardized method is used for assessing both risks. The results can later be compared directly regardless of the source of the requirements.

2. FSA Requirements Under Part 105

33 CFR 105 Subpart C provides guidance on the format and content of a FSA. Although the acronym "SVA" is not used in the regulations, the description of the purpose, use, as well as the content of a FSA clearly indicates that FSA is the USCG's version of a SVA. While the MTSA regulations do not refer to existing industry or government SVA methodologies, the contents of a FSA include the same issues that are addressed in API/NPRA, AIChE (CCPS) SVAs and their equivalents.

The FSA report requirements under § 105.305 (d), include:

- On-Scene Survey;
- Existing Security Measures;
- Vulnerabilities;
- Security Measures to Address Vulnerabilities;
- Key Facility Operations; and
- List of Identified Weaknesses.

A key component of the FSA is the on-scene survey. This is directly analogous to the security survey in a SVA performed using the API/NPRA and CCPS methodologies. The comprehensive security survey documentation in these industry methodologies has been sufficient for demonstrating that a complete survey was made for sites covered by the MTSA regulation.

Based on the information provided to the FSA performer(s) and the results of the one-scene survey, the threats, consequences, and vulnerabilities must be analyzed, and recommendations for inclusion in the Facility Security Plan (FSP) to reduce the vulnerabilities. Then list of events that must be considered in a FSA include:

Regardless of the methodology used, the USCG requires certain threats to be considered as shown in Figure 1.

Figure 1
MTSA Part 105 Facility Security Scenarios and Countermeasures

Threat	Examples for Petroleum or Chemical Facility
(2) Possible security threats, including but not limited to: (i) Damage to or destruction of the facility or of a vessel moored at the facility;	1. Attack by external adversary via water with small craft and explosives while ship is moored at the dock or of facility 2. Attack by external adversary via land by foot across perimeter using force or stealth and explosives while ship is moored at the dock or of facility 3. Attack by insider or external adversary via land by vehicle through perimeter using force or stealth and explosives while ship is moored at the dock or of facility 4. Adversary aboard ship or supplier to ship or visitor or during crew change for ship
(ii) Hijacking or seizure of a vessel moored at the facility or of persons on board;	1. Attack by external adversary via water with small craft while ship is moored at the dock or of facility 2. Attack by external adversary via land by foot across perimeter using force or stealth while ship is moored at the dock or of facility 3. Attack by insider or external adversary via land by vehicle through perimeter using force or stealth while ship is moored at the dock or of facility 4. Adversary aboard ship or supplier to ship or visitor or during crew change for ship
(iii) Tampering with cargo, essential equipment or systems, or stores of a vessel moored at the facility;	1. Attack by external adversary via water with small craft while ship is moored at the dock or of facility 2. Attack by external adversary via land by foot across perimeter using force or stealth while ship is moored at the dock or of facility 3. Attack by insider or external adversary via land by vehicle through perimeter using force or stealth while ship is moored at the dock or of facility 4. Adversary aboard ship or supplier to ship or visitor or during crew change for ship
(iv) Unauthorized access or use including the presence of stowaways;	1. Attack by external adversary via water with small craft while ship is moored at the dock or of facility to offload unauthorized persons 2. Attack by external adversary via land by foot across perimeter using force or stealth while ship is moored at the dock or of facility to enter the facility as ship crew or to board the ship 3. Attack by insider or external adversary via land by vehicle through perimeter using force or stealth while ship is moored at the dock or of facility to enter the facility as ship crew or to board the ship 4. Adversary aboard ship or supplier to ship or visitor or during crew change for ship
(v) Smuggling dangerous substances and devices to the facility;	1. Attack by external adversary via water with small craft while ship is moored at the dock or of facility and waterborne attachment or relay of contraband or device 2. Attack by external adversary via land by foot across perimeter using force or stealth while ship is moored at the dock or of facility and attachment or relay of contraband or device 3. Attack by insider or external adversary via land by vehicle through perimeter using force or stealth while ship is moored at the dock or of facility and attachment or relay of contraband or device 4. Adversary aboard ship or supplier to ship or visitor or during crew change for ship

Figure 1
MTSA Part 105 Facility Security Scenarios and Countermeasures

Threat	Examples for Petroleum or Chemical Facility
(vi) Use of a vessel moored at the facility to carry those intending to cause a security incident and their equipment;	1. Attack by external adversary via water with small craft while ship is moored at the dock or of facility to offload unauthorized persons 2. Attack by external adversary via land by foot across perimeter using force or stealth while ship is moored at the dock or of facility to enter the facility as ship crew or to board the ship 3. Attack by insider or external adversary via land by vehicle through perimeter using force or stealth while ship is moored at the dock or of facility to enter the facility as ship crew or to board the ship 4. Adversary aboard ship or supplier to ship or visitor or during crew change for ship
(vii) Use of a vessel moored at the facility as a weapon or as a means to cause damage or destruction;	1. Attack by external adversary via water with small craft while ship is moored at the dock or of facility to offload unauthorized persons and steal the vessel 2. Attack by external adversary via land by foot across perimeter using force or stealth while ship is moored at the dock or of facility to enter the facility as ship crew or to board the ship and steal the ship 3. Attack by insider or external adversary via land by vehicle through perimeter using force or stealth while ship is moored at the dock or of facility to enter the facility as ship crew or to board the ship 4. Adversary aboard ship or supplier to ship or visitor or during crew change for ship and onboard ship are weapons or adversary uses ship to directly ram or release contents
(viii) Blockage of entrances, locks, and approaches; and	1. Attack by external adversary via water with small craft while ship is moored at the dock or of facility and explosives to cause destruction and oil spill or fire or physical blockage or ramming vessel into key location to cause blockage 2. Attack by external adversary via land by foot across perimeter using force or stealth while ship is moored at the dock or of facility and explosives to cause destruction and oil spill or fire or physical blockage or ramming vessel into key location to cause blockage 3. Attack by insider or external adversary via land by vehicle through perimeter using force or stealth while ship is moored at the dock or of facility and explosives to cause destruction and oil spill or fire or physical blockage or ramming vessel into key location to cause blockage 4. Adversary aboard ship or supplier to ship or visitor or during crew change for ship and explosives to cause destruction and oil spill or fire or physical blockage or ramming vessel into key location to cause blockage
(ix) Nuclear, biological, radiological, explosive, and chemical attack;	More likely explosive or chemical attack, possibly using the hazardous properties of the cargo as the weapon (similar to (i) above)

The FSA report is an appendix in the required Facility Security Plan (FSP) for MTSA covered facilities.

The FSA must be revalidated (although that term is not used) on a five-year basis when the FSP is submitted to USCG. Other changes and events can also trigger a FSA update.

The overall conclusion, based on similarities in requirements and in approval of FSAs from covered facilities, is that a complete SVA performed using the API/NPRA or CCPS methodology or an equivalent whose report includes well-documented descriptions of how each step of the SVA was completed, particularly the security survey portion of the SVA, should suffice for a FSA.

3. API/NPRA SVA Methodology

One possible risk-based approach is to use the security vulnerability assessment (SVA) methodology published by the American Petroleum Institute and the National Petrochemical and Refiners Association.[2] This is a robust security vulnerability assessment methodology that was purposely developed for assessing homeland security risks of energy infrastructures. It has the advantage of providing a structured approach to the analysis and including the necessary steps of an asset characterization, threat assessment, vulnerability assessment, consequence assessment, and a risk assessment as shown in Figure 2.

The American Petroleum Institute (API) and the National Petrochemical & Refiners Association (NPRA) developed the first edition of this Security Vulnerability Assessment Methodology available to the petroleum and petrochemical industry in 2003. The information contained in the document was developed in cooperation with government and industry, and is intended to help refiners, petrochemical manufacturers, and other segments of the petroleum and petrochemical industry maintain and strengthen security of their personnel, facilities, and operations, and to thereby enhance the security of the nation's infrastructure.

API decided that it would be beneficial to use a single methodology for both facility and transportation risks. As such, the API/NPRA SVA Methodology was seamlessly expanded to include transportation risks simply by applying the methodology to the issue and adapting the technique to the need. The exercise proved that the methodology was scalable to the problem of transportation risks. The end result was simply a presentation of the concept against three transportation risks – a pipeline, a trucking operation, and a rail operation. The second edition, incorporating these changes, was published in 2004.

[2] "Security Vulnerability Assessment for the Petroleum and Petrochemical Industries", American Petroleum Institute, August, 2004.

Figure 2
API/NPRA Security Vulnerability Assessment Methodology

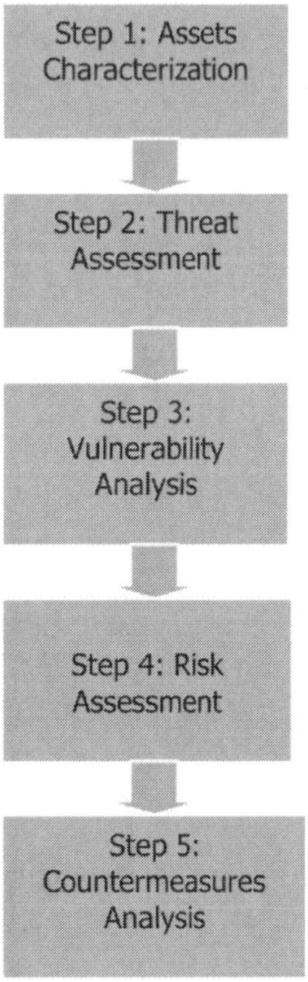

1.1 Identify critical assets
1.2 Identify critical functions
1.3 Identify critical infrastructures and interdependencies
1.4 Evaluate existing countermeasures
1.5 Evaluate impacts
1.6 Select targets for further analysis

2.1 Adversary identification
2.2 Adversary characterization
2.3 Target attractiveness

3.1 Define scenarios and evaluate specific consequences
3.2 Evaluate effectiveness of existing security measures
3.3 Identify vulnerabilities and estimate degree of vulnerability

4.1 Estimate risk of successful attack
4.2 Prioritize risks

5.1 Identify and evaluate countermeasures options
5.2 Prioritize potential enhancements

As part of the methodology, the risks identified can be portrayed against a risk decision matrix such as the example one shown in Figure 3. Either the likelihood is expressed as the absolute likelihood of expected successful attack or, if the likelihood of attack is considered to be a probability of one, it represents the vulnerability (likelihood of success given an attack) vs. consequences.

Figure 3
Example WSA Risk Matrix

LIKELIHOOD \ SEVERITY	5	4	3	2	1
5	High	High	High	Med	Low
4	High	High	Med	Med	Low
3	High	Med	Med	Low	Low
2	Med	Med	Low	Low	Low
1	Med	Med	Low	Low	Low

Severity

Code	Description
1	Very Low
2	Low
3	Med
4	High
5	Very High

Likelihood

Code	Description
1	Very Low
2	Low
3	Med
4	High
5	Very High

4. Transportation Security Assessment

An element of the approval process needing improvement was guidance on the assessment of the safety and security aspects of the transportation of LNG by tankers to onshore terminals. The U.S. Coast Guard recently published guidance (NVIC 05-05) requiring a Waterway Suitability Assessment (WSA) to solve this issue. The latter was developed in conjunction with federal and state agencies, as part of the approval process for securing a Letter of Recommendation (LOR) from the U.S. Coast Guard in order to transport LNG via ship through state and federal waters to an onshore LNG terminal.

Most of the public attention on LNG terminal siting has focused on the potential worst case consequences of LNG releases and fires. A limited analysis such as that is not a complete risk assessment as one must understand both the potential consequences as well as the likelihood of intentional and accidental events. Instead, the WSA process involves examining all of the factors for assessing safety and security risks, including the measures taken by industry and government to manage these risks. By framing LNG in the context of risk instead of basing it solely on potential consequence, an informed judgment on the risk to society can be made. The results can then be used to compare the risk of LNG operations to other industries and other societal risks posed to the public.

The scope of a WSA is to:

- *Address* transportation of LNG from a tanker's entrance into U.S. territorial waters, through its transit to/from LNG terminal (receiving) facility, and including operations at the vessel/facility interface;
- *Address* navigational safety issues and port security issues introduced by proposed LNG operations;
- *Identify* relevant safety and security issues from a broad viewpoint of impact to the entire port, as well as provide a detailed review of specific points of concern along a tanker's proposed transit route.

While the WSA is a major step forward in assessing safety and security risks, in practice the process leaves the applicant with some challenging decisions that the guidance does not fully answer. The NVIC guidance was not meant to be prescriptive in its approach. This paper is intended to share some of those experiences and challenges to stimulate a professional discussion on how the WSA process could be improved.

5. Objectives, Scope, and Approach of the WSA

A primary objective of the WSA is to identify the federal, state, local and private sector resources needed to carry out the mitigation measures identified in the WSA. The WSA also requires the identification of the resources currently available and the mechanism by which funding will be provided for additional public resources needed for the safety and security of the LNG marine and terminal operations in that port.

The WSA planning processes come under the jurisdiction of the USCG. An applicant proposing a new LNG terminal or modification of an existing LNG terminal must submit a Letter of Intent (LOI) to the appropriate USCG Captain of the Port (COTP). The COTP then prepares a Letter of Recommendation (LOR) as to the suitability of the waterway in question for LNG marine traffic. A proposed LNG terminal project can not proceed without an LOR deeming the waterway suitable for marine traffic.

With the issuance of NVIC 05-05, a WSA is now required prior to an LOR being issued. The WSA process is solely under the direction of the U.S. Coast Guard's Marine Safety Office. According to the NVIC, the purpose of a WSA is "...to ensure that full consideration is given to safety and security of the port, the facility, and the vessels transporting LNG." The WSA identifies credible safety hazards and security threats to LNG transportation in that port and waterway and identifies appropriate risk mitigation measures.

The WSA process includes consideration of both safety and security. The safety portion of the WSA is most likely well understood at existing ports and includes issues previously considered for other forms of marine transportation. It is the security part that is less familiar to the participants and as such poses the greatest challenge. Growing public concern for the security of LNG transits make this evaluation even more important. In particular, the two areas of most difficulty have been the assessment of potential impacts and making judgments on risk.

The participants of the WSA process include the USCG and port pilots with existing knowledge of the physical characteristics and existing traffic patterns on the waterway. For safety, they consider the change in the port safety environment due to the introduction of the LNG ship transportation including the potential for groundings, collisions and allisions. The WSA includes information on the commercial traffic within the waterway, recreational vessel usage of the waterway, and the time of day when the waterway is at its peak usage. It also includes physical considerations such as bridges, natural or man-made hazards, underwater pipelines, as well as important or significant icons, such as parks, monuments, etc., along the transit to the terminal and nearby the berth at the LNG terminal.

For LNG terminal projects and related shipping, NVIC 05-05 calls for the involvement of a cross-section of the public officials responsible for the safe transit of LNG vessels bound for a U.S. port. The COTP may also involve existing ad hoc committees (e.g, Area Maritime Security Committee (AMSC), comprising law enforcement, emergency responders, other industrial users of the port, etc.) to participate in the process.

As part of the WSA, an analysis is prepared by the project applicant, drawing on internal and/or external security expertise, to determine potential risks of safety hazards and security threats to the vessel, the public and property along the transit route and at the terminal berth. The analysis includes measures that should be employed to mitigate such risks. Another stakeholder team (e.g., AMSC) reviews the analysis for completeness and accuracy on behalf of the USCG. Subsequently, the COTP, with input from stakeholders, prepares a waterway management plan for use when an LNG vessel calls on that port. The plan should be developed with an eye toward flexibility as it must be able to change to meet the specific operational requirements of each transit as well as changes in MARSEC threat levels.

The results of the recent Department of Energy (DOE) funded study by Sandia National Laboratories on the risks of intentional and accidental spills of LNG on water provide the basis for analyzing the consequences within the WSA. While this study was mostly a consequence analysis, it did recommend that a risk-based approach be used for the analysis of these situations. Consequence-based zones of concern are to be overlaid on tanker routes to assess potential consequences, vulnerabilities, likelihood of attack, and mitigation measures.

The various stakeholders may have differing opinions on the vulnerabilities identified and the degree and type of risk management measures required. Part of the WSA's purpose is to raise and discuss these issues and reach an agreement on the best, site-specific approach for managing the risks.

6. Benefits of WSA Process

The WSA process will have a positive impact on the various proposed projects and existing terminals since it addresses the need for conducting a comprehensive safety and security risk assessment in a systematic and participatory manner. One ramification of not having a standard WSA process is the potential for a wide variety of unanswered public concerns and biased and unscientific studies conducted without the benefit of full subject matter expert and stakeholder input.

The WSA guidance provides for a robust process for reviewing safety and security issues specific to a proposal. Appropriate stakeholder involvement and technical expertise is involved, and the process tackles tough issues in the proper environment. The WSA considers infrastructure dependencies and impacts along the route of the tanker; and it systematically addresses risk measures against standardized attack modes and vulnerabilities to provide a better basis for USCG and FERC decision-making.

7. Issues/Challenges with the WSA Process

Based on the authors' experience in conducting WSA studies soon after the release of the NVIC, the authors can report that the expected benefits were realized. More important, however, is the reporting of challenges in implementing the NVIC guidance with the hope that future WSA studies will benefit from lessons learned. Challenges include the identification of vulnerabilities, the determination of consequences and likelihood, and the evaluation of risk by the USCG. All of these challenges were met by supplementing the guidance and supporting studies to develop a workable solution.

These issues include the following:

8. Definition of a risk-based approach

The use of the guidance in the NVIC is encouraged. The NVIC describes a risk-based framework but is flexible on the exact approach used. One of the references for the NVIC is the Risk-Based Decision Making, COMDTINST M16010.3 (Series) and USCG 'Guidelines for Risk-Based Decision-Making 3rd edition', which provide further guidance on establishing a risk-based approach.

The analysis should include the following key steps:

A. Port Characterization
B. Characterization of LNG Facility and LNG Tanker Route
C. Risk Assessments (Safety and Security)
D. Risk Management Strategies
E. Resource Needs for Safety, Security and Response
F. Conclusions and Recommendations

The NVIC itself does not define a specific risk-based approach other than to state that "Applicants are encouraged to conduct a risk-based assessment. The approach considers risk as the combination of the consequences of an undesired event and the associated probability of that event occurring. Fundamentally, this means asking What can go wrong?, How severe can the consequences be?, and, What is the likelihood that this will occur? By answering these three questions the stakeholders can effectively characterize the risk."

It goes on to state "The applicant may use any assessment methodology deemed appropriate. However, it is recommended that the applicant use a methodology that meets generally accepted risk-based decision-making industry standards and that the assessment is as objective and transparent as possible. The Risk Assessment portion of the WSA looks at the conditions that could result in a release of LNG. The events that could trigger a release may be accidental (collisions, groundings, spills, etc.) or intentional (terrorist act, sabotage, etc). The accidental releases should be considered in a safety assessment that looks at the probability and consequences of various incidents. For the unique case of intentional releases a security assessment is performed."

9. API NPRA SVA Methodology for Transportation Risks

The security risks associated with the transportation of assets is a complex issue but a necessary one to analyze in a security vulnerability analysis. Given that most SVA's have focused on the fixed facility assets rather than the entire value chain of the operation, this is a ripe area for security evaluation. The American Petroleum Institute (API), recognizing the need for a methodology for evaluation of transportation risks, recently expanded their SVA guidelines to include consideration of this risk.

The authors have successfully used the API SVA methodology for both safety and security considerations required under the WSA. The WSA involved the integration of this method along with the guidance provided in the NVIC and from the Sandia report to develop a workable approach. Each segment along the tanker route inbound and outbound is characterized by describing the maritime conditions, length of route, vessel speed, physical obstructions and hazards, and other required considerations.

The Sandia study is used along with NVIC 05-05 Enclosure 3 Risk Management Quick Reference Tool to define:

- the Risk Factors including Zones of Concern (Zones 1-3)
- the Attack Vectors and Accident Types
- and the Risk Mitigation Strategies recommended by the USCG for consideration for each attack vector and accident type.

The potential impacts of an LNG event on port infrastructure, marine traffic, workers, visitors, roadways, and the public were considered by overlaying the Sandia zones and assessing the population or activity within the zones. If a segment could produce a Zone of Concern that impacted one of the four Risk Factors (Transit near heavy marine traffic, Transit near medium population areas, Transit near high population areas, or Transit near critical infrastructure & key assets), then those impacts were summarized and considered during the risk ranking process.

10. Interpretation of the Study Zones of Concern

There are numerous references in the NVIC to the Sandia study, as this was meant to be a definitive resource for the issues of spills of LNG on water from tankers. The guidance provides thorough direction for industry to follow, but in practice it doesn't address all of the issues needed to assess consequences for a specific WSA study since that wasn't the objective of the report. As a result, applicants are required to interpret the results of the Sandia study or add to them to complete the WSA. Due to this there is the potential for some local interpretation leading to national inconsistencies. This will have to be addressed by the USCG during the oversight of the processes.

The Zones are explained in Enclosure 11 of NVIC 05-05 (Figure 4).

Figure 4
Summary of "Zones of Concern" for Intentional LNG Spills

Zone 1: This is the area with the most severe consequences around the LNG tanker, where an LNG spill could pose a severe public safety and property hazard and could damage or significantly disrupt critical infrastructure and key assets located within this area. Zone 1 is considered to extend about 500 m (0.3 miles) for an intentional breach of an LNG tanker. Risk management strategies should address vapor cloud dispersion and fire hazards. The most rigorous deterrent measures should be considered when major critical infrastructure elements, such as population or commercial centers, lie within Zone 1. These measures should include such things as vessel security zones, waterway traffic management, and establishment of positive control over vessels. Coordination among all port security stakeholders is essential. Incident management and emergency response measures should be carefully evaluated to ensure adequate resources (i.e., firefighting, salvage) are available for consequence and risk mitigation.

Zone 2: This is an area with less severe consequences than Zone 1 and is considered to extend from 500 m (0.3 miles) to 1,600 m (1 mile) for an intentional breach of an LNG tanker. Risk management strategies should address vapor cloud dispersion and fire hazards. When major critical infrastructure elements occur within Zone 2, risk management strategies that should be considered include incident management and emergency response measures that ensure areas of refuge (enclosed areas, buildings) are available, the development of community warning procedures, and education programs to ensure that communities are aware of precautionary measures.

Zone 3: This is an area with the least likelihood of severe consequences and is considered to extend from 1,600 m (1 mile) to a conservative maximum of 3,500 m (2.2 miles) from the LNG tanker, in the unlikely event that 3 cargo tanks were breached and a vapor cloud disperses without an initial ignition. Risk management strategies should address the vapor cloud dispersion hazard. When major critical infrastructure elements occur within Zone 3, risk management strategies that should be considered include incident management and emergency response measures that ensure areas of refuge are available and community education programs should be considered to ensure that people know what to do in the unlikely event of the release of a vapor cloud without initial ignition.

Some of the areas where additional supporting information is required are as follows:

11. Understanding the Zones of Concern

The Zones were not produced to derive specific levels of consequences at all possible locations for WSA studies so each proposed LNG terminal site should be evaluated for its specific vulnerabilities and consequences using these as guidance only. Zones 1 and 2 were to be considered as circles representing isotherms. Zones 1/2 were based on a nominal result from considering up to 1 tank volume released. Zone 1 was 37 kW/M2 and Zone 2 was 5 W/m2. Zone 3 was the distance to the isopleth of the Lower Flammability Level (LFL), but while it may be possible to go in any of 360 degrees direction, it is assumed to be a unidirectional plume shaped cloud.

In practice there was some confusion on Zone 3 and how to evaluate risk from it including flash fire risks. Sandia stated that "large, unignited LNG vapor releases are unlikely. If they do not ignite, vapor clouds could spread over distances greater than 1600 m from a spill. For nominal accidental spills, the resulting hazard ranges could extend up to 1700 m. For a nominal intentional spill, the hazard range could extend to 2500 m. The actual hazard distances will depend on breach and spill size, site-specific conditions, and environmental conditions." Zone 3 was a vapor dispersion case from up to 3 tanks released and was anticipated to be a plume resulting from an unignited release. Zone 3 was 100% of the Lower Flammable Limit for methane. Zone 3 should be considered a vapor cloud directed towards a populated area and evaluated as to confinement, likelihood of ignition, and effects.

Figure 5 illustrates the concept. These zones are to be overlaid along the tanker route to determine potential consequences.

Figure 5
Illustration of
Zones of Concern
(Intentional Release)
(NVIC 05-05, Enclosure 11)
(Not to Scale)

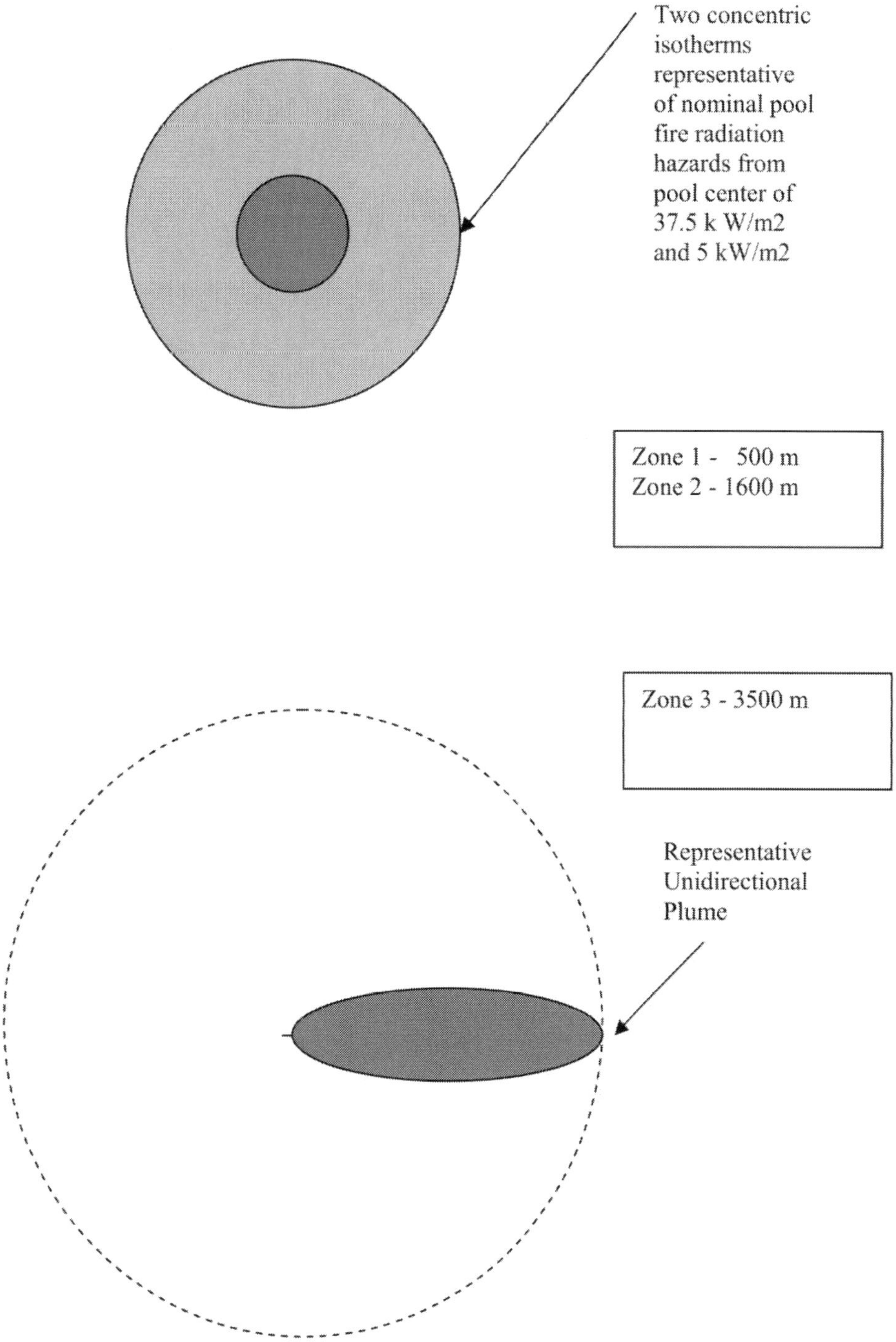

Sandia concluded that releases from intentional acts against an LNG tanker would create a larger potential consequence than from accidental causes as shown in Figure 6. The NVIC guidance (Enclosure 3) provides recommended Risk Management Strategies against these Zones, but note that they do not distinguish between safety and security zones. The authors referred to the original Sandia report and considered the more limited impacts consistent with their conclusion for accidental consequences rather than to use the intentional zones for both categories of events.

Figure 6
Guidance for Impacts on Public Safety from LNG Breaches and Spills

EVENT	POTENTIAL SHIP DAMAGE AND SPILL	POTENTIAL HAZARD	POTENTIAL IMPACT ON PUBLIC SAFETY*		
			High	Medium	Low
Collisions: Low speed	Minor ship damage, no spill	Minor ship damage	None	None	None
Collisions: High Speed	LNG cargo tank breach and small - medium spill	Damage to ship and small fire	~ 250 m	~ 250 – 750 m	> 750 m
Grounding: <3 m high object	Minor ship damage, no breach	Minor ship damage	None	None	None
	Intentional breach and medium to large spill	Damage to ship and large fire	~ 500 m	~ 500 m – 1600 m	> 1600 m
Intentional Breach	Intentional, large release of LNG	• Damage to ship and large fire	~ 500m	~ 500 m – 1600 m	> 1600 m
		• Vapor cloud dispersion with late ignition	~ 500 m	> 1600 m	> 2000 m

* Distance to spill origin, varies according to site
Low – minor injuries and minor property damage
Medium – potential for injuries and property damage
High – major injuries and significant damage to property

The three zones are characterized in Figure 7. Note that there is an implied low likelihood of delayed ignition and flash fire risks making the 'likelihood' of Zone 3 impacts lower given an attack is assumed than Zone 1 or 2.

Figure 7
Summary of Zones of Concern (derived from NVIC 05-05, Enclosure 11 and Sandia report)

Zones of Concern (NVIC 05-05, Enclosure 11)	Type of Initial Event	Extent of Impacts (m)[3]	Hazards	Level of Concern	Likelihood of Impacts Given Initial Event Occurs
1	Intentional single tank to up to 3 tank breach (5 m2 hole size)	391 – 529 (Nominal 500 m)	Thermal radiation from resulting pool fire (conductive, convective, and radiative) leading to severe public safety and property hazards and could damage or significantly disrupt critical infrastructure and key assets	≥ 37.5 kW/m^2	Likely
2	Intentional single tank to up to 3 tank breach (5 m2 hole size)	1305 – 1652 (Nominal 1600 m)	Fire (radiative) leading to less severe public safety and property hazards	37.5 to 5 kW/m^2	Likely
3	Intentional release from single and 3 tanks (both assumed 5m2 hole size)	Single tank 2450m 3 Tanks 3614m Nominal 3500 m)	Vapor cloud (flash fire and possible confined space overpressure) disperses without an initial ignition leading to even less severe public safety and property hazards (down wind impacts as compared to radial for Zone 1 and Zone 2.	Within flammable envelope of dispersed cloud to LFL endpoint	Very Unlikely due to need for delayed ignition and consequences dependent on wind direction. Overpressure requires confinement

[3] Table 41, Sandia SAND2004-6258, pg. 144.

12. Consequence Scales

The guidance referenced in the NVIC did not specify that consequences needed to be quantified, but in practice it was helpful to frame the risks across broad categories of consequences such as potential fatalities or injuries, economic impacts, and environmental impacts.

Sandia did not publish a prescribed factor for estimated fatalities, injuries, or damages within each zone, so the applicant would have to draw those conclusions if more detailed results are desired. Such factors as ability to shelter for Zone 2 should be considered, which is very difficult to estimate, but for which some assumptions can be made. Probit equations can be applied to more realistically estimate the impacts within a zone.

To facilitate the rough estimate of consequences a scale can be developed for the study as shown in Figure 8.

Figure 8
Example WSA Consequence Matrix

Severity (S)	Fatality/ Injuries	Environmental	Offsite Dollar Loss	Operational Disruption to Vessel	Disruption to Terminal	Economic Impact on Port
5	Major fatalities	Major environmental impact onsite or offsite; large-scale contamination to public waterway	>Z $B	>1 year	>1 year	Closure of >Z days
4	Possible large number of fatalities	Very large environmental impact onsite and/or large offsite impact	Y-Z $B	6 months - 1 year	6 months - 1 year	Closure of Y-Z days
3	No fatalities	Environmental impact onsite and/or minor offsite impact	X-Y $B	3 months to 6 months	3 months to 6 months	Closure of port for <Y days
2	Major injuries	Minor environmental impacts to immediate incident site area only	W-X $M	1 week to 3 months	1 week to 3 months	
1	Minor injuries only onsite	No environmental impacts	<$20M	less than 1 week	less than 1 week	

13. Nominal vs. Specific Consequence Estimates

The guidance from Sandia was not intended to provide site specific consequence information. Zones of Concern were derived from the work to assist with the application of the WSA process.

For example, Sandia reported that LNG cargo tank hole sizes for most credible threats range from two to twelve square meters but that nominal sizes for intentional threats are five square meters. Sandia developed three Zones of Concern to address the pool fire and vapor cloud hazards that may result from accidental or intentional results. The Zones of Concern were developed along the concept of providing nominal hazard areas although each attack vector results in a specifically different outcome. As a result the WSA study is limited to a nominal result if the guidance is followed directly.

One possible solution to this problem is to consider more specifically the pool fire radiation vs. distance, so that the applicant could produce a scenario-specific vulnerability vs. consequence estimate. This could assist teams by allowing for more measured impacts to be estimated for specific infrastructure or public areas. In fact Sandia encouraged site-specific use of the model and discouraged blanket use of the nominal zones as representative of the potential consequences at all sites.

14. Specific vs. Generalized Attack and Consequence Pairings

Since the DOE report that explained the attack vector and weapon size vs. hole size and orientation is classified information, the applicant is encouraged to work with the USCG to understand the specific actions that are required to cause given outcome. As a result the guidance from NVIC 05-05 is general, meaning it isn't easily possible to correlate attack mode and scenario to consequences other than to assume that up to the same three zones are possible for any attack. Some assumptions could be made to derive more specific consequence estimates.

15. Generalized Hazard Zones vs. Specific Hazards

The zones were not necessarily intended for generalized risk assessment. Sandia reported that 'current LNG spill and dispersion modeling and analysis techniques have limitations. Variations exist in location-specific conditions that influence dispersion, such as terrain, weather conditions, waves, currents, and the presence of obstacles'. Therefore, they provided guidance on the general range of hazards for potential spills rather than suggest a specific, maximum hazard guideline.

Within a given zone there is no detailed process for consequence analysis prescribed. There is no guidance on how to estimate damages or numbers of injuries or fatalities. Their report gave no guidance on estimating economic or cascading impacts to port infrastructure. These issues lead to the need for the WSA team to develop several processes to semi-quantify, in a rough sense at least, the impacts.

16. Stakeholder Involvement

The WSA is to address the safety and security issues of the various stakeholders. It accomplishes that goal, but mostly by proxy due to the nature of the sensitive homeland security issues involved and the expertise required. The public is not directly involved – there is a representative committee formed to address the WSA process. Sandia recommended that the "risk identification and risk management processes should be conducted in cooperation with appropriate stakeholders, including public safety officials and elected public officials. Considerations should include site-specific conditions, available intelligence, threat assessments, safety and security operations, and available resources." The information is protected by Security Sensitive Information (SSI) requirements, which precludes public disclosure. As a result there is a representative committee formed and there will always be a level of trust required of officials managing the process. FSAs, on the other hand, are generally done without the same level of involvement of the public that the WSA guidance recommends.

17. Suggested Improvements

The analysis of an LNG operation's security risks is not complete unless the entire value chain is evaluated. It would be beneficial if there was a common vulnerability assessment process so that the same scales and assumptions were used. For transportation this may include ship, truck, or pipeline transportation of LNG or associated materials, and may include consideration of both facility and public risks.

A common methodology was proven to be useful for the evaluation of both facility and transportation risks. As such, the API/NPRA SVA Methodology was seamlessly expanded to include transportation risks simply by applying the methodology to the issue and adapting the technique to the need.

Also, applicants attempting a WSA are encouraged to fill in the site specific factors to result in a more specific and complete analysis. The paper described several areas where the guidance provided by NVIC 05-05 doesn't give a complete solution and where an additional risk management framework is suggested. The API/NPRA SVA methodology again assisted in structuring the WSA process.

Acknowledgements

The authors wish to thank Lieutenant Peter Gooding, Chief of the Waterways Management Division for the United States Coast Guard at Sector Los Angeles - Long Beach, and Mike Hightower, Sandia National Laboratories, for their helpful review and comments on this paper.

ITM Syngas: Ceramic Membrane Technology for Lower Cost Conversion of Natural Gas

Christopher M. Chen[1], Douglas L. Bennett[1], Michael F. Carolan[1], Christopher F. Miller[1], James J. Steppan[2], William E. Waldron[1]*

[1.] Air Products and Chemicals, Inc., 7201 Hamilton Blvd, Allentown, PA, 18195 USA
[2.] Ceramatec, Inc., 2425 South 900 West, Salt Lake City, UT, 84119 USA
** Correspondence should be addressed to C. M. Chen, email: chencm@airproducts.com, tel: +1 (610) 481-3315*

Copyright © 2006 Air Products and Chemicals, Inc.

Prepared for Presentation at the AIChE Spring National Meeting
Orlando, FL April 25, 2006

Introduction

The ITM Syngas Team, led by Air Products and including Chevron, Ceramatec, and other partners, in collaboration with the U.S. Department of Energy, is developing Ion Transport Membrane (ITM) technology for the production of synthesis gas. ITM ceramic membranes are fabricated from non-porous, multi-component metallic oxides and operate at high temperatures (typically over 700 °C) with exceptionally high oxygen flux and selectivity. A conceptualization of the ITM Syngas technology is shown in Figure 1. Oxygen from low-pressure air permeates, as oxygen ions, through the ceramic membrane and is consumed through chemical reactions, thus creating a chemical driving force that pulls oxygen ions across the membrane at high rates. The oxygen reacts with natural gas in a partial oxidation process to produce a hydrogen and carbon monoxide mixture (synthesis gas).

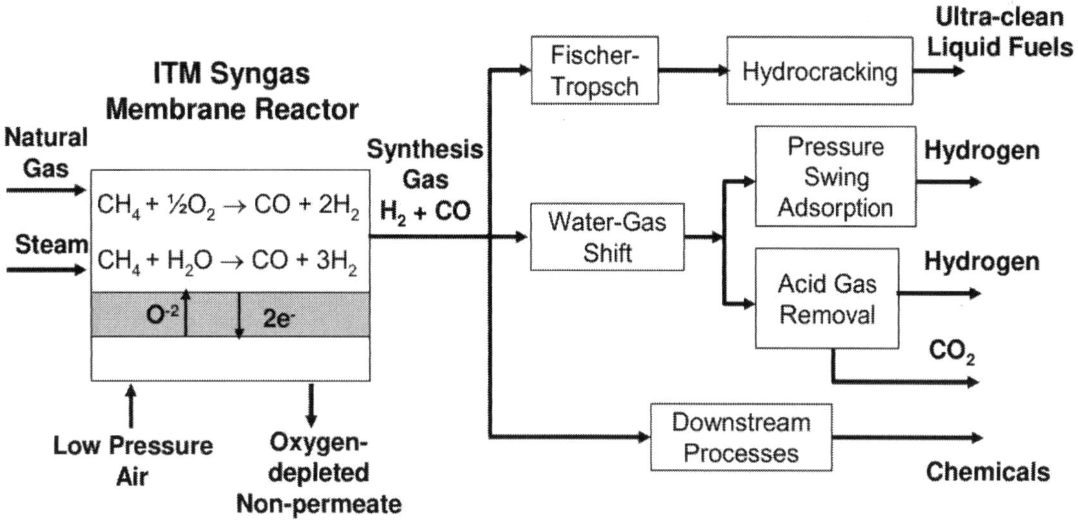

Figure 1. Conceptual ITM Syngas process showing multiple synthesis gas applications

Applications

The ITM Syngas process is a breakthrough technology that combines air separation and high-temperature synthesis gas generation processes into a single ceramic membrane reactor, with significant savings in the cost of synthesis gas production. Because synthesis gas is a feedstock for a range of different processes, ITM Syngas represents a technology platform that has numerous applications, such as gas-to-liquids; hydrogen; ultra-clean fuels, including liquid transportation fuels; and chemicals such as methanol (Figure 1).

Process designs and cost estimates were developed for single-train Fischer-Tropsch Gas-to-Liquids plants producing 6,000 to 18,000 barrels/day. The capital cost of the synthesis gas generation process was projected to be 30% lower for ITM Syngas compared to conventional technology employing oxygen-blown Autothermal Reforming (ATR) with a cryogenic Air Separation Unit (ASU). The 6,000 barrel/day GTL plant was evaluated for a Floating Production Storage and Offloading (FPSO) application, and footprint reductions of up to 40% were identified compared to ATR/ASU.

Applications to co-produce hydrogen and carbon dioxide, such as use of the hydrogen as a decarbonized fuel for centralized or distributed power applications, were also evaluated. ITM Syngas processes producing 150 to 760 MMSCFD (million standard cubic feet per day) of fuel-grade hydrogen were compared to conventional processes based on ATR/ASU. Economic evaluation of the ITM Syngas process showed the potential for 30% capital cost savings in the synthesis gas generation process area and over 20% capital cost savings for the overall synthesis gas/H_2/CO_2 production process. The ITM Syngas process also had a higher thermal efficiency of 74% compared to 71% for the oxygen-blown ATR process. The capital cost and efficiency advantages for ITM Syngas are due to the combination of oxygen separation and high temperature synthesis gas production into a single unit operation and the ability to use low-pressure air as the oxidant source.

Membrane Design and Manufacture

The economics of the ITM Syngas process are most attractive when a high-pressure natural gas feed and a low-pressure air feed are used to produce a synthesis gas product at a high pressure that matches downstream process pressures. This arrangement avoids the expenses associated with compressing the air or the synthesis gas. Ceramic membranes for this application must withstand the stresses generated by the large pressure difference between the high-pressure natural gas/synthesis gas on one side of the membrane and the low-pressure air on the opposite side.

Within the ITM Syngas development program, a planar ceramic membrane was designed and fabricated that is able to achieve high oxygen fluxes while withstanding the thermo-mechanical stresses encountered during operation.[1] The membrane design incorporates micro-channel features to reduce the thermo-mechanical stresses that occur within the membrane assembly because of the pressure loads and the exothermic oxidation reactions. The micro-channel features also provide for efficient distribution of the gases to the membrane surfaces: high-pressure natural gas/synthesis gas flowing outside the membrane and low-pressure air flowing through the micro-channels inside the membrane. Figure 2 illustrates the structure of ITM Syngas micro-channeled planar membrane.

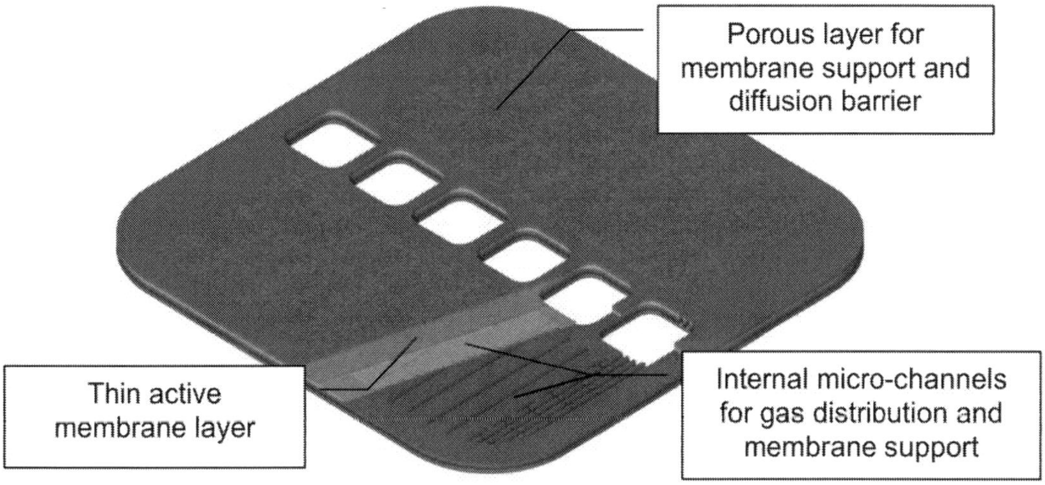

Figure 2. Cut-away view of the ITM Syngas ceramic membrane design showing the multi-layer structure.

The planar membrane design enables very efficient packing of membrane surfaces within the reactor volume. An ITM Syngas module is assembled by stacking membranes vertically. Figure 3 shows a submodule simulant built from 4 full-size ceramic membranes. Ceramic spacers are placed between the individual membranes to create channels for the flow of natural gas and synthesis gas between wafers. Various sizes of modules consisting of pilot-scale membranes or of full-size membranes have been joined and sealed using a ceramic sealant suitable for operating temperatures up to 1100 °C [2].

Figure 3. Full-size ITM Syngas ceramic membranes shown as part of a 4-membrane submodule simulant

The primary challenge in commercializing micro-channel devices has been the ability to mass produce the micro-channel components in the preferred materials at a commercially viable cost. The ITM Syngas planar membranes were designed with a multi-layer structure that can be manufactured using commercial-scale production and automation equipment.

In order to attain dimensional tolerances, a Laminated Object Manufacturing (LOM) method was developed. This LOM process begins with ceramic powders within a specified process window for surface area and particle size. These powders are blended with a mix of organic binders, plasticizers and solvents to form a flowable ceramic slip. This slip is then cast onto a polyester film that dries and cures into a flexible, ceramic-filled plastic tape. This process is called tape casting. Once cured, these tapes can be featured by numerous machining technologies, such as CNC laser machining and punching. The featured, plastic layers are pressure-laminated and bonded together. Once these layers are laminated, they are heated to remove the organics and to densify the ceramic powders. The densification step is called sintering.

Test Results

A laboratory Process Development Unit (PDU) was constructed at Air Products and is used to test the pilot scale planar ceramic membranes. The PDU operates at nominally 24 MSCFD (thousand standard cubic feet per day) of synthesis gas capacity, and is used to evaluate membrane performance under full process operating conditions.

PDU tests with pilot-size membranes have included operation under synthesis gas simulating the conditions of a commercial ITM Syngas reactor including temperatures between 825 and 900°C and syngas pressures between 215 and 425 psig. Membranes demonstrated the ability to operate through a range of temperature, pressure, and composition conditions. Figure 4 shows that the ITM Syngas program achieved fluxes at the commercial flux target under simulated commercial conditions.

Recent tests included the successful demonstration of ceramic to ceramic seals [2,3] and ceramic to metal seals [4] at full process conditions. The ceramic to ceramic seals are used to join the sintered ceramic components together to form a single module. The robustness of the ceramic to ceramic seals was demonstrated by the modules remaining leak tight as they were cycled through the range of temperature, pressure and gas composition. The ceramic to metal seals are used to join the ceramic modules to metal tubes supplying the air feed and withdrawing the spent air from the modules. These seals also remained leak tight against the full pressure differential of up to 410 psi at temperatures up to 900 °C.

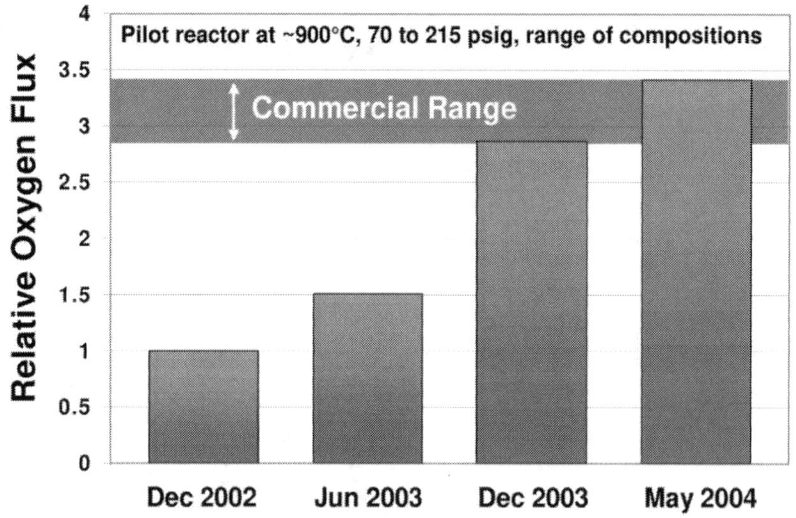

Figure 4. Achievement of oxygen fluxes at the commercial target was demonstrated first in December 2003. Later tests demonstrated fluxes above the commercial target.

To mitigate contamination from reactor components in the PDU, proprietary guard beds were placed upstream of ITM Syngas membranes and proprietary coatings were used on the air- and syngas-side flow paths. These changes to the PDU design significantly reduced membrane contamination and improved membrane performance. Figures 5 and 6 illustrate the improved membrane performance where the oxygen flux through the membrane was higher and more stable during the initial startup compared to tests without the proprietary coatings and guard beds. The improved membrane performance was directly correlated to the reduced amount of contamination on the membrane surface.

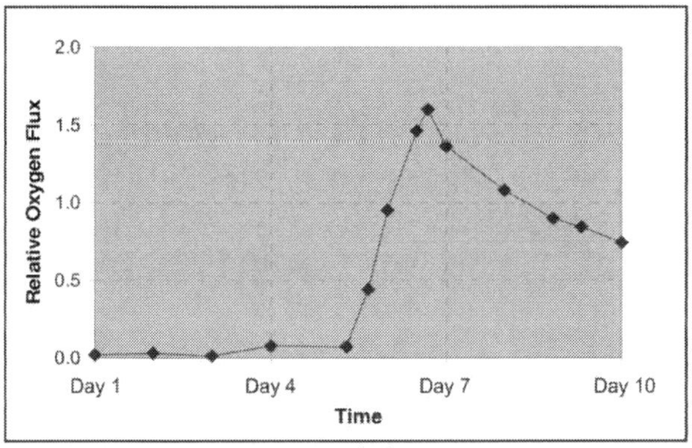

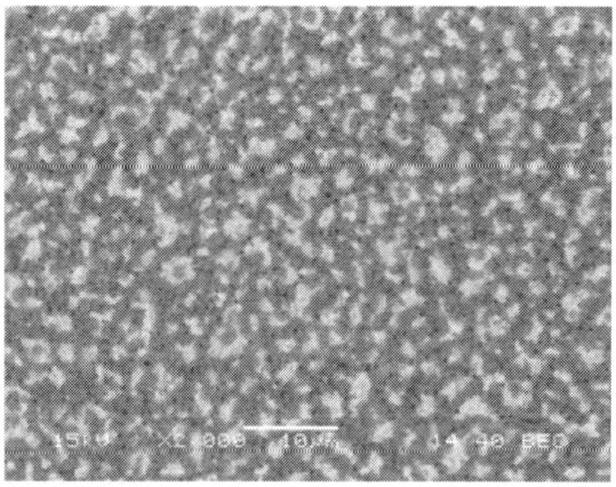

(a) (b)

Figure 5. (a) Oxygen flux decay was observed during initial startup of the PDU when not using proprietary coatings and guard beds. (b) Scanning electron microscopy of the airside surface of the membrane revealed contamination (i.e., dark phases).

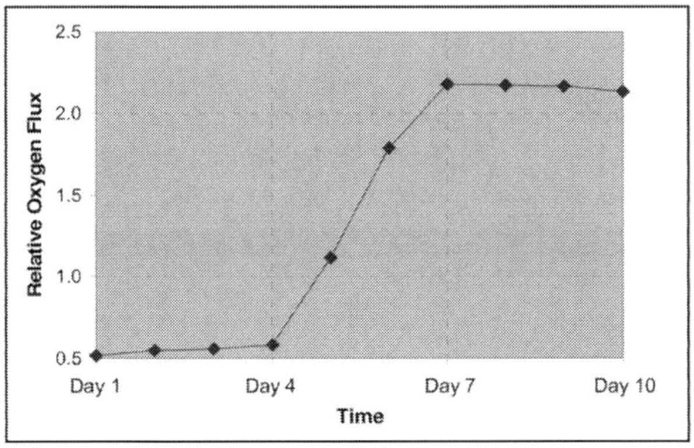

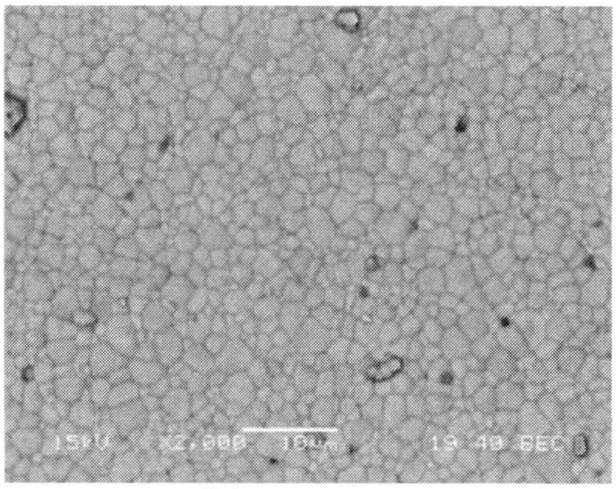

(a) (b)

Figure 6. (a) Stable and high oxygen fluxes were observed during initial startup of the PDU when using proprietary coatings and guard beds. (b) Scanning electron microscopy of the airside surface of the membrane revealed a contaminant-free surface.

Conclusions

Significant progress has been made in designing and manufacturing planar ceramic membranes, and in demonstrating targeted membrane oxygen fluxes, which are necessary steps toward commercializing the ITM Syngas technology. The potential economic benefits of the technology continue to be very promising.

Acknowledgement

The authors gratefully acknowledge the work of ITM Syngas team members at Air Products and Chemicals Inc., Ceramatec Inc., Chevron Energy Technology Co., Norsk Hydro, Eltron Research Inc., SOFCo-EFS Holdings LLC, Penn State University, University of Pennsylvania, and Pacific Northwest National Lab. The co-funding by the U.S. DOE (DE-FC26-97FT96052) is also gratefully acknowledged.

References

[1] US patent application US 2004/0186018 A1, M. Carolan et al., "Planar ceramic membrane assembly and oxidation reactor system"
[2] US Patent application US 2004/0182306 A1, D. Butt et al., "Method of forming a joint"
[3] US Patent application US 2004/0185236 A1, D. Butt et al., "Method of joining ITM materials using a partially or fully-transient liquid phase"
[4] US Patent 6302402 B1, S. Rynders et al., "Compliant high temperature seals for dissimilar materials"

Disclaimer

This paper was written with support of the U.S. Department of Energy (DOE) under Cooperative Agreement DE-FC26-97FT96052. The Government reserves for itself and others acting on its behalf a royalty-free, nonexclusive, irrevocable, worldwide license for Governmental purposes to publish, distribute, translate, duplicate, exhibit and perform this copyrighted paper.

Neither Air Products and Chemicals, Inc. nor any of its contractors or subcontractors nor the DOE, nor any person acting on behalf of either:

1. Makes any warranty or representation, express or implied, with respect to the accuracy, completeness, or usefulness of the information contained in this report, or that the use of any information, apparatus, method, or process disclosed in this report (Technology) may not infringe privately owned rights; or
2. Assumes any liabilities with respect to the use of, or for damages resulting from the use of such Technology.

Reference herein to any specific commercial products, process, or service by trade name, trademark, manufacturer, or otherwise, does not necessarily constitute or imply its endorsement, recommendation, or favoring by the DOE. The views and opinions of authors expressed herein do not necessarily state or reflect those of the DOE.

Metal Promoted Binary Oxides of Ceria and Zirconia for Low Temperature Water-Gas Shift

Gary Jacobs, Sandrine Ricote, Yaying Ji, Patricia M. Patterson and Burtron H. Davis, Center for Applied Energy Research, University of Kentucky, 2540 Research Park Drive, Lexington, KY 40511

Abstract

Mixed oxides (e.g., solid solutions) of ceria and zirconia, promoted with noble metal, like Pt or Au, offer a synergism on the reducibility over metal/ceria or metal/zirconia catalysts alone. For example, by temperature programmed reduction, we observed a shift of more than 50°C to lower temperature in the peak assigned to reduction of the oxide surface shell. This is an important step for activating the Type II bridging OH groups on the surface. A series of mixed oxides of ceria and zirconia were prepared and characterized by a number of techniques, including TPR, XRD, EXAFS, and infrared spectroscopy, and tested for the low temperature water-gas shift reaction, an important reaction for the purification of hydrogen for fuel cell applications. The feed utilized contained CO, H_2O, H_2 and CO_2, in order to simulate the low temperature stage of a fuel processor. Increasing the Zr content, activity was found to pass through a maximum, and then decrease, at a Ce:Zr ratio of 1:1. Results are explained in terms of the number of active sites on the catalyst surface, and the rate of decomposition of surface formate intermediates, the proposed rate limiting step of the mechanism. Inhibition of the WGS rate due to carbon dioxide was also examined.

Introduction

Ceria is an important component in automobile catalytic converter catalysts for buffering the partial pressure of oxygen. To facilitate bulk to surface oxygen transport, zirconium is typically doped into the ceria cubic matrix. Ceria-zirconia binary solid solutions have also been reported to exhibit far greater oxygen storage capacity (OSC) than ceria alone. These characteristics are also of interest from the perspective of the water-gas shift reaction [1].

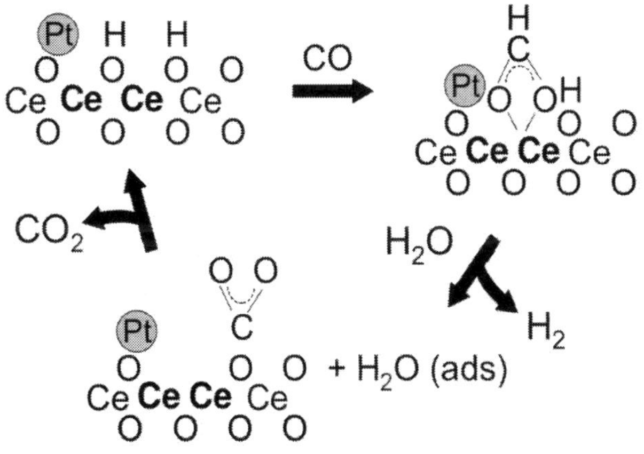

where Ce = Ce^{4+} and **Ce** = Ce^{3+}

In the catalysis community, there is general agreement that reduced defect sites are important sites in the water-gas shift mechanism over metal promoted, ceria-based catalysts. There are two proposed mechanisms for how these active sites operate. One proposed mechanism for water-gas shift is a surface formate mechanism [2,3]. In this mechanism, reduction of the surface of ceria is facilitated by addition of a metal promoter [4], resulting in the formation of bridging OH groups at Ce^{3+} defect sites. One possibility for their formation is the dissociation of H_2O at oxygen vacancies generated during the surface reduction process [2]:

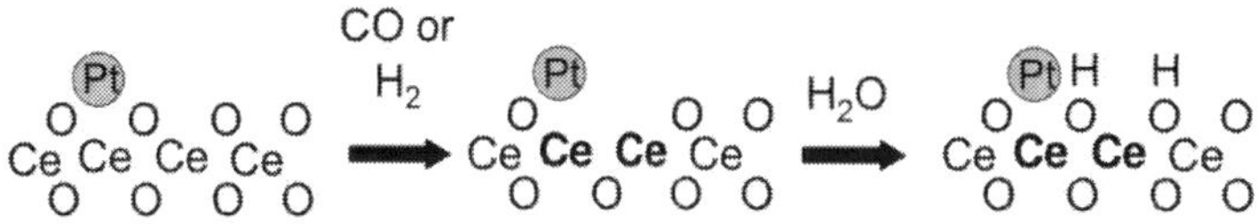

However, another possibility is direct formation of the bridging OH groups via H_2 dissociation on the metal and spillover to the oxide, as follows [3]:

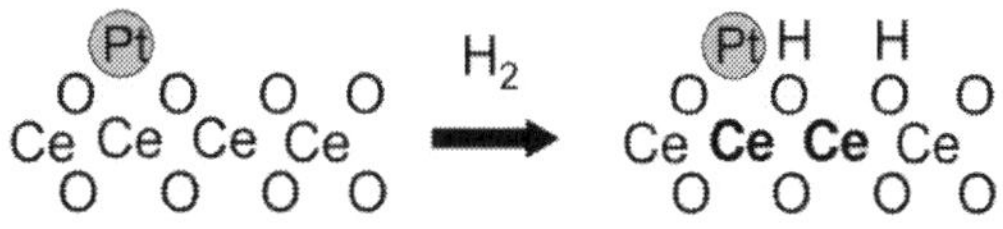

The role of the metal has been suggested to facilitate the shuttling of hydrogen to and from the surface of the ceria and the formate intermediates [3], important for (1) the generation of the active site bridging OH groups and (2) the removal of H_2 during water-assisted surface formate decomposition [3], which is proposed to be the rate-limiting step based on the results of isotopic studies [5].

Alternately, a ceria-mediated redox process [1] has been proposed to operate for low temperature water-gas shift. In this mechanism, CO adsorbed on the metal is suggested to react at the interface with ceria, reducing the ceria surface from the Ce^{4+} to the Ce^{3+} oxidation state. It has been proposed that H_2O then reoxidizes the ceria, liberating H_2 in the process.

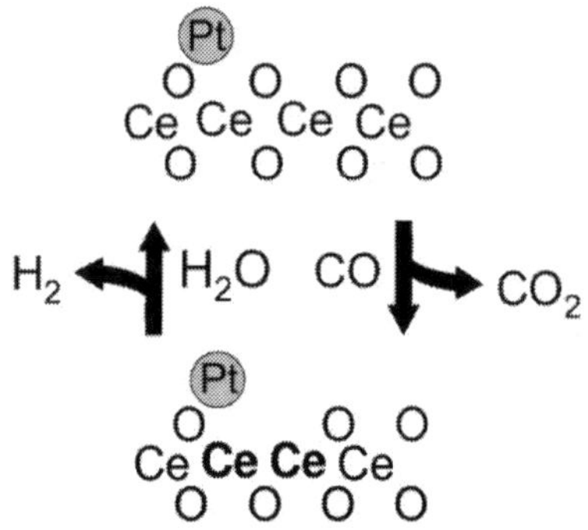

where Ce = Ce^{4+} and **Ce** = Ce^{3+}

Here, a number of spectroscopic techniques are used to gain insight into the impact of doping Zr to Pt/ceria catalysts. To estimate the surface composition of Ce relative to Zr, x-ray photoelectron spectroscopy (XPS) was utilized. To examine whether or not a binary oxide solid solution was formed, extended x-ray absorption fine structure spectroscopy (EXAFS) was employed. The surface reduction processes were monitored by a combination of temperature programmed reduction (TPR) and x-ray absorption near edge spectroscopy (XANES). To examine the impact of zirconia addition on the intensity of formates present on the catalyst surface, as well as their decomposition rate in the presence of steam, diffuse reflectance infrared Fourier transform spectroscopy was carried out using an in-situ reactor cell. Finally, activity studies were conducted using a fixed bed reactor system. The feed composition was selected to mimic typical conditions expected in the low temperature shift stage of a fuel processor for hydrogen production/purification for fuel cell applications. Therefore, a high H_2O:CO ratio was used, and H_2 and CO_2 were included.

2. Experimental

Along with preparation of ceria and zirconia, four ceria-zirconia binary oxides were also prepared, including $Ce_{0.90}Zr_{0.10}O_2$, $Ce_{0.75}Zr_{0.25}O_2$, $Ce_{0.50}Zr_{0.50}O_2$, and $Ce_{0.75}Zr_{0.25}O_2$. The precursor salts used were cerium nitrate and zirconyl nitrate, which were completely dissolved in aqueous solution that was mildly heated. Urea was used to slowly release the precipitating agent, the hydroxide ion, a technique that generally leads to a more homogenous precipitation, and significantly higher surface area materials relative to standard precipitation with ammonium nitrate alone [4]. After filtration, washing, and drying, the ceria was calcined. Tetraammine platinum (II) nitrate was the metal promoter precursor.

BET surface area measurements were conducted with a Micromeritics Tristar 3000 gas adsorption analyzer. In each test, approximately 0.25-0.30 g of sample (either binary oxide or Pt promoted catalyst) was used. Nitrogen adsorption was carried out at its boiling temperature.

Temperature programmed reduction (TPR) was conducted on binary oxide and Pt promoted catalysts using a Zeton-Altamira AMI-200 unit, equipped with a thermal conductivity detector (TCD). The system was modified to allow heating to as high as 1150°C. Argon was used as the reference gas, and 10%H_2 (balance Ar) was flowed through the catalyst at 30 cm^3/min as the temperature was increased from 50 to 800°C at a ramp rate of 10°C/min.

XANES spectra at the Ce L_{III} and Zr K edges, and EXAFS spectra at the Zr K edge, were recorded at Beamline X-18b at the National Synchrotron Light Source (NSLS) at Brookhaven National Laboratory (BNL) in Upton, New York. After the cell was purged for a long period with inert gas at a high flow rate to ensure air removal, the samples were treated in situ using a hydrogen/helium mixture (150 cm^3/min H_2 and 500 cm^3/min He). Samples were scanned in the temperature range of room temperature to 300°C. Scans were obtained in transmission mode to confirm the temperature for surface shell reduction of the partially reducible oxide. XANES spectra were background-subtracted and normalized prior to comparison. Standard data reduction methods were utilized to process the EXAFS results including edge calibration, background subtraction, normalization, truncation, conversion to k-space, background subtraction in k-space using appropriate splines to determine Chi(k). Finally, an appropriate k-range was selected, and the Fourier transform was carried out. The Fourier transform magnitude spectra were compared qualitatively.

The XPS experiments were monitored with a Thermo VG Scientific Minilab. Mg K alpha X-ray (Kinetic energy 1273 eV) were used with a dual cathode X-ray gun. The pressure was maintained in analysis chamber generally between 9 and 5 E-10 Torr. The data were collected using a hemispherical electron analyzer. The spectra were first base line corrected. Then, for both, ceria and zirconia, the area for the 3d3/2 and the 3d5/2 peaks were calculated, and for oxygen and carbon, the area of the 1s peak was calculated. Scofield factors were employed to estimate the different atomic ratios of interest.

For IR spectroscopy measurements, a Nicolet Nexus 870 was used, equipped with a DTGS-TEC detector. A chamber fitted with ZnSe windows was utilized as the WGS reactor for in-situ measurements. During a run the catalyst was first activated at 300°C in hydrogen and then cooled to 130oC in nitrogen. Formates were generated on the surface by adsorption of CO, whereby the flows were maintained at 3.75 cm^3/min CO and 135 cm^3/min N_2. After, a mixture of H_2O:N_2 was flowed at 63:135 cm^3/min. Decomposition of the formate was recorded with time to provide an indication of the impact of the Zr dopant on the formate decomposition rate.

Steady state catalytic activity and carbon dioxide sensitivity measurements conducted using a fixed bed reactor system consisting of a 0.5 in stainless steel tube with an internal thermocouple located inside the catalyst bed. Experiments were conducted using 200 mg of catalyst diluted with 1.8 g of silica beads. Conditions were chosen to reflect those in a fuel processor following the high temperature shift stage, including H_2 and CO_2 in the feed.

3. Results

Table 1: BET surface areas and atomic surface ratios from XPS

Oxide Description	BET SA (m^2/g)	Atomic ratios from XPS			
		Ce/O	Zr/O	(Ce+Zr)/O	Zr/Ce*
CeO_2	109.8	0.47	-	0.47	0.00 (0.00)
$Ce_{0.90}Zr_{0.10}O_2$	134.5	0.38	0.12	0.50	0.32 (0.11)
$Ce_{0.75}Zr_{0.25}O_2$	129.1	0.35	0.15	0.50	0.43 (0.33)
$Ce_{0.50}Zr_{0.50}O_2$	118.6	0.34	0.16	0.50	0.46 (1.00)
$Ce_{0.25}Zr_{0.75}O_2$	141.9	0.09	0.41	0.50	4.88 (3.00)
ZrO_2	124.0	-	0.48	0.48	-

* nominal bulk ratios.

Table 2: Conversion vs. temperature. $H_2O:CO:H_2:CO_2:N_2$ = 26.1%:3.0%:29.8%:11.2%:29.9%. Total gas flow 250 cm^3/min.

Catalyst description	225°C	250°C	275°C	300°C
1%Pt/CeO_2	13.8	26.0	43.3	61.3
1%Pt/$Ce_{0.90}Zr_{0.10}O_2$	13.7	33.9	57.4	70.9
1%Pt/$Ce_{0.75}Zr_{0.25}O_2$	16.2	39.3	54.3	67.9
1%Pt/$Ce_{0.50}Zr_{0.50}O_2$	16.7	41.3	58.5	73.6
1%Pt/$Ce_{0.25}Zr_{0.75}O_2$	17.5	33.3	47.1	64.3
1%Pt/ZrO_2	7.0	16.5	26.9	42.7

Table 3: Link between formate decomposition rate and water-gas shift rate

Catalyst description	formate 0.15-life at 130°C (min)	first order assumed k (min^{-1})	relative site from initial formate area	relative rate using 50%Zr case as basis	relative rate based on X_{CO} (250°C) using 50%Zr as basis
1%Pt/CeO_2	17.5	0.11	1.00	0.50	0.63
1%Pt/$Ce_{0.90}Zr_{0.10}O_2$	9.1	0.21	0.73	0.70	0.82
1%Pt/$Ce_{0.75}Zr_{0.25}O_2$	8.2	0.23	0.75	0.80	0.95
1%Pt/$Ce_{0.50}Zr_{0.50}O_2$	6.4	0.30	0.73	1.00	1.00
1%Pt/$Ce_{0.25}Zr_{0.75}O_2$	8.0	0.24	0.70	0.76	0.81
1%Pt/ZrO_2	18.0*	0.11	0.48	0.23	0.39

Table 4: Catalyst response to added CO_2. Initial conditions $H_2O:CO:H_2O:CO_2:N_2$ = 30.8%:3.5%:35.1%:0%30.6%. Final: $H_2O:CO:H_2O:CO_2:N_2$ = 30.8%:3.5%:35.1%:8.0%:22.6%. T range: 235 to 275°C. Total gas flow: 212 cm^3/min.

Catalyst description	X_{CO} CO + H_2O + H_2 + N_2	X_{CO} CO + H_2O + H_2 + CO_2	Drop, % Initial	X_{CO} CO + H_2O + H_2 + N_2	Recovery, % Initial
1%Pt/CeO_2	38.9	24.8	63.7	32.4	83.2
1%Pt/$Ce_{0.50}Zr_{0.50}O_2$	84.6	60.3	71.3	79.9	94.4
1%Pt/$Ce_{0.50}Zr_{0.50}O_2$	35.3	22.8	64.6	32.2	91.3
1%Pt/ZrO_2	34.4	23.9	69.4	29.9	87.0

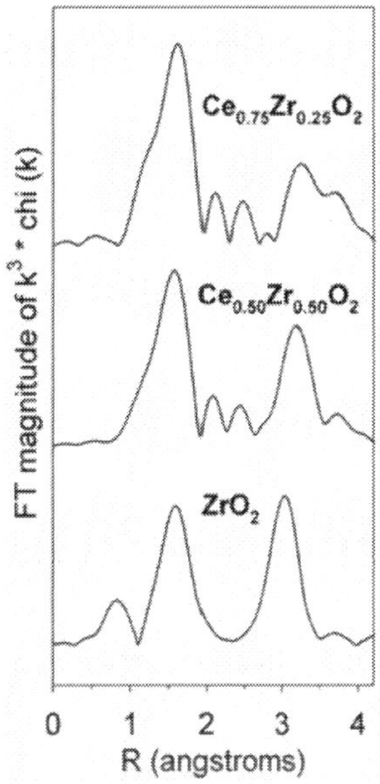

Figure 1: EXAFS FT magnitude spectra. K-edge of zirconium

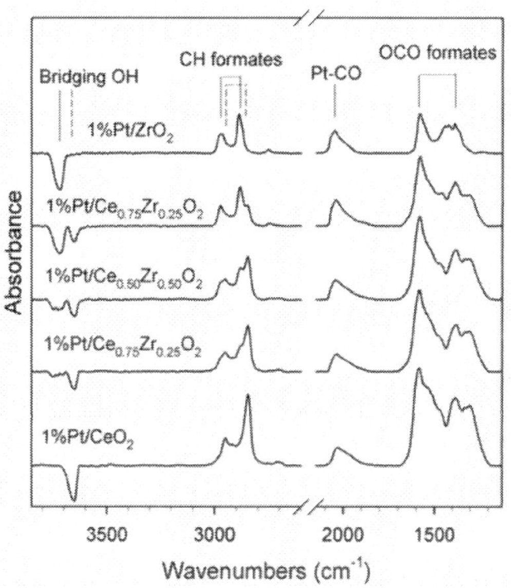

Figure 2: DRIFTS spectra for adsorption of CO to H_2 treated catalysts.

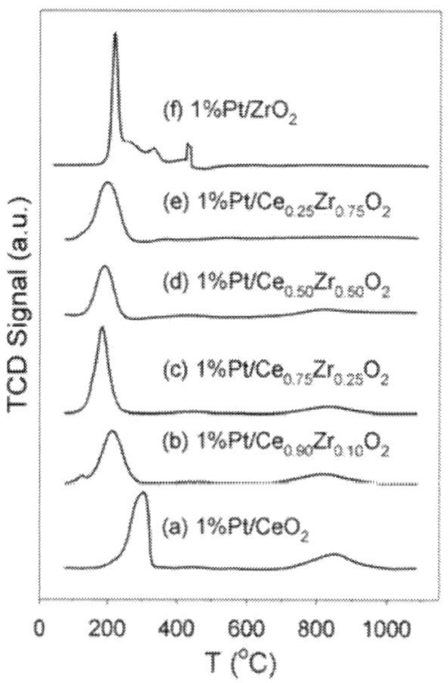

Figure 3: H_2 TPR profiles.

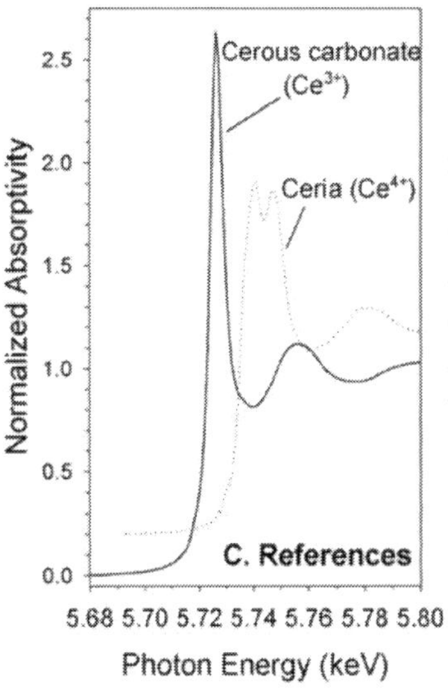

Figure 4: XANES reference spectra for Ce^{3+} and Ce^{4+}.

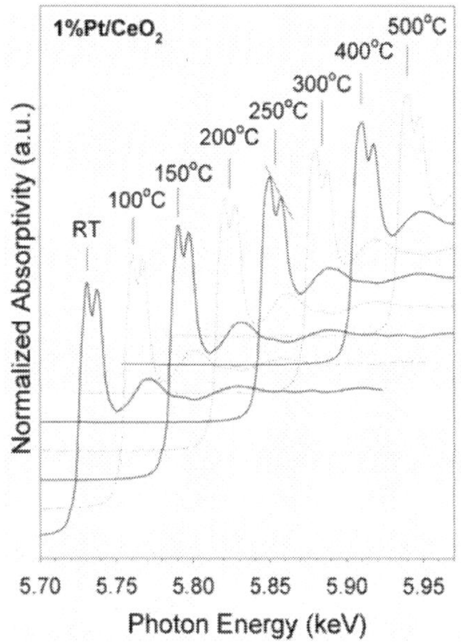

Figure 5: XANES spectra for Pt/ceria in H_2. Partial reduction of ceria at ~250°C.

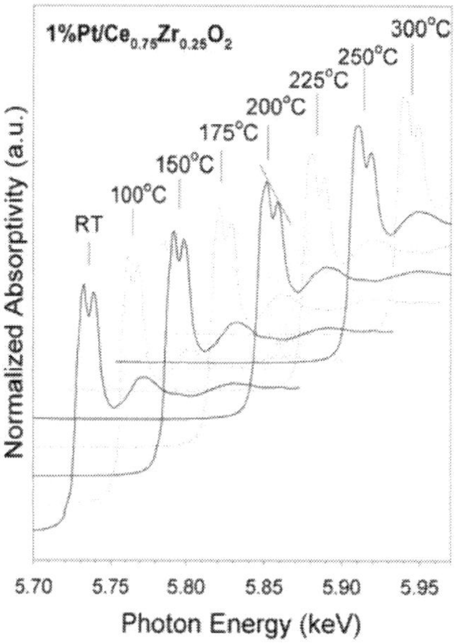

Figure 6: XANES spectra fro Pt/ceria-zirconia in H_2. Partial reduction of ceria at ~200°C.

4. Discussion

Resulting surface areas of the oxides were in the range of 110 – 140 m^2/g, as shown in Table 1. The surface atomic ratios are also provided, and detailed spectra are available elsewhere [6]. Although the surface atomic ratios are close to what is expected in the case of the formation of a solid solution in some cases (e.g., $Ce_{0.75}Zr_{0.25}O_2$, $Ce_{0.25}Zr_{0.75}O_2$), in other cases, the surface is enriched with one particular element over the other (e.g., $Ce_{0.90}Zr_{0.10}O_2$, $Ce_{0.50}Zr_{0.50}O_2$). The EXAFS Fourier transform magnitude (Zr K edge) spectra (Figure 1) suggest that at least a fraction of the ceria forms a solid solution with the zirconia, as a distinct peak arises in the

Fourier transform magnitude spectra at high Ce content at 3.7 Angstroms, corresponding to the coordination of Zr (absorber) to Ce (backscatterer).

DRIFTS spectra are recorded in Figure 2. First, the catalyst was activated in H_2 treatment at 300°C prior and cooled to 130°C prior to CO adsorption. The spectra indicate that CO reacts with the Type II bridging OH groups on the Ce-rich (~3650 cm^{-1}) and Zr-rich (~ 3700 cm^{-1}) surface phases, and this reaction produces surface formate species on the surface. The position of the formate bands on ceria are slightly offset from those on ceria, with the main C-H stretching band situated at 2840 cm^{-1} for ceria, and 2880 cm^{-1} for zirconia). The formate OCO asymmetric and symmetric stretching bands are also evident. The total intensity of formates decreases with increasing Zr content, suggesting that the total number of active bridging OH groups actually decrease.

For unpromoted ceria, surface reduction occurs in the range of 450-500°C. However, addition of a metal promoter, like 1%Pt, facilitates the reduction of the surface shell of ceria to ~250-275°C. In this way, the active bridging OH groups are generated at the lower temperature. Interestingly, as shown in Figure 3, the surface reduction step is further promoted by doping Zr into ceria. At 25-50% Zr doping, Figure 3 shows that the temperature is shifted to about 175-200°C. So, while DRIFTS indicates that the total number of bridging OH groups decreases with Zr doping (negative effect), the temperature at which they form is lower (beneficial effect). The promoting impact of Zr on the reduction of ceria was further tested using XANES (Figures 4-6). When partial reduction of ceria takes place (Figures 5,6), the spectra resemble a combination of the Ce^{4+} and Ce^{3+} oxidation states (Figure 4). Clearly, the partial reduction of ceria occurs at a lower temperature (~175-200°C) for the Zr doped Pt/ceria catalyst, as opposed to the higher temperature (250-275°C) for undoped Pt/ceria, in agreement with TPR results. XANES results for the Zr edge, not shown, indicate that Zr remains largely in the Zr^{4+} oxidation state, which can account for a declining number of surface defects (reduced sites) in the oxide with Zr content.

The net result on catalytic activity is summarized in Tables 2-4. Testing of the catalyst in the fixed bed reactor (Table 2) indicated that CO conversion passed through a maximum with Zr loading, the best catalyst being that with a 50% Zr doping level. Setting this catalyst as a basis, relative rates for WGS and formate decomposition are assessed (Table 3). Total sites were ranked on the basis of initial formate band intensity, while their 0.15-life was assessed from transient formate decomposition tests in H_2O at 130°C. Good agreement is observed between relative WGS and formate decomposition rates, suggesting that H transferred from formate may be facilitated with doping of Zr. Zr-doped catalysts were also found to exhibit a better recovery in CO_2 addition experiments (Table 4). This is likely due to the suppression of carbonate buildup with the presence of Zr.

Conclusions

A measurable, but limited, benefit in WGS rates are realized by doping Zr atoms into Pt/ceria catalysts. While addition of Zr to Pt/ceria decreases the temperature for surface shell reduction of ceria, as monitored by TPR and XANES, the total number of defect sites (bridging OH group active sites) decreases, since Zr remains largely in the Zr^{4+} oxidation state. The net result is that CO conversion (and formate intermediate decomposition rate) passes through a maximum at 50% Zr-doping levels.

References

1. T. Bunluesin, R. Gorte, G. Graham, Appl. Catal. B 15 (1998) 107.
2. T. Shido and Y. Iwasawa, J. Catal. 141 (1993) 71.
3. G. Jacobs, P. Patterson, U.M. Graham, A.Crawford, A. Dozier, B.H. Davis, J. Catal. 235 (2005) 79.
4. Y. Li, Q. Fu, M. Flytzani-Stephanopoulos, Appl. Catal. B 27 (2000) 179.
5. G. Jacobs, A. Crawford, B.H. Davis, Catal. Lett. 100 (2005) 147.
6. S. Ricote, G. Jacobs, M. Milling, Y. Ji, P.M. Patterson, B.H. Davis, Appl. Catal. A, *in press*.

Efficiency Study in Small Scale Hydrogen Production by Methane Steam Reforming in High Temperature Catalytic and Low Temperature Plasma Processes

T. Pham, T. Hoang, L.L. Lobban, R.G. Mallinson

School of Chemical, Biological and Materials Engineering
Institute for Natural Gas Utilization Technologies
University of Oklahoma
100 E. Boyd St, T-335, Norman, Oklahoma 73019

Prepared for Presentation at Gas to Chemicals session for 2006 AIChE Spring meeting
April 23rd -27th , 2006

Copyright © *T.Pham, T.Hoang, L.L. Lobban, R.G. Mallinson*

March 2006
Unpublished

AIChE shall not be responsible for statements or opinions contained in papers or printed in its publications

Abstract

Catalytic methane steam reforming is relatively efficient for large scale systems with integrated heat recovery. However, for smaller scale systems in distributed production, optimized heat recovery will be less complete, limiting the efficiency that may be achieved. Low temperature plasmas use electricity with limitations on electrical power generation efficiency, but with small amounts of unrecoverable heat production. For distributed production, the efficiency difference between the two processes may be small compared to other costs and complexities.

I. Hydrogen role in the overall U.S energy

The overall consumption energy level of the U.S was 99.74 quadrillion Btu (1 Quad=10^{15} Btu) in 2004. Fossil fuels were consumed as high as 85.65 Quads. One third of the total energy consumption went to transportation (27.70 quads) which includes conventional fuels such as gasoline and diesel. With a current hydrogen production of 9 million tons per year, which is equivalent to 1 Quad, hydrogen is projected to be produced to 10 quads by 2010[2] to gradually replace one third of the transportation energy demand, which has been solely depended on conventional fuels. In the United States, hydrogen is a major feedstock for ammonia synthesis and refining industry such as desulfurization and hydrotreating processes. The projection of one-quad hydrogen production by 2010 is a challenging task requiring various resources and novel technologies to overcome production cost barriers. Large scale production of hydrogen has been intensively focused on steam reforming of natural gas, which is a fossil fuel. Getting hydrogen from natural gas, although a mature industrial process, is still an efficiency loss process. To get 1 quad of hydrogen would require 1.2 quad of natural gas. Other resources such as solar energy are also considered as an abundant source to produce hydrogen. However, while the required area to install solar panels is justified, the investment on the solar panels is very high. The return on investment is extremely low. But solar energy has still been under study to lower cost and increase overall efficiency. Also looking in the energy flow diagram in Figure 1, most of the energy supply came from fossil fuels (56 Quads ~ more than 50% of supply) while renewable energy contributed only 6.12 Quads. Solar energy has a long way to go if it is to be used to supply hydrogen for transportation sector. Other resources to produce hydrogen such as electrolysis, biomass have also been studied but with similar position with solar energy[3], the supply side needs to be much higher to meet the demand side of 10 Quads of hydrogen by 2010. It is possible to have all the resources pulled together to meet this objective; however, large scale production of hydrogen is thus far limited to the supply of natural gas.

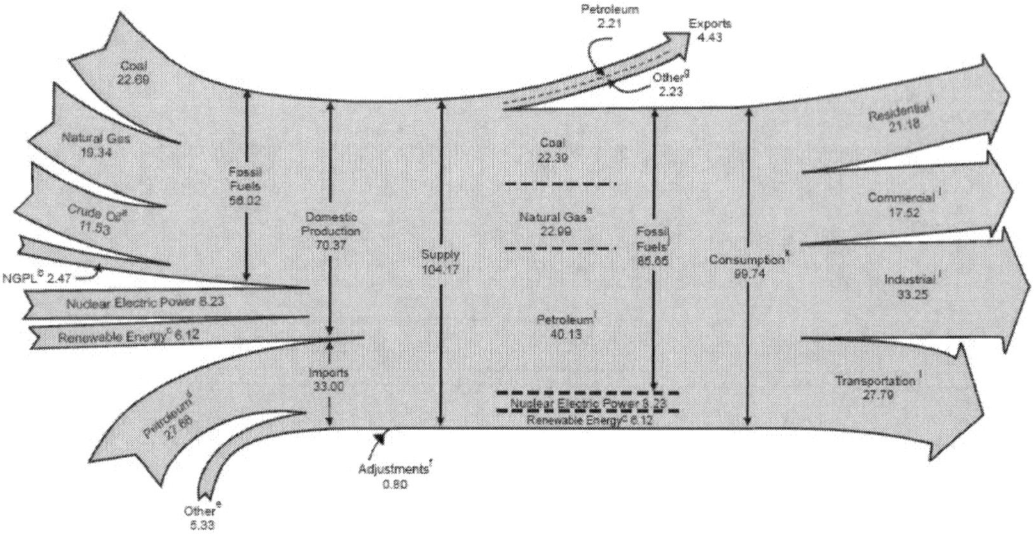

Figure 1: U.S energy flow in 2004 (EIA, 2004)

II. How efficiently is Hydrogen produced?

Efficiency definition:

Efficiency of hydrogen production can be roughly defined by the heat content of the product (hydrogen) relative to the energy input of natural gas on low heating value basis. This is a very direct definition for any hydrogen production process. Within the hydrogen production process, by product such as steam can be credited and reported in the efficiency value. With the consideration of by product, the efficiency becomes the overall efficiency of the process. The efficiency of large scale production of hydrogen is relatively high at 78.1% on LHV basis[3]. This efficiency is conservative in that no credit for steam export is reported. There are two types of efficiency: practical and ideal efficiencies, where the practical efficiency is the real operating efficiency of a plant and the ideal efficiency is the maximum efficiency dictated by thermodynamics. The ideal efficiency is defined as follows[3]:

$$\eta_{ideal} = \frac{\text{LHV product/mol}}{\text{LHV methane/mol}} \bullet \frac{\text{mol product}}{\text{mol methane}} \qquad (1)$$

If based solely on thermodynamics, the ideal efficiency for steam reforming can be as high as 90-120%. The lower end assumes one mole of methane gives three moles of hydrogen while the higher end assumes one mole of methane yields four moles hydrogen with water gas shift reaction. According to Rostrup-Nielson[4], steam reforming of natural gas is a very efficient process. The thermal efficiency of the furnace of a tubular reformer is as high as 95%, and most of the heat input to the process (up to 90%) is recovered from the flue gas and the product. So, why is the practical efficiency only 78.1%, which is much lower than the ideal efficiency? The reason is that the ideal efficiency is dictated by thermodynamics and it is not concerned with any complexity according to the ideal efficiency equation (1). Also, the ideal efficiency assumes complete conversion of methane to hydrogen. The practical efficiency, on the other hand, takes into account other factors such as power consumption, conversion, compression and heat recovery. The practical efficiency of steam reforming is affected by the conversion of methane to hydrogen product and the energy input since this is a highly endothermic process. Increasing conversion produces more hydrogen, however; this requires much more heat input. The normal operating temperature of steam reforming is at 800-900°C and requires furnace for combustion of natural gas to generate the heat input for the reaction. At the temperature range 800-900°C, the methane conversion is up to 95%. Increasing temperature further could increase methane conversion. However, the energy input to raise the temperature is not linear with the methane conversion after 95% conversion point. Thus, operating higher than 900°C hurts the overall efficiency of the process.

Another factor beside conversion is the hydrogen selectivity. Hydrogen selectivity not only depends on the methane converted but also on the CO conversion from the water gas shift reaction. Hydrogen can be further produced if CO is shifted in the water gas shift reaction with water to convert to CO_2. The water gas shift reactions are carried out in two stages including high and low temperature shift, respectively. By doing two stage shift, the equilibrium and catalyst use can be optimized. CO is desired to be fully converted to hydrogen. However, due to thermodynamic equilibrium limitation, CO can only be converted to a 0.3 % composition in the stream after two-stage water-gas-shift. The rest of the CO is converted back to methane in the methanator to prevent catalyst poisoning. For modern plants, pressure swing adsorption purification can be applied to give high purity hydrogen (>99%). However, in conventional plants, although the initial methane conversion is high; the methanation reaction consumes some of the produced hydrogen and decreases the overall methane conversion. The efficiency suffers a decrease of up to 5% due to the stoichiometric ratio of H_2 to CO of 3 to 1 in the methanation reaction that takes up some of the H_2 produced in the first place.

Finally, energy consumption accounts for a major difference in the ideal efficiency and practical efficiency. The ideal efficiency does not take into account the methane feedstock used in the fired heater to drive the reaction. For every mole of hydrogen produced, the practical efficiency, on the other hand, accounts for both the methane used in the reaction and the methane fed to the fired heater. This makes a substantial difference since a great amount of methane is needed for combustion to generate high temperature heat. Proll simulation showed that without considering the methane fed to the combustion furnace in the efficiency equation, the ideal efficiency on low heating value basis is 110%, which is similar to the ideal efficiency of 120% of one mole of methane producing four moles of hydrogen with complete water-gas-shift reaction. On the other hand, if the energy value of the methane fed to the furnace is taken into account, the efficiency is only 74% on low heating value basis. This efficiency of 74% has not been accounted for the steam export credit. The efficiency could be as high as 80% if steam credit was taken into account. Proll simulation showed that for conventional steam reforming of natural gas in distributed size plants, if complete heat integration was installed and maintained the same way as in the centralized plants, the efficiency of the two would be very comparable. However, the capital investments on the furnace, reactors, and heat exchangers are quite high and would make the process uneconomical since the production rate of hydrogen is low for distributed plants. The economy of scale made it infeasible to invest in a complete integrated distributed plant that produces low throughput hydrogen ~500 kg/day. Thus, in economic sense, it would not be profitable to construct such a plant to operate at high temperature at 800-900°C to produce hydrogen via steam reforming of natural gas. A complete integrated plant at small scale has efficiency somewhat lower than the large scale steam reforming plant due to lower the economy of scale; however, the investment cost is so high for such a small scale that it is not economical at all to construct it. Other alternatives have been looked at with the objective of reasonable cost and efficient production of hydrogen at small scale. Low operating temperature might be preferred and the runoff heat might be recovered. A compact design, quick startup process is also desired for small scale hydrogen production. Plasma conversion of methane, therefore, has an edge that meets the above criteria. The by-product of this process is un-converted methane and some C_2 hydrocarbons that can be sold un-separately for its heating value. The hydrogen can be separated using Pressure Swing Adsorption to meet high purity requirement. Since this is a small scale process, the operating pressure can be kept at atmospheric condition.

For the plasma oxidative system, a revised efficiency is defined with respect to the hydrogen product, the methane feed and other hydrocarbon products, and power consumption based on a lower heating value basis

$$efficiency = \frac{LHV\,of\,Hydrogen}{LHV\,of\,CH_{4\,fed} - LHV\,of\,CH_{4\,unreacted} - LHV\,of\,C_2\,products + power\,consumption} \quad (2)$$

The energy unit of hydrogen, methane, C_2 and power are in kWt on thermal basis. Power consumption was considered with 55% efficiency assumption converting from electrical to equivalent thermal power. This efficiency takes into account the credit of the products, which include un-reacted methane and C_2 products. Water vapor, which is formed from the oxidation of hydrogen and oxygen, was not accounted for and considered as an energy waste both on the heating value and hydrogen balance. Recovering the hydrogen or accounting for the heating value of the water loss might increase the efficiency. This efficiency is practical for any independent plants that use natural gas as a feedstock and convert it to hydrogen or other valuable products.

III. Discussion of efficiency in oxidative plasma conversion

From the maximum efficiency obtained at 48% from partial oxidation system, extrapolations were made to construct the following figure showing the relationship of efficiency and power consumption at different yields of hydrogen.

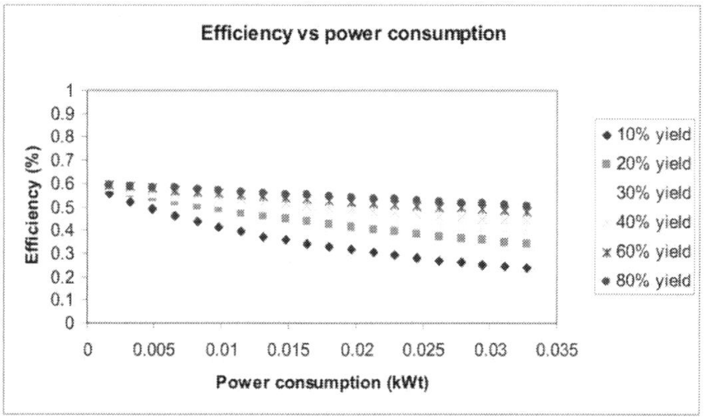

Figure 2: Efficiency varies with power consumption at different yields of hydrogen for a partial oxidation of methane using high voltage AC corona discharge

It was observed that even at no power consumption (assumed zero power consumption if a free power source were available), the efficiency is at 60% on a lower heating value basis of the product and feed gases. For plasma system, although the highest conversion observed was 68%, it was still a lot lower than 95% methane conversion in conventional steam reforming. Also, the hydrogen selectivity was only 41%, very low compared to conventional steam reforming. Thus, the hydrogen yield in plasma system is only as high as 28%. In Figure 1, as the power consumption increases, the efficiency decreases. This means additional power input results in waste heat. Methane conversion increases very little with more power input due to thermal effects if the overall temperature is greater than 600°C, and efficiency decreases. The methane conversion differential is little compared to the power consumption differential. At higher yield of hydrogen, the efficiency drops off to a lesser extent than at low yield. The increase in power input does not relate linearly with the increase in methane conversion. According to the efficiency definition:

$$efficiency = \frac{LHV\ of\ Hydrogen}{LHV\ of\ Methane_{fed} - LHV\ of\ Methane_{unreacted} - LHV\ of\ C_2\ products + power\ consumption}$$

It can be rearranged for methane in the denominator to give:

$$efficiency = \frac{LHV\ of\ Hydrogen}{LHV\ of\ Methane_{converted} - LHV\ of\ C_2\ products + power\ consumption} \qquad (3)$$

The rearranged efficiency equation is useful in analyzing the power consumption influence on the methane conversion, hydrogen yield and the overall efficiency. Suppose the power consumption was increased twice to push more methane conversion, if the power increase was linear with methane conversion, on the denominator, the LHV of $methane_{converted}$ would increase twice, and the LHV of C_2 products would presumably increase twice. Thus, the whole denominator increased twice. On the denominator, the hydrogen would also increase twice since hydrogen yield directly relates to methane conversion. Both the numerator and the denominator increased twice when power was increased twice, but they cancel out and result in the same ratio and efficiency stays the same. Upon the assumption that power increase was linear with methane conversion, the efficiency stayed the same. It was also reasonable to rule out that a certain increase in power would result in a substantial increase in methane conversion. Thus, a lot of power is required to gain some increase in methane conversion. As power increases, the efficiency decreases; to what extent is determined by the hydrogen yield as seen in Figure 2.

IV. Analysis:

a. Limitation:

The highest efficiency we obtained of 48% is based on partial oxidation of methane. With oxygen in the feed, the lowest eV/molecule of methane converted was observed (2eV/molecule). While steam reforming and autothermal conditions require higher power input eV/molecule (8-12eV/molecule). Thus, oxygen is needed as a feedstock. While pure oxygen is expensive, a cryogenic separation unit is capable to produce relatively pure oxygen from air with reduced cost compared to oxygen import option. But either way, this is a necessary cost to pay. Within the same system, if we carry out steam reforming reaction, although the water feed is free, it takes electric energy to vaporize the water to vapor phase before mixing with methane feed. The consumed energy to vaporize water to steam accounts for higher overall consumption energy eV per molecule of methane converted. Based on experimental results, the combined reactions of steam reforming and partial oxidation also consumes higher energy than partial oxidation alone. Although theoretically, the exothermic heat released by the partial oxidation reaction can compensate for the endothermic heat required for the steam reforming reaction with a balanced mixing ratio of the feeds, in plasma system, this thermal effect is minimal because it still takes energy to initiate the partial oxidation reaction and the reactor volume is not perfectly adiabatic as for large scale plants so not all the heat is transferred to the steam reforming reaction, which also requires energy to vaporize the water to steam. Thus, energy or power consumption is a sensitive factor that greatly affects the efficiency.

b. Improvements:

With the current partial oxidation system, the power consumption was lowest compared to steam reforming or combined steam reforming and partial oxidation. The hydrogen yield is also highest with highest methane conversion. The lowest energy consumption is due to the introduction of oxygen into the system with relative mole ratio of 2 to 1 to methane. Oxygen provides energetically active species that help initiating methane dissociation to methyl radicals that lead to high conversion of methane with the lowest energy consumption. Hydrogen selectivity could be higher but the secondary combustion reactions inhibit this. The by-products, therefore, are carbon monoxide, carbon dioxide and water vapor. Water is formed via hydrogen bonding with oxygen resulting in net loss of hydrogen. The direction to improve the hydrogen selectivity is to minimize the water formation or the secondary combustion reaction. CO formation can be shifted to H_2 and CO_2 under water-gas-shift reaction, but the CO_2 is also desired to be sequestrated. Other by-products in the system include C_2 hydrocarbons such as C_2H_2, C_2H_4, and C_2H_6 with trace amount. These can be left intact with the un-reacted methane and can be exported for their heating value. One way to look for improvement is to keep the C_2 amount as low as possible so the amount of CO and H_2 can be maximized.

According to the defined efficiency for the partial oxidation of methane under plasma conversion, the efficiency depends on the hydrogen yield, methane conversion, and power consumption. These variables are interrelated to each other. Hydrogen yield depends on methane conversion and methane conversion is also affected by power consumption. If power input is high, methane conversion is high and hydrogen yield is also high. But hydrogen yield, on the other hand, also depends on hydrogen selectivity and hydrogen balance. Current hydrogen balance is as low as 47%, the major loss of hydrogen is due to water formation. If hydrogen is recovered from water, the efficiency will increase. For example, if hydrogen balance is 90% and hydrogen selectivity is 85% (43% increase from current selectivity of 42%), the overall hydrogen yield will be 59% that is twice the current yield at 28%. The corresponding efficiency at 59% yield is 53%.

Increasing methane conversion might not linearly result in hydrogen yield because hydrogen might be wasted to form water due to shift in selectivity. Power increase is not linear with methane conversion. To a certain degree of conversion, the methane conversion curve levels out to a plateau shape (not straight line) with the same amount of power input into the system. More power input does not result in a desired increase in methane conversion which is limited. The electrical power input after a certain threshold ends up in waste heat and thus contributes to the thermal effect that leads to plasma cracking. To improve efficiency, the power consumption needs to be optimized; higher power results in higher conversion but not necessarily better efficiency while lower power might result in an optimum efficiency. We will explore the efficiency sensitivity based on power input according to the efficiency equation along with other variables.

An improvement in hydrogen selectivity will lead to higher efficiency since hydrogen is directly in the numerator of the efficiency equation. The wasted hydrogen to form water needs to be recovered or restrained from reacting with oxygen in secondary reactions. This can be done catalytically but more research on the pathways needs to be conducted. Water can be quantified before and after reaction to determine the exact amount of hydrogen loss. Another way to look at it is the hydrogen balance over all the species. Hydrogen balance needs to increase higher than the current value to ensure hydrogen is not lost to water formation. Increasing hydrogen yield by increasing methane conversion is another way to increase efficiency. However, as mentioned above, increasing conversion requires intensive power input which results in efficiency loss. So, increasing hydrogen selectivity without disturbing the optimal conversion is more preferable. Further research can also be done on shifting the C_2 products to hydrogen and CO and CO can be further shifted to H_2. Ideally, it is desired to convert methane directly to hydrogen bypassing the formation of C_2 products. However, C_2 products are formed under a complicated coupling radical mechanism under plasma discharge and they can be exported for their heating value when computing the overall efficiency. Among the C_2 products, ethylene has highest economic value (as a chemical) and thus, it might be desirable to maximize ethylene production. In our system, acetylene is the predominant product and next is ethylene and ethane. The C_2 selectivity depends on the electric field which relates to power and reactor configuration. It is possible to produce all acetylene and ethylene can be subsequently produced via separate hydrogenation of acetylene with Pd catalyst.

c. Plasma and thermal effect distinction

Thermodynamically, the operating temperature needs to be at least at 600^0C for methane to slowly start decomposing into other products. However, in the high voltage plasma discharge, methane can readily convert to other products at lower temperature ~ 200-300°C shown by Le[6] at al. works with an IR camera. At this low temperature, the methane chemistry cannot be due to thermal effects because thermally it takes much higher temperature for methane to have any conversion to occur. In the partial oxidation system, the feed of methane and oxygen are preheated with oven at 500^0 C and this temperature is still lower than the thermal temperature of methane cracking. Therefore, all the chemistry happens due to the plasma effect, not thermal. However, upon increasing power to increase further conversion of methane, the power eventually results in raising the temperature inside the plasma and if the temperature reaches 600°C, the thermal effect couples with plasma effect and both contribute to further conversion of methane. The thermal cracking of methane at high temperature results in coke formation with hydrogen produced via direct methane pyrolysis reaction $CH_4 \rightarrow C + 2H_2$. However in the oxygen rich environment the carbon can be burned or oxidized to form carbon dioxide, which contributes to the cost of separation and sequestration and decreases yield. Plasma effect, on the other hand, results in methane conversion to carbon oxides with CO being the predominant over CO_2. CO can then be shifted to produce additional H_2 and eventually ends up as CO_2. However, for the same amount of CO_2 produced either by thermal or plasma effect, the amount of hydrogen produced from plasma is twice that from thermal route. Thus, CO_2 emissions per kg of H_2 produced is less with plasma effects than with thermal effects via pyrolysis route of methane.

Since we are still operating within a temperature range of 200-500°C, the plasma effect is still predominant over the thermal effect. Upon increasing power input or longer residence time with the temperature within 200-500°C range, the efficiency decreases due to waste heat. However, once temperature goes up outside the plasma effect range (>600°C), the thermal effect starts creeping in and increases the rate of methane dissociation with respect to Arrhenius rate law. Hydrogen might be produced more at the high temperature. But simultaneously, the combustion effect with oxygen might also take away the hydrogen to form water. Depending on the comparative rates of methane dissociation and water formation, the efficiency curve might increase as power input increases. However, if the water formation rate is higher than the methane dissociation rate, the efficiency will decrease. This is purely thermal and combustion effect that is not related to plasma effect. It is desirable to keep the operating temperature as low as possible so heat recovery can be minimized, and plasma effect is the predominant factor.

Further power increase will increase the efficiency due to more methane conversion via thermal effects when the temperature is greater than 600°C. Thus hydrogen yield increases and efficiency increases. However, the efficiency will increase slowly with temperature and will not reach the highest original point since more hydrogen is produced but also consumed in water formation. So how is efficiency increased? With recent power measurement experiments, the actual power consumption is only 30% of the power reported. The 70% loss is to waste heat and to resistive windings in the transformer and not if measured on high voltage side.

References:

1. EIA, *Annual Energy Review,* 2004 (Report No. DOE/EIA-0384(2004)):p.1-34
2. NREL, *Hydrogen Program,* 2001 (Report No. DOE/GO-102001-1450(2001)):p.1-2
3. National Academies Press, *The Hydrogen Economy: Opportunities, Costs, Barriers, and R&D Needs* , 2004
4. Rostrup-Nielson, J.R., *Catalysis Today* **21** (1994), p. 257-267
5. Rostrup-Nielson, J.R., *Catalysis Today* **18** (1993), p. 305-324
6. Le, H., *M.S Thesis Dissertation*, 2003, University of Oklahoma, p. 73-75
7. Spath, P.L., Mann, M.K., *Life Cycle Assessment of Hydrogen Production via Natural Gas Steam Reforming*, 2001 (Report No. NREL/TP-570-27637), p. 5-33
8. Zhao, G.B., John, S., Zhang, J.J., Wang, L., Muknahallipatna, S., Hamann, J.C., Ackerman, J.F., Argyle, M.D., Plumb, O.A., *Methane Conversion in Pulsed Corona Discharge Reactors* , 2005, unpublished Reviewed Paper.

OXIDATIVE COUPLING OF METHANE ENHANCED BY THERMALLY OPTIMIZED REACTORS AND INTERMEDIATE SEPARATION

Ames Kulprathipanja, Alexander S. Lindquist, Joseph C. Poshusta, and Jerry L. Martin
Mesoscopic Devices, LLC, Broomfield, CO, USA

Introduction

While the natural gas reserves in the world are very large (approximately 13,000 trillion cubic feet), a substantial fraction of those reserves cannot be economically exploited because either their yield is too low to justify the expense of building a natural gas pipeline, or they are located too far from markets. One method of enabling the use of these stranded natural gas wells is to first convert the methane in natural gas to a more transportable liquid form. These liquids can then be transported by less expensive methods such as liquid pipelines, trucks and ocean barges depending on the location of the well.

Background

Few direct methane conversion schemes (i.e. no syn-gas intermediates) have been designed that offer simple and economical production of transportable liquid fuels from the abundance of stranded natural gas fields in the world. In one method, the production of C_2 hydrocarbons (i.e. ethylene + ethane) through the oxidative coupling of methane (OCM) has failed to find commercial success due to low overall yield (conversion x selectivity). In particular, today's catalysts exhibit either high selectivity (>70%) coupled with low conversion (<5%) or high conversion (>75%) with low selectivity (<15%) (Hutchings, 1992).

Most of the research conducted on the oxidative coupling of methane deals with increasing selectivity and yield of the OCM catalyst. However, many studies have come to the conclusion that economical practice of OCM is not possible without both high selectivity catalysts and a novel reactor or process design to achieve high yield. The need for a novel reactor design is due to the limitations of present catalysts to achieve both high selectivity and high conversion through a single pass (Kalenik, 1992). Su and co-workers have shown through theoretical modeling that the limit to ethylene yield by OCM using conventional, packed bed, continuous feed operation is approximately 28% (Su, 2003). For oxidative coupling to be economically competitive with other processes, researchers have shown the processes must achieve concurrent C_2 selectivities and CH_4 conversions of approximately 85 and 35%, respectively (Matherne, 1992 and Kuo, 1992). Emphasis must be made that high yield alone is not sufficient for OCM feasibility; the process must be able to achieve high yield with high selectivity in a single pass.

Two process designs have been published in Science and patented, one by Tonkovich and one by Jiang, that report overall C_2 yields of approximately 50 and 85%, respectively (Tonkovich, 1993 and Jiang, 1994). Both process designs use adsorbents to collect the C_2s to minimize further oxidation of the products to CO_2 and H_2O. Though both process designs achieve the yields deemed necessary for economical practice of OCM, the complexity of the process designs, which consists of numerous switching valves and beds or gas recycle, have prohibited the technologies from reaching commercialization. Jiang's process, which can achieve 85% in batch mode, only achieves 49% yield in continuous-flow steady state mode with a C_2 selectivity and CH_4 conversion of 76 and 65%, respectively. Tonkovich's process obtains a C_2 selectivity and CH_4 conversion of 80 and 65%, respectively. Furthermore, both these processes use large amounts of carrier gas, which dilutes the desired product. The carrier gas is used to sweep the product from the adsorbent while a thermal swing is applied during product desorption. Diluting the ethylene is a main disadvantage of the above processes and decreases the overall value of the product.

Rotating Sorbent Bed Reactor

To overcome these problems, Mesoscopic Devices has designed a novel multi-bed system with optimized thermal integration that may enable commercial practice of oxidative coupling with low cost and high yield. In addition to increasing conversion and selectivity for the oxidative coupling of methane, the rotating sorbent bed reactor (RSBR) can be applied to other oxidation processes resulting in similar benefits.

Mesoscopic Devices' design builds on the processes described above by increasing system single-pass product selectivity and methane conversion through a multi-bed arrangement of thermally optimized reactors and separation units. Furthermore, system complexity of the multi-bed arrangement is minimized with the use of a simple rotating valve. The major advantages of the RSBR are the following:

- only one moving part reducing system complexity
- no carrier gas diluting the desired product
- high throughput
- no coke formation
- no gas recycle
- easily scalable
- compact assembly
- continuous operation

Process Design

Mesoscopic Devices' OCM process is based on 4 key elements: 1) thermally optimized reactors, 2) intermediate product separation units, 3) a single rotating valve, and 4) a unique multi-bed arrangement that incorporates all of the above elements. The reactors are based on a thermally optimized design capable of achieving higher selectivity and production rates compared to traditional reactors in a single pass. No coke formation is observed in our thermally optimized reactors. **Table 1** presents data obtained from a traditional packed bed reactor compared to our thermally optimized design for the OCM reaction at less than 15 ms catalyst contact time. The short residence time allows for high throughput of product and/or small reactors. The same reaction conditions were used for both sets of data shown in **Table 1** except for the reactor geometry. The thermally optimized design increases selectivity by decreasing unselective secondary oxidation. In a similar manner, the thermally optimized reactor also decreases CH_4 conversion to undesired products including coke formation. Important to note is that the thermally optimized reactor may be able to achieve higher conversion and selectivity with a more active catalyst. Both conversion and selectivity in **Table 1** are defined on a carbon basis. Furthermore, even though the traditional reactor geometry achieves a higher overall yield in a single pass, the selectivity is significantly lower than necessary to be economically practical. With the use of the RSBR and the thermally optimized reactor design, high selectivity can be combined with high conversion.

Table 1. Thermally optimized reactor compared to traditional packed bed reactor at the same operating conditions.

	Thermally Optimized Reactor	Traditional Packed Bed Reactor
C_2 Selectivity (mol%)	75	43
CH_4 Conversion (mol%)	10	99
C_2 Yield (mol%)	8	43
C_2H_4 Selectivity (mol%)	49	29
C_2H_4 / C_2H_6	2.2	2.5
Residence Time (ms)	12	12

To increase overall yield, the desired ethylene product is separated from the by-products and unreacted feed in an adsorption unit. Our proprietary regenerable adsorbent has high ethylene breakthrough and saturation capacity and are capable of removing ethylene from a mixture of OCM products (C_2H_4, C_2H_6, CO_2, CO, H_2, N_2, CH_4, H_2O). **Figure 1** shows several breakthrough curves using an adsorbent bed with an inlet feed from the exit of a thermally optimized reactor. Notice, the ethylene is strongly adsorbed as it breaks through after 100 min. On the other hand, the unreacted methane and by-products, except for CO_2, are either weakly adsorbed or not adsorbed at all. This allows the RSBR to concentrate the desired ethylene product and continue reacting the methane to achieve higher overall conversion. The combination of the thermally optimized reactors and separation units also increases product selectivity in the RSBR by reducing over-oxidation of the desired ethylene product. The ethylene product is collected and beds regenerated continuously in the multi-bed arrangement. More importantly, the product collection in the RSBR is completed without the use of carrier gases, which dilute the desired product. Unlike the other two OCM processes described earlier by Jiang and Tonkovich, the ethylene collected in the RSBR is more concentrated and has higher purity. The undiluted ethylene product requires minimal downstream separation and therefore has higher value.

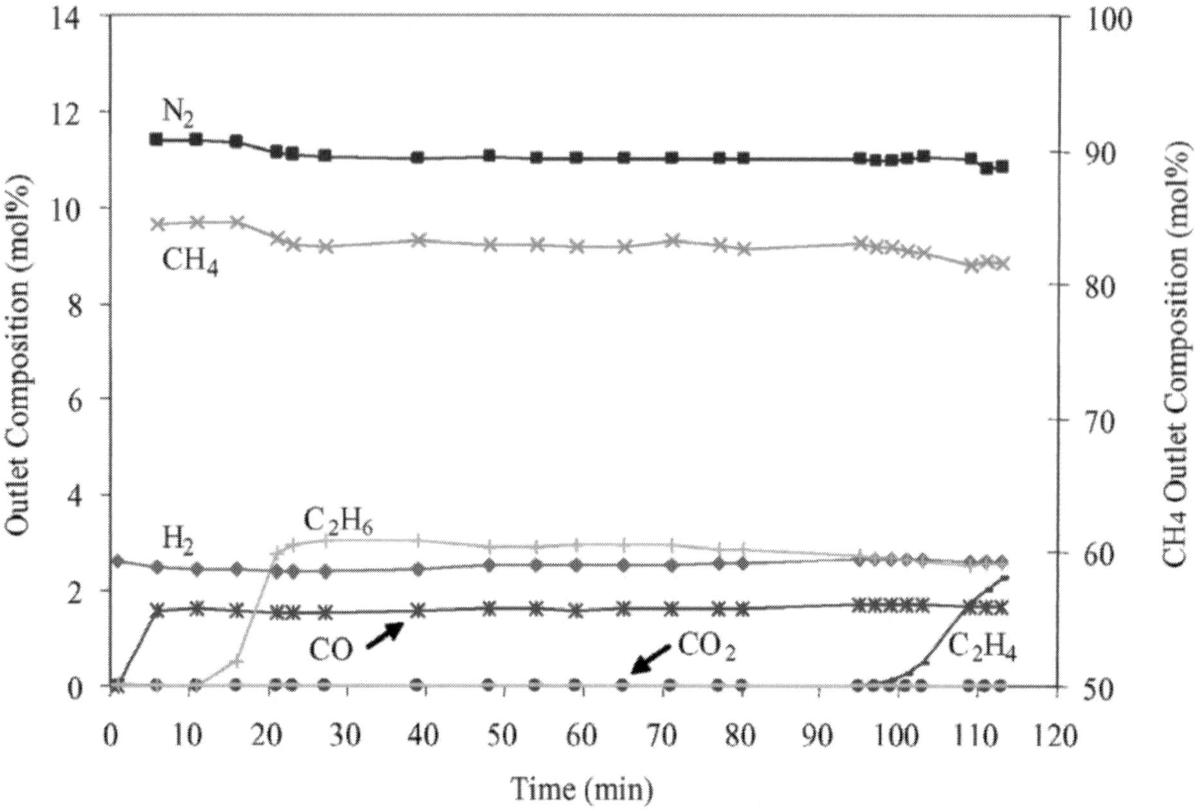

Figure 1. Breakthrough behavior using a proprietary sorbent with an inlet feed from our thermally optimized reactor of 10% N_2, 2.5% C_2H_4, 2.5% C_2H_6, 77% CH_4, 1% CO, 1% CO_2, and 2% H_2, 3% H_2O.

To increase methane conversion and overall product yield, the aforementioned series of unit operations is implemented in a multi-bed system where all of the necessary fluid flow switching is controlled by a single rotating valve (**Figure 2**). The rotating valve operates with a single moving part eliminating numerous mechanical switching valves, which are expensive and potentially unreliable. The valve switches the adsorption units between two phases, ethylene adsorption and desorption, to maximize methane conversion, product selectivity and sorbent efficiency. Mesoscopic Devices has concentrated tremendous efforts on optimizing design, robustness, control, operation and simplicity of the rotating valve assembly. Using a combination of hard ceramic surfaces, expected operating lifetime for the valve assembly is several years of continuous operation. Earlier generations of this valve design exceeded the equivalent of one year of continuous operation in accelerated life tests without measurable increase in leak rates. Driven by a stepper motor, the valve completes a single rotation in minutes.

Figure 2. Rotating valve used to complete all necessary fluid flow switching.

Concept Demonstration

We have built and tested a 2 bed system to prove feasibility of our rotating sorbent bed reactor concept. Our small scale system has shown that we can increase overall methane conversion using 2 beds as well as maintain our high selectivity by intermediate separation of the desired ethylene product (**Table 2**). The CH_4 conversion data is based on product formation since the presence of the water gas shift and methanation reactions convert CO and H_2 to CH_4. Mass balances confirmed the method of calculating conversion.

Table 2. Performance data from our 2 bed system with reactors and sorbent beds scaled to produce 1 kg/day.

	After 1st Reactor	After 2nd Reactor
C_2H_4 Selectivity (mol%)	49	49
CH_4 Conversion (mol%)	8	14
C_2 Yield (mol%)	6	11

We have also built an 8 bed system that operates with a rotating valve capable of producing 1 kg per day of ethylene and are in the process of demonstrating large scale feasibility. In the 8 bed process, we expect the ethylene selectivity and conversion to be approximately 75% and 30%, respectively, giving a process yield of 23%. The 8 bed system consists of both adsorption and product collection phases that contain 4 beds each. Therefore the output of the 8 bed system is expected to be 2 times the data reported above for the 2 bed system, which did not have a product collection phase. We are using the 8 bed system to determine necessary scaling parameters for industrially important scales. The parameters will be used to predict size, performance and cost in our scaling models.

Acknowledgements

This work was supported by a grant from the U.S. Department of Energy (contract no. DE-FG02-03ER83872). The content of this paper does not necessarily reflect the position of the Government, and no official endorsement should be inferred.

References

1. Hutchings, Graham J. and Michael S. Scurrel. "Studies of the mechanism of the oxidative coupling of methane using oxide catalysts". *Methane Conversion by Oxidative Processes: Fundamentals and Engineering Aspects, edited by E.E. Wolf,* Van Norstrand Reinhold Catalysis Series, New York 1992.

2. Kalenik, Zbigniew and Eduardo E. Wolf. "The role of gas-phase reactions during methane oxidative coupling". *Methane Conversion by Oxidative Processes: Fundamentals and Engineering Aspects, edited by E.E. Wolf,* Van Norstrand Reinhold Catalysis Series, New York 1992.

3. Su, Yee San, Jackie Y. Ying, and William H. Green, Jr. "Upper bound on the yield for oxidative coupling of methane". *Journal of Catalysis,* 218 (2003) 321-333.

4. Matherne, J.L. and G.L. Culp. "Direct conversion of methane to C_2's and liquid fuels: process economics". *Methane Conversion by Oxidative Processes: Fundamentals and Engineering Aspects, edited by E.E. Wolf,* Van Norstrand Reinhold Catalysis Series, New York 1992.

5. Kuo, James C.W. "Engineering evaluation of direct methane conversion processes". *Methane Conversion by Oxidative Processes: Fundamentals and Engineering Aspects, edited by E.E. Wolf,* Van Norstrand Reinhold Catalysis Series, New York 1992.

6. Tonkovich, Anna Lee, Robert W. Carr and Rutherford Aris. "Enhanced C_2 yields from methane oxidative coupling by means of a separative chemical reactor". *Science,* 262 (1993) 221-223.

7. Jiang, Y., I.V. Yentekakis and C.G. Vayenas. "Methane to ethylene with 85 percent yield in a gas recycle electrocatalytic reactor-separator". *Science,* 264 (1994) 1563-1566.

Development and Evaluation of the Gasification Technology as Alternative Power Generation

Emilio Manzanares-Papayanopoulos, Manuel Fernández-Montiel, Agustín Moisés Alcaraz-Calderón, Jorge A. Altamirano-Bedolla and César A. Romo-Millares*

*Gerencia de Procesos Térmicos, Instituto de Investigaciones Eléctricas, Reforma No. 113, Col. Palmira, CP62490, Cuernavaca, Morelos, México. * e-mail: emp@iie.org.mx*

ABSTRACT

The technology of gasification transforms any fossil fuels such as natural gas, coal, liquid or solid refinery residues or heavy oils into a synthesis gas, also called syngas, for subsequent utilisation as combustible gas. Also, biomass and wastes can be used as feed fuel.

In the past years, the Instituto de Investigaciones Electricas (IIE) has been working on economical and technical evaluation studies of the gasification technology as an alternative for power generation.

Despite being considered as an expensive technology compared to other technologies for power generation, the gasification technology as Integrated Gasification Combined Cycle is quite attractive in many ways since it offers a clean path to power generation; sulphur and heavy metals can be removed from the syngas and the CO_2 can be sequestered and stored in special places, avoiding the traditional contamination to the atmosphere.

Nowadays, the main energy companies in Mexico, PEMEX and CFE, are interested in the potential that the gasification technology offers for power generation in the following 5 to 10 years. In that sense, IIE began to develop an ambitious research and technological development project to gather the knowledge of the technology and to gain expertise, in particular on the topics related to (1) the thermodynamics and kinetics of the chemical reactions inside of the gasification reactor, (2) hot clean-up syngas systems, (3) liquid residual feedstock systems, and (4) modelling processes using Computer Fluid Dynamic (CFD) tools.

The project has been planned in various stages aiming to design a research pilot plant gasifier. In this first stage, a study of the behaviour of the syngas production in function of the composition of the fuel, temperature and pressure has been carried out using a computational commercial program with a general model for gasification. The fuels used are bituminous coal and refinery residues such as the liquid vacuum distillation residues and the petroleum coke. It is important to notice that all the feedstock are Mexican fuels, in which the refinery residues have relative high sulphur and metals concentration and the bituminous coal is from MICARE mine, at the north of the country, with relative high ash concentration. The results obtained of the syngas behaviour and its comparison with those reported in the open literature allowed us to modify parts of the experimental prototype gasification system design.

In a second stage, a gasification system at laboratory level will be constructed. Also, a CFD models for the gasification of Mexican feedstocks will be developed. The experimental results will validate the computational models.

I. Introduction

In México, about 81% of the electric power generation is base on fossil fuels (CFE-Subdirección de Distribución, 2004a and 2004b). Furthermore, the electric power generation for the next 10 to 20 years is expected to continue with the use of such kind of fuels (SENER,2002).

Mainly, the expansion of the electric power generation, around the world and also in Mexico, has been based on combined cycle power plants. In addition, the volatility of the natural gas price and the worldwide environment concerns of clean power generation point out a clear path to explore new technologies, such as the gasification technology, for electric power generation.

The Instituto de Investigaciones Electricas (IIE), as technological research institute consulting of the main energy companies in Mexico, focus his effort in the gasification technology as Integrated Gasification Combined Cycle (IGCC) for power generation (Fernández-Montiel, *et.al.*, 2003; Fernández-Montiel and Alanis, 1993 and 1994, Fernández-Montiel,1994). The gasification technology offers flexibility of the fuel used and with a required pollution reduction systems allow the clean power generation (Higman and Der Burgt, 2003; Rezaiyan and Cheremisinoff, 2005).

The aim of the study presented is the analysis of the syngas production behaviour in function of fuel compositions, temperatures and pressures. Also, the comparison of the energy produced in function of the fuel composition at the same operational parameters is presented.

II. Simulation Methods of the Gasification Process

The analysis of the syngas production behaviour in function of fuel compositions, temperatures and pressures were carried out using the computational program ThermoFlex. However, the energy production was performed with the software package GT PRO, both from Thermoflow Inc. Company.

The simulation models used are described below. The fuels used in the analysis were liquid residuals and petcoke from Mexican refineries, mixtures of Mexican coal, and also a mixture of imported coal. The analysis composition of the fuels is described below.

Feedstock Characterization

The liquid residuals used in the study were from Tula Refinery, Salamanca Refinery, Cd. Madero Refinery, Minatitlan Refinery and Salina Cruz Refinery, all of them from PEMEX. The heating values and weight composition analysis are listed in Table 1.

Table 1. Heating values and weight composition analysis of the refinery liquid residuals.

Property	Tula Refinery	Salamanca Refinery	Cd. Madero Refinery	Minatitlan Refinery	Salina Cruz Refinery
HHV kj/kg	41,533	41,680	40,839	41,755	41,559
LHV kj/kg	39,281	39,362	38,676	39,448	39,271
Composition					
C wt %	83.42	83.96	82.82	84.16	83.66
H wt %	10.32	10.24	9.91	10.57	10.10
S wt %	5.13	4.77	4.99	5.01	5.07
O wt %	1.032	0.08	2.137	0.178	1.078
Ash wt %	0.098	0.95	0.143	0.082	0.092

The solid fuels used in the simulation were three different mixtures of coals. The first mixture was the one used in the Mexican Coal Power Plant Rio Escondido, which is a mixture of Mexican coals from MICARE and SABINAS regions. The second one, used in the Mexican Coal Power Plant Carbon II, in which is a mixture of Mexican coal from also MICARE and SABINAS regions, and imported coal form USA, in particular from Texas, Nuevo México y Colorado regions. Finally, the third mixture was the one used in the Petacalco Coal Power Plant, in which the coal mixture used in the power generation is imported coal from Australia, China, Colombia and Canada. The heating values and weight composition analysis are listed in Table 2.

Table 2. Heating values and weight composition analysis for both the Coal used in the Mexican power plants and the coke produced in the Mexican refineries.

Property	Rio Escondido Coal Mixture 1	Rio Escondido Coal Mixture 2	Carbon II Coal Mixture 1	Carbon II Coal Mixture 2	Petacalco Coal Mixture	Coke 70% Maya - 30% Isthmus
HHV (kj/kg)	19,603	18,778	21,231	19,791	29,649	31,298
LHV (kj/kg)	18,749	17,983	20,348	19,017	28,925	30,617
Composition						
C wt %	47.98	46.27	52.38	49.18	72.98	88.39
H wt %	3.96	3.68	4.09	3.59	3.36	3.12
S wt %	1.32	1.30	1.03	1.07	0.40	6.70
O wt %	4.48	4.44	4.02	2.31	6.99	0.49
N wt %	0.21	0.26	0.19	0.28	1.15	0.96
H_2O wt %	2.52	2.68	2.53	3.60	2.48	--
Ash wt %	39.53	41.37	35.76	39.97	12.64	0.34

In addition of the three different mixtures of coals used in the simulation as solid feedstock, the design composition of coke for the Mexican refineries using 70% Maya and 30% Isthmus mixture was considered (Martínez, 2000). The heating values and weight composition analysis are also listed in Table 2.

Schematic Models

Two simulation schematic models were used. The first model was performed to study the syngas composition behaviour as a function of pressure and temperature. The pressures were 2, 15 and 30 bars and the temperatures were 1000 °C, 1250 °C and 1500 °C. The model, from ThermoFlex program, was represented by a Gasifier with a separation air unit as shown in Figure 1.

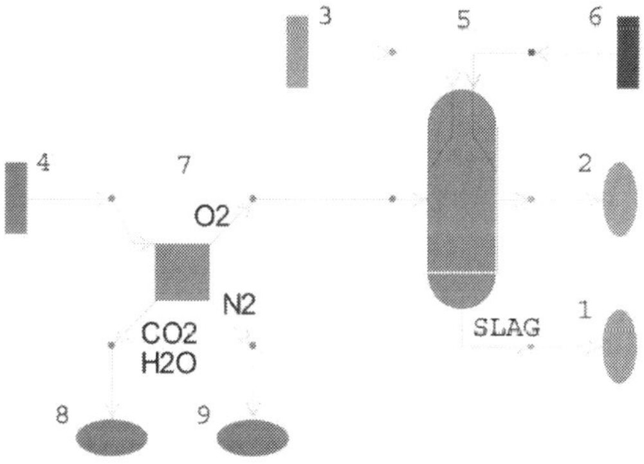

Figure 1. ThermoFlex schematic model used by the study of the syngas composition behaviour for Mexican feedstock.

The second model used in the simulation was to compare the energy produced in function of the fuel composition at the same operational parameters. The model, from GT-PRO program, was represented in a similar way than the first model with the addition of clean syngas system and a gas turbine as shown in Figure 2.

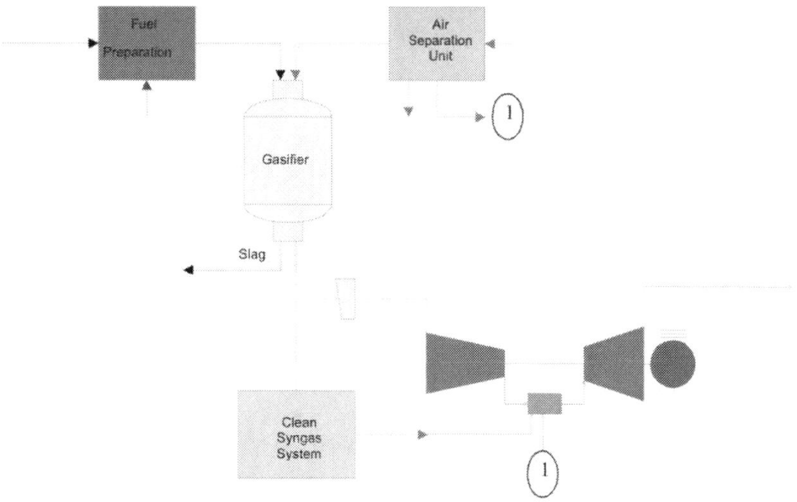

Figure 2. GT-PRO schematic model used by the study energy produced in function of the fuel composition at the same operational parameters.

III. Simulation Results and Discussion

The results of the simulations for the study of the syngas composition behavior were divided in three different cases. The first case analyzes the syngas composition behavior of the liquid residual from Mexican refineries as wet-feedstock in the gasification process. The second case evaluates the syngas composition behavior of solid wet-feedstock, as coal and petcoke, in the gasification process. Finally, the third case studies the syngas composition behavior of a chosen solid feedstock with dried basis in the gasification process.

***1**st* ***Case**.* The syngas composition behaviors of the liquid residual from the Mexican refineries as a function of pressure and temperature are presented in Table I in the appendix. The premises taken for the simulation are listed in Table 3.

Table 3. Premises for the simulation of the syngas composition behavior.

Property	Liquid Residuals from refineries & coke	Coal
Ambient Temperature (°C)	15.6	15.6
Relative Humidity (%)	60	60
High Site Plant over Sea Level (m)	0	0
H_2O in Gasification Process (%)	40	40
Atmospheric Pressure (bar)	1.013	1.013
Ash (%)	6	--
O_2/fuel Ratio (%w / %w)	80	80

The trend behaviors shown of the syngas composition for the five liquid residuals feedstock used from the Mexican refineries are practical similar. Therefore, the data from Tula and Madero refineries were selected as representation of all refineries for graphics purposes. These syngas composition behaviors were expected since the feedstock compositions were quite similar between the five liquid residuals as shown in Table 1 above.

In Figure 3, the syngas compositions from the liquid residual from Tula Refinery and Madero Refinery are plotted. As shown in Figure 3, the main syngas composition production, about 90%, correspond to H_2 and CO components. At low temperature, the H_2 and CO compositions drop when the pressure is increased. However, the other syngas components are favored, as shown in Figure 3(a) and 3(d). Nonetheless, when the gasification process temperature is high, about 1500 °C, there is not change in the syngas composition for all components as a function of the pressure, see Figure 3(C) and 3(f).

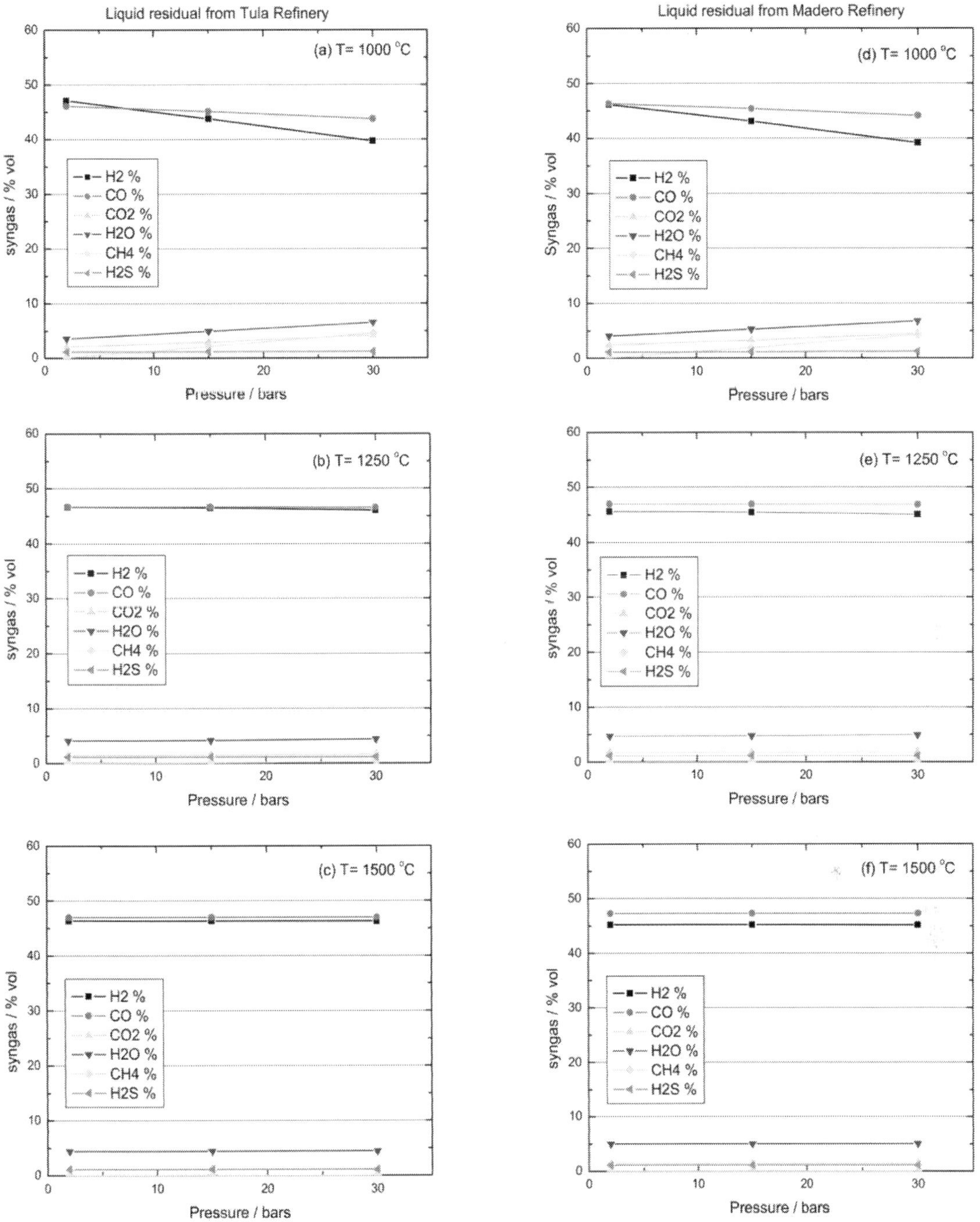

Figure 3. Syngas behaviors of liquid residuals from (a), (b) and (C) Tula refinery and (d), (e) and (f) Madero refinery.

2nd Case. The syngas composition behavior of the solid wet-feedstock as a function of pressure and temperature is presented in Table II in the appendix. The premises taken for the simulation are listed in Table 3. It is important to notice that for the coal simulation, ash was not used for these kinds of feedstocks since there is a relative high ash component, as shown previously in Table 2.

The trend behaviors shown of the syngas composition for the solid wet-feedstock in the gasification process are not quite similar. However, the H_2 composition is lower about 15% points than the one given for the liquid residuals. Nevertheless, three different trends can be point out in the CO, CO_2 and H_2O compositions. One trend for the Petacalco coal mixture, a second trend for the Rio Escondido and Carbon coal mixtures, and a third one for the coke. These different syngas composition behaviors were expected since the feedstock compositions, shown in Table 2, were different between the coals and coke.

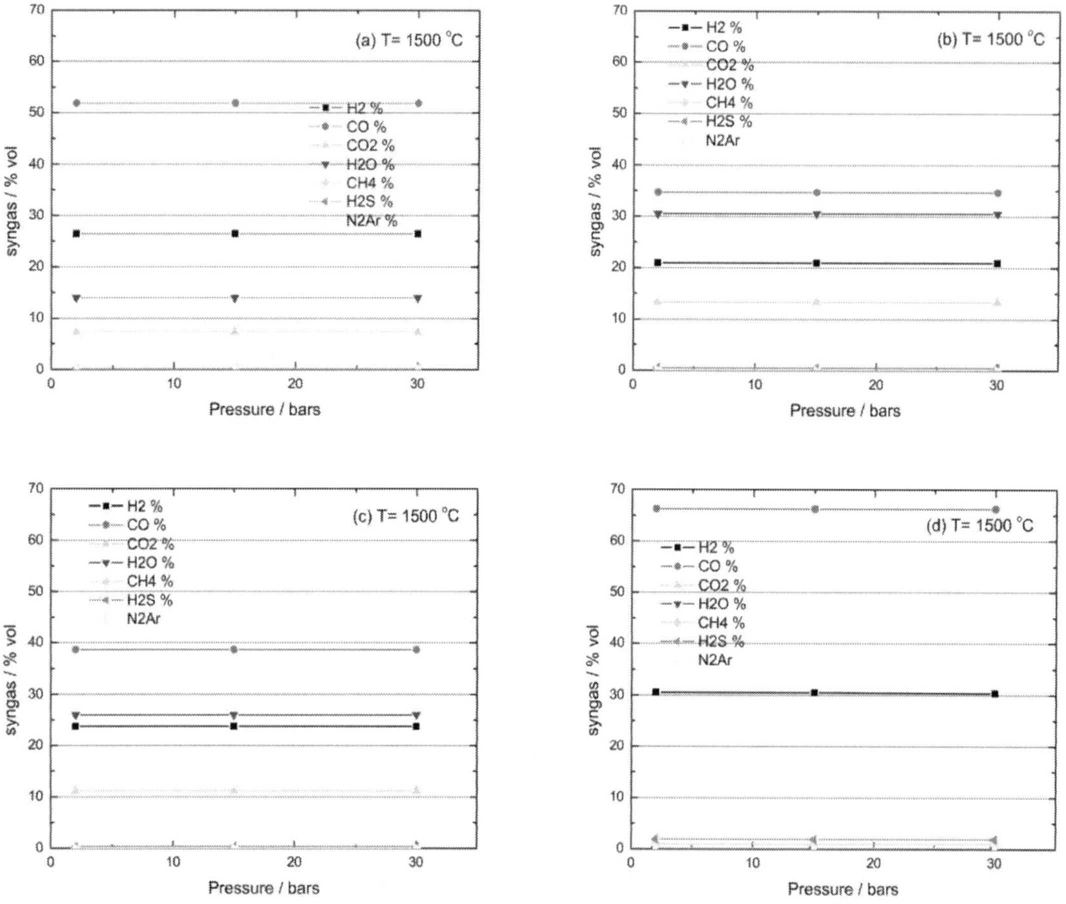

Figure 4. Syngas behavior of solid feedstock from (a) Petacalco coal mixture, (b) Rio Escondido coal mixture 1, (C) Carbon II coal mixture 1, and (d) coke.

In Figure 4, the syngas composition behaviors for the coal mixtures with the high heat value reported in Table II are plotted. The three different trends behaviors described above are shown in the Figure. Petacalco coal mixture has a high concentration of CO, see Figure 4(a) than those for Rio Escondido and Carbon II coal mixtures, see Figure 4(b) and 4(c). However, due to the high C concentration of the coke feedstock, the CO concentration of the syngas from the coke gasification is higher than the rest of the samples, as shown in Figure 4(d). The syngas composition behavior of the coke is, in a sense, similar to those for the liquid residuals feedstock described earlier. In this case, the main syngas composition, corresponding to H_2 and CO, is about 96 %, see Figure 4(d).

3rd Case. The syngas composition behavior of solid dry-feedstock as a function of pressure and temperature is presented in Table III in the appendix. The premises taken for the simulation are reported in Table 3, but without H_2O for the gasification process. As in the second case, ash was not used for these kinds of feedstocks since there is a relative high ash component concentration, as shown previously in Table 2.

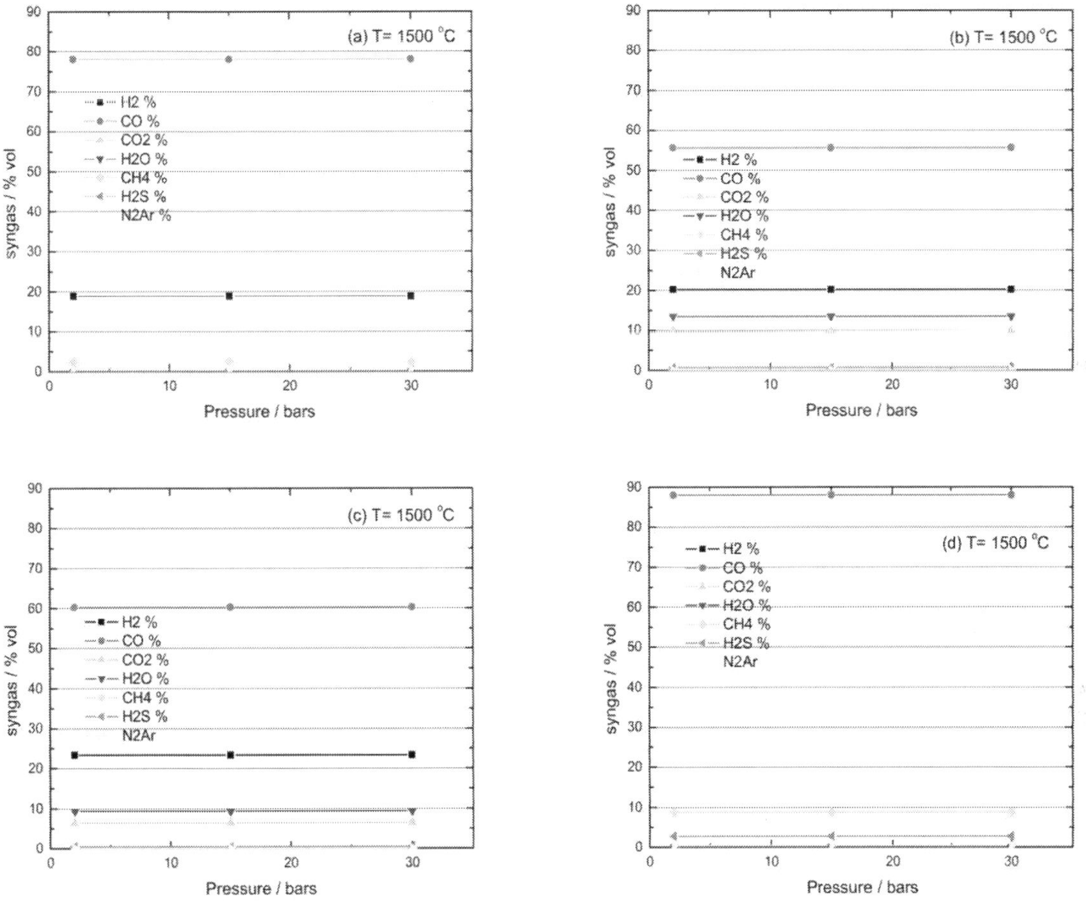

Figure 5. Syngas behavior of solid dry-feedstock from (a) Petacalco coal mixture, (b) Rio Escondido coal mixture 1, (C) Carbon II coal mixture 1, and (d) coke

The trend behaviors exhibited by the syngas compositions for the solid dry-feedstock were expected for the coal mixtures, see Figure 5(a), 5(b) and 5(c). However, the syngas behavior composition of the coke was unexpected, as shown in Figure 5(d). The syngas behavior composition expected is base on the composition of the feedstock reported earlier in Table 2, and the results reported in the open literature (Phillips and Mahagaokar, 1991; Mahagaokar and Krewinghaus, 1990a and 1990b).

The syngas behavior of the coke reported here is due to the lack of O_2 in the operational conditions during the simulation to give preference to the gasification reactions that favored the H_2 production, and then to focus the CO production.

A second part of the simulation was to compare the energy produced in function of the fuel composition at the same operational parameters. Here, two gas turbines were selected from a library of the computational program GT-PRO. One of them was a General Electric gas turbine model 7241FA, and the second gas turbine was Siemens KW501F. It is important to notice that it is not possible to compare the turbines due to 1) it was not the aim of the study and 2) they have not got the same operational characteristics.

The gross power obtained was as a function of temperature at a fixed pressure given for the gas turbine operational requirements. Table 4 shows the results obtained for a Rio Escondido coal mixture, Salamanca liquid residual refinery and a coke as wet-feedstock of the gasification system.

Table 4. Simulation results of the gross power and efficiency of the syngas of different wet-feedstock as a function of temperature.

T (°C)	Component	Rio Escondido Coal Mixture 1 Pressure (bar) (GE7241FA) 28.2	Rio Escondido Coal Mixture 1 Pressure (bar) (Siemens KW501F) 30.93	Salamanca Refinery Pressure (bar) (GE7241FA) 28.2	Salamanca Refinery Pressure (bar) (Siemens KW501F) 30.93	Coke Pemex Pressure (bar) (GE7241FA) 28.2	Coke Pemex Pressure (bar) (Siemens KW501F) 30.93
1000	Gross Power (kW)	190,923	216,045	219,574	249,983	199,250	226,490
	Gross Efficiency (%)	43.69%	44.73%	42.02%	43.45%	42.82%	44%
1250	Gross Power (kW)	180,853	203,469	212,321	242,350	192,311	217,761
	Gross Efficiency (%)	44.37%	45.21%	42.49%	43.87%	43.26%	44.33%
1500	Gross Power (kW)	169,118	188,818	205,512	233,898	184,673	208,244
	Gross Efficiency (%)	45.24%	45.80%	42.91%	44.21%	43.76%	44.69%

The results obtained shows that the power generation decreases when the temperature is increased for all the wet-feedstock reported in Table 4. However, the efficiency of the turbine keeps its value due to the additional O_2 needed in the gasification model is given externally by the air separation unit. Therefore, the global system efficiency will decrease.

As a second simulation performance on the power generation was carried out for a dry-feedstock for two fuels. Table 5 shows the gross power obtained as a function of temperature at a fixed pressure given for the gas turbine operational requirements.

Table 5. Simulation results of the gross power and efficiency of the syngas of different dry-feedstock as a function of temperature.

T (°C)	Component	Rio Escondido Coal Mixture 1 Pressure (bar) (GE7241FA) 28.2	Rio Escondido Coal Mixture 1 Pressure (bar) (Siemens KW501F) 30.93	Coke Pemex Pressure (bar) (GE7241FA) 28.2	Coke Pemex Pressure (bar) (Siemens KW501F) 30.93
1000	Gross Power (kW)	211,328	248,869	215,814	246797
	Gross Efficiency (%)	40.59%	42.15%	41.66%	43.08%
1250	Gross Power (kW)	215,675	246,791	213,392	243,782
	Gross Efficiency (%)	41.99%	43.40%	41.79%	43.18%
1500	Gross Power (kW)	210,129	239,831	211,329	241,216
	Gross Efficiency (%)	42.33%	43.68%	41.90%	43.27%

The results obtained for the Rio Escondido coal mixture 1 using the GE7241FA gas turbine as a function of temperature do not present the same gross power behavior than both the coke fuel reported in Table 5 and the wet-feedstock gross power behavior shown in Table 4. The analysis of this behavior should be more deeply taking into account more parameters and results given by the model using the GT-PRO program. Further simulations with different operational conditions are required for better understanding of the phenomenon.

Iv. Conclusions and Future Work

The technology of gasification is still under study around the world. In this particular study, the understanding of the syngas composition behavior as a function of the feedstock composition, temperature and pressure was the main goal.

The simulation results present here open the discussion of the possibility to use the gasification technology as an attractive alternative for power generation since the crude found around the world is heavier and also the tendency to grow of the natural gas price. However, more studies and research on the gasification technology and all its complementary systems, i.e. a separation air unit, feedstock systems, clean syngas system, CO_2 sequestering and storage system, syngas turbine systems, etc., required in a gasification power plant should be performed.

There are several operational and physical parameters that affect the syngas production in a gasification process. The work reported here is just a begging of a systematic study to understand the black box represented by a gasifier. The understanding of the chemical reactions, chemical kinetics, and both mass and energy balances can favored the syngas prediction with much real approach to the experimental gasification results. More simulations are needed with different operational parameters conditions.

A second stage after to gather the knowledge of the technology, should be the design and construction of an experimental gasification pilot plant to perform both an investigation of the gasification of local feedstock and to model the process by CFD tools.

VI. References

CFE-Subdirección_de_Distribución (2004a). National Electric Sector Statistics 2004 (Estadísticas del Sector Eléctrico Nacional 2004).

CFE-Subdirección_de_Distribución (2004b). Operational Report 2004 (Informe de Operación 2004

Fernández-Montiel, M., Alcaraz-Calderón, A. M. y García-Chávez, L.; "Estudios y Desarrollo de Alternativas de Gasificacion para el Sector Eléctrico", Informe 12585, Instituto de Investigaciones Eléctricas, Diciembre 2003, México.

Fernández-Montiel, M; "Reporte de actualizacion sobre el estado que guardan las tecnologias de gasificacion aplicadas a la generacion de energia electrica", Informe 44256, Instituto de Investigaciones Eléctricas, Octubre 1994, Mexico.

Fernández-Montiel, M. y Alanís, F.J.; "Estudios de Factibilidad para Plantas de Gasificacion de Carbon y de Combustoleo Integradas a Ciclos Combinados", Informe 44288, Instituto de Investigaciones Eléctricas, Marzo 1994, México.

Fernández-Montiel, M. y Alanís, F.J. "Reporte del Estado Actual de las Tecnologias de Gasificacion Aplicadas a la Generacion de Energia Electrica", Informe 44207, Instituto de Investigaciones Eléctricas, Octubre 1993, México.

Higman C. and Der Burgt M. Van, "Gasification", Elsevier, 2003.

Mahagaokar U., KrewinghausA.B.; "Shell Coal Gasification Plant No.1 – Recent Results on Domestic Coals", Presented at POWER-GEN '90, Orlando, Florida, 1990a.

Mahagaokar U., KrewinghausA.B.; "Shell Coal Gasification Process – Recent Performance Results on Drayton, Buckskin, Blacsville No.2 and Pyro No. 9 Coals", Presented at 1990 International Joint Power Generation Conference & Exposition, 1990b.

Martínez, E., " Estudio de tecnologías potencialmente utilizables para la combustión de coque de la refinería de Cadereytia", Master Thesis, 2000, Faculty of Engineering, UNAM.

Phillips J.N., Mahagaokar, U. "Shell Coal Gasification Process Recent Results on Low Ah Feedstock, Including Petroleum Coke", Presented at Eighth Annual Pittsburgh Coal Conference, 1991.

Rezaiyan J. and Cheremisinoff, N. P.; "Gasification Technologies: A Primer for Engeneers and Scientists", Taylor & Francis, 2005.

SENER, Prospectiva del Sector Eléctrico 2002-2011, Dirección General de Formulación de Política Energética, 2002.

APPENDIX

Table I. Simulation results of syngas composition for the liquid residual from the Mexican refineries as a function of pressure and temperature.

T (°C)	Composition	Tula Refinery Pressure (bar)			Salamanca Refinery Pressure (bar)			Madero Refinery Pressure (bar)			Minatitlán Refinery Pressure (bar)			Salina Cruz Refinery Pressure (bar)		
		2	15	30	2	15	30	2	15	30	2	15	30	2	15	30
1000	H_2 (vol %)	47.08%	43.74%	39.69%	48.02%	44.20%	39.96%	46.11%	43.08%	39.18%	47.87%	44.14%	39.93%	47.72%	43.97%	39.76%
	CO_2 (vol %)	2.06%	2.98%	4.21%	1.63%	2.65%	3.90%	2.39%	3.26%	4.47%	1.71%	2.71%	3.95%	1.69%	2.69%	3.94%
	CH_4 (vol %)	0.07%	2.10%	4.62%	0.09%	2.43%	5.07%	0.05%	1.89%	4.30%	0.08%	2.36%	4.99%	0.08%	2.38%	5.00%
	H_2S (vol %)	1.11%	1.16%	1.21%	1.05%	1.10%	1.16%	1.10%	1.14%	1.19%	1.07%	1.12%	1.18%	1.12%	1.17%	1.23%
	H_2O (vol %)	3.60%	4.94%	6.51%	2.89%	4.42%	6.06%	4.07%	5.28%	6.78%	3.02%	4.52%	6.14%	2.96%	4.45%	6.07%
	CO (vol %)	46.08%	45.09%	43.77%	46.32%	45.21%	43.86%	46.27%	45.36%	44.08%	46.25%	45.16%	43.81%	46.43%	45.34%	44.00%
	Total	100.00%	100.00%	100.00%	100.00%	100.00%	100.00%	100.00%	100.00%	100.00%	100.00%	100.00%	100.00%	100.00%	100.00%	100.00%
	LHV (kJ/kg)	15,869.06	15,593.12	15,277.88	16,300.52	15,980.74	15,648.88	15,491.64	15,245.32	14,945.07	16,226.27	15,914.91	15,585.84	16,209.14	15,896.82	15,568.75
	HHV (kJ/kg)	17,314.91	17,038.96	16,723.73	17,775.23	17,455.45	17,123.59	16,899.74	16,653.41	16,353.17	17,696.99	17,385.63	17,056.56	17,671.35	17,359.04	17,030.96
1250	H_2 (vol %)	46.63%	46.47%	46.03%	47.71%	47.50%	46.95%	45.56%	45.43%	45.06%	47.53%	47.33%	46.80%	47.40%	47.20%	46.66%
	CO_2 (vol %)	1.48%	1.51%	1.59%	1.16%	1.19%	1.29%	1.73%	1.76%	1.83%	1.21%	1.25%	1.34%	1.20%	1.23%	1.33%
	CH_4 (vol %)	0.00%	0.09%	0.34%	0.00%	0.12%	0.44%	0.00%	0.08%	0.29%	0.00%	0.12%	0.42%	0.00%	0.12%	0.42%
	H_2S (vol %)	1.11%	1.11%	1.12%	1.05%	1.05%	1.06%	1.10%	1.10%	1.10%	1.07%	1.07%	1.08%	1.12%	1.12%	1.13%
	H_2O (vol %)	4.11%	4.18%	4.38%	3.28%	3.37%	3.62%	4.67%	4.73%	4.90%	3.43%	3.52%	3.76%	3.36%	3.45%	3.69%
	CO (vol %)	46.67%	46.64%	46.55%	46.80%	46.76%	46.65%	46.94%	46.91%	46.83%	46.75%	46.71%	46.60%	46.93%	46.89%	46.78%
	Total	100.00%	100.00%	100.00%	100.00%	100.00%	100.00%	100.00%	100.00%	100.00%	100.00%	100.00%	100.00%	100.00%	100.00%	100.00%
	LHV (kJ/kg)	15,892.55	15,880.09	15,845.93	16,324.34	16,307.57	16,263.66	15,515.42	15,505.21	15,476.63	16,249.93	16,234.11	16,192.22	16,232.72	16,216.70	16,174.42
	HHV (kJ/kg)	17,338.40	17,325.94	17,291.79	17,799.05	17,782.28	17,738.37	16,923.51	16,913.31	16,884.73	17,720.65	17,704.82	17,662.93	17,694.94	17,678.91	17,636.63
1500	H_2 (vol %)	46.34%	46.32%	46.29%	47.48%	47.46%	47.42%	45.22%	45.21%	45.18%	47.29%	47.27%	47.23%	47.16%	47.14%	47.10%
	CO_2 (vol %)	1.18%	1.18%	1.18%	0.92%	0.92%	0.93%	1.38%	1.38%	1.39%	0.96%	0.97%	0.97%	0.95%	0.95%	0.96%
	CH_4 (vol %)	0.00%	0.01%	0.03%	0.00%	0.01%	0.04%	0.00%	0.01%	0.02%	0.00%	0.01%	0.03%	0.00%	0.01%	0.03%
	H_2S (vol %)	1.11%	1.11%	1.11%	1.05%	1.05%	1.05%	1.10%	1.10%	1.10%	1.07%	1.07%	1.07%	1.12%	1.12%	1.12%
	H_2O (vol %)	4.41%	4.41%	4.43%	3.51%	3.52%	3.54%	5.02%	5.02%	5.04%	3.68%	3.69%	3.71%	3.60%	3.61%	3.63%
	CO (vol %)	46.97%	46.97%	46.96%	47.04%	47.04%	47.03%	47.28%	47.28%	47.28%	47.00%	47.00%	46.99%	47.17%	47.17%	47.16%
	Total	100.00%	100.00%	100.00%	100.00%	100.00%	100.00%	100.00%	100.00%	100.00%	100.00%	100.00%	100.00%	100.00%	100.00%	100.00%
	LHV (kJ/kg)	15,900.59	15,899.71	15,897.05	16,330.93	16,329.72	16,326.09	15,524.48	15,523.77	15,521.61	16,256.79	16,255.66	16,252.23	16,239.46	16,238.31	16,234.85
	HHV (kJ/kg)	17,346.44	17,345.56	17,342.90	17,805.64	17,804.43	17,800.80	16,932.58	16,931.86	16,929.70	17,727.50	17,726.37	17,722.95	17,701.67	17,700.53	17,697.06

Table II. Simulation results of syngas composition for solid feedstock used in Mexican power plants as a function of pressure and temperature.

T (°C)	Composition	Petacalco Coal Mixture Pressure (bar)			Rio Escondido Coal Mixture 1 Pressure (bar)			Rio Escondido Coal Mixture 2 Pressure (bar)			Carbon II Coal Mixture 1 Pressure (bar)			Carbon II Coal Mixture 2 Pressure (bar)			Coke Pressure (bar)		
		2	15	30	2	15	30	2	15	30	2	15	30	2	15	30	2	15	30
1000	H_2 (vol %)	29.59%	29.09%	27.89%	25.54%	25.45%	25.19%	23.84%	23.78%	23.59%	28.11%	27.95%	27.50%	25.11%	25.01%	24.73%	30.21%	26.27%	22.95%
	CO_2 (vol %)	10.40%	10.68%	11.36%	17.88%	17.94%	18.10%	19.38%	19.42%	19.55%	15.48%	15.57%	15.84%	18.10%	18.16%	18.35%	0.10%	1.41%	2.68%
	CH_4 (vol %)	0.01%	0.30%	1.02%	0.00%	0.05%	0.20%	0.00%	0.04%	0.15%	0.00%	0.10%	0.36%	0.00%	0.06%	0.22%	1.17%	3.41%	5.38%
	H_2S (vol %)	0.12%	0.12%	0.12%	0.49%	0.50%	0.50%	0.50%	0.50%	0.51%	0.37%	0.37%	0.37%	0.42%	0.42%	0.42%	1.92%	2.00%	2.07%
	N_2Ar (vol %)	0.40%	0.40%	0.41%	0.09%	0.09%	0.09%	0.12%	0.12%	0.12%	0.08%	0.08%	0.08%	0.10%	0.10%	0.10%	0.31%	0.33%	0.34%
	H_2O (vol %)	10.79%	10.94%	11.28%	25.90%	25.94%	26.05%	27.72%	27.75%	27.83%	21.64%	21.71%	21.90%	24.13%	24.17%	24.28%	0.08%	0.96%	1.62%
	CO (vol %)	48.69%	48.47%	47.92%	30.10%	30.04%	29.87%	28.45%	28.40%	28.27%	34.32%	34.23%	33.96%	32.15%	32.08%	31.90%	66.21%	65.62%	64.97%
	Total	100.00%	100.00%	100.00%	100.00%	100.00%	100.00%	100.00%	100.00%	100.00%	100.00%	100.00%	100.00%	100.00%	100.00%	100.00%	100.00%	100.00%	100.00%
	LHV (kJ/kg)	10,051.16	10,017.92	9,938.64	6,903.58	6,897.95	6,881.73	6,353.67	6,349.68	6,338.02	7,945.38	7,934.99	7,906.08	7,001.24	6,995.17	6,977.73	13,900.76	13,660.05	13,463.15
	HHV (kJ/kg)	10,903.74	10,870.51	10,791.23	7,957.94	7,952.32	7,936.09	7,386.87	7,382.88	7,371.22	8,993.71	8,983.31	8,954.40	7,994.65	7,988.57	7,971.13	14,655.53	14,414.81	14,217.92
1250	H_2 (vol %)	27.62%	27.61%	27.57%	22.68%	22.68%	22.67%	20.94%	20.94%	20.93%	25.40%	25.39%	25.38%	22.25%	22.25%	22.25%	30.52%	30.10%	30.53%
	CO_2 (vol %)	8.42%	8.42%	8.44%	15.01%	15.02%	15.02%	16.47%	16.47%	16.47%	12.76%	12.76%	12.77%	15.24%	15.24%	15.25%	0.00%	0.10%	0.00%
	CH_4 (vol %)	0.00%	0.01%	0.03%	0.00%	0.00%	0.00%	0.00%	0.00%	0.00%	0.00%	0.00%	0.01%	0.00%	0.00%	0.01%	1.00%	1.22%	1.00%
	H_2S (vol %)	0.12%	0.12%	0.12%	0.49%	0.49%	0.50%	0.50%	0.50%	0.50%	0.37%	0.37%	0.37%	0.42%	0.42%	0.42%	1.91%	1.92%	1.91%
	N_2Ar (vol %)	0.40%	0.40%	0.40%	0.09%	0.09%	0.09%	0.12%	0.12%	0.12%	0.08%	0.08%	0.08%	0.10%	0.10%	0.10%	0.31%	0.31%	0.31%
	H_2O (vol %)	12.77%	12.78%	12.79%	28.76%	28.76%	28.77%	30.62%	30.62%	30.63%	24.36%	24.36%	24.36%	26.98%	26.98%	26.99%	0.00%	0.12%	0.00%
	CO (vol %)	50.67%	50.67%	50.66%	32.96%	32.96%	32.96%	31.35%	31.35%	31.35%	37.04%	37.04%	37.03%	35.00%	35.00%	35.00%	66.25%	66.23%	66.25%
	Total	100.00%	100.00%	100.00%	100.00%	100.00%	100.00%	100.00%	100.00%	100.00%	100.00%	100.00%	100.00%	100.00%	100.00%	100.00%	100.00%	100.00%	100.00%
	LHV (kJ/kg)	10,090.80	10,090.04	10,087.75	6,958.07	6,957.95	6,957.59	6,407.66	6,407.58	6,407.33	7,998.74	7,998.52	7,997.83	7,054.77	7,054.64	7,054.26	13,920.13	13,896.28	13,920.64
	HHV (kJ/kg)	10,943.39	10,942.63	10,940.34	8,012.43	8,012.31	8,011.95	7,440.86	7,440.77	7,440.53	9,047.06	9,046.84	9,046.15	8,048.18	8,048.05	8,047.67	14,674.90	14,651.04	14,675.41
1500	H_2 (vol %)	26.42%	26.42%	26.42%	20.95%	20.95%	20.95%	19.18%	19.18%	19.18%	23.77%	23.77%	23.77%	20.53%	20.53%	20.53%	30.53%	30.49%	30.39%
	CO_2 (vol %)	7.22%	7.22%	7.22%	13.29%	13.29%	13.29%	14.71%	14.71%	14.72%	11.13%	11.13%	11.13%	13.52%	13.52%	13.52%	0.00%	0.01%	0.03%
	CH_4 (vol %)	0.00%	0.00%	0.00%	0.00%	0.00%	0.00%	0.00%	0.00%	0.00%	0.00%	0.00%	0.00%	0.00%	0.00%	0.00%	1.00%	1.01%	1.06%
	H_2S (vol %)	0.12%	0.12%	0.12%	0.49%	0.49%	0.49%	0.50%	0.50%	0.50%	0.37%	0.37%	0.37%	0.42%	0.42%	0.42%	1.91%	1.91%	1.91%
	N_2Ar (vol %)	0.40%	0.40%	0.40%	0.09%	0.09%	0.09%	0.12%	0.12%	0.12%	0.08%	0.08%	0.08%	0.10%	0.10%	0.10%	0.31%	0.31%	0.31%
	H_2O (vol %)	13.97%	13.97%	13.97%	30.49%	30.49%	30.49%	32.38%	32.38%	32.38%	25.99%	25.99%	25.99%	28.70%	28.70%	28.70%	0.00%	0.01%	0.04%
	CO (vol %)	51.87%	51.87%	51.87%	34.69%	34.69%	34.69%	33.11%	33.11%	33.11%	38.67%	38.67%	38.67%	36.72%	36.72%	36.72%	66.25%	66.25%	66.25%
	Total	100.00%	100.00%	100.00%	100.00%	100.00%	100.00%	100.00%	100.00%	100.00%	100.00%	100.00%	100.00%	100.00%	100.00%	100.00%	100.00%	100.00%	100.00%
	LHV (kJ/kg)	10,114.35	10,114.30	10,114.16	6,990.83	6,990.80	6,990.78	6,440.24	6,440.21	6,440.20	8,030.59	8,030.56	8,030.52	7,087.01	7,086.98	7,086.95	13,920.64	13,918.68	13,913.15
	HHV (kJ/kg)	10,966.94	10,966.89	10,966.75	8,045.19	8,045.16	8,045.14	7,473.45	7,473.41	7,473.40	9,078.91	9,078.88	9,078.84	8,080.42	8,080.39	8,080.36	14,675.41	14,673.45	14,667.91

Table III. Simulation results of syngas composition for solid dry-feedstock used in Mexican power plants as a function of pressure and temperature.

T (°C)	Composition	Petacalco Coal Mixture Pressure (bar)			Rio Escondido Coal Mixture 1 Pressure (bar)			Carbon II Coal Mixture 1 Pressure (bar)			Coke Pressure (bar)		
		2	15	30	2	15	30	2	15	30	2	15	30
1000	H_2 (vol %)	18.80%	16.84%	14.60%	23.62%	23.33%	22.58%	25.85%	25.24%	23.88%	0.00%	0.00%	0.00%
	CO_2 (vol %)	0.03%	0.79%	1.73%	13.18%	13.38%	13.90%	8.94%	9.29%	10.10%	0.00%	0.00%	0.00%
	CH_4 (vol %)	2.46%	3.55%	4.82%	0.00%	0.18%	0.63%	0.01%	0.37%	1.18%	8.82%	8.82%	8.82%
	H_2S (vol %)	0.17%	0.17%	0.17%	0.67%	0.68%	0.68%	0.49%	0.50%	0.50%	2.75%	2.75%	2.75%
	N_2Ar (vol %)	0.54%	0.55%	0.57%	0.12%	0.12%	0.12%	0.10%	0.10%	0.11%	0.45%	0.45%	0.45%
	H_2O (vol %)	0.01%	0.29%	0.56%	10.16%	10.22%	10.36%	6.83%	6.95%	7.22%	0.00%	0.00%	0.00%
	CO (vol %)	78.00%	77.82%	77.55%	52.25%	52.10%	51.72%	57.78%	57.55%	57.01%	87.98%	87.98%	87.98%
	Total	100.00%	100.00%	100.00%	100.00%	100.00%	100.00%	100.00%	100.00%	100.00%	100.00%	100.00%	100.00%
	LHV (kJ/kg)	12,564.29	12,458.72	12,340.14	9,069.68	9,051.30	9,003.95	10,368.25	10,329.65	10,243.46	12,321.94	12,321.94	12,321.94
	HHV (kJ/kg)	13,024.63	12,919.06	12,800.48	9,728.79	9,710.42	9,663.07	11,029.99	10,991.38	10,905.19	12,652.65	12,652.65	12,652.65
1250	H_2 (%)	18.87%	18.79%	18.55%	21.54%	21.53%	21.52%	24.30%	24.28%	24.24%	0.00%	0.00%	0.00%
	CO_2 (%)	0.00%	0.03%	0.10%	11.09%	11.09%	11.11%	7.37%	7.37%	7.39%	0.00%	0.00%	0.00%
	CH_4 (%)	2.42%	2.47%	2.59%	0.00%	0.00%	0.02%	0.00%	0.01%	0.03%	8.82%	8.82%	8.82%
	H_2S (%)	0.17%	0.17%	0.17%	0.67%	0.67%	0.67%	0.49%	0.49%	0.49%	2.75%	2.75%	2.75%
	N_2Ar (%)	0.54%	0.54%	0.54%	0.12%	0.12%	0.12%	0.10%	0.10%	0.10%	0.45%	0.45%	0.45%
	H_2O (%)	0.00%	0.02%	0.06%	12.24%	12.24%	12.25%	8.40%	8.40%	8.41%	0.00%	0.00%	0.00%
	CO (%)	78.00%	78.00%	77.99%	54.33%	54.33%	54.32%	59.35%	59.35%	59.33%	87.98%	87.98%	87.98%
	Total	100.00%	100.00%	100.00%	100.00%	100.00%	100.00%	100.00%	100.00%	100.00%	100.00%	100.00%	100.00%
	LHV (kJ/kg)	12,567.89	12,563.60	12,551.82	9,107.23	9,106.85	9,105.71	10,398.27	10,397.40	10,394.76	12,321.94	12,321.94	12,321.94
	HHV (kJ/kg)	13,028.23	13,023.94	13,012.16	9,766.35	9,765.97	9,764.83	11,060.00	11,059.13	11,056.49	12,652.65	12,652.65	12,652.65
1500	H_2 (%)	18.87%	18.87%	18.85%	20.25%	20.24%	20.24%	23.32%	23.32%	23.31%	0.00%	0.00%	0.00%
	CO_2 (%)	0.00%	0.00%	0.01%	9.80%	9.80%	9.80%	6.39%	6.39%	6.39%	0.00%	0.00%	0.00%
	CH_4 (%)	2.42%	2.42%	2.43%	0.00%	0.00%	0.00%	0.00%	0.00%	0.00%	8.82%	8.82%	8.82%
	H_2S (%)	0.17%	0.17%	0.17%	0.67%	0.67%	0.67%	0.49%	0.49%	0.49%	2.75%	2.75%	2.75%
	N_2Ar (%)	0.54%	0.54%	0.54%	0.12%	0.12%	0.12%	0.10%	0.10%	0.10%	0.45%	0.45%	0.45%
	H_2O (%)	0.00%	0.00%	0.01%	13.54%	13.54%	13.54%	9.37%	9.37%	9.38%	0.00%	0.00%	0.00%
	CO (%)	78.00%	78.00%	78.00%	55.63%	55.63%	55.63%	60.33%	60.33%	60.33%	87.98%	87.98%	87.98%
	Total	100.00%	100.00%	100.00%	100.00%	100.00%	100.00%	100.00%	100.00%	100.00%	100.00%	100.00%	100.00%
	LHV (kJ/kg)	12,567.97	12,567.69	12,566.86	9,130.42	9,130.38	9,130.31	10,416.54	10,416.48	10,416.32	12,321.94	12,321.94	12,321.94
	HHV (kJ/kg)	13,028.31	13,028.03	13,027.20	9,789.54	9,789.50	9,789.43	11,078.27	11,078.21	11,078.05	12,652.65	12,652.65	12,652.65

128f Optimization of Amins Mixture `S Formulation by Neural Networks in Gas Sweeting

iman Farasat[1], Mohammad Samipoor giri[2], and Mojtaba Shariati Niasar[2]. (1) Chemical Engineering, Sharif University of Technology, Tehran, Iran, (2) Faculty of Engineering, Tehran University, Tehran, Iran

Generally natural gas has impurities such as H2S and CO2 (known as acid gases) that causes many problems. Different methods are used for removal of these impurities.The common method is absorption of the acid gases by amine solutions. One amine, can not absorb both acid gases preferentially. By using amine mixtures, absorption of acid gases from natural gas is possible and more adequate.But because of the complexity of absorption and reaction equations, general mathematical methods can not solve the partial differential equations. That is due to several unknowns parameters By using Neural Network method one can overcome these problems and the solution of unknowns parameters becomes possible, by a more simple method respect to similar methods. By using of Neural Network method the formulaes of amine mixture can be optimized. Neural Network has the capability of solving, nonlinear equations.

CFD Study on Natural Gas Fluidized Bed Combustors

Felix Severino de Farias Junior [a], Rubens Maciel Filho [a], Sergio Lucena [b], Jornandes Dias da Silva [b]

[a] *State University of Campinas (UNICAMP), CP 6066, Campinas, CEP 13081-970, SP, Brazil, Phone: 55 + 19 + 3788-3971, Fax: 55 + 19 + 3788-3965*

[b] *Federal University of Pernambuco (UFPE), Rua Prof°. Artur de Sá, S/N, Cidade Universitária, Recife, PE, Brazil, CEP 50670-090, Phone: 55 + 81 + 3274-7274*

Abstract

The increasing availability of high performance and parallel computer environments has stimulated depth studies of complex reactor systems in chemical engineering. Using computational fluid dynamics approach, better description and comprehension of the most important phenomena taking place in the system can be accomplished, resulting in more efficient reactor and process design. Fluidized bed is a relevant system in many applications in process industries. Fluidized bed combustors are important alternatives to conventional ones, due to the fact that they promote large contact area between gas and solids and intense gas mixing throughout the bed, leading to cleaner burning and less pollutants emission, because the relatively low operating temperatures and isothermal profiles. In this work, computational fluid dynamics simulation of a natural gas fed fluidized bed was carried out. The simulated reactor bed was composed by inert particles (sand) with 350 μm. A two-stage kinetics model for methane combustion was used. Natural gas, the cleanest fossil fuel, was used in the simulated fluidized bed combustor, minimizing pollutant emissions even more. Dynamic simulations were developed with disturbances imposed into process input variables. The behavior of the simulated combustor was analyzed, and then compared to experimental data from literature.

Keywords: CFD, natural gas, fluidized bed, fluid dynamics.

1. Introduction

The increasing production and availability of natural gas in several parts of the world are creating better conditions and opportunities for the use and development of new technologies and applications for its consumption. Among the possible applications, it is detached the use of fluidized bed combustors in substitution to conventional ones. In the last years, a lot of research and investments have been made in this technology, due to the increase of the environment restrictions and constraints about the emission of pollutants and consequent need for cleaner alternative technologies.

Fluidized bed systems present interesting thermal properties for its application as combustors. The large contact area between gas and solids promotes high rates of energy generation during the solid fuel burning (Kuo, 1986) and very efficient thermal energy exchange between gas and solid particles. The Fluidized Bed Combustor (FBC) has diverse applications: generation of thermal energy, regeneration of catalysts, CO_2 production in large scale for other process feeding purposes, food processing, treatment of solid residues (drying, overhauling, calcination and incineration), among others. The adjustment of the flows of air and fuel gas allows the control of the working temperature in relatively low levels, 850-900°C (Pre et al, 1998). In such low temperatures, the emissions of NOx are small, which represent an important advantage of the fluidized bed system over other conventional combustors. Natural gas is yet a relatively clean fuel when compared with common liquid fuels; in fact, it is the cleanest fossil fuel. The choice of natural gas as fuel in a fluidized bed combustor contributes even more for low formation of pollutants. With all these characteristics combined, the development of the technology of natural gas fluidized bed combustors assures great environment and economic importance.

In this work, a natural gas fluidized bed combustor simulation is carried out using computational fluid dynamics approach thought MFIX, a multiphase CFD code developed at the U.S. Department of Energy - National Energy Technology Laboratory (NETL) (Syamlal et al., 1993).

2. Kinetic Model

A natural gas combustion kinetic model can easily extend to more than 150 elementary equations. However, when simulating dynamics of combustors and reactors, a simpler mechanism approach is often desirable, because its much better computational performance. Considering a global two stage kinetic model, the methane combustion can be expressed by the following mechanism:

$$CH_4 + \frac{3}{2}O_2 = CO + 2H_2O$$

$$CO + \frac{1}{2}O_2 = CO_2$$

Dryer and Glassman (1973) considered a model of two stages for the oxidation of the methane in a reactor with turbulent flow and obtained the following kinetic rate expressions:

$$R_{CH_4} = 10^{13} e^{(-49600/RT)} [CH_4]^{0.7} [O_2]^{0.8} \qquad (1)$$

$$R_{CO_2} = 10^{14.35} e^{(-45200/RT)} [CO][H_2O]^{0.5} [O_2]^{0.25} \qquad (2)$$

This two stages model allows that the CO and CO_2 profiles along the bed be obtained, estimating the nonlinear behavior of the two steps global reaction.

3. CFD Model

A fluidized bed computational model was developed based on the fluidized bed described by Pre et al. (1998) using the Multiphase Flow with Interphase eXchanges (MFIX) code. The simulated reactor consisted of a 180mm diameter and 800mm pipe with a perforated plate distributor with 1.8% porosity (free area). Sand with 350 μm average diameter and 1400 kg/m3 specific mass was used as solid phase. A mesh grid of 18x50 points was considered for a 2D Cartesian coordinates simulation. Axial symmetry was considered. The bed was assumed to be isothermal and operating at 900

°C. Chemical reaction, momentum and mass balances were considered using the following general equations (Syamlal et al, 1993):

3.1. Momentum balances

Momentum balance for gas-phase:

$$\frac{\partial}{\partial t}(\varepsilon_g \rho_g \vec{v}_g) + \nabla \cdot (\varepsilon_g \rho_g \vec{v}_g \vec{v}_g) = \nabla \cdot \overline{\overline{S_g}} + \varepsilon_g \rho_g \vec{g} - \sum_{m=1}^{M} \vec{I}_{gm} + \vec{f}_g \tag{3}$$

Momentum balance for solid-phase:

$$\frac{\partial}{\partial t}(\varepsilon_{sm} \rho_{sm} \vec{v}_{sm}) + \nabla \cdot (\varepsilon_{sm} \rho_{sm} \vec{v}_{sm} \vec{v}_{sm}) = \nabla \cdot \overline{\overline{S_{sm}}} + \varepsilon_{sm} \rho_{sm} \vec{g} + \vec{I}_{gm} - \sum_{\substack{l=1 \\ l \neq m}}^{M} \vec{I}_{ml} \tag{4}$$

3.2. Mass balances

Continuity equation for the gas-phase:

$$\frac{\partial}{\partial t}(\varepsilon_g \rho_g) + \nabla \cdot (\varepsilon_g \rho_g \vec{v}_g) = \sum_{n=1}^{N_g} R_{gn} \tag{5}$$

Continuity equation for the solid-phase:

$$\frac{\partial}{\partial t}(\varepsilon_{sm} \rho_{sm}) + \nabla \cdot (\varepsilon_{sm} \rho_{sm} \vec{v}_{sm}) = \sum_{n=1}^{N_{sm}} R_{smn} \tag{6}$$

Species conservation equations (CH4, O2, CO, CO2, H2O, N2):

$$\frac{\partial}{\partial t}(\varepsilon_g \rho_g X_{gn}) + \nabla \cdot (\varepsilon_g \rho_g X_{gn} \vec{v}_g) = R_{gn} \tag{7}$$

$$\frac{\partial}{\partial t}(\varepsilon_{sm} \rho_{sm} X_{smn}) + \nabla \cdot (\varepsilon_{sm} \rho_{sm} X_{smn} \vec{v}_{sm}) = R_{smn} \tag{8}$$

The resulting partial differential equations system was solved through the finite volume technique. The convective flux discretization uses Superbee method and the iterative algorithm is a modified extension of SIMPLE (Syamlal, 1998).

4. Results and Discussion

4.1. Simulation

The sand used as inert solid is recognized as a group B solid of Geldart classification. Under this condition, classical fluidization theory predicts 10cm diameter gas bubbles, fast ascending gas velocity in bubbles and a slower one in the emulsion phase. These characteristics were observed in the simulation results too, according to Figure 1. In Figure 1.a, a snapshot of bed void fraction distribution is represented. Figure 1.b shows an axial gas velocity profile with scalable gas velocity vectors, illustrating gas recirculation and dispersion phenomena. From Figure 1-b, it is possible to observe higher gas velocities in bubbles than in emulsion (larger vectors) and counter-current streams in the emulsion phase. This gas flow pattern promotes intense mixing and temperature homogenization along the dense phase of the bed, promoting an isothermal profile in a real bed.

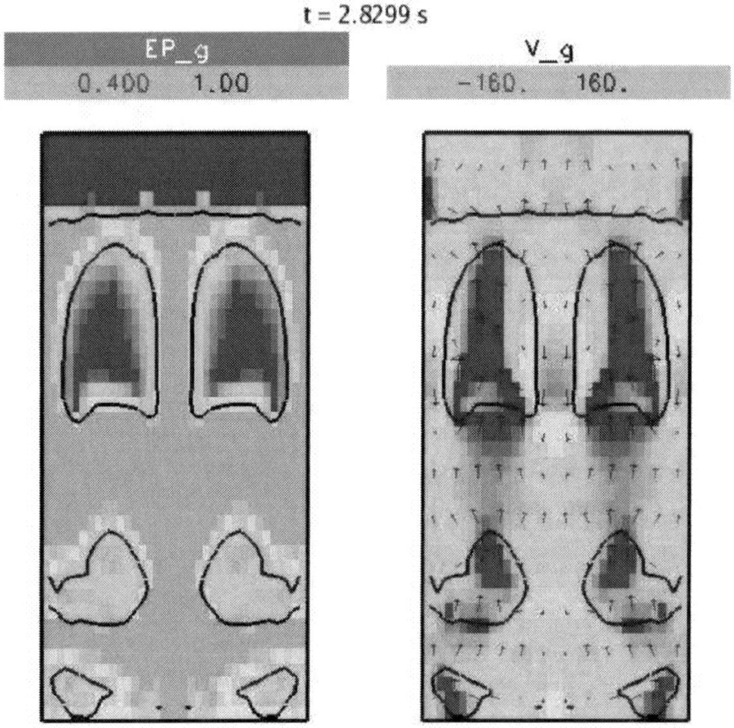

Figure 1 - a) Void Fraction profile b) Axial gas velocity profile

In Figure 2, the dynamic oscillatory behavior of the fluidized bed for the carbon monoxide component after a step in methane feed is shown. The oscillation amplitudes are related to bubble sizes and positions and gas circulations patterns at each time sample.

4.2. Experimental Validation

Steady state simulation results were compared to data from literature (Pre et al., 1998). Average values of component molar fraction profile were calculated over a range of five seconds in dynamic simulation results and are compared to experimental data in Figure 3.

The results obtained by simulation presented faster methane kinetics at the entrance of the reactor. This can be attributed to the isothermal assumption of gas temperature along the reactor and to the limitations of the simplified kinetic mechanism adopted. Brien et al. (2003) point out that the present uncertainty in the kinetics law is a known source of errors in simulation of the combustion of the natural gas. It is expected that better results can be obtained with a better kinetic law and through a non isothermal simulation.

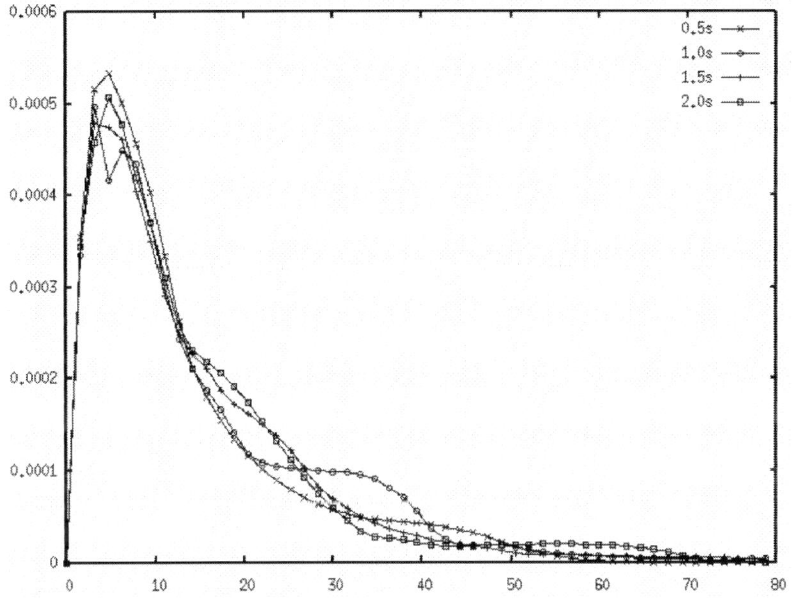

Figure 2 - Carbon Monoxide dynamic response profile after step in methane feed

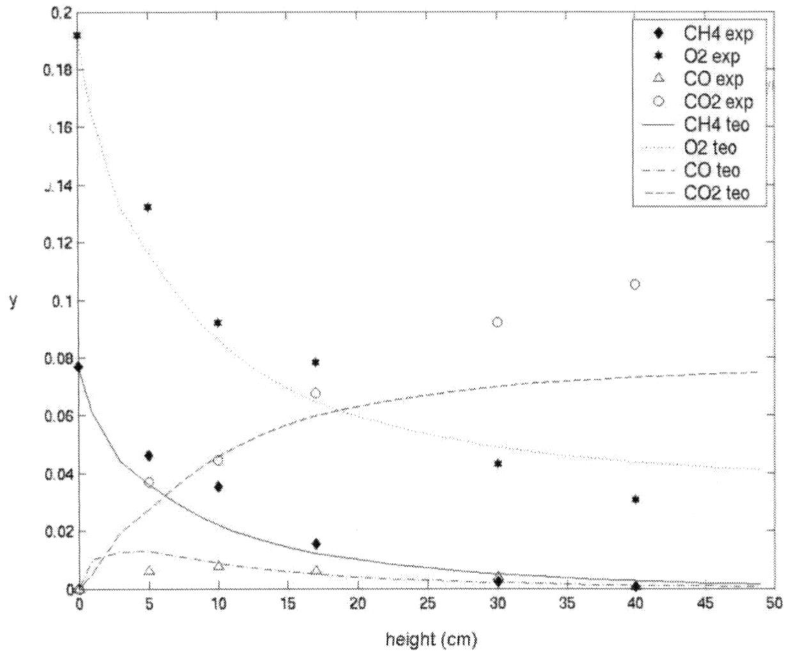

Figure 3 – Components molar fraction profiles

5. Conclusions

Despite of the simplifications assumed (isothermal bed, relatively small mesh grid and two stages methane combustion kinetics mechanism), simulation results were able to produce good predictions, both qualitatively and quantitatively for the fluidized bed. This shows the suitability of the proposed approach.

In this work, sand was used as inert particle which is a cheaper alternative to limestone. However, the limestone has an interesting property of sulfur adsorption that can be explored in future simulation works.

6. Acknowledgments

The authors wish to thanks UFPE/ANP PRH-28 program, FINEP and PETROBRAS for the financial support.

References

1. Dryer, F.L.; Westbrook, C.K.; 1973. Chemical kinetic modeling of hydrocarbon combustion. Prog. Energy Combust. Sci., 1984, Vol 10, pp 1-57. Pergamon Press Ltd.
2. Kuo, L.K. 1986. Principles of Combustion. 1 ed. New York, John Wiley & Sons, 601-614.
3. Moustoufi, N.; Cui, H.; Chaouki, J.; 2001. A comparasion of two- and single-phase models for fluidized-bed reactors. Ind. Eng. Chem. Res. 40:5526-5532.
4. O'Brien, T. J.; Syamlal, M.; Guenther, C.; 2003. Computational fluid dynamic simulations of chemically reactive fluidized bed processes. Third International Conference on CFD in the Minerals and Process Industries. CSIRO, Melbourne, Australia. 469-474.
5. Patankar, S.V. 1980. Numerical Heat Transfer and Fluid Flow. 1 ed. Series in Computational Methods in Mechanical and Thermal Sciences. Hemisphere Publishing Corporation.
6. Pre, P.; Hemati, M.; Marchand, B.; 1998. Study on natural gas combustion in fluidized beds: modelling and experimental validation. Chemical Engineering Science, 53:2871-2883.
7. M. Syamlal, W. Rodgers, T.J. O'Brien, MFIX Documentation: Theory Guide, Technical Note, DOE/METC-94/1004, 1993.
8. M. Syamlal, MFIX Documentation: Numerical Technique, Technical Report, DE-AC21-95MC31346, 1998.
9. Multiphase Flow with Interphace Exchanges (MFIX). National Energy Technology Laboratory (NETL). U.S. Department of Energy. http://www.mfix.org

RasGas Makes Extensive Use of Process Operator Training Simulators in LNG Operations

James M. Dawson*, Ahmet Pekediz[§], John W. Womack[¶]
* Ras Gas Company Limited, P.O. Box 24200, Doha, State of Qatar
[§] Honeywell Process Solutions, 300-250 York Street, London, Ontario, Canada
[¶] 32 Markham Rd, Princeton, NJ 08540

AIChE Spring Meeting, April 2006
6th Topical Conference on Natural Gas Utilization
Orlando, Florida, April 23-27, 2006

ABSTRACT

Ras Gas Company Limited (Ras Gas) operates a major LNG production facility in the state of Qatar. Ras Gas has made significant use of high quality Process Operator Training Simulators (POTS) since before operation of the first LNG Train began in 1998. The primary use is for Distributed Control System (DCS) training of central control room Operators. However, RasGas have captured significant, and unexpected, benefits from faster plant start-up as result of POTS development enabling (a) early identification of some control difficulties, and (b) identification of minor process problems. The paper explains the training use of POTS made by RasGas, and provides examples of these other benefits.

INTRODUCTION

Ras Gas Company Limited (Ras Gas) operates a major LNG production facility in the state of Qatar. The majority of the operation involves treatment of natural gas received from the North Field in Qatar, Liquefaction to produce lean or rich LNG, and transportation of gas to the market. Four LNG trains are now operating at their Ras Laffan site; Trains 5 is under construction, and Trains 6 & 7 are in the engineering phase.

Ras Gas have made significant use of high quality Process Operator Training Simulators (POTS) since before operation of the first LNG Train began in 1998. POTS primary use is for training of the Distributed Control System (DCS) central control room (CCB) operators. Each simulator includes duplicates of the Invensys DCS operator stations which are used in the central control room. The POTS' DCS Operator Stations are linked to a simulation computer system which includes a reasonably accurate first-principles dynamic model of the process being operated, a full emulation of the Process Emergency Shutdown System (PESD), an Instructor station, and supporting software.

The following summarizes POTS system installed at the Training Department

- In 1997 a full POTS system was purchased for Train-1 pre-start-up training.
- In 2001 another POTS system was purchased for Train-3 pre-start-up training.
- In 2004/5 the original POTS computer systems were upgraded and a new system was purchased for the Ras Gas Expansion Project pre-start-up training.

Table 1 lists the specific processes for which POTS process models have been custom built with the proprietary dynamic simulation algorithms from Honeywell.

Table 1: Process Models used by Ras Gas

Train-1 Project	Train-3 Project	Train-4 Project
OffShore Facilities	Acid Gas Treating	Acid Gas Injection
OnShore Inlet Facilities	Sulfur Plant	LNG Train-4 (with NGL)
Acid Gas Treating	LNG Train-3 with Refrigerant Fractionation	NGL Recovery Unit
Sulfur Plant	Dehydration & Selexol	Helium Extraction Unit
LNG Train-1		NGL Off-Plot facilities
Refrigerant Fractionation		
Utility Boiler		
Storage & Loading Facility		

Ras Gas' experience has shown that successful POTS deployment requires

- An Operating Company plan to use the training simulator (intent; resource allocation; actual funding; support of acceptance testing; training plan and simulator maintenance/ support plan)
- A simulator development and acquisition plan (definition of simulator scope and schedule; contractor and/or owner project team; vendor selection; liaison among POTS vendor, engineering contractor and DCS/PESD vendor; Factory and Site Acceptance Testing procedures)
- An effective vendor with staff experience in modeling & DCS/PESD technology.
- Liaison with EPC & DCS/PESD vendor for timely provision of information.

Ras Gas has found that the benefits from successful POTS implementation include:

- Effective pre-start-up training of Operations staff (as expected).
- Certification of Panel Operators, including Refresher Training (as expected).
- Faster plant start-up because many DCS/PESD design and implementation shortcomings are found and corrected during factory

acceptance tests (FAT) and site acceptance test (SAT) of the POTS-system. This benefit had not been anticipated for the Train-1 activity, but was known to be significant for subsequent training simulator procurements.

- Faster plant start-up because of early identification and correction of small process problems (some of these were identified by dynamic simulation studies carried out as part of the POTS acquisition plan; others were identified during model acceptance test (MAT), FAT and SAT of the POTS process models)

Quantifying the benefits from use of POTS is difficult. However, commissioning team members assert that use of these simulators speeded start-up by "several days" for each plant. In addition, Ras Gas' Operations Department have enthusiastically supported POTS for new plants because of training and other benefits which they have observed.

OPERATOR TRAINING

When Ras Gas first started operations it had a new staff hired from throughout the world, with thirty-seven different nationalities employed. To smooth the start up as much as possible Ras Gas used the Operator Training Simulator described above. All newly hired Panel men and Senior Operators were put through a one-week training session using models of the units they were going to operate. Other operators were run through a shorter session to familiarize them with moving throughout the DCS screens to find the information related to their units.

Since start-up of the first LNG train Ras Gas has been expanding rapidly, reaching a peak in 2005 for the number of new units to be started up. To meet the process training requirements there has been consistently high usage of the simulator and the instructor's time. Ras Gas has provided POTS-based operations training for all necessary personnel before scheduled plant startups. During a current slowdown in new-plant commissioning activity, "refresher training" of personnel and training of new hires is resuming.

Quantitatively: the Training Department has developed fourteen different training courses using the OTS models. A total of some 200 delivered courses have trained almost 300 operations staff, using more than 1500 man-days.

Ras Gas is currently using most of the delivered POTS process models. This training assists Operations management in evaluating the performance of their current Panel Operators and also in evaluating other personnel for promotion into panel positions.

Expansion of the Ras Gas facilities is continuing, generating a constant requirement for training. All models are kept running and up-to-date (with regard to changes in the real-world control schemes) through use of a Maintenance contract with Honeywell, the vendor of all presently installed POTS equipment and models.

SCOPE of OPERATOR TRAINING SIMULATOR MODELS

For each model the scope is defined, early in the POTS-project, by marking all in-scope equipment on a set of Process & Instrumentation Diagrams (P&IDs). Model scope then includes all marked equipment and piping plus all controls and process logic associated with that equipment. Work on POTS usually begins 12-to-15 months before delivery date: small changes in P&IDs are expected, and remain in-scope. High quality emulation of DCS controls and PESD are required, but emulation of special purpose controls (e.g. compressor surge controls from CCC or the GE Mark-VI gas turbine controls) are included with somewhat less detail. In order to simulate the effect of air temperature and/or feed compositions, boundary conditions such as plant feed and ambient conditions are set or changed by the POTS instructor. In addition there are usually a few logic inputs required from out-of-scope equipment. In general, process-related packaged equipment, such as lube oil system of a pump or compressor, is left out-of-scope, or a lower fidelity simulation is used in the model. Also, first-out-alarm systems are not included in the POTS system.

As an example of the level of detail used, consider the model which supports start-up of the Frame-7 single shaft gas turbine used to drive the main propane refrigeration compressor for the Train-3 LNG plant. This gas turbine was designed for electric power generation, and does not produce significant power below about 92% of the design speed. However, the refrigeration compressor's power demand becomes significant well below that speed. Since the standard gas turbine "starter motor" cannot deliver enough power, its output is augmented by a variable frequency electric motor drive, the "helper motor". It is considered necessary to model the start-sequence in some detail. An attempt to start the compressor from an initial pressure which is too high leads to a "stall" condition in which the system fails to reach normal operating speed. The POTS provides an effective way to teach this limitation to panel operators.

After the operator hits the "start" button the Mark-V control system moves the gas turbine through a pre-programmed start-up sequence. Initially the "starter" motor spins the turbine up to about 600 rpm and holds that speed for pre-set period while compressed air purges the combustion path. Then fuel is admitted and gas turbine speed is ramped up to minimum governor speed following a fixed time-speed schedule imposed by Mark-V. During this ramp-up the gas turbine actually absorbs power until speed is about 90% of design speed, and both compressor power and accelerating torque are provided by the "starter" motor and the "helper" motor.

A block diagram of the simulation-approach used to represent the gas turbine start-up for the propane compressor is shown below in Figure 1. Compressor speed is calculated from a dynamic torque-balance by the SHAFT-module of the dynamic model. The three "multiplier" blocks marked X in a circle in Figure 1, use start-up governor controller output signals to modulate the maximum torque output from various drivers in order to control speed. The pre-set speed-vs-time relationship supplies a varying setpoint to the speed controller, NIC. The two "relay" blocks, RY1 and RY2, implement split range control of speed by using first the GE starter motor output and then the ABB Helper motor output to regulate speed.

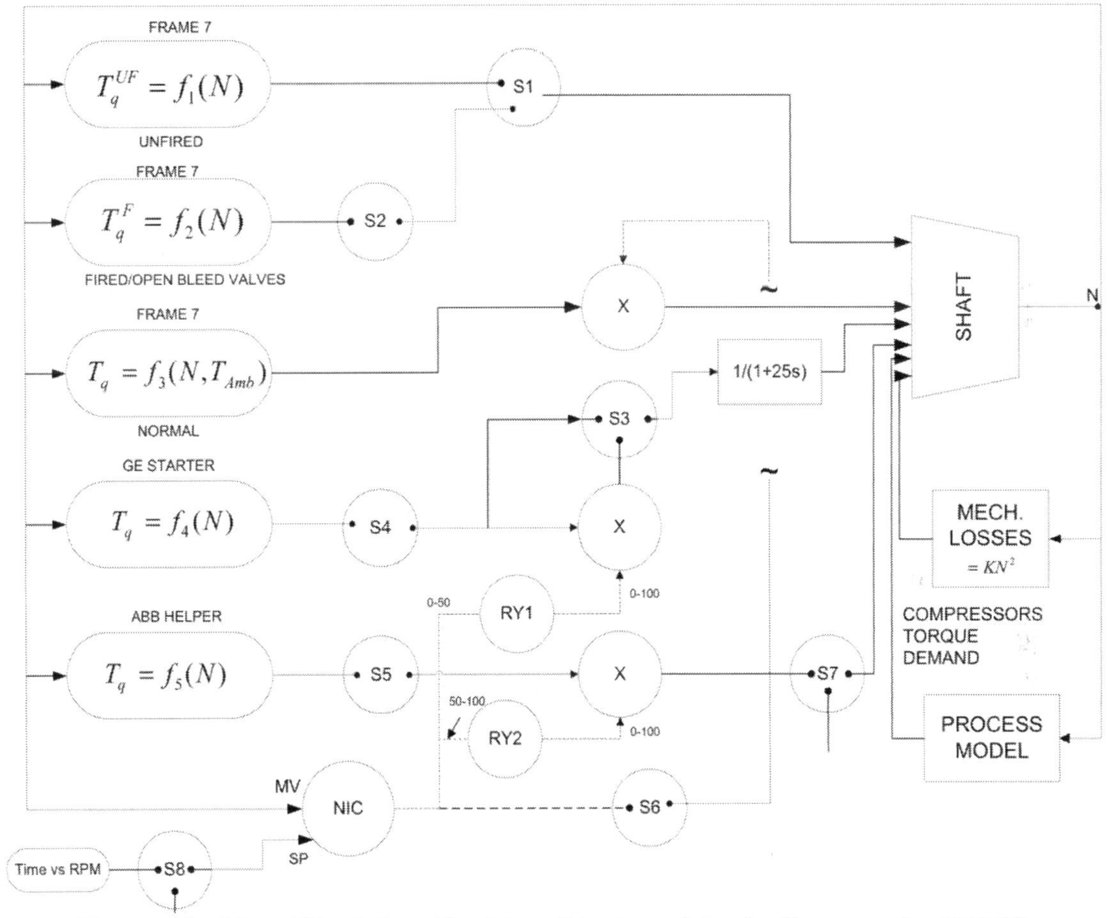

Figure 1: Simplified Gas Turbine Start-up block diagram for POTS

Switches have the following functions

S1	Turbine fired/unfired	Selects between Unfired Frame 7 torque and Fired Frame 7 (w.Open bleed valves) torque. Changes state at Time = 600 sec during startup. S1 is also used to apply "breaking torque" during TRIP simulations when Frame 7 is running at high speed during an unfired condition.
S2	Bleed valves open	Switch closed from Time = 0 until shaft speed N=91.5% (3294 rpm) and bleed valves close.
S3	Purge Cycle	During the Purge cycle, from Time = 0 sec until Time = 60 sec after N reaches 612 rpm, this switch connects full Starter motor torque with the Keonig clutch. After Purge cycle this switch connects the controller-modulated Starter torque to the clutch.
S4	Starter Motor On/Off	GE starter motor turns ON at Time = 0 sec and remains ON until speed reaches N= 2160 rpm (60%).
S5	ABB Helper On/Off	ABB Helper motor turns ON when speed reaches N=1008 rpm, and off when speed reaches the "empowerment speed", N=3294 rpm.
S6	Normal Governor On/Off	S6 closes when the "empowerment speed", 91.5%/3294 rpm, is reached; enabling the normal governor function of controlling speed by regulating fuel supply to the gas turbine. When S6 is not closed a zero signal is sent to the multiplier.
S7	Minimum Governor	After minimum governor is reached the user is allowed to set helper motor power through S7. This Power is set in MW.
S8	Minimum Governor	After minimum governor is reached the user is allowed to change the speed through S8 from a DCS block (95% to 101%). The set point change, increment or decrement, is limited to 96RPM per minute. If kickback is initiated from the logic, then speed ramps down through S8 at 96RPM per min.
NIC	Mark 6 Governor	Speed governor is modeled as a conventional P + I controller, with measured variable range 0-to-4200 rpm. During "normal operation" (after speed reaches 3420 rpm) the speed setpoint is limited to the range 95-to-101% (3420-to-3636 rpm). During "start-up" this controller's setpoint varies in a pre-determined fashion: (a) From T=0 to 600 seconds, speed setpoint = 360 rpm (10%), (b) From T=600-to-1200 speed setpoint varies as described in the narrative, (c) For T>1200 sec, speed setpoint = 95%/3420rpm unless it is changed manually. For the start-up, the tuning constants for the speed governor were set at GAIN=15 and INTEGRAL=1

HELP with COMMISSIONING of CONTROL SYSTEMS

For each training simulator purchase Ras Gas specified that POTS be delivered on-site, ready for training use, six months before the planned plant start-up, thus allowing lead-time for both basic and refresher training of key Operations personnel. The critical path item in this schedule is always the delivery date for the post-FAT configurations for the "real plant" distributed control system and emergency shutdown system. Holding to the schedule needs close liaison between DCS and POTS vendors, and this has been achieved over four separate projects.

Factory acceptance tests for DCS configuration and for the PESD system are carried out separately in the instrumentation vendor's shop, after which these configurations are used for FAT of the various POTS plant models. Detailed configuration and testing of DCS and PESD are carried out by different engineers. During DCS and PESD FAT the operating company's acceptance testing teams are often different, and these tests are completed without use of a dynamic process model.

In contrast to that, the POTS integrates the post-FAT configurations of DCS and PESD with a dynamic process model. Tests carried out during the POTS Factory Acceptance Test include a process start-up, adjustment of process throughput and product quality, a process shutdown, key "emergency failure" situations, and some random testing. The POTS test team uses the latest available Operating Manuals as a guide for these activities. During the FAT of every POTS process model the test team has discovered "dozens" of items which require correction prior to the start-up of the "real"-plant. While some of these corrections are minor, many are not; some are obvious and some are subtle. There are significant cost savings to making these corrections BEFORE the same problems are discovered during the actual plant start-up.

Examples of problems identified during POTS FAT range from the DCS configuration using the "direct action" for a simple control loop instead of "reverse action" (simple to spot and correct, even during plant start-up) to a trip-recovery procedure, implemented as described in an Operating Manual, which caused shut-down of product-supply to a major sales gas customer (this was not simple to identify, and took weeks of study and reviews to develop satisfactory revisions to DCS configurations and PESD logic).

Other examples are (a) incorrect alignment of "steps" between an automated DCS start-up procedure and the logic protection, PESD/sequence code, which was to implement it, (b) control strategies which fail to achieve their objectives, and (c) automated sequences of operations which "lock-up" if the Operator takes some unexpected action or fails to follow instructions correctly.

DESIGN SHORTCOMINGS IDENTIFIED BY POTS ACCEPTANCE TESTING

One of the POTS process models includes a closed loop cooling water system which was isolated from the site-wide cooling water system because of contamination risks. The model was built to match "approved for construction" P&IDs represented by Figure 2, and during acceptance testing for the POTS a problem was recognized.

Three pumps are normally operating and the fourth is spare. The spare pump should be "lined-up" ready for immediate operation, with the minimum-flow control-loop in Automatic using the correct setpoint. However, with the spare pump off-line, there is no flow measured by this controller, so it moves the recycle valve fully open. Figure 2 shows that flow from the three on-line pumps will then pass through the open recycle valve back to the storage tank, which was not the desired control strategy. The obvious correction is to move the check valves in the discharge line for each pump to downstream of the recycle flow take-off line, but they had already been installed. After considering several ways to use the DCS to correct this problem it was decided to move the pump discharge check valves. Use of the POTS model exposed this problem early enough that it could be corrected before start-up.

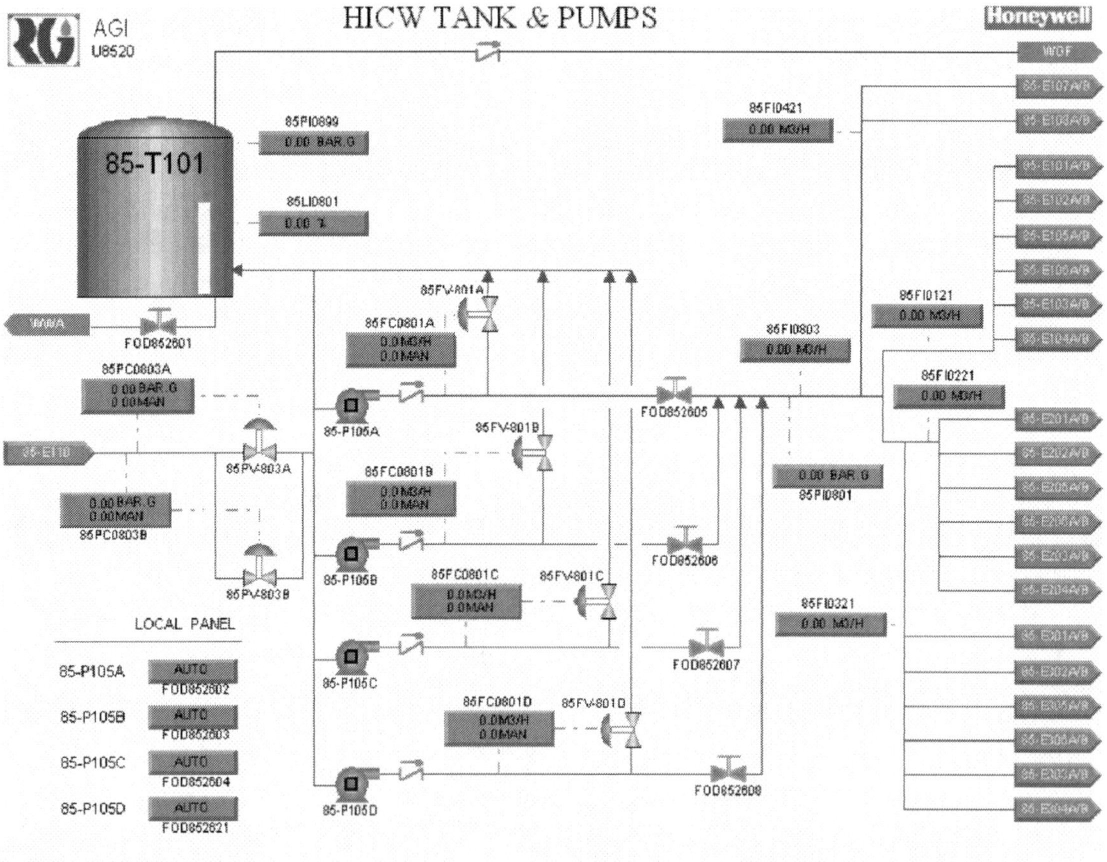

Figure 2: Cooling water pumps and protective minimum flow controls

ANOTHER EXAMPLE

POTS models are designed to represent re-start of each plant unit. However, due to length of time required to start each model, some are not scoped to include the initial commissioning or manual pre-start operations such as drying and defrosting. During the POTS FAT for the NGL Off-Plot facilities a number of DCS and PESD logic problems were identified. These drew attention to difficulties with the initial start-up and cooldown of the huge product storage tanks. Once it was on-site, the POTS model was utilized by the commissioning team to assess the effectiveness of alternate procedures for the initial plant start-up.

CONCLUSION

Process Operator Training Simulators have been used effectively by Ras Gas training programs. In addition, Ras Gas has derived significant, unexpected, benefits from their use of high-fidelity dynamic-model-based operator training simulators. This experience increases the validity of the argument for use of first-principles models throughput the life of a processing facility. Early on, models drive the choice of equipment to implement the design: after that, continuously improving model(s) can answer "what-if" questions throughout the life of the plant.

BIOGRAPHY OF SPEAKER

Ahmet Pekediz is a Sr. Application Analyst with Honeywell Process Solutions in London-Ontario's Office. He has been working in Dynamic Process Simulation applications related to Engineering Studies and Plant Operator Training for over 6 years. Examples of completed projects are H-Oil process, Olefin Catalytic Polymerization Unit, Hydrogen Plants, Sulfur Recovery Units, Acid Gas Injection and Compression Units, Natural Gas Refrigeration and Liquefaction units. He holds a PhD degree in Chemical Engineering from the University of Western Ontario.

138c Performance Characteristics of LNG Tandem Expanders

Hans Kimmel, Ebara International Corporation, Sparks, NV 22222 and Chen-Hwa Chiu, Fuels and Petrochemicals Division/Group 16, Chevron Energy Technology Company, 1500 Louisiana Street, Houston, TX 77002.
In 1996 the first LNG Expanders went into operation at a liquefaction plant in Malaysia and ever since every new liquefaction plant has been equipped with LNG expanders. It is a frequently discussed trend that the size of LNG liquefaction plants is continuously increasing and consequently also the size of each plant component is increasing as well. For the past 10 years the average size of LNG expanders has doubled and the traditional design of the liquid expanders has satisfied the specified performance requirements without major changes. To comply with the further projected increase in the size of liquefaction trains the design of LNG expanders has to be modified. The key objective for the performance of LNG expanders is high efficiency over a wide range of mass flow variations. Further design objectives are reliability, safety, weight and size. These objectives are best achieved with a new design arranging two LNG expanders in a tandem configuration. The paper presents the basic design and the typical performance characteristics of LNG Tandem Expanders and analyzes the differences and advantages compared to currently designed LNG expanders.

Pressure induced Non-Linear Oscillations in Two-Phase LNG Pipe Flow

Andrew Kimmel
University of Nevada, Reno
Department of Physics
Reno, Nevada

AIChE Spring National Meeting 2006
6th Topical Conference on Natural Gas Utilization
Orlando, Florida, April 23-27, 2006

Abstract

Two-phase fluid flows are often subject to pressure induced oscillations. Due to its compressibility the vapor bubbles act as a spring with an asymmetric non-linear characteristic. The volume of the vapor bubbles increases or decreases differently if the pressure fluctuations are compressing or expanding. Consequently, compressing pressure fluctuations in a two-phase pipe flow cause less displacement in the direction of the pipe flow than expanding pressure fluctuations. The displacement depends on the ratio of LNG liquid to vapor, the ratio of pressure fluctuations over average pressure and on the exciting frequency of the pressure fluctuations.

The paper describes the governing non-linear equation for two-phase fluid oscillations and presents steady state solutions for free and for pressure induced oscillations. Resonance criteria and effects of LNG two-phase flows on plant safety and operation are discussed.

Introduction

Two-phase LNG is a mixture of liquefied and gaseous natural gas. It forms in LNG liquefaction plants during the condensation of the saturated natural gas or during the expansion of the already condensed pressurized natural gas. The volumetric ratio r between gaseous and liquid natural gas is defined as

$$r = \frac{V_g}{V_g + V_l}$$

with $r = 0$ for pure liquid and $r = 1.0$ for pure gas. V_g is the volume of gas and V_l is the volume of liquid. With very good accuracy, the liquefied natural gas can be assumed to be incompressible when compared to the compressible gaseous natural gas. Gaseous natural gas follows approximately the ideal-gas equation of state with the gas volume V_g inversely proportional to the gas pressure p at constant temperatures. Pressure changes Δp and volumetric changes are related through the following equation

$$\frac{p + \Delta p}{p} = \frac{V_g}{V_g + \Delta V_g}$$

and can also be written in the form:

$$\Delta p = -p \frac{\Delta V_g}{V_g + \Delta V_g}$$

Within a controlled volume of two-phase fluid defined as $V_g + V_l = 1$ the volumetric gas content is equal to

$$V_g = r$$

and for very large Δp the gas condenses. Therefore the maximum volumetric change is $\Delta V_g = -r$ for increasing pressures (compression) and $\Delta V_g = +\infty$ for decreasing pressures (expansion) with $p + \Delta p \rightarrow 0$ approaching vacuum.

$$\Delta p = -p$$

The density of the gaseous natural gas is negligibly small compared to the density ρ of the liquefied natural gas. Therefore the oscillating mass m within the controlled volume $V_g + V_l = 1$ depends only on the volume of the liquid V_l.

$$m = \rho\, V_l = \rho(1-r)$$

For pipe flows with constant cross sectional area the axial displacement z of the fluid is linearly dependent to the volumetric change

$$z \approx \Delta V_g$$

Free oscillations in two-phase flow

Pressure changes Δp act as a force in the axial direction of the pipe. The compressible gaseous volume and the mass of the liquid volume form an oscillating system. Due to the non-linear characteristic of the ideal-gas equation, the compressible gaseous volume acts as an asymmetric non-linear spring and generates axial forces that are non-linear to the displacement z as a function of the time t.

This oscillating system can be modeled in the case of free vibrations by the following second order differential equation

$$(1-r)\rho \frac{d^2 z(t)}{dt^2} + p \frac{z(t)}{r + z(t)} = 0$$

In case of cylindrical pipe flows r has the dimension of a length and a dimensionless displacement $y(t)$ is introduced:

$$y(t) = \frac{z(t)}{r}$$

$$r(1-r)\rho\frac{d^2y(t)}{dt^2}+p\frac{y(t)}{1+y(t)}=0$$

For very small oscillating displacements $y(t) \ll r$ the oscillating system can be considered as a linear harmonic oscillator

$$\frac{d^2y(t)}{dt^2}+f_N^2y(t)=0$$

with the natural frequency $f_N=\sqrt{\frac{p}{r(1-r)\rho}}$ that is introduced to the non-linear differential equation for free oscillations in two-phase pipe flows

$$\frac{d^2y(t)}{dt^2}+f_N^2\frac{y(t)}{1+y(t)}=0$$

This non-linear equation has a first integral

$$\frac{dy(t)}{dt}=\pm f_N\sqrt{2}\sqrt{\Phi-y(t)+\ln\frac{1+y(t)}{1+\Phi}}$$

with the initial condition of $y(t)=\Phi$ for $\frac{dy(t)}{dt}=0$ and Φ is the positive amplitude (expansion) of the two-phase oscillation and Ψ is the negative amplitude (compression) of the two-phase oscillation. Using the Lambert $W(x)$-function, Valluri[(1)], the expansion amplitude Φ and the compression amplitude Ψ are related through the following equation:

$$\Psi=-1-W(-\frac{1+\Phi}{e^{1+\Phi}})$$

The ranges of the expansion and compression amplitudes are

$$0\le\Phi\le\infty \text{ and } -1\le\Psi\le 0$$

Analyzing pressure waves in two-phase mixtures Henry[(2)] recorded the differences in amplitudes and frequencies of expansion and compression waves.

Fig.1 demonstrates the asymmetric oscillations in a two-phase pipe flow. For a given absolute pressure difference the absolute displacement $\Delta x(t)$ is smaller for compression than for expansion.

Forced oscillations in two-phase flow

In the case for a forced vibration with harmonic excitation the following differential equation describes the pressure-induced oscillation

$$\frac{d^2y(t)}{dt^2} + f_N^2 \frac{y(t)}{1+y(t)} = Gf^2\cos(ft)$$

The exciting force Gf^2 depends on the exciting amplitude *G* and the square of the exciting frequency *f*.

This highly non-linear differential equation cannot be solved by analytical methods, and numerical or approximation methods have to be applied. Common methods like Perturbation, Variation of Parameter, Harmonic Balance [Tse[(3)]] or Ritz-Galerkin [Rao[(4)]] to generate approximate analytical solutions for the steady state are not practical to apply due to the discontinuity at *y(t) = -1*, and numerical methods generate only non-algebraic solutions.

A different method using Fourier coefficients for the periodic error is applied to generate an approximate analytical solution for the steady state case of pressure induced oscillations in two-phase pipe flow.

If *y(t)* is the exact solution then let *x(t)* be an approximate solution for the above described differential equation.

$$x(t) \approx y(t)$$

Then *E(t)* is the function of the error for this approximate solution *x(t)*.

$$\frac{d^2x(t)}{dt^2} + f_N^2 \frac{x(t)}{1+x(t)} - Gf^2\cos(ft) = E(t)$$

For periodic functions of the error *E(t)* can be expanded in a Fourier series

$$E(t) = \frac{1}{2}a_0 + \sum_{k=1}^{\infty}[a_k\cos(kft) + b_k\sin(kft)]$$

The Fourier coefficients a_k and b_k are determined by the following integration:

$$a_k = \int_0^{2\pi} E(t)\cos(kft)d(ft)$$

$$b_k = \int_0^{2\pi} E(t)\sin(kft)d(ft)$$

By setting certain a_k and certain b_k equal to zero, the error for the *k*-th term of the Fourier series *E(t)* becomes zero. The approximate solution *x(t)* is then correct for the *k*-th term of the function *E(t)*.

Approximate solution for the oscillating system

For the given differential equation with forced oscillation through harmonic excitation $Gf^2\cos(ft)$ the approximate solution *x(t)* corresponds to the exciting periodic function but with a constant term S and different amplitude *A*.

$$x(t) = S + A\cos(ft)$$

This is part of a Fourier series with $a_0=2S$ and $a_1=A$ and all other Fourier coefficients equal to zero. By substituting *x(t)* into the differential equation the Fourier coefficients a_0 and a_1 of the error *E(t)* can be determined by integration:

$$E(t) = -(A+G)f^2\cos(ft) + f_N^2\frac{S + A\cos(ft)}{1 + S + A\cos(ft)}$$

$$a_0 = \int_0^{2\pi} E(t)d(ft)$$

$$a_1 = \int_0^{2\pi} E(t)\cos(ft)d(ft)$$

The integration of these equations leads to the following functions:

$$a_0 = a_0(A,S)$$
$$a_1 = a_1(A,S)$$

Setting both functions $a_0(A,S)=0$ and $a_1(A,S)=0$ produces approximate solutions for *A* and *S* as a function of the frequency *f* of the exciting force. The solutions are in implicit form with *f* as a function of *A*, and *S* as a function of *A*.

$$S = -1 + \sqrt{1+A^2}$$

$$\frac{f}{f_N} = \sqrt{2}\sqrt{\frac{-A+\sqrt{1+A^2}(-1+\sqrt{1+2A(A+\sqrt{1+A^2})})}{A(A+G)(A+\sqrt{1+A^2})}}$$

Review of the approximate solution

The approximate solution corresponds to the true solution through the following relation: $\{ S + A \approx \Phi \}$ and $\{ S - A \approx \Psi \}$.

Fig.2 shows the approximate solution for a characteristic two-phase oscillation with $\{A=1\}$. The positive expansion amplitude has a maximum displacement of $\{S + A = \sqrt{2}\}$ and the negative compression amplitude has a maximum displacement of $\{S - A = -2 + \sqrt{2}\}$. The dashed line indicates the constant $\{S = -1 + \sqrt{2}\}$.

Fig.3 demonstrates typically the asymmetric response for free and for forced oscillations in two-phase flow. The dotted line indicates the asymmetric free oscillation. The frequency f of the free oscillation is a function of the initial displacement or amplitude and is always smaller than the natural frequency f_N for very small oscillations.

For a certain frequency f the expansion amplitude is related to the compression amplitude by the Lambert function, as shown above. The full line demonstrates the asymmetric response for the forced oscillation. The response includes typical non-linear characteristics, like more than one stable solution, amplitude jump phenomenon [Rao[(4)]] and phase switching, if the exciting frequency is in the range $\{0 \le f \le f_N\}$.

Fig.4 shows a typical amplitude response for the first harmonic $\{ A \cos ft \}$ of the approximate solution. The dotted line represents the amplitude for free oscillations and the full line for forced oscillations. If the exciting frequency is above the natural frequency $\{ f > f_N \}$, then the responding amplitude A is negative, which means that the fluid system oscillates 180^0 out-of-phase with the excitation.

This is also the case for frequencies to the right side of the dotted line. For very high frequencies the responding amplitude A approaches the negative value of the exciting amplitude G. For frequencies between zero and the dotted line, the fluid system oscillates in-phase with one or two possible stable solutions for the amplitude A. The oscillating system eventually switches between in-phase and 180^0 out-of-phase response, also changing then the magnitude of the amplitude A.

Fig.5 shows the amplitude response spectrum and Fig.6 the absolute amplitude response spectrum for different values of the exciting amplitude G.

Conclusions

- The natural frequency of oscillating two-phase fluids depends on the volumetric ratio between gas and liquid. For the ratio 0.5, which means equal volumes for gas and liquid, the natural frequency has a minimum value.
- The frequency of free oscillating two-phase fluids decreases with increasing initial displacement amplitude.

- Pressure induced forced vibrations in two-phase fluids generate complicated non-linear response patterns and resonance characteristics if the exciting frequencies are smaller than the frequencies of free oscillations.
- Low frequency responses are therefore typical for oscillations in two-phase pipe flows.
- Due to the complicated non-linear effects of two-phase LNG pipe flows the oscillations and vibrations of two-phase flows have to be closely monitored and observed to maintain plant safety and reliable plant operation. Particular exciting frequencies below the natural frequency of the two-phase fluid should be avoided.

Acknowledgement

The author is indebted to the encouragement and support regarding the publication of this paper received from Dr. Chen-Hwa Chiu, Chair of the Sixth Natural Gas Utilization Conference at the 2006 AIChE Spring National Meeting in Orlando, Florida.

References

1. Valluri, S.R. et al "Some applications of the Lambert W-function" Canadian Journal of Physics Vol.78, No 9, pg. 823-831, Sept. 2000
2. Henry, R.E. "Pressure Wave propagation in two-phase mixtures" Heat transfer Minneapolis, Chemical Engineering Progress Symposium Series No.102, 1970, Volume 66, American Institute of Chemical Engineers, New York.
3. Tse, Francis S. et al. "Mechanical Vibrations", Allyn and Bacon Inc., Boston, MA 02210, 1978, ISBN 0-205-05940-6
4. Rao, S.S. "Mechanical Vibrations" Pearson Prentice Hall, New Jersey 07458, 2004, ISBN 0-13-048987-5

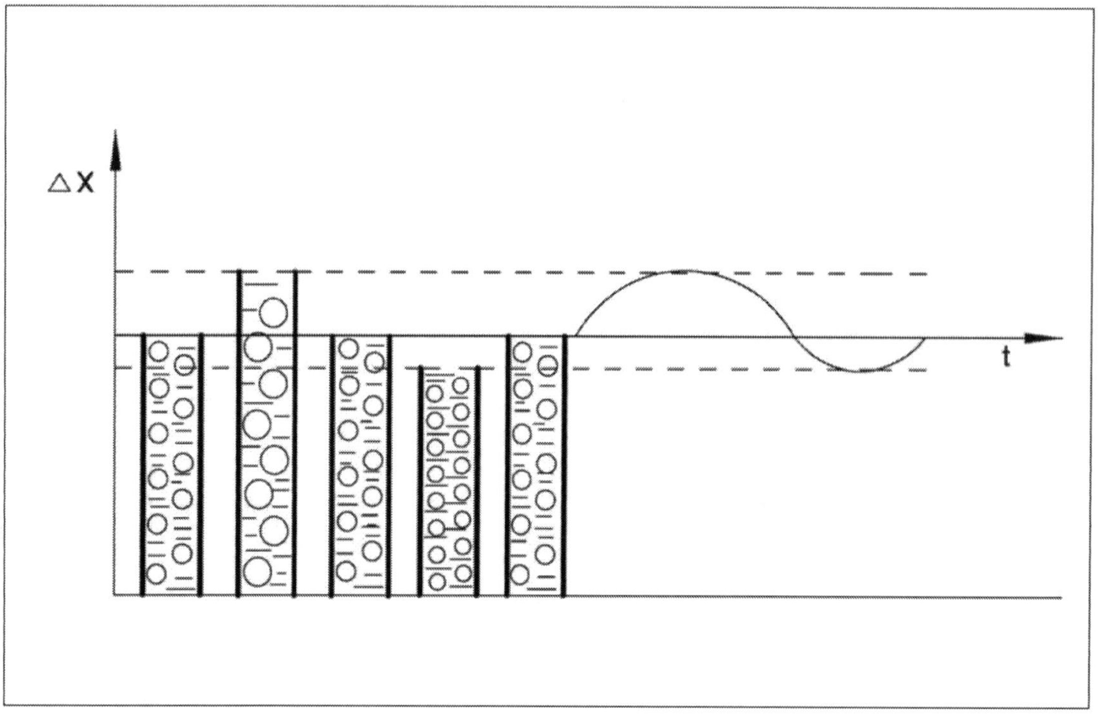

Figure 1

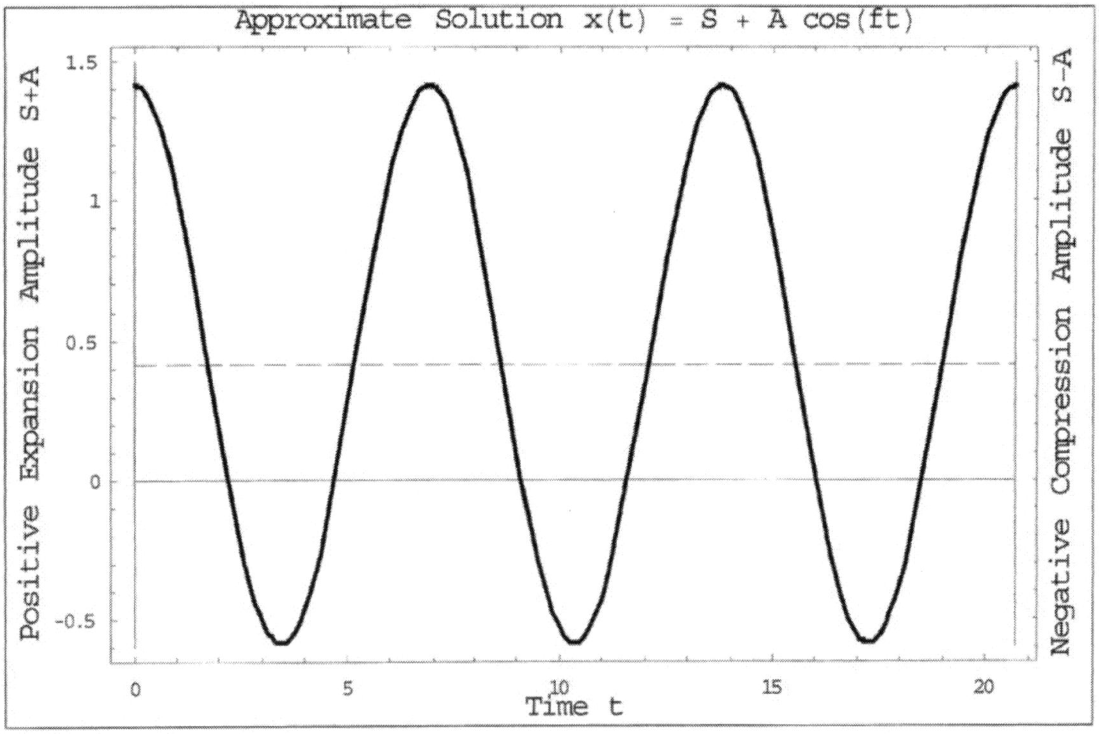

Figure 2

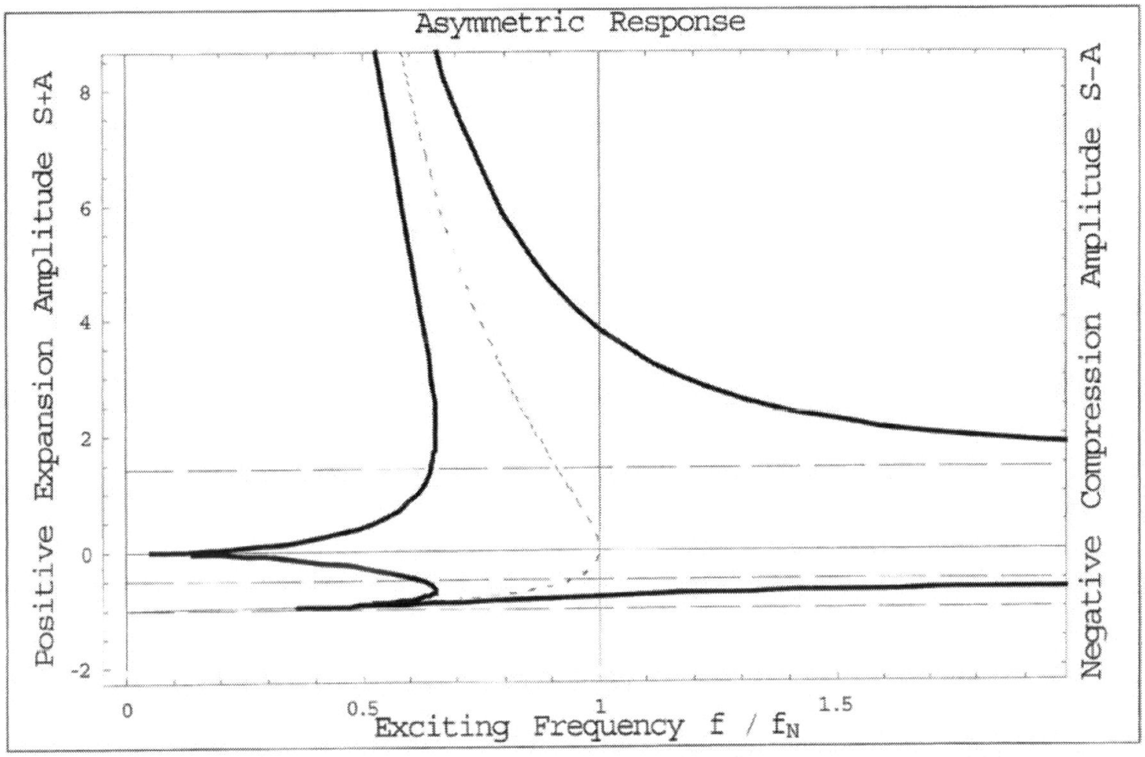

Figure 3

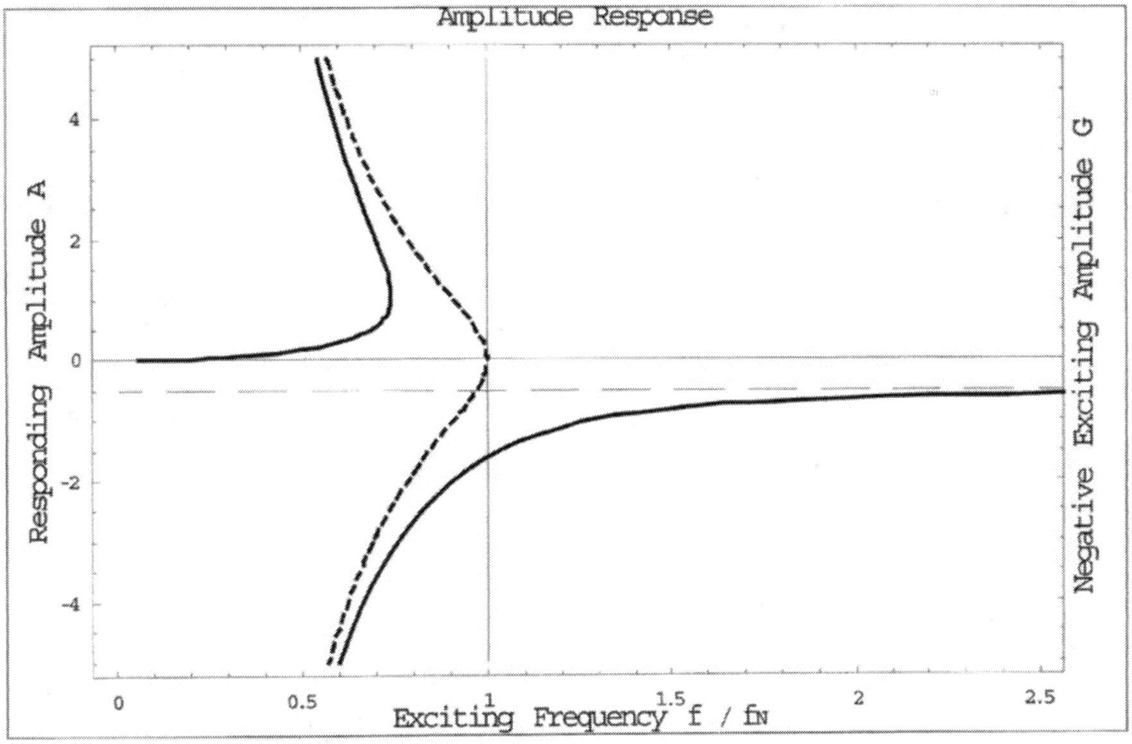

Figure 4

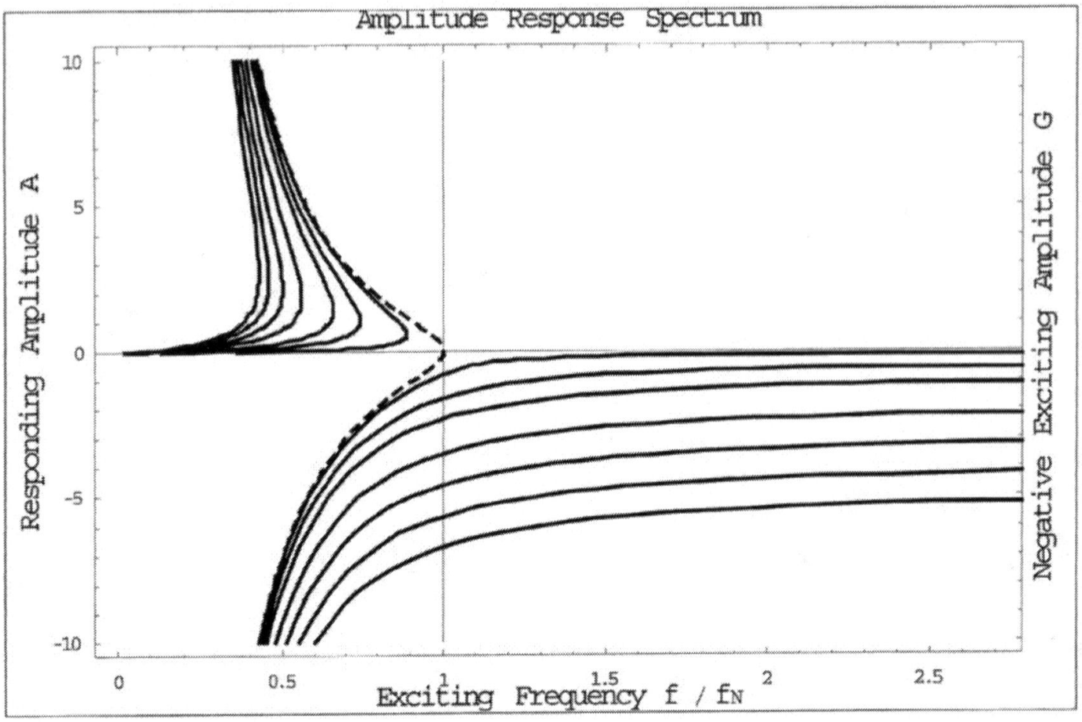

Figure 5

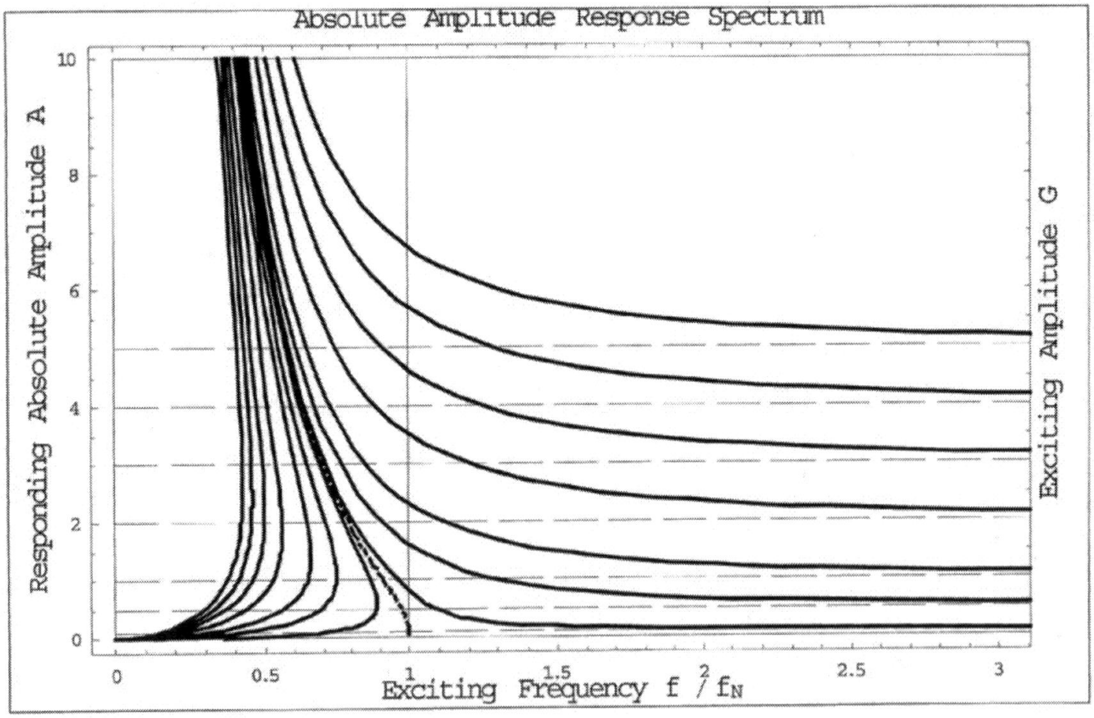

Figure 6

Biography

Andrew Kimmel

Andrew Kimmel is graduating at the University of Nevada, Reno, College of Science, in Physics and Mathematics. His research contributions include analyses of non-linear oscillations, and steady state solutions for non-linear vibrations with asymmetric spring characteristic. He is a member of the German Physical Society and participates in the German research group for non-linear phenomena in physics.

How Big Is the Area required for LNG Base Load Plant And their Design Considerations

J.C. Kuo
Chevron Energy Technology Company
1500 Louisiana Street
Houston, TX 77002

AIChE Spring Meeting, April 2006
6th Topical Conference on Natural Gas Utilization
Orlando, Florida, April 23-27, 2006

1.0 Introduction

The purpose of this paper is to outline the general considerations and requirements for developing the layout of the LNG base load plant. The paper also presents some equations for determining a quick estimate of the LNG plant plot size required based on correlations developed between the LNG plant production and plot space requirements for some of the existing LNG plants around the world. The information used to develop the correlation is based on the author's personal collection while working on some LNG projects.

2.0 Basis of the LNG Plant Layout [6.1,6.2,6.3,6.4,6.5]

The bases of the LNG plant layout are:

2.1 To ensure the plant is safe, easy to construct, operate, maintain and meet the necessary insurance codes, local laws and regulations.

All local statutory laws and regulations must be followed. Additionally, codes, guidelines or standards set by individual oil companies or chemical companies should be met. The following organizations provide guidelines for fire protection and safety in chemical plants including LNG plants. These guidelines directly influence the size of the plot space due to the spacing requirement.

2.1.0	ANSI	(American National Standards Institute)
2.1.1	API	(American Petroleum Institute)
2.1.2	CFR	(Code of Federal Regulations)
2.1.3	EEMUA	(Engineering Equipment and Material User Assoc.)
2.1.4	EN	(European Norm Standard)
2.1.5	IP	(Institute of Petroleum)
2.1.6	IRI	(Industrial Risk Insurers)
2.1.7	NFPA	(National Fire Protection Association)
2.1.8	OIA	(Oil Insurance Association)
2.1.9	OSHA	(Occupational Safety and Health Administration)

The documents listed below are commonly referred to in planning for LNG Plant Layouts:

- API RP 521-Guide for Pressure Relieving and Depressuring Systems.
- API RP 2510-Design and Construction of LPG Installations.
- API 752-Management of Hazards Associated with Location of Process Plant Buildings.
- API RP 505-Recommended Practice for Classification of Locations for Electrical Installations at Petroleum Facilities ..."
- 49CFR Part 193-Liquefied Natural Gas Facilities.
- 33 CFR Part 127-Waterfront Facilities Handling Liquefied Natural Gas and Liquefied Hazardous Gas.
- EN1473/1160-Installation and Equipment for Liquefied Natural Gas.
- NFPA 30-Flammable and Combustible Liquids Code.
- NFPA 59A-Standard for the Production, Storage and Handling of Liquefied Natural Gas (LNG).

2.2 To provide adequate separation between major facilities, LNG tanks and parallel process trains so that a loss at one train facility would not cause physical damage and loss of production to the other train.

Location and separation of major areas such as process trains, utilities, offsite, LNG storage and loading shall consider the following:

- Major areas shall be located considering prevailing winds, grading of site, toxic, and flammable vapour gas dispersion such as CO_2, H_2S, LNG, LPG and hydrocarbon.

- As far as possible, process units, LNG storage tanks, and other flammable liquid storage tanks (e.g. condensate, refrigerants) should be located so that prevailing winds will not carry flammable vapors to ignition sources such as furnaces, heaters, welding shops, and flares. Such facilities should therefore be located perpendicular to the prevailing winds (cross winds) at the LNG plant site. Please refer to the Illustration on Figure 1.

- Disposal of toxic vapors from the flare should also be considered relative to the prevailing winds.

- Offices, shops, and other similar potential ignition sources relative to the process units and storage areas should be considered.

- The site grading such that a storage or process area spill will not migrate towards ignition sources, process areas, or control areas.

- Parallel trains are better to be identical instead of mirror images.

- The prevailing wind direction and site drainage should be considered when determining the location of control buildings. Control buildings that are exposed to potential explosions or fires must be designed to withstand potential overpressures explosion and fire effects.

- Electrical power lines, main switchgear, and transformers should be located such that they are not exposed to damage due to the malfunctioning of processing equipment, storage, or loading/unloading facilities.

- Pressurized horizontal product tanks containing natural gas liquids (NGL) or similar volatile liquids should be oriented such that the long axis does not point toward nearby equipment or structures.

Finally, consideration must be given to the subsequent Process Trains without restricting the operation of the first Process Train. This is because most of the new LNG plants have multiple Process Trains.

2.3 To Ensure Adequate Spill Containment

2.3.1 LNG Storage Tanks

LNG storage tanks should be designed with proper containment and remote from process units with common remote impound for a 10 minute design spill.

Minimum separation distance between LNG tanks and other facilities will be set by the thermal radiation limits given in NFPA 59A for a tank top fire on a contained tank. Spill containment will conform to NFPA 59A and all other codes, standards, and recommended practices. Additionally, separation between the tank and the Outer Fence Line may be controlled by vapour cloud dispersion.

2.3.2 Atmospheric Storage Tanks

Atmospheric storage tanks (e.g. diesel) should be located such that spills do not impact on the security of process units, other critical facilities, administration or other occupied buildings and will be dike. Tanks shall be spaced in accordance with related code and standards.

2.3.3 Pressure Storage

Pressure storage of hydrocarbons (e.g. refrigerants) in spheres and bullets shall be remote areas, outside the dike enclosure of atmospheric flammable liquid storage.

2.4 To permit safe evacuation of people inside the plant and those living close to the fence-line during a hazardous event.

This is to segregate from toxic, highly reactive, or otherwise hazardous materials and high risk facilities to reduce the potential for event escalation.

- Careful consideration should be given to the location of facilities in relation to property lines and neighbors in order to minimize potential exposures. Exposures from fire, explosions, spills, gas releases, vehicle traffic, etc. should be evaluated in both directions (e.g. exposure to the facility and exposure to neighbors from the facility).
- In addition to the citing criteria noted above, potential escalation of minor events should be limited through the selective inclusion of the flowing into the design:

  - Fire and gas detection initiating alarms
  - Passive fire protection
  - Fire water and other fire suppression systems
  - Installation of Emergency Shutdown and Emergency Depressuring Valves
  - Use of barriers or structures designed for explosion overpressure

2.5 To meet thermal radiation levels and flammability contours from gas or LNG pool fires.

2.6 To limit blast loads in the event of ignition of flammable clouds to acceptable levels.

Additionally,

- To minimize capital and operating costs while satisfying the above safety and maintenance requirements.
- To minimize the engineering and construction bulk material quantities given that all safety, operability and environmental objectives are met.

3.0 General design considerations and requirements for special areas and equipments [6.1,6.2,6.3,6.4,6.5]

Required Studies

First, the plant layout usually needs to accommodate the results of the following studies.

(1) Process Hazard Review,
(2) Hot Air Recirculation Study (Very critical for air cooling),
(3) Vapour and Toxic Gas Dispersion Studies,
(4) Thermal Radiation Exclusion Zones,
(5) Noise Contour Studies, and

(6) Waste Treatment Studies to avoid pollution into the sea and the latest geotechnical and topographical site data.

Process Train Units

The process and utility units should be preferably divided into discrete areas with their size and grouping of equipment based on:

- commonality of process, operation and maintenance
- acceptability failure consequences in a single fire or explosion
- consideration of size of venting and flaring system
- operational and emergency access

Main Control Room

By providing adequate separation distance between the Main Control Room and the process or storage units, the Main Control Room should be protected from explosion or fire. Alternatively, if adequate separation distance is not practical, the building shall be built to be blast resistant.

Other Plant Buildings

Substations, according to function, may be located within or near process and utility areas. These buildings, not frequented by plant personnel, will not be designed for blast resistance. The Marine Terminal Building should be designed in such a way as to minimize blast potential. Consideration shall be given to personnel occupancy levels and the criticality of the systems installed within them.

Utilities and Power Generation

Utilities and power generation facilities should preferably be segregated from process train units and storage of hydrocarbons.

- Consideration of firewater system sizing. Fire risk area size must be consistent with sizing of flare and firewater system headers
- As far as possible, areas should be rectangular to permit fire fighting from at least two opposite sides. Consideration should also be given to limit the longer dimension by including additional fire fighting access ways.

Access Roads

Within the perimeter of process units, access roads of at least 15 meters wide shall be provided for fire fighting and maintenance. Only a 6 meters wide portion of each access road shall be paved. Access roads separating units shall be designed to accommodate trucks, maintenance equipment and mobile cranes.

Fired Equipments

Where practical and economical, fired equipment (e.g. incinerator, hot oil furnace, gas heater for dehydration regeneration heater) should be grouped together in an isolated area. Fired equipment shall be separated by 50 meters from other process equipment handling flammable products and should be located at the edge of process unit with an access way for maintenance and fire-fighting.

Air Intakes

Air intake for equipment such as furnaces, heaters, gas turbines, ventilation, pressurization, or air conditioning, cooling of large motors, plant air, and air separation plants should be from a safe area. In determining a safe area, consideration should be given to normal and emergency exhausts and vents and possible recirculation patterns around buildings and air coolers. Consideration should also be given to including gas detection in the air inlets to such equipment.

Air for all control rooms should be drawn from a safe location. The air intake stack shall be equipped with a combustible gas detector.

Air Coolers

Layout and positioning, orientation, and height of air coolers shall minimize air recirculation given the prevailing site ambient conditions.

Security Stacking and Equipment

Where heat exchangers are stacked, the top exchanger centerline would be better not to be more than 3.5 meters above the ground or operating platform. Stacking of equipment in more than three levels should be avoided; the possibility of an intense fire surrounding and above any stacked equipment must be considered. Equipment handling flammable material should not be located over pumps or any other equipment containing flammable material above the lesser of its auto-ignition temperature or 316 C (600 F).

Refrigerant Compressor Shed

The compressor shed (covering refrigerant compressors and their drivers) shall be open on all sides.

Flare Systems

Location, layout, and height of flares should consider factors such as acceptable thermal radiation, noise, and dispersion of flammable and toxic vapors. An agreed-to heat radiation and dispersion model shall be used. Separation distance between the flare riser and the LNG Plant Fence should follow accordingly with the company or API standards.

A clear zone around flares which is less frequented by plant personnel should be based on the recommended radiation level of API RP 521. There should be enough distance between flares to allow maintenance of one set of flares while the other flares are in operation.

4.0 Plot plan size Requirements [6.1,6.2,6.3,6.4,6.5]

4.1 Parameters Affecting the Plot Plan Size

Different liquefaction processes such as APCI, COP or Shell DMR processes will have some differences in plot sizes. However, the impact is not significant for this estimation purpose because the liquefaction unit is only part of the total process train. Nevertheless, some of these important parameters affecting the plot plan size are listed below:

4.1.1 Gas Composition. If the feed gas has high impurities such as CO_2, H_2S, COS, CS_2, O_2, Aromatic (BETX) or Mercury then it is necessary to add the LNG plot space to include the larger and/or additional gas treating and LPG product treating. Then more plot space is required. Also if the feed gas has high concentration of heavy hydrocarbon, then larger and/or additional condensate or LPG product storage is required. More plot space is also required.

4.1.2 Client or country requirement. Each client and country specifies their own regulations, guidelines, standards or codes to follow. For instance, some LNG plants in one country need spare flare knock out drum, spare flare as well as spare condensate stabilizer column in order to maintain the plant on stream factor and to satisfy the statutory inspection requirement. The plot plan will be larger.

4.1.3 There are other parameters which may affect the size of the LNG plot plan. These are climate, flare type (Ground or Elevated type), site conditions including topography and soil conditions, landfill, foundation design, earthquake criteria, elevation, contour, subterranean water, surface drainage, proximities of existing facilities available infrastructure and limited construction site accommodations.

4.2 Plot Plan Size

The total LNG plot plan was divided into four areas.

(1) The LNG Process Train area,
(2) The Utilities System area,
(3) The Storage Tank area, and
(4) The Supporting area. This may include construction lay down, construction worker camp, permanent dorm, parking lot, material off loading facilities, and temporary buildings.

For a very quick and rough estimate for all the four above mentioned areas and/or the individual area, please refer to the followings.

Total four areas (Total LNG plant) - (6.30-12.50 hectares per MTPA)

(1) The LNG Process Train area,	(1.15 to 2.30 hectares per MTPA)
(2) The Utilities System area,	(2.00 to 3.70 hectares per MTPA)
(3) The Storage Tank area and	(1.15 to 2.30 hectares per MTPA)
(4) The Supporting area	(2.00 to 4.20 hectares per MTPA)

For up to 20.0 MTPA grass root LNG plant, 9.0 hectares per MTPA is a "rule of thumb" number for the initial total area estimation. Larger plant will need smaller area.

MTPA is million metric tons per annual.
One Hectares =100 meters * 100 meters.

For more accurate estimation, it is recommended to use the following correlations.

Based on plot plans of many of those existing LNG plants, some correlations between the capacity of the LNG plant feed gas rate and the required LNG plant plot space were developed for Process train area. Utilities System area is based on the ratio of the Process train area.

4.2.1 LNG Process Train area

The LNG process area includes the unit of

- Feed gas reception,
- Gas treating,
- Gas dehydration,
- Mercury removal,
- LNG liquefaction,
- Condensate stabilization and
- Fractionation.
- Refrigerant storage
- Substation and/or the local control room
- Access road around the Process Train

The LNG process plot space required for the gas turbine and air cooling is larger than that of steam turbine and water cooling. The use of gas turbines, particularly with the waste heat recovery unit, has a major impact on the space required. Because the temperature of the exhausted flue gas is very high, the ductworks are considered to be a source of ignition. There are some limitations on the minimum distance between the ductworks and other equipment. The orientation of the turbine ductworks and waste heat recovery unit in the plot space is another factor; there are some limitations in the minimum distance between air coolers and furnace, fire heater or pumps. However, the impact is slightly less severe as with gas turbine.

For an estimation of the above Process train area required, the following equations could be used for train size up to 5 MTPA. The LNG process plant plot space required is in hectares and feed gas rate is in million standard cubic feet per day MMSCFD.

4.2.1.1 Gas turbine and air cooling plant;
Hectares = 0.130*(MMSCFD**0.6)-1.22

4.2.1.2 Steam turbine and water cooling plant;
Hectares = 0.115*(MMSCFD**0.6)-0.93

4.2.1.3 Gas turbine and water cooling plant;
As 4.2.1.1 and minus 10.0%.

4.2.1.4 Steam turbine and Air cooling plant;
As in 4.2.1.2 and add 15.0%.

Additional area is required for any LNG plant which has more than those aforementioned units such as the followings

- Slug catcher (finger type),
- Feed gas compression,
- MEG (Mono Ethylene Glycol) regeneration and storage,
- CO_2 compression and injection.
- Domestic gas processing
- Sulfur plant

4.2.2 Utilities System Area

The LNG utilities area includes all the process services required such as

- Instrument air and Plant air,
- Nitrogen,
- Fuel gas,
- Demineralized water,
- Steam/Condensate system,
- Power generation, Emergency power & UPS
- Cooling water,

- Heat medium,
- Fire water,
- Flare and blow down,
- Drain systems,
- Diesel,
- Waste water/Effluent treatment,
- All sterilized area due to radiation,
- Access road.

The LNG utilities area required for the steam turbine and water cooling is larger than that of gas turbine and air cooling. There are more equipment involved for steam generation and condensate return.

For an estimation of the Utilities system area required, it is recommended to use the following ratio.

4.2.2.1 Gas turbine and air cooling
Use 140% of Process train area for utilities area

4.2.2.2 Steam Turbine and Water Cooling
Use 160% of Process train area for utilities area

4.2.2.3 Gas Turbine and Water Cooling
Use 140% of Process train area for utilities area

4.2.2.4 Steam Turbine and Air Cooling
Use 160% of Process train area for utilities area

4.2.3 Storage Tank Area

4.2.2.5 LNG storage tank
Hectares = 0.0096*((Meter**3)**0.6) -7.69

4.2.2.6 Condensate storage tank
Hectares = 0.0082*((Meter**3)**0.6) -2.39

Meter**3 is the volume of the storage tank.
The LNG storage tank is based on the double containment tank or full containment tank. More plot space is required for the single containment tank because larger dike area required. However, not many plants install the single containment tank nowadays.

Additional area is required for any LNG plant which has more than those aforementioned storage tanks such as the export storage for Propane, Butane, LPG or others.

4.2.4 The Supporting Area

It is recommended to use the similar plot space as the utility area. However, it should be noted that the area of airport is not included herein.

5.0 Acknowledgements

The author would like to thank his colleagues from various oil companies and engineering contractors for some of the data and information used in this paper. The author would also like to thank Chevron for providing permission to publish this paper.

6.0 References

6.1 IRI (Industrial Risk Insurers) General Recommendation For Spacing

6.2 Equipment spacing design guidelines from major oil companies and Engineering Contractors.

6.3 Minimum safety separation distances between units guidelines from major oil companies and Engineering Contractors

6.4 Calculating safety distance in LPG storage and handling installations guidelines from major oil companies and Engineering Contractors

6.5 In house studies and data base from major oil companies and Engineering Contractors

FIGURE 1

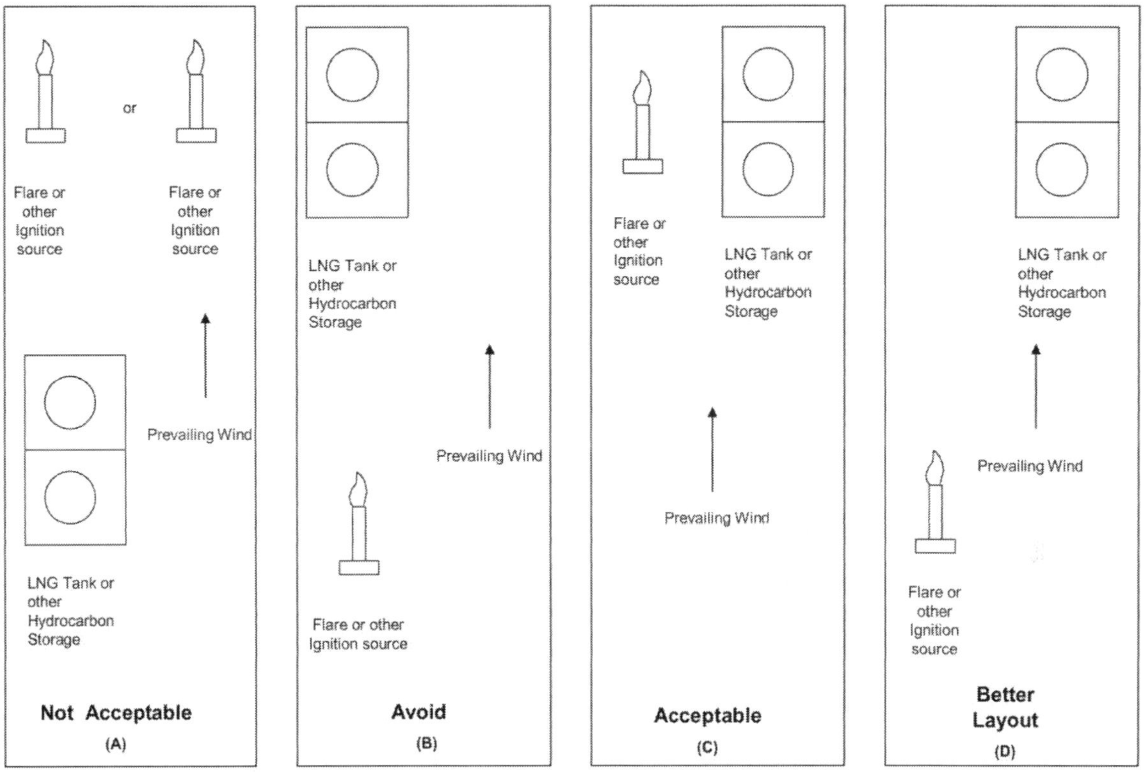

Biography

J. C. Kuo is Senior Technical Advisor for Chevron, He has more than 32 years experience in gas processing, gas treating and LNG. He has BS from Chung Yuan Christian University, Taiwan and MS from the University of Houston all in Chemical Engineering. He also holds an MS in Environmental engineering from Southern Illinois University, Carbondale. He is a register professional engineer (Texas) and member of AIChE.

LNG Rollover: Converting a Safety Problem to Tank Loading Operational Asset

Yanjun Wang, Ben Cormier and Harry H. West
Mary Kay O'Connor Process Safety Center
Chemical Engineering Department
Texas A&M University
College Station, Texas 77843-3122

AIChE Spring Meeting, April 2006
6th Topical Conference on Natural Gas Utilization
Orlando, Florida, April 23-27, 2006

Abstract

Rollover following density stratification in LNG storage tanks was considered a potential safety problem that required careful monitoring. This phenomenon was first observed during several unexpected rollover events in the 1970s. However, today density stratification is routinely used to reduce high LNG boil-off gas rates, particularly when tank filling through the top loading nozzles is used for heavier LNGs. Thus boil-off gas compressor and pre-heater costs can be reduced both during and after unloading LNG Tankers.

LNG rollover predictive models are widely used tools in conjunction with internal tank travelling temperature and density instrumentation to predict and update the behaviour of LNG stratification. Several proprietary and commercial rollover software models are identified and the potential use of Computational Fluid Dynamics (CFD) models for the non-circular tanks proposed for various off-shore applications is described herein.

Introduction

LNG consists of mainly methane (85%~98%), higher hydrocarbons and sometimes a small amount of nitrogen and carbon dioxide. Liquefied Natural Gas (LNG) is usually stored in storage tanks at temperature of about 115K and pressures slightly above atmospheric pressure...

LNG stratification and subsequent rollover can cause a rapid increase in boil-off rate and possibly a tank overpressure if the resultant boil-off rate is beyond the capacity of pressure relief system. It has been studied extensively in 1970s and 1980s as a potential safety problem. However, stratification is now routinely used to suppress evaporation and achieve reduced boil-off rates (Williams, 1994; Uznanski and Versluijs, 2004). This operational mode not only preserves the valuable LNG but can also reduce the capital and utility costs of boil-off gas compressor and reliquefaction.

Real-time monitoring of the stratification and a reliable model to predict the occurrence and severity of rollover phenomenon are essential to ensure safety while achieving reduced boil-off rates and costs. LNG rollover predictive models are widely used tools in conjunction with internal tank travelling temperature and density instrumentation to predict and update the behaviour of LNG stratification. This paper provides a brief review of the stratification and rollover phenomena, the rollover incidents reported in the open literature, and the benefits associated with using stratification intentionally during operation and storage. Several proprietary and commercial rollover software models are identified and the potential use of Computational Fluid Dynamics (CFD) models for the asymmetric and non-circular tanks proposed for several off-shore applications is described in this paper.

Stratification

LNG is stored in well insulated double-wall storage tanks with proper pressure control, pressure relief, and reliquefaction systems. Under normal conditions, LNG in a static storage is in a thermodynamic equilibrium state. Heat leaks from the tank wall, roof, and bottom is released through the quiet boiloff of the LNG at the free surface. A natural, convective circulation is thermally induced as shown in Figure 1. LNG near the bottom and along the walls is warmed by the heat leak and becomes relatively buoyant and rises along the tank wall, then it cools down providing the latent heat for the evaporation at the free surface, sinks back toward the bottom and mixes with the bulk contents, thus most of the bulk is at an essentially uniform composition and temperature except for thermally induced density and temperature differences maintaining the convection.

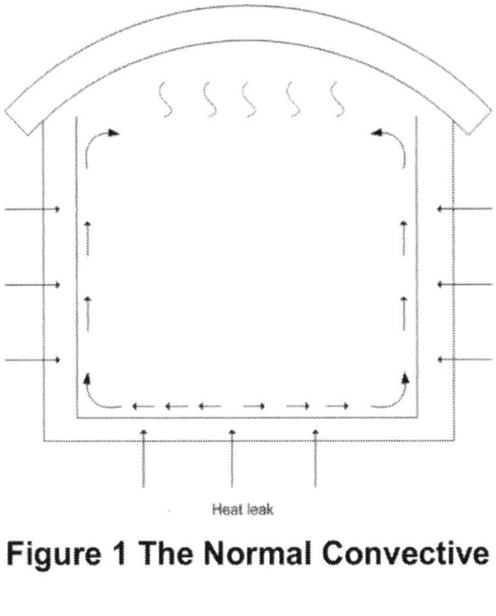

Figure 1 The Normal Convective Circulation in a LNG Storage Tank

Under certain circumstances, LNG stratification with clear interface might result from inhomogeneous density distribution in the storage tanks. In a stratified tank, the upper layer is lighter than the lower layer(s), which is temporarily stable if the convection current in the lower layer is not buoyant enough to penetrate interface between two layers. In this case, the lower layers will develop a separate convective circulation which is confined in the lower layer only, as shown in Figure 2.

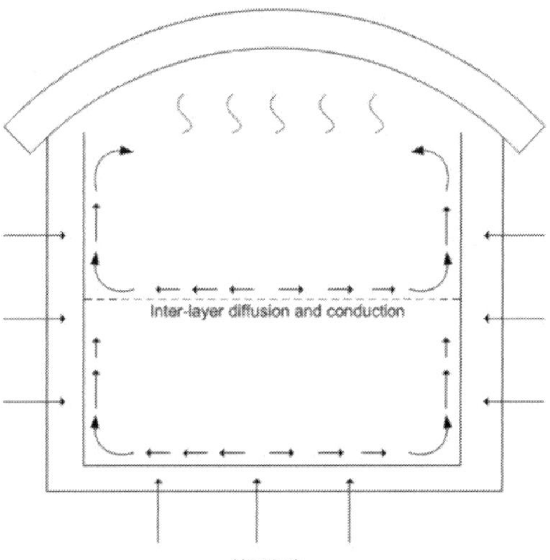

Figure 2 The Convective Circulation in a Stratified LNG Storage Tank

As shown in Figure 2, convective current in the bottom layer is not able to penetrate the interface between the layers; therefore heat exchange between the two layers is restricted to conduction through the interface. Heat leak into the bottom layer will be contained in the bottom layer and not dissipated at the free surface. The accumulation of this heat will eventually lead to a rapid mixing of these two layers upon density equalization and cause a massive boiloff. The formation of stratification and its evolution to rollover are discussed in the following context.

<u>Formation of Stratification</u>

LNG consists of mainly methane, higher hydrocarbons and sometimes small amount of nitrogen and carbon dioxide. The stratification could be operationally induced or spontaneous depending on the composition and density of LNG in the tank and incoming cargos.

The addition of LNG with different densities to partially filled LNG tanks has been known to lead to the formation of stratified layers. The density difference can be due to different sources of the receiving LNG or the weathering of LNG in the tank.

When LNG is bottom filled into a tank contains a lighter and colder heel than the cargo, the initial tank heel will be displaced upwards with only minimal mixing with the off-loaded cargo due to the density difference.

Many peak shaving tanks weather for months, or even years which makes it denser and warmer due to preferential evaporation of lighter content. When fresh LNG is top fed into a storage tank with a denser heel, it will stay on top of the denser heel with minimal heat and mass transfer with the existing heel. If additional mixing is not provided or not sufficient, stratification will form in the tank.

It was pointed out that stratification can also occur spontaneously if the LNG contains a substantial amount of nitrogen (Chatterjee and Geist, 1976; Chatterjee and Studer, 1983). Nitrogen is heavier than methane but more volatile. Hence, the convective current reaches the free surface and flashes, becomes lighter and stays at the top of the tank. As this proceeds, the lighter layer on top grows in depth. This lighter layer will grow with time and finally reach a thickness where the denser fluid no longer has sufficient buoyancy to penetrate and form stratification.

Evolution of Stratification

When layers of different densities exist in an LNG tank, the less dense upper layer can convect normally and dissipate heat by boiloff at the free surface, as shown in Figure 1. With the preferential evaporation of lighter content (mainly methane), the density of the top layer tends to increase, together with the saturation temperature under constant pressure.

In the case of the lower layer, the presence of a top layer suppresses the evaporation from it, and the convective current is confined and much weaker. Heat trapped in the bottom layer either from initial superheat from the cargo or subsequent tank heat leak can only be released through heat and mass transfer at the interface between the two layers. If the heat leak into the bottom layer from environment is greater than the heat loss across the interface, thermal energy will be accumulated in the bottom layer, and the temperature increases. The density of the bottom layer will decrease as a result of thermal expansion.

When the densities of two layers approach each other, the interface becomes increasingly unstable and the core flow in the upper layer is able to break into the lower layer. Eventually, the two layers mixes with a sudden increase in the vapor evolution and sometimes tank pressure.

Rollover

As the equalization of the densities of the two layers is reached, they mix rapidly to form one uniform layer. This phenomenon is termed as rollover. Rollover is usually accompanied by a sudden increase in vapor evolution and tank pressure to release the heat trapped in the lower layer.

Flow visualization experiments show that mixing of two stratified layers generally involves two stages in sequence: migration of the interface and rapid mixing between the remaining liquids (Shi et al., 1993; Bates and Morrison, 1997).

Physical processes involved during rollover include the boundary layer penetration at the wall, entrainment mixing at the interface, and boiling off of the bottom layer. As the interface becomes less unstable, the convection flow along the wall of the bottom layer penetrates into the upper layer and the core flow from the upper layer breaks entrains into the bottom layer.

The other phenomenon that might contribute in the rollover occurrence is the onset of the boiling of the bottom layer. The ullage pressure and the hydrostatic head of the upper layer suppress vapor evolution from the bottom layer. As the heat from the base and the wall leaks into it, its temperature increases. The bottom layer will start to boil when the temperature reaches the saturation point and lead to dispersion of the stratified layers.

Rollover Incidents

Rollover occurs with excessive vapor evolution and tank overpressure. While the threat to the mechanical integrity of the tank is remote, the loss of valuable gas through pressure relief system and venting is undesirable. For extreme cases, vented LNG may lead to a flammable concentration at ground level under certain atmospheric conditions, and cause subsequent fire or explosion. There are a number of rollover incidents reported, although the quality of the data available from these incidents varies and many are not complete.

The first rollover incident reported occurred at Le Havre, France (Williams, 1995). Eight days after charging of a denser cargo into a partially filled storage tank, excessive boiloff occurred and it took three days for it to return to the normal boiling off rate.

The most dramatic and well documented rollover was experienced at a LNG terminal tank in La Spezia, Italy (Sarsten, 1972). A warmer and denser cargo, which had been weathered in a ship for an extended time period was bottom filled into a tank containing a colder and lighter heel. Eighteen hours after the loading, this particular tank experienced a sudden increase in pressure causing a discharge from the tank safety valves and tank vent. The total amount of LNG vented during the occurrence of rollover is estimated to be 410,000 lbs. The resultant peak pressure is about 710mm of water, which is 210mm of water above the tank design pressure but relieved promptly by the pressure relief valves. The roof of the tank was slightly damaged. Although no ignition took place, the cold vapors traveled downwind beyond the fence of the facility. The specific process parameters of the La Spezia rollover were well recorded by the available instrumentation. The La Spezia incident has become the breakthrough event for validation of simulation works (Chatterjee and Geist, 1972; Germeles, 1976, Heestand et al., 1983; Williams, 1995).

Chatterjee and Geist (1972) used two other rollover incidents together with La Spezia to validate their model. One of them is similar to the La Spezia case except that the initial temperature and compositional difference was less. The other case is a scenario where light LNG was slowly added from a liquefaction plant on top of a heavy heel existing in the tank.

Chatterjee and Geist (1976) reported a known yet unpublished case of rollover due to high N_2 content (3~4%) at a peaking shaving plant of Citizens Gas and Coke Utility in Indiapolis, IN. A sequence of rollovers occurred in a storage tank at a liquefaction plant progressively at 20 to 25 days intervals. The peak boil-off rates were only 3 to 4 times the normal boil-off rate, which is milder than most of the filling induced rollovers.

A 1986 technical paper by Acton and Van Meerbeke reviewed 41 rollover incidents from 22 plants between 1870-1982 (Acton and Van Meerbeke, 1982). Most of the 41 LNG rollover incidents were mild and only a few cases achieved peak vapor evolution rates 20 times more than the normal evaporation rate.

A rollover occurred at the British gas storage facility at Partington, England (Baker and Creed, 1995). About 1900 tons of lighter fresh LNG was top filled into a storage tank with 18,650 tons of richer LNG. Approximate 150 tons of gases were vented during a two hour period.

Simulation Models for Rollover

Reliable mathematical models simulating the dynamics of stratification and subsequent rollover can be used to predict its occurrence and severity. Such models are powerful tools both in the prevention of rollover and the intentional stratifications to reduce LNG loss.

As discussed in the evolution of stratification, composition and temperature changes in both layers tend to minimize the density difference of these two layers, which reduces the stability of the interface. Physical phenomena involved in the development of stratification to rollover include: composition change due to boiloff from the upper layer, heat and mass transfer between layers, and heat leak into the bottom layer. Density variations due to the above physical processes in a stratified LNG storage tank can be described if heat and mass balances are made for each stratum. Literally all the simulation models developed to date were based on mass and heat balances for each layer. The differences among the models lie in the estimation of physical properties, estimation of the transfer coefficients, and the stability/mixing criteria of the layer interface.

A reliable model of rollover requires accurate estimation of physical properties over the concerned range of composition including density, viscosity, heat capacity, thermal conductivity, saturation temperature, latent heat of vaporization, and the equilibrium constant of each component, etc. Historically, heating value has been correlated well with the density of LNG especially when nitrogen concentration is low. Density changes due to compositional changes and thermal expansion are presented by Hashemi and Wesson (1971).

The stability for stratified LNG layers has not been studied explicitly. The stability conditions of a series of layers of saline solution were presented by Huppert (1971). The mixing criterion is usually the equalization of density while some models also used the equilibration of temperature and composition as well.

Hashmi and Wesson (1971) developed a model to compute the boiloff rate from stored LNG. The model was based on the concept that evaporation occurs with a small temperature difference between the bulk liquid and the surface.

$$\dot{m} = 0.328\frac{\rho_0 C_p}{H_v}\left(\frac{g\alpha k^2}{\rho\mu C_p^2}\right)^{1/3}(T_n - T_s)^{4/3}$$

Thereafter, simulation models for rollover have been developed by Chatterjee and Geist (1972 and 1976), Germeles (1976), Heestand et al. (1983), Shi et al. (1993), and Bates and Morrison (1997) to estimate the change of temperature and composition with time for each layer in a stratified tank.

Chatterjee and Geist (1972) developed a model for LNG assuming an equi-molal counter-diffusion across the layers of different densities in LNG storage tanks. The stratification is modeled as multiple homogeneous layers of given heights and uniform physical properties separated by sharp interface. Their criteria for mixing requires both temperature and composition equilibration which implies density equilibration. However, a density equilibration when rollover occurs does not necessary require temperature and composition equilibration.

Realizing the spontaneous stratification due to the presence of nitrogen, Chatterjee and his coworkers (Chatterjee and Geist, 1976) later modified their original model to include the effects of nitrogen. The original binary mixture is now replaced with a tertiary mixture of methane, ethane, and nitrogen. Turner's correlation (Turner, 1965) is used to model the mass transfer at the interface instead of equi-molal counter-diffusion.

Assuming LNG as a binary system of methane and a fictitious solute equivalent to all impurities, Germeles (1976) developed a dynamic model for LNG stratification evolution based on a double-diffusive convective layer interface. Turnur's experimental work with stratified saline solutions (Turner, 1965) were borrowed and adjusted to estimate the heat and mass transfer coefficients between the layers.

Both Germeles model and CG model assume that heat and mass transfer between adjacent layers are by molecular diffusion only and constant layer heights are used. Heestand et al. revised and expanded Germeles model to incorporate more constituent components and time-varying layer heights (Heestand et al., 1983). Heat transfer coefficient is estimated for a fully developed turbulent flow pattern.

A table summarizing the principal features of the CG model, Germeles model, and Heestand model is presented by Heestand et al. (1983).

All the above mentioned models are one dimensional. The stratification is modeled as n homogeneous layers of given heights and uniform physical properties separated by sharp interface. Shi et al. (1993) developed a 2-dimensional numerical model to simulate the mixing of two stratified layers subjected to a uniform lateral heat flux. Based on their visualization experiments with liquid Freon, migration of interface and entrainment mixing across the interface are considered. However, in their model, loss of mass due to surface evaporation was ignored.

Bates and Morrison (1997) considered two stages from the evolution of stratification to rollover based on experimental observations. The model developed by Heestand et al. (1983) was used to simulate the first phase which is quiescent and stable. Experimental correlations are required to calculate the penetrative convection in the second phase, which is yet available at this time. The transition point between the two phases was decided using Turner's correlation.

Tokyo Gas developed its own simulation model based on Germeles model. LNG Stratification and subsequent rollover is divided into three phases depending on the vertical velocity of the interface. Kyushu Electric Power developed a simplified model of stratification based on their experiments. The details of these two models are not published in the open domain.

Korea gas is sponsoring a research project titled as *analysis of rollover phenomenon and testing research of boil off on the surface of LNG tank* (http://www.parcon.uci.edu/ OLD_WEBSITE/TABLES/research.html#78). No further information is available regarding this research.

Experimental Work for Rollover

A few experiments have been carried out to improve the understanding of stratification and rollover processes and to validate the models developed.

Rollover tests carried out in an inground LNG tank using LNG from two different sources confirmed the gradual migration of the interface and more rapid movement just before the rollover occurs (Sugawara, Y. et al, 1983).

Small scale LNG rollover experiments have been carried out by Kawasaki Heavy Industries, Ltd. and Osaka Gas Co., Japan (Muro et al., 1986; Sugawara et al., 1984). The container used to conduct the experiment is a vacuum insulated, double wall, stainless steel tank with an outer diameter of 580mm and a height of 1m. Their experimental results validated the use of Hashemi-Wesson's expression for LNG boiling off rate. Heat transfer and mass transfer across the interface were also measure and correlated with the stability factor.

Gaz de France, Shell research Ltd, Osaka Gas, Tokyo Gas, and CFP-Total collaborated on an experimental study of stratified LNG. A series of experiments were performed in a 500m3 tank at Nantes Cryogenic Testing Station from 1987 to 1990. The evolution of stratification is observed to have four distinct phases. Heat and mass transfer between the layers in the first phase is minimal and can be ignored. During the second phase, convective penetration occurs at the layer interface and reduces the density difference further. Rollover occurs at the third phase where rapid mixing of the two layers takes place upon equalization of densities. During the fourth phase, the LNG slowly reestablishes an equilibrium state. Their experimental data are available commercially.

Laboratory work to study the mixing of two stratified layers under different heating conditions have been carried out by Nakano et al. (1982) and by Morrison and Richardson (1990) using liquid Freon and butane/pentane mixtures, respectively. Their results confirmed the migration of the interface with only lateral heating leak. However, when heat leak is only from the bottom, the layer height remains unchanged prior to the rapid mixing between the layers. When both side and base heating are applied, the outcomes depend on the relative ratio of the bottom heating to the lateral heating and are not so clear cut.

Prevention of Rollover

Rollover occurs with excessive vapor evolution and tank overpressure which may result in tank damage or at least loss of valuable fuel. Certainly we would like to prevent rollover to avoid any undesired damage or loss.

Most of papers on LNG rollover are concerned with the preventions of rollover as a safety hazard. Historically, rollover prevention has been focused on the prevention of stratification and ensuring of well mixing. A number of preventive measures have been raised including:

- Restricting variations in the incoming LNG composition;
- Providing capability of both top filling and bottom filling and developing procedures to select filling configuration based on the density of the cargo and the heel in the tank;
- installation of jet mixing nozzles or distributors;
- Lower the tank operating pressure;
- Circulation pump which sucks from the bottom and discharges at the top;
- Providing adequate pressure relief capability to handle the maximum vapor evolution;
- Oversizing storage tanks and vapor handling system capable of handling the potential rollover.

All of these solutions require excessive investment for additional equipment and overdesigning of vapor handling system and/or additional operating and utility costs.

Actually, rollover can be effectively prevented by removing the stratified LNG before rollover takes place. This is the principal prevention method when stratification is maintained intentionally during normal operation and used to suppress evaporation and achieve reduced boil-off rates.

To use such method reliably, we must be able to predict the incubation time for rollover accurately with the help of advanced models and real time stratification monitoring. LNG sendout must be scheduled carefully before the occurrence of rollover. If a sendout can be performed promptly, other preventive measures must be in place to avoid rollover.

Benefit of Operation with Stratification

Artificially imposed stratification during normal operation not only reserves the valuable LNG but also reduces the capital and capital and utility costs of boil-off gas compressor and reliquefaction.

The boil-off rate of LNG in a static tank can be estimated using Hashmi and Wesson's expression (Hashmi and Wesson, 1971). The evaporation rate depends on the tank absolute pressure and the states of compressor as well. Higher absolute pressure reduces LNG boil-off rate. The running compressor results in decreases in the tank pressure which facilitate the boiloff of LNG. When the compressor is turned off, the tank pressure rises quickly to the saturation pressure, and the boil-off rate is reduced.

Typically, a 60,000m^3 well insulated tank with 25,000 tons of LNG inside will lose an average of 12.5 tons of gas daily with a relatively stable average temperature and pressure (Bates and Morrison, 1997).

During unloading operations, most of the vapors in the storage tank that are displaced due to the incoming cargo are reliquefied and returned to the tank. Excess vapor are vented through a tank vent. While displacement vapor always occurs during a loading operation, the rate of flash vapor generation depends on the filling configuration (bottom or top) and the form of the stratification in the tank.

Top filling dissipates the superheat that originally exists in the cargo or is gained from the transit. Flash vapor generated during a top filling is more than a bottom filling. That is essentially why top filing induced rollover is much milder than the bottom filling induced rollover (Drake, 1976). Flash vapor may not be generated as much during bottom loading especially if a heaving LNG forms a stratum under a lighter layer. The upper layer produces hydrostatic head to suppress vapor evolution from the bottom layer.

Stratification is deliberately created when the tank heel is displaced upwards with only minimal mixing with the incoming LNG. In this case, the normal vaporization at storage tank pressure had been temporarily suppressed by the static pressure of an initial low vapor pressure heel, and boil-off rate can be reduced significantly (Uznanski and Versluijs, 2004).

Taking La Spezia incident as an example, the displaced vapor is about 100,000lb, and the flash vapor during filling is 25,000 lbs, which is significantly less than the estimated flash vapor evolution of 400,000 lbs during normal off-loading (Sarsten, 1972).

Conclusion

In operating an LNG storage tank with artificial stratification, it would be required to monitor created stratification. The following measurements are essential:

- Measuring density differences between feed and tank liquid analytically or indirectly by measuring heating values;
- Measuring temperature distribution across the height of the tank;
- Monitoring tank vaporization rates.

A reliable model to predict the occurrence and severity of rollover phenomenon is a must to ensure safety while achieving reduced boil-off rates and costs. LNG rollover predictive models are widely used tools in conjunction with internal tank travelling temperature and density instrumentation to predict and update the behaviour of LNG stratification. Better and more precise designs of the tank together with its vapor handling system and pressure control system can be achieved using a more accurate prediction of boil-off rates, thus reducing the initial investment and subsequent operating costs.

Several proprietary and commercial rollover software models are identified in this work and the potential use of Computational Fluid Dynamics (CFD) models for the non-circular tanks or asymmetric heat leak for rollover applications are under investigation.

References

Acton, A., R. van Meerbeke (1972), Rollover in LNG storage tank – an industry view, *LNG8 conference*, Los Angeles, 738-742

Baker, N., M. Creed (1995), Stratification and rollover in liquefied natural gas storage tanks, ICHEME Symposium Series, No. 139, 621-634

Bates, S., D. Morrison (1997), Modelling the behavior of stratified liquid natural gas in storage tanks: a study of the rollover phenomenon, *Int. J. Heat Mass Transfer.* 40 (8), 1875-1884

Chatterjee, N. J. Geist (1972), the effect of stratification on boil-off rates in LNG tanks, *Pipeline Gas Journal,* September, 40-45

Chatterjee, N. J. Geist (1976), Spontaneous stratification in LNG tanks containing nitrogen, *ASME* paper 76 –WA/PID-6

Chatterjee, N. D.W. Studer (1983), Improved computer program to predict roll-over in LNG with and without nitrogen, *AICHE symposium series*, 38-45

Drake, E.M. (1976), LNG rollover – update, *Hydrocarbon processing*, January, 119-122

Drake, E.M., N.J. Geist, K.A., Smith (1973), Prevent LNG 'rollover', *Hydrocarbon processing*, March, 87-90

Germeles, A. (1976), A model for LNG tank rollover, *Cryog Eng* 21, 326-336

Hashemi, H. and H. Wesson (1971), Cut LNG storage costs, *Hydrocarbon processing*, August, 117-120

Heestand, J. C.W. Shipman, J.W. Meader (1983), A predictive model for rollover in stratified LNG tanks, *AICHE Journal* 29 (2), 199-207

Huppert, H.E. (1971), On the stability of a series of double diffusive layers, *Deep-sea Research* 8, 1005-1021

Morrison, D., A. Richardson (1990), An experimental study on the stability of stratified layers and rollover in LNG, *Proceeding of the Low temperature Engineering and Cryogenic Conference*, Session 01, paper 1

Muro, M., M. Yoshiwa, Y. Yasuda, T. Miyata, Y. Iwata, Y. Yamazaki (1986), Experimental and analytical study of the rollover phenomenon using LNG, *Proceedings of the International Cryogenic Engineering Conference*, 633-637

Nakano, R. et al. (1982), An experimental study of the mixing of stratified layers using liquid Freon, *GASTECH VI*, paper 7

Sarsten, J.A. (1972), LNG Stratification and rollover, *Pipeline Gas Journal* 199 (11), 37-39

Shi, J., C. Beduz, R. Scurlock (1993), Numerical modeling and flow visualization of mixing of stratified layers and rollover in LNG, *Cryogenics* 33 (12), 1116-1124

Sugawara, Y., A. Kubota, S. Muraki (1984), Rollover test in LNG storage tank and simulation model, Advances in cryogenic 29, 805-811

Takao, S., U. Narusawa (1980), An experimental study of heat and mass transfer across a diffusive interface, *Int J Heat Mass Transfer* 23, 1283-1285
Turner, J.S. (1965), the coupled turbulent transports of salt and heat across a sharp density interface, lint. *J. Heat Mass Transfer* 8, 759-767
Uznanski, D., P. Versluijs (2004), Advanced LNG storage tank management, GASEX 2004 Conference & Exhibition, SUNTEC, Singapore, May 31 – June 3
Williams, T.A. (project manager) (1995), Rollover – an update of current technology, topical report for Gas Research Institute, Project Technical Liaison Associates, Inc, 20119 Stuebner Airline, Suite C, Spring, Texas 77379

Biography

Dr. Harry West has over 30 years of experience in the areas of LNG and its safety, chemical process safety analysis, environmental hazard assessment, and regulatory compliance for petroleum, natural gas, LNG & chemical industries. He has participated in several hundred forensic assignments and provided expert testimony in various cases, ranging from civil litigation and international arbitration to regulatory compliance hearings. Dr. West's background includes the following highlights:

- Chemical Engineering Bucknell University (BS); University of Oklahoma (PhD 1968).
- Registered professional engineer (PE), Texas & Pennsylvania.
- Elected member of the National Academy of Forensic Engineers.
- Author of over 80 technical papers on system safety analysis, hazards assessment, process safety management and process safety automation.
- Co-author of AIChE-sponsored text, "Guidelines for Safe Process Operations and Maintenance."

Unlocking the Potential of Advanced Process Control in LNG Plants

Kees den Bakker
Team Manager Advanced Process Control
Shell Global Solutions
Process Control and Optimisation department
Amsterdam, The Netherlands

AIChE Spring Meeting, April 2006
6th Topical Conference on Natural Gas Utilization
Orlando, Florida, April 23-27, 2006

Introduction

Advanced Process Control (APC) can be described as a model based, predictive technology that can automatically control online a total unit or LNG train. Within the oil and gas industry APC was firstly applied in refining but over the past decade, also has become widely recognized as a value-adding technology in LNG processing.

Shell Global Solutions has been implementing and supporting Advanced Process Control (APC) projects in several LNG plants worldwide, including ones in Oman, Brunei, Malaysia, Nigeria, Indonesia and Australia. This presentation focuses on work carried out with some of the LNG producers.

Discussions

The production process in LNG plants is sensitive to a wide range of disturbances, notably, the variations in the air ambient temperature and feed conditions / composition. APC technology reacts constantly, and very quickly, by making automatically countless small changes to the process parameters.

With such tight control, a process may be operated closer to its limits. In fact, APC continuously pushes the process against its constraints while keeping at the same time a proper distance of the safeguarding (trip) and relief settings which leads to financial rewards, see Figure 1.

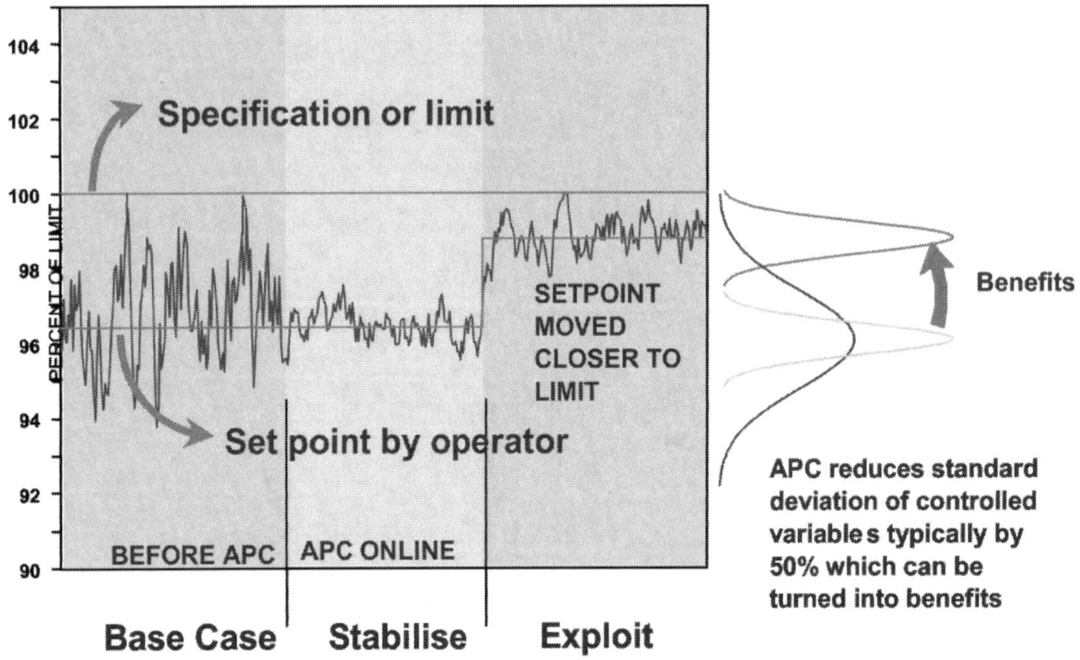

Figure 1 - APC stabilises process and moves operations closer to limits

APC in LNG processing has been applied to a variety of units. Results noted are:

- 1 to 3% higher throughput,
- Higher efficiency,
- More LPG or NGL,
- Less flaring.

APC stabilizes plant operations resulting in less upsets and higher plant availability. Also, operations between various shifts become more consistent. APC is flexible to different objectives e.g. under certain conditions LNG production is maximized where under other conditions efficiency is maximized.

The levels of benefits achieved depend on local conditions such as cooling-system type, compressor driver type and product prices. Gas-turbine-driven sites tend to enjoy higher benefits because of the strong impact of air ambient temperature on gas-turbine efficiency.

APC covers one whole LNG train including:

- liquefaction sections (main cryogenic heat exchangers [MCHE] and refrigerant loops)
- fractionators
- scrubber columns
- stabilisers

Figure 2 shows the areas in the LNG liquefaction process where APC has been applied.

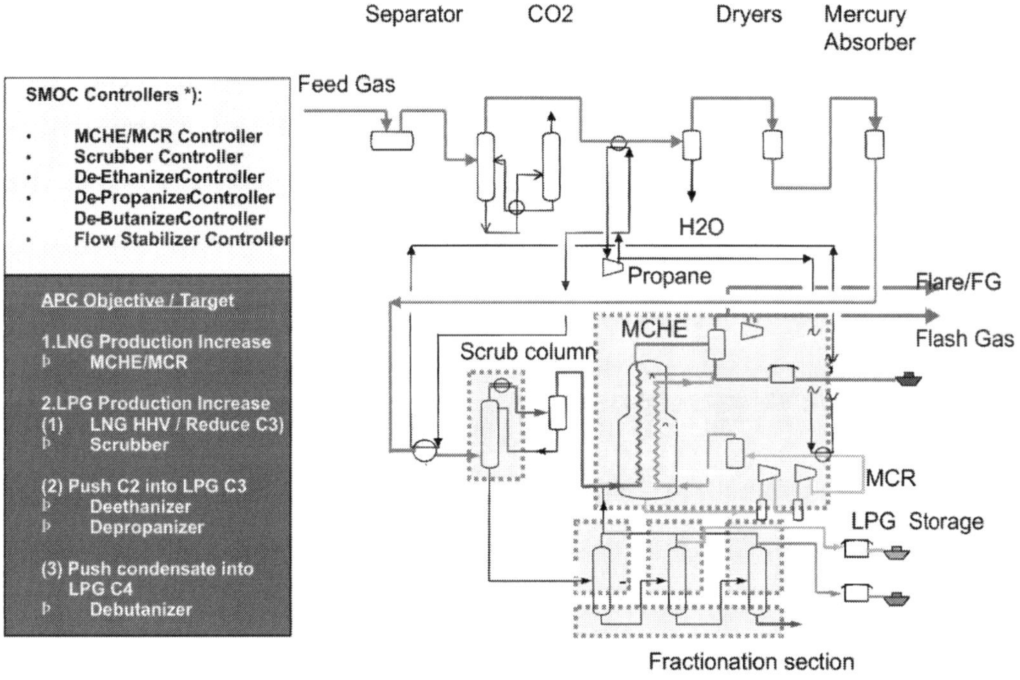

) SMOC = Shell Multivariable Optimising Controller

Figure 2 - LNG APC applications

Recent developments in Shell Global Solutions look at APC over a number of trains where the fuel gas balance is optimized and overall efficiency is further enhanced to further boost the benefits.

Shell affiliates own patent rights on certain APC applications on the MCHE, compressors and cooling cycles.

Case 1 : APC increases LNG production

Figure 3 shows an example where the production of one LNG train with steam driven compressors was increased with 2.4 % by applying APC compared to its base line without APC.

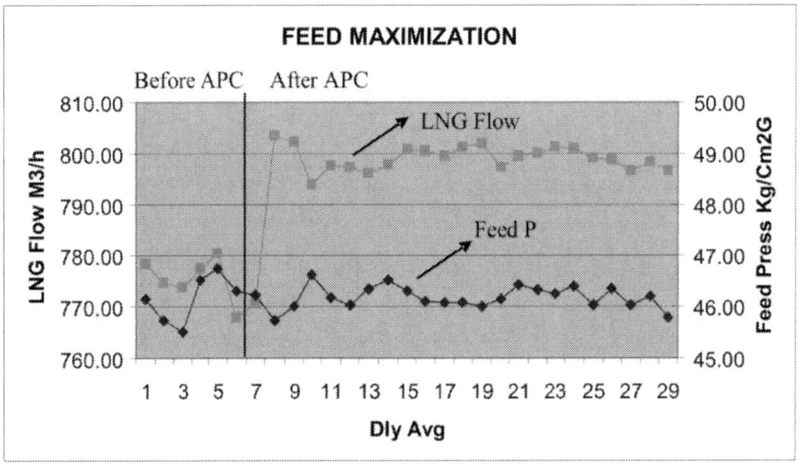

Figure 3 – LNG production increase

Case 2 – Operational aspects

APC projects typically also deliver advantages for operations. For one of the projects the following feedback was obtained from operations:

Before APC	After APC
o Units difficult to control due to interactions o During upsets various controllers taken onto manual o Inconsistencies between laboratory and analyzers o Unit control requires significant operator attention	o Continuous control every and each minute. o High uptimes o Online analysers are now used for APC.

Biography of Speaker

Kees den Bakker is Team Manager Advanced Process Control in Shell Global Solutions in Amsterdam, the Netherlands. He holds a Master of Science degree in Electrical Engineering (specialisation control) from Delft University of Technology. He joined Shell in 1984 and worked in various positions in Germany, the Netherlands and Brunei. He now works in Amsterdam where he manages a team of APC specialists that implements projects and services around the world.

Kees den Bakker
Manager Advanced Process Control Team
Shell Global Solutions International BV
P.O. Box 38000, 1030 BN Amsterdam, The Netherlands

Tel: +31 20 630 3337 **Fax:** 2900 **Other Tel:** +31-(0)655136174
Email: kees.denbakker@shell.com
Internet: http://www.shell.com/globalsolutions

Overview of Fischer-Tropsch Products and Their Upgrading to Useful Products

Mingsheng Luo, Stephen C. LeViness, Gary Jacobs and Burtron H. Davis*
*Center for Applied Energy Research, University of Kentucky, 2540 Research Park Drive, Lexington, KY 40511; *Syntroleum Corp., 1900 North 161st East Ave., Tulsa, OK 74116*

Abstract

Historically and at present only two catalyst systems are available for large-scale commercials plants: one based on cobalt and one based on iron. Two operating windows are available: low and high temperatures. The iron catalyst produces gaseous and gasoline range products when operated in the high temperature range, usually in fluid catalyst bed reactors. In the low temperature range, the iron catalyst can produce 50-75% wax products. The cobalt catalyst produces high boiling, waxy products when operated in the low temperature range but attempts to operate in a high (above about 230°C) results in the production of too much methane to be a viable option. The gaseous and liquid products are highly olefinic when operating with iron at high temperatures. Reforming to obtain a higher octane gasoline and alkylation of gaseous products to produce gasoline and/or diesel range products are required. Currently the low-temperature iron FT catalysis produces smaller quantities of products that can be used as chemicals than the high temperature operation; however, when large-scale production of low-temperature FT products with either an iron or cobalt catalyst is used, the wax will have to be converted to lower boiling range products. Differences in processing the high molecular weight products from the iron and cobalt FT catalysts will be considered.

Introduction

As indicated, two general process options, based on the temperature utilized for Fischer-Tropsch Synthesis (FTS), are available: a high temperature (300-350°C) or a low temperature (180-270°C) operation. The exothermic nature of FTS makes it difficult to operate the high-temperature process except in a fluidized bed reactor. While some preliminary work was done on this concept, the first major effort in this area was undertaken by a group headed by Hydrocarbon Research Inc. (HRI) at Brownsville, Texas in the U.S. and by the group now known as Sasol in South Africa. The HRI effort concentrated on the high-temperature process while Sasol utilized both low- and high-temperature processes. The HRI effort concentrated on the fixed-fluid bed reactor operation while the Sasol effort utilized the circulating bed reactor, a process concept similar to the fluid catalytic cracking process developed during WWII by Standard Oil (now Exxon Mobil). Since HRI encountered some operating problems, Sasol was glad that they had chosen the circulating fluid bed reactor, even though they encountered many early operational problems [1]. However, after operating the circulating fluid bed reactor for about forty years, Sasol converted all their production using this type of reactor (about 90% of the total production) to fixed fluidized bed reactors because of several significant operating advantages [2]. For the high-temperature operation one is restricted for the catalyst and the choice is based on iron because the cobalt catalyst produces far too much methane to be utilized in this process.

A variety of reactors have been utilized for the low temperature operation but only two are seriously considered today: the fixed bed reactor and the slurry bubble column reactor. Because of the density and low activity of the catalyst used by the US Bureau of Mines in a slurry reactor, they utilized circulation of a solvent with a boiling range that matched the operating reactor temperature to keep the catalyst suspended [3]; this resembles the later HRI H-Oil and H-Coal (direct coal liquefaction) processes except the catalyst particles were larger. Another variation was the foam process, developed in Germany during the 1940s, where the catalyst particles were suspended in a lower-density foam rather than a liquid. The cobalt catalysts were utilized in fixed bed reactors by the Germans during the 1933-1945 period. Sasol utilized many tubes of about 5 cm diameter contained within a shell which utilized steam generation to remove heat and this type of reactor is still utilized by Sasol today for iron catalysts. Shell began commercial operation of the tubular reactors in Malaysia about 1990 and continue to use this reactor type today in a commercial operation with a production capacity in the 15,000 bbl/day range.

High-Temperature Process

This process is conducted so that essentially all of the products must be in the gas phase at the reaction temperature, otherwise the heavy products will remain on the catalyst causing particle agglomeration and the development of particles too large to be fluidized at the gas flow rate employed for the reaction. Initially, two reactor types were used: (1) the circulating fluid bed reactor that evolved at Sasol that resembled the reactor used for fluidized catalytic cracking (Fig. 1c) and (2) the fixed bed fluidized catalyst reactor that was utilized at the Brownsville, Texas plant (Fig. 1d). Initially, startup problems were encountered for both types of fluid-bed reactors but eventually both types were operated successfully. The Brownsville plant was shut down before it had been operated for extended times because the price of natural gas increased to the point that it was uneconomical to operate the plant. Sasol solved the initial problems and operated these reactors for nearly 40 years (Fig. 1c); however, they eventually tried the fixed fluid bed reactors and its successful operation caused them to quickly replace all of the circulating to fixed fluidized bed reactors (Fig 1c and 1d). Many operating advantages are obtained with the fixed fluid bed reactor [4].

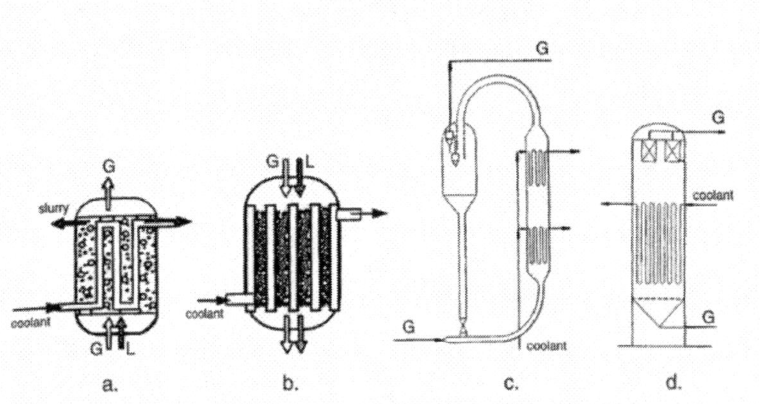

Figure 1. Possible reactors for FTS. (a) slurry, bubble column reactor, (b) multitubular trickle bed reactor, (c) circulating- and (d) fluidized-bed reactor.

The fixed bed reactor operated by Shell is typical of that utilized for low temperature FTS and is operated commercially at about a 15,000 bbl/day using a supported cobalt catalyst (Fig. 1b). In South Africa Sasol operates the fixed bed, slurry and fluid bed reactors to produce about 150,000 bbl/ day FTS products and PetroSA produces nearly 50,000 bbl/day FTS products using the older Sasol-type circulating fluid bed reactors. The Sasol production is based primarily on coal while the PetroSA production is based on natural gas. Sasol is set to bring on line about 35,000 bbl/day FTS plant in Qatar during 2006 that will utilize a slurry reactor and natural gas.

FTS Products to Commercial Products

The petroleum refining industry has had experience with waxy crudes for many years. For example, in its early years, the Atlantic Refining Co. processed significant amounts of Pennsylvania crudes which were very paraffinic in nature. In fact, during the 1950-1960 period they shipped significant quantities of paraffin wax to Europe in heated tankers. An example of this processing is contained in a Shell patent [5] that applies a common processing to both petroleum derived waxes and to waxes produced synthetically in processes such as FTS.

Shell workers considered the production of lubricating oil from a range of products [5]. In this patent they wrote, "...The hydrocarbon base oil should be aromatic-free and consist essentially or wholly of isoparaffins having a molecular weight of at least 350 and a pour point of at least -40°F. Oils of this type can be prepared by any suitable means, of which the preferred process is the production of isoparaffinic oils by isomerization of paraffin wax as described in U.S. Patents 2,668,866 and 2,668,790, as well as by any other suitable methods. The waxes from which the aromatic-free isoparaffinic lubricating oil are prepared can be any normally solid hydrocarbon wax. The wax will usually be derived from natural mineral sources such as petroleum oil, shale oil, oil from tar sands, gilsonite, ozokerite, or the like, but synthetic wax produced by the Fischer-Tropsch synthesis or as a by-product of other processes may also be isomerized....Essentially, the paraffin wax is isomerized in the vapor, liquid, or trickle phase in the presence of Pt, Pd, Ni, Co, Mo oxide or sulfide catalyst...on suitable carriers or support such as alumina, activated bauxite, and the like." A range of operating process conditions were specified.

The above could enable one to relate the conversion of FTS products to useful products to the vast processes and literature for the conversion of petroleum; however, it must be recognized that modifications of these processes may be required to take into account the differences in the chemical components of the various natural and synthetic feedstocks. This is a more exact comparison for high-temperature FTS products which contain aromatics in addition to paraffinic components.

Early efforts to upgrade FTS reaction products into fuels, lubricants, and useful materials has been reviewed [6].

High-Temperature FTS Product Upgrading

The products of the high temperature process are in the transportation fuels range and in those products which are gases at room temperature; for example, with an alpha value of 0.75 the average molecular weight of the products is C_4. Thus, the upgrading of high-temperature FTS products provides what appears to be unique upgrading opportunities since the carbon number distribution of the liquid products do not have to be altered. The liquid products fall in the gasoline and in the diesel, kerosene, and jet fuels boiling range so that the only processing that is needed is to convert their original properties to meet the requirements for their use as fuels. For the heavier products, such as diesel fuel, this requires isomerization to meet the properties required, such as gel point. For the gasoline range materials this becomes a more demanding task with the prohibition of using TEL for octane improvement. While isomerization of the n-paraffins to meet the specifications of each product stream (diesel, kerosine and fuel oil) is relatively easy, upgrading the gasoline to the required 87-93 RON range is not an easy task. The isomerized naphtha range paraffins usually have octane ratings in the 60-70 range. A significant improvement in the octane rating comes from the COD (conversion of olefins to diesel) process. The COD process uses a silicalite (a high-silica ZSM-5 type catalyst) catalyst to convert C_2-C_4 alkenes to high octane alkylate boiling in the gasoline range and in the diesel range [7]. The alkenes in the gas range ($C_2 - C_4$) may be converted to highly branched, high-octane paraffins. In addition, isomerization of the C_4 fraction permits the production of an alkylate using acid catalysis that has an octane in the 90+ range.

Low-Temperature FTS Product Upgrading

Unlike high-temperature FTS, the dominant products of the low-temperature FTS can be waxy materials with boiling points above that of the heavier transportation fuels. Thus, cracking, hydrocracking and/or hydro-isomerization can be utilized to effect the upgrading of the waxy products to transportation fuels, chemicals and/or lubricants. Today, there is little need to upgrade the wax since wax commands a price that is about four times that of transportation fuels; however, the market for the high-value wax is limited and a doubling of current wax production has the potential to saturate the market.

The light gases from the iron catalyst are predominantly alkenes (ca. 80%) so that oligomerization of these as described above for the high-temperature case is possible; however, in this case the volume of liquids obtained will be limited by the small fraction of the C_2-C_4 products that are produced. Thus, the important processing will be the conversion of the wax products to transportation fuels and chemicals.

The simplest way to obtain suitable products for transportation fuel could be the isosynthesis . Early efforts to accomplish this were carried out in Germany and the early work on the isosynthesis has been reviewed [8]. In spite of much work in Germany and in many countries at various time intervals, little success has been made in moving from the laboratory to larger units. This is caused primarily because the catalysts that effect the direct synthesis to isoalkanes in a single operation requires very high pressures and temperatures and, to date, only low activity catalysts have been found to be selective for the iso-products.

Another approach to directly produce higher branching in the products is to use bifunctional catalysts or dual catalysts in a single reactor. An early, if not the first, attempt at this was by workers at Gulf Oil [9]. Thus, a cobalt Fischer-Tropsch catalyst was physically mixed with a catalyst that effected alkene isomerization and/or cracking, such as a silicalite-1 catalyst, so that FTS could be carried out as well as secondary reactions of alkene and/or alkane isomerization and cracking. In the 1970-1980 period workers at Mobil oil attempted the bifunctional catalytic conversion using an iron FTS catalyst mixed with ZSM-5 isomerization-cracking catalyst in an attempt to make gasoline range products directly. However, Mobil abandoned this approach because the ZSM-5 catalyst did not have sufficient activity and/or life at the low temperatures that were optimum for the iron FTS catalyst and the iron catalyst produced too much methane and/or deactivated too rapidly at the higher temperature that was optimum for the ZSM-5 catalyst. Thus, Mobil modified their approach to that of dual reactor processing, using the first reactor for FTS and the second reactor system for converting the FTS products to gasoline range products [10]. Since the Mobil work, efforts continue to combine the FTS and upgrading into one reactor and publications continue to appear on this topic today.

An approach to get around the rapid catalyst aging is the use of supercritical conditions. Early work on this utilized solvents with such a low critical temperature (e.g., CO_2, CH_4, etc) that the density was essentially that of the gas phase synthesis. However, more recent work has focused on the use of high-density solvents, such as n-pentane, n-hexane or mixtures of these two. Unfortunately, these experiments are demanding and many of the experimental results were obtained during runs lasting only a few hours. Claims have been made for significant deviations from the normal ASF product distribution, caused mainly because of olefin reincorporation during the synthesis [e.g., 11]. Runs that extend for about four weeks and the use of alkenes labeled with ^{14}C did not provide evidence for reincorporation of alkenes to significantly alter the product distribution from that obtained at normal pressures [12].

In summary, it does not appear that success has been attained in utilizing one catalyst bed containing catalysts with two functions using either normal FTS pressures or those required for operation at high-density supercritical conditions.

A number of options are available for processing the wax product. One approach is to practice steam cracking which will produce primarily low molecular weight alkenes. This could be an appropriate process since steam cracking is currently utilized for this purpose although frequently with lower molecular weight fractions than FTS wax. However, there is a limited market for the alkene products and the introduction of several commercial scale plants will saturate the alkene market, making this approach impractical for the long term. However, less severe thermal cracking to make liquid products is still a viable option that is currently receiving serious consideration [e.g., 13].

The production of high quality diesel was recently reviewed and will not be repeated here [14,15]. The approach considered is to simply hydrocrack the heavy wax to produce high quality diesel, and some low quality naphtha range material; the final product was about 80% diesel, 15% naphtha and 5% C_1-C_4 . The diesel was high quality but it was noted that the naphtha would require extensive isomerization and Pt-reforming to yield a high octane gasoline. At this time, hydrocracking is the major route to upgrading the heavy wax fraction to transportation fuels.

In summary, the processing of low and high temperature FTS products are similar and usually require both oligomerization and hydrocracking (Fig. 2). For high temperature FTS, oligomerization will dominate in volume whereas for low temperature hydrocracking or hydro-isomerization will be the dominant process.

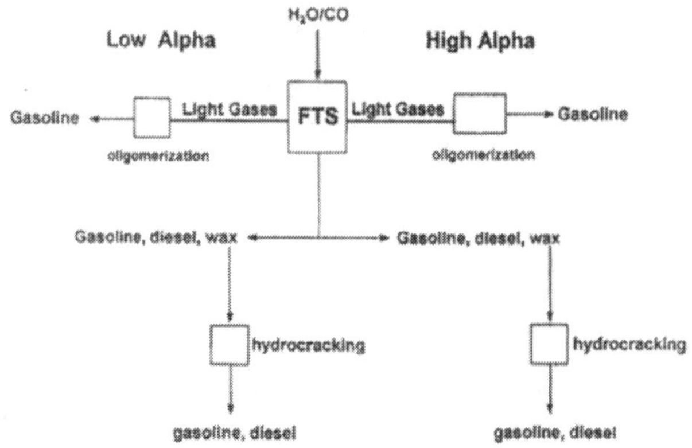

Figure 2. Comparison of processing the products from low- and high-temperature FTS.

References

1. M. Dry, personal communication.
2. A.P. Steynberg, T. Shingles, M.E. Dry; B. Jager; Y. Yukawa, SASOL commercial scale experience with synthol FFB and CFB catalytic FT reactors; Int. Conf. on Circulating Fluidized Beds (1991) 527.
3. H.E. Benson, J.H. Field, D. Bienstock, R.R. Nagel, L.W. Brunn, C.O. Hawk, J.H. Crowell and H.H. Storch; Development of the Fischer-Tropsch oil-recycle process, Bureau of Mines, Bulletin 568, 1957.
4. D.J. Duvenhage and T Shingles; Synthol reactor technology development; Catal. Today 71, 301-305 (2002).
5. H. Diamond and C. L. Mahoney, U.S. Patent 2,794,781, June 4, 1957.
6. R. Freeks, AIChE Spring Natl. Mtg., New Orleans, La, April 2, 2003, paper 86d.
7. The COD Process, Sud-Chemie Group bulletin.
8. H. Pichler and K.-H. Ziesecke, Bureau of Mines Bulletin 488, US Gov. Printing Office, Washington, 1950.
9. H. Beuther, C. L. Kibby, T. P. Kobylinski and R. B. Pannell, U.S. Patent 4,399,234, August 16, 1983.
10. Kuo, J. C. W.; Two-Stage process for conversion of synthesis gas to high quality transportation fuels; Final Report; DOE/PC/60019-9; October 1985.
11. K. Yokota and K Fujimoto, Ind. Eng. Chem. Res., 30, 95-100 (1991).
12. G. Jacobs, K. Chaudhari, D. Sparks, Y. Zhang, B. Shi, R. Spicer, T. K. Das, J. Li and B. H. Davis, Fuel, 82, 1251 (2003)
13. S. E. Henderson, D. R. Johnson, S. J. Miller and R. R. Krug, U.S. Patent 2005/0131082, June 16, 2005.
14. M. E. Dry, J. Chem. Technol. Biotechnol., 77, 43 (2002).
15. L.P. Dancuart, R. de Haan, A. de Klerk, Stud. Surf. Sci. Catal., 152 (2005) pp 482-532.

Fischer-Tropsch Synthesis with Ultrafine Iron-Based Catalyst: Nano-Scale Growth of Particles

Amitava Sarkar, U. M. Graham, J. K. Neathery, R. L. Spicer, and Burtron H. Davis.*
Center for Applied Energy Research, University of Kentucky, 2540 Research Park Drive, Lexington, KY 40511-8479.

1. INTRODUCTION

Increasingly stringent environmental regulations for "clean-fuels" (low-sulfur, low-aromatics) make Fischer-Tropsch synthesis (FTS) an environmental-friendly route for the production of middle-distillatess and a wide variety of hydrocarbons and oxygenates from coal, natural-gas or biomass-derived synthesis gas (SG).[1] Iron-based catalysts are preferred for FTS utilizing a low H_2/CO ratio SG derived from coal or biomass because of their excellent activity for the water-gas shift reaction. The use of iron-based catalysts is also attractive in view of lower cost, lower methane selectivity, high olefin-selectivity, lower sensitivity towards poisons, and flexible product slate. When the operation favors long chain growth, a large amount of waxy material is produced that can be processed to produce premium fuels and chemical feed-stock. In removing wax, suspended particles must be separated and recycled back to the reactor to maintain the required conversion and to avoid catalyst loss.[2] Catalyst/wax separation is also necessary for wax upgrading. The recommended solid catalyst content of the wax is in the 2-5 ppm range or even lower.[2]

The use of iron catalysts in the most economical slurry bubble column reactors (SBCR) has been limited by their high rate of attrition to ultrafine particles leading to catalyst loss, high slurry viscosity and difficulty in wax/catalyst separation. Phase-transformations during activation/FTS play an important role in determining the structural integrity or attrition resistance of the catalyst particles. Sequential phase-modification during the activation process (with CO or SG) from hematite to magnetite and finally to Fe_xC_y (iron carbides) have been reported.[3-6] The chemical conversion of Fe_3O_4 to Fe_xC_y induce a volumetric change due to significant difference in densities and resulting shear lead to formation of small crystallites of Fe_xC_y which split off rapidly to form ultrafine particles.[5,7-8] The conversion of Fe_3O_4 to Fe_xC_y is dynamic and reversible depending upon the environment. Under oxidizing conditions (e.g., at high water and CO_2 partial pressures), layers of surface Fe_3O_4 can form on carbide kernels and ultimately pure Fe_3O_4 crystals can be formed.[6] Thus the possibility of either fragmentation or growth of particles depending on the process condition exists. Kuo et al.[9] reported an increase in particle size during the slurry FTS which is contrary to what generally reported in literature. The particle size of most of catalysts used for attrition studies are in micrometer range. The use of submicron particles in SBCR is advantageous for efficient control, utilization of high reaction exotherm and enhanced mass transfer rate of gaseous reactants to catalyst surface. A detailed knowledge of the change of particle size of submicron particles is necessary for application in SBCR. The objective of the present study is to monitor the morphological and chemical change of ultrafine iron catalyst particles (3-5 nm diameters) during the activation/FTS in a CSTR by HRTEM, EELS, and Mössbauer spectroscopy. This information can be utilized for design and optimization of a filtration unit for wax separation.

2. EXPERIMENTAL

A commercial ultrafine iron oxide ($\alpha - Fe_2O_3$) catalyst (NANOCAT®, Mach I, Inc.; particle size, 3 nm; surface area, 250 m^2 g^{-1}; bulk density, 0.05 g cm^{-3}) was used for all experiments. Polywax 850 (polyethylene fraction with average MW of 850) purchased from Baker Petrolite, Inc., was used as start-up solvent. Experiments were conducted in a CSTR equipped with a turbine impeller, a gas-inlet line, and a vapor outlet line with a fritted filter (2 μm) placed external to the reactor. A tube fitted with a fritted filter (0.5 μm) extends below the liquid level of the reactor and is used to maintain a constant level. A dip-tube was used to collect catalyst slurry at increasing times. 54.7 g catalyst was added to 300 g melted (150°C) Polywax 850. The temperature was raised to 270°C at a rate of 1°C min^{-1}. Catalyst was activated with CO at a SV of 3.0 sl h^{-1} g Fe^{-1} at 270°C and 175 psig for 24 h. The FTS then started with a H_2/CO ratio of 0.7 and a SV of 3.0 sl h^{-1} g Fe^{-1}. The conversions were calculated from GC analysis of the exit gas. The products collected in three traps maintained at 200, 100 and 0°C, were analyzed by GC.

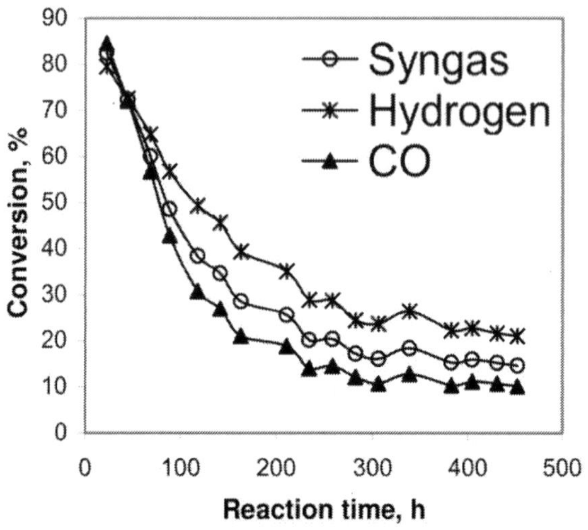

Figure 1: Plot of conversion of synthesis gas, H_2 and CO against time during the FTS (270°C, 175 psig, 750 rpm) using iron nano-catalyst.

The slurry samples were diluted with hot (about 70°C) o-xylene to remove the wax. Although it was not possible to completely remove the wax by this method, the leftover wax did not interfere with TEM analysis and acted as a protective cover for the air-sensitive particles. An optimum ratio of o-xylene to catalyst was used to make a slightly turbid suspension for TEM analysis. A drop of the suspension was placed onto a lacey carbon film on 200 mesh copper grid and loaded into the microscope.

Mössbauer spectra were collected in a transmission mode by a standard constant acceleration spectrometer (MS-1200, Ranger Scientific). A radiation source of 30 mCi ^{57}Co in Rh matrix was used and spectrums were obtained using a gas detector. The solid catalyst slurry was mounted in Plexiglass compression holders. Samples were investigated at room temperature as well as at liquid helium temperature, typically over a velocity range of $\pm$ 10 mm s^{-1}. For the low temperature measurements, the samples were placed inside a vibration free closed cycle cryostat. The structural analysis of the samples was done by least-squares fitting using user defined functions within the PeakFit® program. Errors in the determined percent of Fe values are about $\pm$3-5 % depending on the complexity of the spectrums.

The particle morphology was analyzed by a field emission analytical transmission electron microscope (JEOL JEM-2010F) operated at an accelerating voltage of 200 kV and equipped with a STEM unit with high-angle annular dark field (HAADF) detector, and a Gatan Imaging Filter (GIF)/PEELS system. The electron beam had a point-to-point resolution of 0.2 nm. EELS spectrums were recorded in TEM imaging mode with an energy resolution of 1 eV and dispersion rate of 0.2 eV/channel. Gatan Digital Micrograph® software was used for image and EELS data processing.

3. RESULTS AND DISCUSSIONS

The initial activity of the ultrafine catalyst was high and showed a gradual decrease during a total synthesis time of over 475 h (Figure 1). The removal of catalyst particles during the synthesis also contributes to the gradual decrease of conversion.

Compositional changes in terms of percent peak area of the different iron species in Mössbauer spectrum are listed in Table 1. The superparamagnetic (spm) phase found at 300 K can be either Fe_3O_4 or iron carbide or a mixture of both phases. Hence, to identify the spm phase, the low temperature (20 K) results are considered. 24 h pretreatment with CO converts Fe_2O_3 to a mixture of 85% $\chi-Fe_5C_2$ and 15% magnetite which is in agreement with reported results.[10] Low temperature Mössbauer results showed that the spm phase was magnetite, and not the iron carbide, which can also be present in the spm phase. The water produced in FTS acts as an oxidant and the oxidation of iron carbide continues with synthesis time and the amount of carbide phase gradually decreases with a corresponding increase in the fraction of the magnetite phase. At the end of 311 h of synthesis time, the catalyst is composed of mainly magnetite (96%) and the conversion of SG drops to a value lower than 15%. Thus, the presence of some fraction of iron carbide is necessary for a catalyst to exhibit high activity in FTS.

Table 1: The composition of the catalyst during the activation/FTS determined from Mössbauer spectroscopy. The CO activation (for 24 h) and FTS was carried out at 270°C and 175 psig. The TOS was counted from the start of CO activation.

Sample # (TOS, h)	Fraction (%) of Fe at 300 K	Fraction (%)of Fe at 20 K
S1 (0 h)	85 in χ-Fe_5C_2 15 in spm	84 in χ-Fe_5C_2 16 in Fe_3O_4
S2 (28.9 h)	48 in χ-Fe_5C_2 33 in Fe_3O_4 19 in spm	50 in χ-Fe_5C_2 50 in Fe_3O_4
S3 (73.2 h)	65 in Fe_3O_4 18 in χ-Fe_5C_2 17 in spm	76 in Fe_3O_4 22 in χ-Fe_5C_2 2 in spm
S4 (167.7 h)	89 in Fe_3O_4 11 in spm	91 in Fe_3O_4 6 in χ-Fe_5C_2 3 in spm
S5 (311 h)	93 in Fe_3O_4 7 in spm	96 in Fe_3O_4 4 in spm

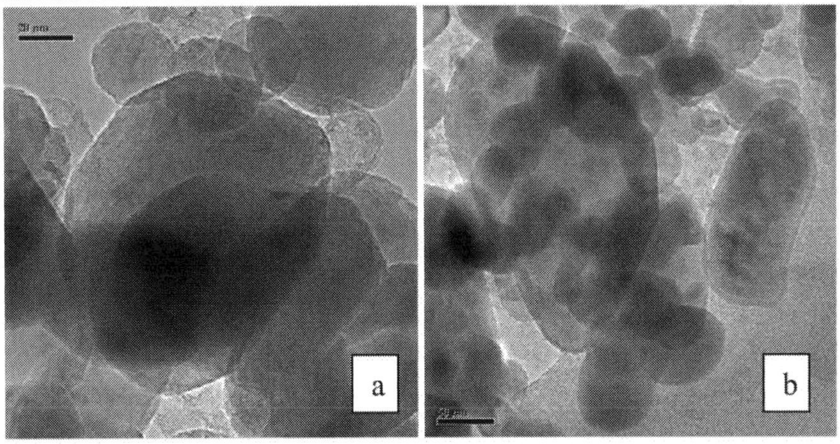

Figure 2: TEM image of particles after (a) 505.6 h & (b) 620.8 h TOS.

Size analysis of irregular shaped particles requires some approximations of representative size. Non-equidimensional particles are often characterised by averaging major dimensions. Since, the particle sizes are measured from the TEM images, Feret diameter, the maximum distance between pairs of parallel tangents to the projected outline of the particle is used. An arithmetic average of the first two large diameters (i.e., first two Feret diameter) has been taken as the "representative particle diameter" for the present study. The size of catalyst particles increased with the synthesis time. The FTS was started with an average particle size of 12.5 nm and after 620.8 h, the average particle size was 61.5 nm. Some particles with a maximum dimension larger than 130 nm was also observed (Figure 2a and 2b). Amorphous carbon rims of 3-5 nm thickness around some of the particles and some hexagonal shaped particles were observed. The bigger particles were formed during FTS (Figure 3). The variation of 'largest observed particle size' with TOS revealed a 10-fold increase of size. The thickness of carbon rims did not grow above 5-6 nm. Hence, the growth of particles is not due to carbon deposition on surface.

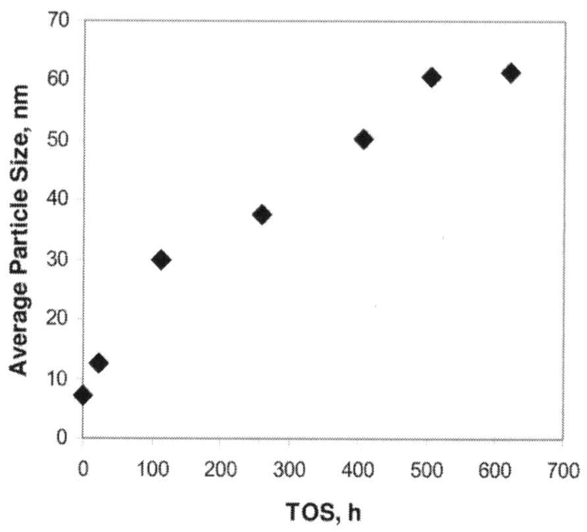

Figure 3: Variation of average particle size with TOS (counted from start of CO activation).

TEM image of the catalyst precursor (Figure 4a) showed a particle size of (4-7) nm. The diffuse and non-sharp rings in the diffraction pattern (inset, Figure 4a) revealed the lack of well-defined crystallinity and inelastic scattering from ultrafine particles. The TEM image of samples activated in CO for 24 h showed a particle size in the range of 8-15 nm. The average diameter of the particles in this sample was 12.5 nm and a 2-3 nm rim of amorphous carbon around the some particles were visible. The diffraction pattern confirmed the presence of loose aggregates of

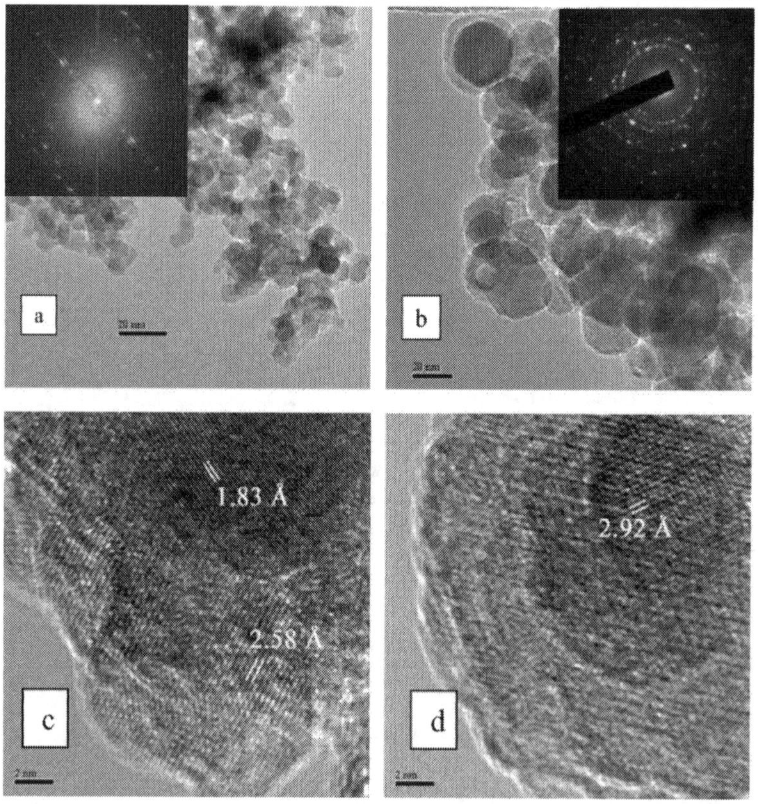

Figure 4: HRTEM image of particles: (a) fresh catalyst; (b, c, and d) CO activated (for 24 h) catalyst.

iron carbides.[11] Carbide particles formed in layers overlapping with each other as found by HRTEM (Figure 4c & 4d). The iron carbides appeared as small crystallite with a d–spacing of 1.83 Å ([511] plane of Hägg carbide). A crystallite with a d–spacing of 2.58Å can be assigned either to the [311] plane of magnetite or $\chi - Fe_5C_2$ (2.65Å for [311] plane or 2.5Å for [002] plane). Some of the crystalline aggregates were composed of an outer layer of iron carbide and an unreduced magnetite core (Figure 4d) with a characteristic d–spacing of 2.92 Å ([220] plane). The formation of such nano-zone was also observed with particles at higher TOS. In Figure 5a (112.6 h TOS), an outer-zone composed of magnetite (with a d–spacing of 2.58 Å for [311] plane) formed by re-oxidation of carbides and a carbide inner-zone (d–spacing of 1.83 Å, [511] plane) are observerd. Similar features were observed in Figure 5b.

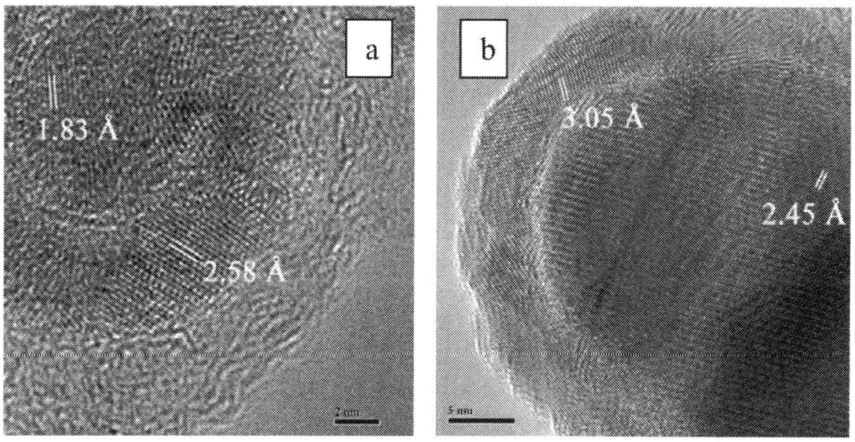

Figure 5: TEM image of particles after (a) 112.6 h; (b) 505.6 h TOS.

It was reported that while using particles of 100-200 micron in size, the carbide phase forms as small nodules on the surface of the magnetite and the phase transformation proceeds slowly into the bulk.[4,6] The volumetric change during the oxide to carbide transformation results the sprouting of carbide buds/nodules on the surface of magnetite crystal. However, such phenomena were not observed with nanometer range particles. The carbides were identified as bulk phase with layers and sometimes with an oxide core. When the dimension of magnetite formed during FTS reaches the micrometer range, some of the carbide phase appears as nodules at top of the magnetite crystal. This was confirmed by the EELS analysis coupled with STEM imaging (Figure 6). The larger particle was pure oxide while the nodule was carbidic in nature. The

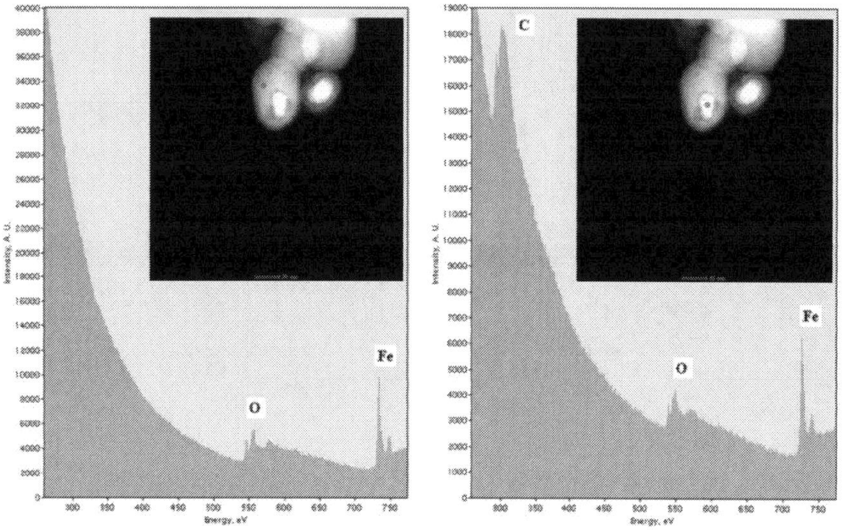

Figure 6: EELS spectrum of catalyst samples collected after 505.6 h TOS. Positions on the sample used to record the spectrum are also shown by red spot in the STEM image (inset).

nature of the carbon was analyzed with the high-energy loss portion of the EELS spectrum at the carbon K-edge. The differences between the spectra of crystalline and amorphous carbon are due to the relative intensity of carbon $1s \rightarrow \pi^*$ and $1s \rightarrow \sigma^*$ core-level transitions which provides a robust measure of the percentage sp^2-bonded carbon. The spectra of both ordered and disordered graphite show a sharp π^* peak followed by a higher energy σ^* peak.[12] For the spectrum for amorphous carbon, however, only has a small π^* shoulder with a sloping σ^* peak. The shape of present EELS spectrum suggests that the carbon phase was crystalline.

4. CONCLUSIONS

The change in size and morphology of ultrafine iron nano-catalyst during the FTS revealed a different phenomenon than commonly observed with micrometer sized iron catalysts. The phase transformation of ultrafine iron oxide to iron carbides does not create any fragmentation or attrition of the particles. This may be due to lower magnitude of stress generated (because of smaller size), by volumetric change in carbidation, and is not sufficient to break the particles. The small particles convert into carbides in a "shrinking core" manner with different nano-zones. The carbide/oxide conversions are dynamic and with the progress of FTS, the carbides get oxidized in an "egg-shell fashion" into pure single crystal oxide phase and grow in size. However, when the dimension of magnetite reaches the micrometer range, the crystalline carbide phase appears to be sprouted on the surface of magnetite single crystal.

5. REFERENCES

1. Steynberg AP, and Dry ME, (eds), Fischer-Tropsch Technology, *Stud. Surf. Sc. Catal.*, 152, Elsevier, Inc.: San Diego, 2004.
2. Zhou PZ, and Srivastava RD, DOE Report, Burn and Roe Services (1991).
3. Huang CS, Ganguly B, Huffman GP, Huggins FE, and Davis BH, *Fuel Sc. Tech. Int'l*, 11 (9), 1289 (1993).
4. Shroff MD, Kalakkad DS, Coulte KE, Köhler SD, Harrington MS, Jackson NB, Sault AG, and Datye AK, *J. Catal.*, 156, 185 (1995).
5. Zhao R, Goodwin JG, Jothimurugesan K, Gangwal SK, and Spivey JJ, *Ind. Eng. Chem. Res.*, 40, 1320 (2001).
6. Li S, Ding W, Meitzner GD, and Iglesia E, *J. Phys. Chem. B*, 106, 85 (2002).
7. Srinivasan R, Xu L, Spicer RL, Tungate FL, and Davis BH, *Fuel Sci. Tech. Int'l*, 14(10), 1337 (1996).
8. Pham HN, and Datye AK, *Cata. Today*, 58, 233 (2000).
9. Kuo JCW, Sanzo FPD, Garwood WE, Gupte KM, Lang CK, Leib TM, Malladi M, Molina T, Nace DM, Smith J, Tarallo N, Kirk JF, DOE Report, Contract DE-AC22-83PC80019 (1985).
10. Rao KRPM, Huggins FE, Huffman GP, Gormley RJ, O'Brien RJ, and David BH, *Energy Fuels*, 10, 546 (1996).
11. Morozova OS, Maksimov YV, Shashkin DP, Shirjaev PA, Matveyev VV, Zhorin VA, and Krylov OV, *Appl. Catal.* , 78, 227 (1991).
12. Fernández A, Prieto P, Quirós C, Sanz JM, Martin JM, Vacher B, *Appl. Phys. Let.*, 69, 764 (1996).

181c Conversion of Synthesis Gas to Gasoline

Finn Joensen, Haldor Topsøe A/S, Nymøllevej 55, DK-2800 Lyngby, Denmark

High and sustaining oil and energy prices have spurred a renewed interest in manufacture of synthetic fuels and chemicals from gas, coal and other sources. The TIGAS (Topsoe Integrated Gasoline Synthesis) process for the conversion of synthesis gas into gasoline was developed and demonstrated on a T/d scale in the mid-eighties using natural gas as feedstock. A subsequent demonstration program based on coal gasification was initiated, but later abandoned due to plummeting oil prices and lack of interest in the market. Today, we are back to the situation where alternative sources for production of liquid fuels and increasing the degree of self-supply is high in demand. The basic principle in the TIGAS process is that of integrating a methanol/dimethyl ether synthesis and the subsequent conversion into gasoline in a single synthesis loop, i.e., without the isolation of intermediate MeOH or DME. Due to the flexibility of the MeOH/DME synthesis a variety of synthesis gas compositions may be applied. For CO-rich gases as obtained by gasification of coal, coke or biomass the integrated process offers a particular advantage due to the high efficiency that may be obtained in the synthesis of the MeOH/DME intermediate at the low (< 1) H2/CO ratios characteristic to gasification. As opposed to methanol synthesis, where a module – the effective H2/CO ratio – adjustment to 2 is required, the ideal stoichiometry with respect to the combined MeOH/DME synthesis is at H2/CO = 1, making DME by far the dominant intermediate. In addition to improving the conversion efficiency less steam is required for module adjustment. Finally, CO2 removal as well as module adjustment may be carried out inside the synthesis loop, minimizing the recycle of unconverted synthesis gas. The virtues of the integrated process and recent developments will be highlighted.

Enhanced Natural Gas and Methane Conversion into Methanol in a Membrane Driven Reactor (Experimental Modeling and Simulative Comparison with Fixed Catalytic Reactors)

Dr. Savvas Vassileiadis (PI), and

Dr. Zoe Ziaka

Chemical, Biochemical, & Materials Engineering Department
ZIVATECH INSTITUTE

26 School St, Princeton, Maine, 04668

ABSTRACT

Modeling and simulative comparison of catalytic permreactors (membrane driven reactors) with fixed plug flow reactors (fixed catalytic reactors) is performed using numerical analysis techniques, for the synthesis gas (syngas) and methanol production processes from methane and natural gas feedstocks. Experimental fixed catalytic reactors are utilized to derive kinetic data for the catalytic methane-steam reforming and the water gas shift reactions using combined optimization-integration techniques. Experiments are also performed with catalytic membrane reactor modules (permreactors) for comparison purposes with the fixed catalytic reactors (the syngas to methanol process). The mass and pressure balances of these reactors were solved numerically to provide the profiles of partial pressures or molar flowrates or fractional conversions and yields of each species along the reactor axial length. Several other reactor parameters were varied in the numerical model to obtain the respective variation of conversion and yield at the reactor exit. Such parameters are the reactor temperature and pressure, space time, and reactor dimensions. The developed numerical reactor models can be also used to predict the reactors operation at different running conditions in order to increase the syngas and methanol product yield.

The methane reforming and shift reactions under analysis are of special interest in syngas and methanol generation and utilization processes. Utilization of improved reactors and reaction systems for methanol production from natural gas is of high importance in the recent fuel, gas, and transportation fuels industry. Methanol is a key fuel and can be directly produced from the synthesis gas which is coming out from the catalytic reforming of methane or natural gas as described in this paper. Use of methanol is mainly in transportation fuels, chemical synthesis, and as fuel in direct methanol fuel cells (DMFC).

Hydrogen is also a main product from the reforming and related conversion reactions. Other hydrogen and syngas utilization processes include feed for electrochemical fuel cells, hydrogen assisted combustion systems, hydrogen rich hydrotreating of heavy oils, distillates and asphaltenes, chemical synthesis of hydrogenated products (e.g., methanol and olefins).

KeyWords: Methanol synthesis, modeling of fuel resources, reforming and shift reactions, syngas generation, permreactors, computational design.

*** Correspondence directed to: Dr. S.Vassileiadis (PI); Tel&fax: 818-892-4238; email: vasil501@peoplepc.com**

1. Introduction

In several of our previous journal communications [1-9], we described and analyzed catalytic processors for steam reforming of methane, natural gas, and coal gasification for use in chemical synthesis and fuel products (such as methanol, gasoline, hydrocarbons etc) and in various types of fuel cells [2,3,5,8,9]. The catalytic reactors or processors can be described by mass, thermal, and pressure balances which after solution of the respective differential equations provide the variation of the species conversions or yields at the reactor exit.

This communication focuses on the numerical analysis and procedures for the design of fixed catalytic reactors and membrane reactors (permreactors) of inorganic type membranes for production of syngas (syngas or synthesis gas is a H_2, CO, CO_2 rich exit gas mixture) and methanol products.

The two main catalytic gas phase reactions taking place and under current consideration are the methane steam reforming and the water gas shift which have been also reported earlier kinetically and catalytically [10-15].

The overall catalytic reaction is endothermic, as can be found by the overall heat balance of the two reactions. We have reported several earlier communications on the methane-steam reforming, water gas shift, and related catalysis processes and experimental operations [1-7,18].

2. Process modeling and experimental analysis

The computational model developed here describes the main mass, thermal, and pressure balance equations in the reformer or shift reactor and is used to quantify the reactor and the reaction variables describing this system [1,2,3,4,5].

The describing steady state-plug flow reactor model equations (fixed catalytic reactor type, FCR) for the two reactions are given as follows:

$$dX^{CH4}/dz = (\pi\, d_T^2 / 4n_{CH4o})\, \rho_B\, R_{CH4} \tag{3}$$

with: $R_{CH4} = -R_1$

$$dX^{CO2}/dz = (\pi\, d_T^2 / 4n_{CH4o})\, \rho_B\, R_{CO2} \tag{4}$$

with: $R_{CO2} = R_2$

$$dX^{CO}/dz = (\pi\, d_T^2 / 4n_{CH4o})\, \rho_B\, R_{CO} \tag{5}$$

with: $R_{CO} = R_1 - R_2$

and for hydrogen species:

$$dX^{H2}/dz = (\pi\, d_T^2 / 4n_{CH4o})\, \rho_B\, R_{H2} \tag{6}$$

with: $R_{H2} = 3R_1 + R_2$

$$dX^{H2O}/dz = (\pi d_T^2 / 4n_{CH4o}) \rho_B R_{H2O} \quad (7)$$

with: $R_{H2O} = -R_1 - R_2$

The thermal balance in the reactor is given as:

$$dT_R/dz = (\pi d_T^2 / 4) (1 / m'c_p) \{ \rho_B [(-\Delta H_r^1) R_1 + (-\Delta H_r^2) R_2] - 4 (U/d_T) (T_R - T_S) \} \quad (8)$$

The pressure balance is given as:

$$- dP_T/dz = 2 f \rho_g u_s^2 / g_c d_p \quad (9)$$

the above equations are complemented by initial conditions:

at $z=0$ (reactor inlet), $X^j=0$, $T_T=To$, $P_T = P_{To}$

The above system of equations is integrated numerically as an initial value problem to provide the reactive/reactant species conversions and yields, the reactor temperature and pressure along the axial length, and to obtain the axial profiles of these parameters and their values at the reactor exit. Detailed analysis of the simulation results of such reforming-water gas shift systems are provided in our previous communications [1,2,3,4,5]. Moreover, the overall reaction rates R_1, R_2 shown above for the methane steam reforming and the water gas shift reactions have been investigated and reported. These rate expressions include kinetic and adsorption constants as previous reaction studies have also reported [1-5]. The reaction rate expressions utilized within the model are similar with those reported initially by Xu and Froment in their well known methane-steam reforming kinetic analysis study [10]. Our reforming reaction experiments utilized methane mixed with various compositions of steam and flown in nickel oxide (-NiO-) type catalysts supported on various supports. The kinetic rate constants for the two reactions above, were derived by using a nonlinear regression technique (optimization type routine) coupled with the integration of differential equations (3-9) above. The reaction data used was obtained by varying the space time of the reactor (total rate of flow in the inlet) and by measuring the exit CH_4 conversion and the CO_2 yield [1,2,3]. The amount of catalyst in the fixed bed reactor was kept constant and the ratio (W_{cat}/F^o_{CH4}) was varied by varying the F^o_{CH4} feed flowrate parameter. By means of the two kinetic equations described above, the methane conversion is correlated with the extent of the methane steam reforming, in the reforming reaction. The carbon dioxide yield is correlated with the extent of the water gas shift, in the second reaction.

Below we elaborate additionally in the method of obtaining the rate constants for the two reactions. The material balance equations (3) to (7) were used to calculate the molar flowrates or fractional conversions/yields (of each component) at the exit of the fixed catalytic reactor (FCR) [10-15]. The calculated quantities were then compared with the experimental ones by using the optimization routine. By minimizing the sum of the squares of the residuals between the two quantities, the best values for the kinetic constants k_1 and k_2 were determined. The numerical method involved was using a modified Levenberg-Marquardt algorithm to generate a series of approximations that search for the minimum point of the response surfaces $\sigma^2 = f(k_1, k_2)$, until the stopping criteria are satisfied. The integration algorithm of equations (3) to (7) is based on implicit integration techniques with variable stepsize to ensure a higher accuracy.

The detailed kinetic analysis study of the fixed catalytic reactor, with the variation of its parameters, is given in earlier communications, (Vasileiadis, S., 2000, 2002, 2004), [1,2,3]. The k_1 and k_2, kinetic Arrhenius type expressions are given as follows:

$$\boldsymbol{k_i = k^o_i \exp(-E_i/RT)}$$

The employment of an inorganic permreactor (a membrane driven reactor) as the main catalytic processor for the conversion of methane or natural gas feedstocks with steam as the reactant produces synthesis gas (syngas) under different process conditions. This makes the new "syngas to methanol process" for syngas production. The use of the tubular permreactor defines two gas exits for this type of membrane driven reactor. The composition of the permeate from the membrane gas is rich in hydrogen while the composition of the main outlet exit gas is rich in CO, CO_2, some H_2, and some unreacted CH_4 and steam. The two exit streams may be combined and fed into the next syngas reactor R2 for methanol synthesis, as shown in Fig.1. This will provide for additional hydrogen in R2 (methanol reactor) for tailored production of alcohols (e.g., methanol) or hydrocarbons (e.g., olefins via Fischer-Tropsch synthesis) [1,9,18]. The above described

FCR design equations are modified accordingly to include the permeation effects of the different components via the membrane reactor walls. The following term has to be added to the right hand side of the equations (3) to (7):

$$- (2\pi/n^{T}_{Ao})\, P_{A,e} [(p^{T}_{A} - p^{S}_{A}) / \ln (r_1/r_2)] \qquad (10)$$

wherein $\boldsymbol{P}_{A,e}$ (gmol/s.cm.atm) is the effective permeability coefficient of any permeating species A via a catalytic or a non-catalytic (e.g., blank) inorganic membrane.

The corresponding mass, temperature, and pressure variation equations of the permreactor are written as well for the gas which permeates via the membrane wall materials and flows in the permeate side of the membrane-separated reactor. We assume that there are no reactions occurring in the permeate side and the reactor operates at steady state condition. In the case of an isothermally kept reactor with no pressure drop in the permeate side, the mass balance for the moles of gas permeating in the reactor shell becomes:

$$dn^{S}_{A}/dz = - (2\pi P_{A,e}) [(p^{T}_{A} - p^{S}_{A}) / \ln (r_1/r_2)] \qquad (11)$$

The detailed model for the permreactor and related operation has been described analytically in earlier communications (Vasileiadis, S., 2000, 2002, 2004) [1-5].

In parallel experimental studies to study the characteristics and performance measures of the inorganic membrane reactor we utilized mesoporous aluminum oxide membranes having a thin top permselective membrane layer (3-5μm thickness, 50% porosity) with about 40-50Å pore diameter [1-5]. The same catalyst was used as in the fixed catalytic reactor experiments.

By using the modified equations for the membrane reactor (by use of equations (10), (11) above) in the modeling procedure, a detailed membrane reactor modeling analysis is obtained.

The solution of the permreactor (membrane reactor) equations is also obtained numerically with the use of the same initial value integration technique as in the case of the fixed catalytic reactor. An implicit integration method for solving ODEs is used again here with a variable stepsize to ensure a higher accuracy of integration. All equations of the permreactor (the fixed bed side/feedside and the permeate side equations) are integrated at the same time at steady state conditions to obtain the profiles of fractional conversions and yields at the reactor two(dual) exits.

The modeling of the permreactor is compared with experimental results obtained by the use of the bench scale experimental membrane reactor as described below.

3. Results and discussion

Fig.1 [18], below shows the flow chart of the steam reforming process using methane or natural gas as feedstocks in the catalytic reactor (R1). The process can utilize a fixed plug flow (catalytic) reactor or a permreactor (syngas to methanol process permreactor) as described above to produce synthesis gas for methanol synthesis and other consecutive synthesis reactions [1-5,8,9,18]. Such synthesis reactions in R2 vessel (syngas mixer) can be of methanol synthesis [1,2,3,8,9] or of olefinic hydrocarbon synthesis via the well studied Fischer-Tropsch synthesis reactions [8,9].

The conversion of syngas to methanol in the described process for use as fuel is of increased and considerable importance, especially for DMFC (direct methanol fuel cells) and as transportation and synthesis fuel.

Fig.2 below [18], shows the modeling of the fixed catalytic reactor (FCR) at various space times as described in the modeling procedure above. Variation of the total feed flowrate in the reactor inlet results at different methane conversions at the exit as shown in the Figure. The modeling results simulate the experimental reactor performance data very well. The experiments were done with an experimental tubular reactor filled with catalyst particles as described above, by using fixed compositions in the reactor feed and by varying the reactor temperature (450-650°C) at a fixed range of temperatures. The data were also used to derive the kinetic constants of the reforming reactions:

$k_1 = k^o{}_1 exp(-E_2/RT)$, with the procedure described above. Fig.3, [18], which is shown consecutively, corresponds to the carbon dioxide yield. The modeling and experimental CO_2 yields are indicative of the extent of the water gas shift reaction and were used to obtain values for the second kinetic constant: $k_2 = k^o{}_2 exp(-E_2/RT)$. These experimentally measured conversions and yields approach the thermodynamic equilibrium calculated values for the upper (higher) space times (for low flowrates in the inlet) as used in the FCR module, and seen in Figs. 2&3.

In the next two Figures 4 and 5, [18], we plot results obtained with both the experimental membrane driven reactor (made by a mesoporous γ-alumina tube, with 40-50Å pore diameter, and 3-5μm thickness for the top most permselective layer) and with the fixed catalytic reactor. The operation of both reactors was varied by varying the reactor pressure (between 1 to 3atm) in the tube-side while the reaction temperature was kept constant at 550°C at all cases. Both the CH_4 conversion and the CO_2 yield were measured for the different reactors and are plotted respectively in Figs.4&5, [18]. The shell-side of the reactor at the outside side of the membrane was kept hermetically sealed by means of high temperature flexible carbon seals. The feed composition in both reactors was kept constant at all cases, as also the reactor space time at about 32.5 g_{cat}.h/gmolCH_4, and the reactor temperature kept constant at 550°C. Further, the membrane reactor experiments are simulated by modeling results using the permreactor model as was described above. In addition, the thermodynamic equilibrium values (conversion and yields) have been calculated and are plotted as well in the same Figures (dotted lines). The thermodynamic equilibrium values are indicative of the maximum conversions and yields which can be achieved by the fixed catalytic reactor operation FCR (i.e., without the use of a membrane in the reactor).

Use of the permreactor (membrane reactor) in this process, increases the reactant conversion and the product yield due to the separation of products and especially this of hydrogen out of the reaction zone. This effect can be seen clearly in both plots, for all the reactor pressure values examined experimentally. The yield to syngas and the yield to methanol in the R2 vessel (syngas to methanol mixer) are higher.

The measured increased shift in conversion and yield in the membrane reactor results at a higher quantity of synthesis gas product for the next synthesis reactor R2 at a lower operation temperature than the counterpart fixed bed (impermeable) reactor (Vasileiadis,S., 2000, 2002, 2004), [1,2,3,4,5, 8, 9]. Process operation at a lower temperature is beneficial for increasing the (R1) and (R2) reactor and catalyst life times and for reducing the required heating load (Btu/hr) of the endothermic reformer.

The economics of the permreactor in methane reforming and syngas production and methanol production processes is projected competitive by using membranes which are stable on stream, have high life times and low manufacturing costs of materials used. High temperature pure or metal enriched ceramics as the aluminas used in these studies are good candidates for the syngas to methanol process. Silicas and titanias can be also suitable candidates as membranes in those inorganic permreactors [1,18]. The use of the permreactor (membrane driven reactor) creates a two outlet reaction system which carries synthesis gas or hydrogen rich gas at different compositions into the methanol synthesis reactor (Fig.1). The permeate stream can be used either in the next methanol synthesis reactor, together with the non-permeate stream, or in a separate utilization via a separate outlet as shown in Fig.1. Such additional utilizations include feed in a fuel cell or power system [2,3,8,9], also in hydrogenation reactions [8,9], and in hydrotreating of heavy crudes, asphaltenes, and tar sands [17]. The use of hydrogen rich gas separated via the membrane can be used as fuel in the anode of fuel cells such as in PEMFCs, Phosphoric acid, Alkaline, MCFCs and SOFCs. The conversion of methane and natural gas via the desrcibed syngas to methanol process for use as fuel is of increased and considerable importance, especially for the DMFCs (direct methanol fuel cells) and as quality transportation and synthesis fuel.

4. Conclusions

Modeling and simulation of fixed catalytic reactors for methane steam reforming and water gas shift involves use of parameters which are calculated from experimental data. The fixed reactor modeling involves both optimization and integration algorithms in order to obtain values for kinetic rate constants and obtain modeling values for fractional conversions and yields at the reactor exit. The kinetic parameters of the fixed bed reactor can be used in a membrane driven reactor (permreactor) which operates at similar catalytic conditions, to model its performance for the methane steam reforming and water gas shift reactions for consecutive methanol synthesis. The use of the permreactor results at higher conversions and syngas yields by means of selective product permeation via the membranes. Both reactors can be designed to deliver syngas in an interconnected synthesis reactor for methanol or olefins synthesis. The ceramic or inorganic permreactor (membrane driven reactor) defines a versatile reaction system with two outlets which can be used effectively for multiple reaction uses (syngas to methanol process). Among these uses the direct production of methanol from syngas is of high interest in the fuel and transportation sectors and in DMFCs (direct methanol fuel cells). The economics of the permreactor is projected better than the fixed catalytic reactor, due to the higher syngas yields and consecutive methanol yields obtained at the same reaction and operating conditions. The permreactor may produce and deliver also hydrogen in the permeate stream which can be used as direct fuel for fuel cells or in hydrogenation and hydrotreating reactions.

References

1) Vasileiadis, S., Ziaka, Z., Environmentally benign hydrocarbon processing applications of single and integrated permreactors, *in Reaction Engineering for Pollution Prevention*, Elsevier Science, **2000,** 247-304.

2) Vasileiadis, S., Ziaka, Z.; "Efficient catalytic reactors-processors for fuel cells and synthesis applications", paper#13, presented and published in: *17th International Symposium on Chemical Reaction Engineering,* Aug. 25-28, **2002**.

3) Vasileiadis, S., Ziaka, Z., "Efficient catalytic reactors-processors for fuel cells and synthesis applications" in *Separation & Purification Technology Journal*, 34, 213-225, Febr. **2004.**

4) "Selective reforming catalysis-catalyst testing and reactor design", PETR Division, #89, 229th ACS National Meeting, SanDiego, **2005.**

5) Ziaka, Z., Vasileiadis, S.; "Catalytic reforming-shift processors for hydrogen generation and continuous fuel cell operation", *IASTED-power and energy systems*, May **2002**, pp.360-365.

6) Vasileiadis, S., Ziaka, Z., *Chem. Eng. Comm.*, **1999**, 176, 247-252.

7) Ziaka, Z., Vasileiadis, S., *Chem. Eng. Comm.*, **1996**, 156, 161-200.

8) Ziaka, Z., Vasileiadis, S., Reactor-membrane permeator process for hydrocarbon reforming and the water gas shift reactions, *USPatent* #6,090,312, **2000**.

9) Vasileiadis, S., Ziaka, Z., Permreactor and separator type fuel processors for production of hydrogen and hydrogen, carbon oxides mixtures, *USPatent* #6,919,062, **2005**.

10) Xu, J., Froment, G.F., Methane steam reforming, methanation and water-gas shift: I. Intrinsic Kinetics, *AIChE J.*, **1989**, 35,1, 88-96.

11) Bodrov, I.M., Apelbaum, L.O., and Temkin, M.I., Kinet. Katal., 5, 4, 614, **1964**.

12) Agnelli, M.E., Ponzi, E.N., and Yeramian, A.A.,. Ind. Eng. Chem. Res., 26, 8, 1707, **1987**.

13) Van Hook, J.P., *Catal. Rev. Sci. Eng.,* **1980**, 21,1,1.

14) Rostrup-Nielsen, J.R., *J. Catal.,* **1984**, 85, 31.

15) Rostrup-Nielsen, J.R., *J. Catal.,* **1974**, 33, 184.

16) Ziaka, Z., Vasileiadis, S. "Multifunctional hydrocarbon processing systems for production of fuels and valuable chemicals", *AIChE annual meeting series,* paper*#361i*, Nov. **2000**.

17) Wen, C.S., Chilingarian, G.V., and Yen, T.F., Properties and structure of Bitumen. Bitumen, Asphalts, and Tar Sands, Elsevier Publ. Co., New York, 7, 155 **1978**.

18) Vasileiadis, S., "Simulative comparison and modeling of a catalytic permreactor and a fixed bed flow reactor for syn-gas production", *IASTED-modeling and simulation,* Febr. 24-26 **2003.**

19) Alkhonifer, J., and Ershaghi, I.: "Detection of Channel Sands and Vertical Shale Continuity Using a new Approach in Interference Analysis of Parallel Horizontal Wells," SPE, #53928 presented at 1999 SPE Latin American and Caribbean Petroleum Engineering Conference, **1999.**

20) Du, C., and Yortsos, Y.C., "A numerical study of the critical gas saturation in a porous medium", Transport in Porous Media, **1999**, 35, 205-225.

181e One-Step Dimethyl Ether Synthesis from Carbon Monoxide-Rich Syngas in a Slurry Reactor

Yongwon Seo, Sung-Ho Jo, Ho-Jung Ryu, Chang-Keun Yi, and Gyoung-Tae Jin. Clean Energy Research Center, Korea Institute of Energy Research, 71-2 Jang-dong, Yuseong-gu, Daejeon, South Korea

Recently, DME (dimethyl ether) has received growing attention due to its potential as an alternative clean fuel and a chemical feedstock. One-step DME synthesis from syngas has many economic and technical advantages over conventional two-step synthesis. In the present study, one-step DME synthesis from CO-rich syngas over the bifunctional catalyst comprising a methanol synthesis catalyst (Cu/ZnO/Al2O3) and a solid acid catalyst (γ-Al2O3) was investigated. Considering the highly exothermic characteristics of DME synthesis, liquid-phase DME synthesis was carried out in a slurry reactor that provides good mixing and efficient heat control. The influence of various reaction parameters such as reaction temperature (230-280 oC), reaction pressure (3.0-7.0 MPa), GHSV (800-6000 ml/g-cat/hr), and content of γ-Al2O3 on CO conversion and DME space time yield (STY) was closely examined at a H2:CO ratio of 0.5 that is very close to coal-derived syngas. The effect of H2:CO ratios (0.5-3.0) on DME synthesis was also investigated. Reaction temperature and pressure showed similar positive effect on DME synthesis. Lower GHSV caused higher CO conversion, but lower DME STY. The excessive content of γ-Al2O3 had an adverse effect on both CO conversion and DME STY. Among various H2:CO ratios, the maximum DME space time yield (13.5 mol/kg-cat/hr) was observed at a H2:CO ratio of 1.0. The results obtained in this study can be used as basic data for design and operation of a fluidized bed DME synthesis reactor.

181f Rcc Algorithm for Automated Attainable Regions Analysis: Methanol Synthesis

Tumisang Seodigeng, B. Hausberger, D. Hildebrandt, and D. Glasser. School of Chemical and Metallurgical Engineering, University of the Witwatersrand, Private Bag 3, Wits, Johannesburg, 2050, South Africa

For reactor network synthesis, attainable regions analysis provides a set of all possible product states that can be realized from given chemical reaction scheme with underlying kinetics and specified feed, permitted fundamental processes and system constraints [1]. This set of all possible system output states that can be achieved by application of all permitted fundamental processes and or combinations thereof is termed the attainable region (AR). The boundary of the attainable region sets the limit to all achievable states and it is therefore of special interest as it is in most cases, where a given objective can be optimised. Even though the general theory of AR analysis has been developed over the years [2,3], there is no known sufficiency conditions that characterise the boundary of the AR. Necessary conditions have been derived, which a nominated boundary has to satisfy in order to be considered a candidate AR. It is this absence of the sufficiency conditions that make identification of AR boundaries an intricate procedure guaranteeing no certainty of the complete or true result.

Feinberg demonstrated by use of mathematical proofs arising from the theory of geometric optimal control that the AR boundary, for systems considering fundamental processes of isothermal reaction and bulk mixing, is outlined by special types of reactors [4,5]. These special reactors that are characterized by stringent mathematical conditions give access to extreme points from which reaction manifolds (representing plug flow reactors) and mixing planes that form the final shape of the AR boundary originate. For reaction and mixing, these special types of reactors are always reactor structures in which simultaneous reaction and mixing occur optimally under governance of some control policy [4,5]. Feinberg derived the mathematical equations that the control policies have to obey in order for a special type of reactor structure to occur on the AR boundary. From these proofs it can be asserted that a candidate AR can be identified primarily by determining the structures that outline the boundary giving access to extreme points from which fundamental process manifolds emanate. This work proposes an automated technique for identifying candidate attainable regions using recursive constant control policy algorithm (RCC). This approach uses iterative application of constant control policies to approximate the optimal state varying optimal control policy that govern the occurrence of a special reactor type on the boundary.

For a system considering fundamental processes of reaction and mixing, the special types of reactors that delineate the outline structure of the AR boundary are the critical continuous flow stirred tank reactor (CFSTR) and the critical differential side stream reactor [4,5]. Mathematically, it can be shown that the CFSTR is a special type of a DSR where the vector representing the combined rate of reaction and bulk mixing diminishes to zero. In this work we propose that the critical CFSTR therefore occur on the AR boundary at discrete points where the vector field of the critical DSR goes through a node. By identifying critical DSR's, the critical CFSTR's will be automatically identified at the DSR nodes. The support of this proposition has been demonstrated with results from case studies. Once the critical DSR's and CFSTR's have been identified, the boundary of the attainable regions can be completed by application of plug flow reactor (PFR) and mixing planes that emanate from critical structures. We have demonstrated the practicality and applicability of the automated RCC technique to identify candidate ARs. The RCC algorithm has been used to solve various case studies ranging from the simple van de Vusse mass action kinetic reaction scheme in three dimensional and computer intensive four dimensional system to several industrial case studies that extend the applicability of the technique to non-isothermal reaction schemes considering fundamental processes of reaction, bulk mixing, and heat transfer such as the generic first order exothermic reversible reaction, ammonia synthesis, methanol synthesis and oxidative dehydrogenation of 1-butene.

Attainable regions analysis is used to identify optimal reactor networks configurations for the examples detailed below that have previously been thought to be impossible. This procedure is carried out in an easy automated manner using the RCC technique. The method is implemented into an easy to use software routine using MATLAB®. Due to the robustness of the underlying RCC algorithm, the software package can be compiled into a custom package that allows the end user to enter reaction scheme, select system variables and permitted fundamental processes as well as control policies. Due to the built-in data tracking capabilities of the RCC method, interpretation of the resulting attainable regions boundary into optimal reactor network structure can also be done automatically.

This paper illustrates how the RCC method can be used to identify optimal reactor structures for methanol synthesis via attainable regions analysis. In the methanol synthesis system, the water gas shift reaction can be carried out separately or can be done along with methanol formation reaction using a WGS active catalyst. By using the RCC algorithm we are able to show the best configuration relative to an economic evaluation. Different reactor structures are considered to incorporate fundamental processes of reaction, bulk mixing, heating and cooling. The results obtained are compared for consistency with the reactor sequencing in current methanol synthesis industrial practice.

Understanding Ambient LNG Vaporizers

James Davis
SNC-Lavalin GDS, Inc.
Houston, TX

AIChE Spring Meeting, April 2006
6th Topical Conference on Natural Gas Utilization
Orlando, Florida, April 23-27, 2006

Abstract

This paper presents an overview of a recent Study conducted to assess the best vaporization technology for use at a specific southern U.S. site. The Client directives led to the examination of ambient vaporization technologies as a primary process and with fired-heaters as the back-up heat source. The Study itself is comprised of a methodical analysis of the underlying thermodynamics, the overall thermal efficiency, a comprehensive assessment of the total system reliability and maintainability and finally economic viability. This paper is a digest of that Study and offers some general observations of the applicability of ambient LNG vaporizers.

Increasing energy costs combined with the regulatory and socio-political environment directed the analysis away from traditional technologies of submerged combustion vaporizers (SCV) and open-rack vaporizers (ORV). Licensors and manufacturers are marketing ambient air LNG vaporization systems as standalone systems or more properly hybrid systems of ambient units coupled with auxiliary fired heaters and/or with waste heat recovery systems. There appears to be considerable promotion of these technologies with frequent comparisons to the more traditional processes. However, without an experience base from which to draw, it is difficult to assess whether to employ ambient units and which system is best suited for a specific application.

Introduction

The Study upon which this summary-level paper is based was initiated by a request for an un-biased assessment of available ambient LNG vaporization systems for use in a southern U.S. location with a further provision that the ambient units be used as the primary heat source vs. acting as backup support to SCV units. The directives included the following requirements:

1. Minimum Life Cycle NPV of Capital and Operating Cost over 20 years;
2. Minimal Energy Cost – concern about further energy price increases
3. Vaporization rate of 1.0 bcf/day with send out at 40 °F @ 1400 psig.
4. No water to be used for vaporization

The use of ambient vaporizers in many industries such as medical, electronics, research for oxygen, hydrogen, helium, and nitrogen has become common place. The use in LNG facilities has become the subject of great study and interest – yet none are installed as the primary source of vaporizing heat.

This report is a summary of our general findings for given ambient conditions – Table 1. A rigorous analysis is currently underway to confirm preliminary conclusions.

Background

This recent interest in ambient systems has arisen to a great extent because of the U.S. Gulf Coast "fish egg" controversy of 2004/ 2005. Early modeling for the EIS calculated a large percentage of the eggs, larvae and plankton crucial for red fish and drum would be killed with the startup of a single LNG terminal utilizing open-rack vaporizers. This controversy essentially halted any consideration of ORV usage along the US Gulf Coast and elsewhere. The Magnuson-Stevens Fisheries Conservation and Management Act is intended to manage open fishing seasons to help rebuild the inventory of key species. For 25 years sampling and records have been kept about certain species but the data collected does not address all species count by location, by depth or by season. The ability to identify species at the egg or larvae stages is difficult even for trained personnel.

The context of the impact of an LNG terminal moving large volumes of water is one year's operation would impact the same as one week of normal "by-catch" from routine fishing operations by the US Gulf Coast fleet (over 600 mm lbs of 'by-catch" were tossed back per year – 100% "kill" assumed). Dr. Paul Boehm, Vice President of Environmental Services of Exponent reportedly said that their analysis has shown the EIS calculations overstate the impact by a significant margin[1].

This active discussion has the Commerce Department looking into the matter because the ability to use water is the lowest cost source of vaporization heat. Dr. Wm. Hogarth, Asst. Administrator for NOAA and National Marine Fisheries has called for more data and study to resolve the lack of data and the need for accurate modeling[2]. This will take time, money and eventual concurrence by scientists, fishermen and regulators.

In the interim, selection of vaporizer heat source is moving away from the permitting challenges associated with water usage as the heat source for open rack vaporizers and toward other proven processes such as submerged combustion vaporizers (SCV). This technology has two significant issues that affect applicability to a specific project:

1) they require 1.2 – 1.5 % of the feedstock as fuel
2) they create an emission source to be permitted.

[1] www.lngfacts.org/newsroom/Eval-OLV.pdf
[2] LNG Express, Vol XVI, No. 3, pg 12-14; Zeus Development Corporation

In addition, they have a waste water stream that must be pH treated prior to discharge. There are ample studies and papers related to the advantages and disadvantages of water-based open rack vaporizers (ORV) and SCV Units – not discussed herein.

Ambient Air Vaporization Study

Other papers[3] discuss the relationship of ambient units to those more traditional processes and even the integration of vaporization with other heat sources such as cooling towers and power plants.

This paper focuses upon a specific request to find the "best" ambient vaporization process that would offer the lowest life-cycle cost to the project and which would meet the objectives listed in Part A. The objective is to avoid existing paradigms and to sift through marketing and promotional information to arrive at a factual comparison of the currently available units being offered in two basic configurations. These are:

1. Process A - direct exchange of LNG with ambient air and in-line trim heater and 100% backup SCV unit
2. Process B - indirect exchange of LNG with air utilizing an intermediate fluid with a 100% backup heater.

Process schematics of both processes are included as Figure 1.

Previous Studies Indicate Ambient Vaporizers are Viable

Other Studies and data provided by vendors indicate ambient units are viable in the scenario in which water is not available or usable and in which the cost of gas is above $5.75. Based upon a multi-client technical and economic vaporization system study performed by the Author[4] in 2004 / 2005 for all vaporization technologies, it was concluded that ambient air vaporizers are economic when considered over a 20 year life cycle given that ORV units are not considered. In that Study, multiple factors were considered including the following:

1. Capital cost – cost for an entire system capable of 1.3 bcf/day with fans, blowers, pumps, heaters, surge vessels, valves, piping, etc.
2. Energy Cost – total energy required for pumps, fans, blowers, heaters (all fuel calculations based upon market value for gas – considering fuel as lost revenue).
3. Maintenance Costs – cleaning; routine maintenance; periodic replacement of parts.
4. Associated Costs – assign a cost of "space" – how much area does each system require to be fully functional.

[3] References
[4] "LNG Vaporization Technology Survey for Onshore & Offshore Locations"; J.F. Davis & D. McInnis, May 2005 (Confidential Report)

Also considered were "soft costs" such as the reliability / availability of the entire system (a quantitative reliability assessment has since been performed) and ease of maintenance (special training, special equipment, shop only repair, etc). A graphical representation is provided as Figure 2.

The NPV calculations were performed using a discount rate of 12% over 20 years. The graphical results in Figure 3 on a "life cycle" basis indicate ambient units perform well. Also, it remains true that water-based vaporization units offer the best balance of economy and low operating cost. The Study also shows that the indirect contact ambient units have an operations cost of 1.2 times that of the single exchange direct contact units. This differential was confirmed within this Study.

Ambient Vaporizer Technology - Basic Thermodynamics

Results of previous work prompted us to exclude the SCV units (due to operations cost) as the preferred vaporizer unit and to focus upon the "best" ambient solution for the environmental conditions at the site. Within the context, budget and schedule of this Study, we did not have the luxury of engaging vendors of ambient systems to provide a Process Design Package. The assessment had to be qualitative in nature and based upon what we could glean from the vendor provided information and from our own design experience and thermodynamic assessment of the systems.

We started with finding the answer to "how do you get the most heat transfer from the air into the LNG?". A cursory review indicated several intuitive observations such as:

a. Multiple Exchanger Approach: From Figure 1 it seems obvious direct contact (Process A) with a single exchanger approach would be able to get more heat into the LNG from a given ambient condition than a system with two approaches (Process B).
b. Latent Heat: Both systems will capture the latent heat of vaporization as the humidity cools. If the system can frost over, the latent heat of fusion (LHF_W = 144 Btu/lb) of the humidity can be used to gain more heat from a given amount of air.
c. Lowest Ambient Operating Temperature: Process A can add heat to the LNG – being heated from -260 ^oF – at an ambient temperature down to 40 ^oF. Process B commences heater operation at about 65 ^oF and shuts down ambient units as outlet air temp approaches 40 ^oF and potential freezing of the fins. If the ambient condition is at 50 ^oF, Process B is full fired heater while Process A has about 25% of the heat being added by fired heater.
d. Space Allocated – The space required for Process A to vaporize

General Operating Procedure – Process A

The individual cell (Figure 4) is very simple and is grouped into a "bank" of 4 cells and a "train" is multiple banks. The units may be "passive" ambient utilizing natural draft for warming; fan-assisted units have shrouded sides providing more capacity for a given cell (more compact layout) at the expense of a forcing fan (about 4300 hp for a 1 bcfd system).

The design parameters set by the vendor indicates that for Process A, the cells are allowed to frost over forcing a "defrost cycle" of 1.75 hours during each 5.5 hour cycle – 3.5 hours "on". Therefore the vaporizer configuration comprises 33% extra exchanger units to accommodate the defrost portion of the cycle. In a typical example, for a 30 °F approach, there would be 14 banks (64 cells) on with 7 banks (28 cells) defrosting. A switching operation would occur every 15 minutes bringing fresh vaporizers on-line. As each vaporizer comes online, it ability to warm begins to decline as frost builds on the outer surface. The decline in heat transfer due to frosting is a third order polynomial but is shown as a straight line for this presentation. This mode of operation does create a "wavy" gas outlet temperature with a period and amplitude that will decrease as more banks are put into the switching order.

Figure 5 shows the average gas outlet temperature for 8 banks and ambient temperature of 70 °F daytime and 60 °F night. Figure 6 shows the outlet temperature of the ambient units for a cold day of 45 °F high and 40 °F low. In this instance, the trim heater would make up the difference. These examples have 8 banks but the actual design basis has 21. Some high-level observations include the following:

Advantages:

(1) Simplicity – the unit is very simple with few moving parts. The ambient have only valves as moving parts.
(2) Proven – the cell technology has been in oxygen, argon, nitrogen, helium and hydrogen service for many years for mission critical applications in medicine, semiconductors and space operations.
(3) Flexibility – cells and banks can be connected to manifolds and operated by computer to optimize the operations. Various combinations of sequencing and adjusting switch times will enable the facility to match the demand for gas into the pipeline.
(4) "Passive" Capability – the cells create a natural draft. Ambient units without fans are effective with the exception more cells (1.5 X) are needed.
(5) Space - the full ambient system will require about 25,000 square feet with units located at ground level with a 3 m dis-engagement zone

Disadvantages:

(1) Switching Valve - the switching operation is typical of PSA units. However, the constant operation of cryogenic ball valves is a maintenance and safety concern.
(2) Space and Orientation to the Wind – as with all ambient units they do take space vs. SCV and must have proper orientation to the prevailing winds.
(3) Cryogenic Application – the units are not in LNG service and each cell requires cryogenic materials for cyclic pressures and temperatures.

General Operating Procedure – Process B

Process B utilizes proven air fin-fan units with belt-driven fans (about 3800 hp for a 1 bcfd unit) drawing air across finned tubes through which an intermediate heat transfer fluid (propane) at moderate pressure is pumped. That fluid then passes through a fired heater (rated at 100% of vaporizer capacity) to the cryogenic exchanger with the LNG. Only this last exchanger must be rated at gas send-out pressures and for cryogenic duty. The fired heater provides the entire heating service when ambient temperatures fall below 40 °F. In some configuration, the fired heater is not in-line but rather on a separate heating medium loop with an exchanger just after the fin-fans. The nature of the fin fan units forces controls to shut down the fin-fan exchangers to preclude formation of ice on the fins.

Advantages:

(1) Proven – the fin-fan components have been in service for many years in LNG liquefaction plants.

(2) Cryogenic Exchanger – only the cryogenic exchanger experiences cyclic temperatures and pressures.

Disadvantages:

(1) Re-Circulation – fans must operate for ambient benefit. The forced fans can produce a looping action with exhausted cool air being drawn back into the inlets. Extensive modeling[5] in the LNG liquefaction plant design has shown careful attention must be given top prevailing winds and adjacent operations.

(2) Dis-engagement Zone – the fans must be located at a level of 10 - 12 m above grade to allow sufficient dis-engagement. A problem in seismic areas (added structural cost)

(3) Fired Heater – the dual approach requirement with two exchangers causes the heater to be engaged at a higher ambient temperature (65 °F) than with the Process A. This means more energy is consumed for a given amount of LNG vaporized (see comments below).

(4) Propane – propane as heat transfer medium causes safety concerns with portions of Client management.

(5) Responsiveness to Demand Changes - thermal inertia

Observations

The absence of good comparative information led to considerable effort in modeling the systems. Within the allotted time frame assumptions had to be made. Process A information provided a breakout of equipment, energy, power, maintenance costs as well as on-stream factors. From this the impact of fuel and energy cost and efficiency can be evaluated and an NPV calculated. The results are offered as normalized data utilizing the SCV as the reference point in Figures 7 and 8.

[5] "predicting Environmental Impacts on Multi-Train LNG Facility using Computational Fluid Dynamics, AIChE fifth Annual topical Conference on Natural Gas , April 2005; Yee, Lin, Mehrota and Diwaker

To the extent possible analysis is made with information available but in some instances only a comparison can be made vs. using absolute numbers.

General Conclusion

Some general conclusions for ambient units include the following:

1) A key factor is overall energy consumption – conclusion represented in Figure 7. Process B requires more energy than does fan-assisted Process A over the course of a year (calculated @ $2.5 mm annually)
2) The overall capital cost, operating cost (maintenance, energy, etc.) and the NPV as calculated for 25 yr @ 12 % is present in Figure 8.
3) Sufficient space is needed for free flow of ambient air around the units and an orientation perpendicular to the prevailing winds is preferred. For a 1 BCFD send out, Process A and Process B require 25,000 sq. ft and 44,000 sq ft respectively.
4) The ambient units perform better when average humidity is high
5) The weight of the units is spread over large area making foundations simple
6) Considerable water will be produced than must be handled properly – process water or natural moisture?
7) The Process A cells benefit more from wind than does Process B
8) The simplicity and flexibility of Process A is appealing

"Lessons Learned" in Evaluating Ambient Vaporizers

The implementation of ambient units is a function of time - they will be utilized. Several LNG terminal sites have selected either one of the processes for secondary vaporization. In some expansions there may be a push to make ambient units primary since the site has considerable SCV /ORV backup.

From our experience, the following suggestions are made if someone is contemplating ambient units. That would include the following:

1) Prepare a Detailed Basis of Design including extensive weather data for the specific site upon which vendors can prepare a detailed technical and commercial offer.
2) Perform a rigorous heat and energy balance of the system across the entire range of anticipated weather.
3) Require detailed breakdown of the installed capital cost, equipment lists and the operating cost with itemized values for fuel gas for each heater, power for pumps and for fans, maintenance cost including cleaning and spares costs.
4) Require detailed breakout of energy costs by season.
5) Require guarantees for availability, energy usage, scheduled downtime. Perform a Quantitative Risk Assessment on the system.
6) Solicit early "buy-in" from insurers, lenders and regulators.

7) Rely upon the warranty, guarantees and financial backing / stability of the vendors. It is these commitments that bankers and lenders and insurers will be assessing as they contemplate funding a terminal project.

Migration to Ambient Units?

So if the ambient units are superior to submerged combustion units, why are they not being adopted? The answer is that they are being adopted – slowly – and in most instances the application is secondary or backup to SCV units. This approach facilitates the permitting of the facilities and yet enables the testing of the units for primary operation – with a likely revision to the permit (the reduction in air emissions should be well received). The LNG industry does not make any changes quickly – Owners, operators, insurers, regulators are all reluctant to step-out with any new technology. The impact upon the $4 -5 billion USD supply chain would be immense if a new and untried device or equipment were to fail and shut down the terminal.

Some of the events that must occur before the ambient units are routinely adopted for primary vaporization duty include the following:

1. Operations - tested and trialed for an extended period to prove durability under repetitive cycles of temperature and pressure. This will take a step-out by an existing Operator. Several expansion sites are considering this technology.
2. Insurers and Regulators: must be convinced safety and risk are not compromised. Someone will need to be first.
3. Sharing of experience – will there be open sharing of the operating data between terminal locations such that all may benefit?
4. Multiple Installations and Experience Base: multiple system installed at various locations will provide the longevity and operations experience to satisfy regulators, insurers and Owners.

ACKNOWLEDGEMENTS

Daniel McInnis – Process Engineer, Paragon Engineering
Jay Kiper – Senior Process Engineer, SNC-Lavalin GDS
Lynn Tressel – Senior Process Engineer, SNC-Lavalin GDS

REFERENCES

1. J.F. Davis; "Salt caverns and the Bishop Process™", OMAE Conference, June 2004, Vancouver; paper 2.5 of LNG Forum
2. J.F. Davis; "Enabling Technologies for Mooring, Unloading, Regasification and Storage for Offshore LNG Receiving terminals", Society of Naval Architects and Marine Engineers, Annual Meeting; Nov. 2004

3. Himmelberger et al.; "Optimizing Energy Consumption for LNG Vaporization" Hydrocarbon Processing, Jan. 2006, pg. 57-64
4. Billman, Himmelberger; "LNG Vaporization Technology: The Environment vs. Energy Sourcing"; AIChE 2005 Spring Meeting – LNG Session, April 2005, "Limitations in LNG Vaporization Process Selection";
5. Cho, Mathur, et al; AIChE Fifth Topical Conference on Natural Gas Utilization, April 2005;

TABLE 1
AMBIENT CONDITIONS

FA VA
CT LU
ORE

	Avg. Summer Daytime High	91 degree F
	Avg. Summer Evening Low	73 degree F
	Avg. Summer Humidity	67 % R.H.
	Avg. Summer Wind Speed & Direction	S.E @ 6 km/hr
	Avg. Winter Daytime High	67 degree F
	Avg. Winter Evening Low	42 degree F
	Avg. Winter Humidity	39 % R.H.
	Avg. Winter Wind Speed & Direction	N.W @ 12 km/hr

FIGURE 1
Process Schematics

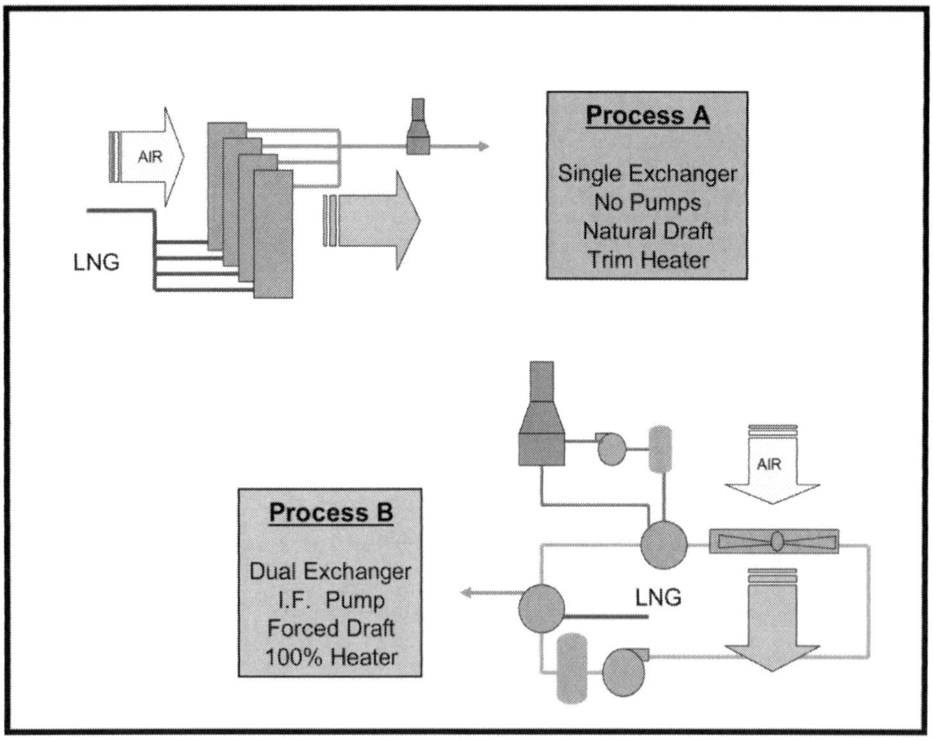

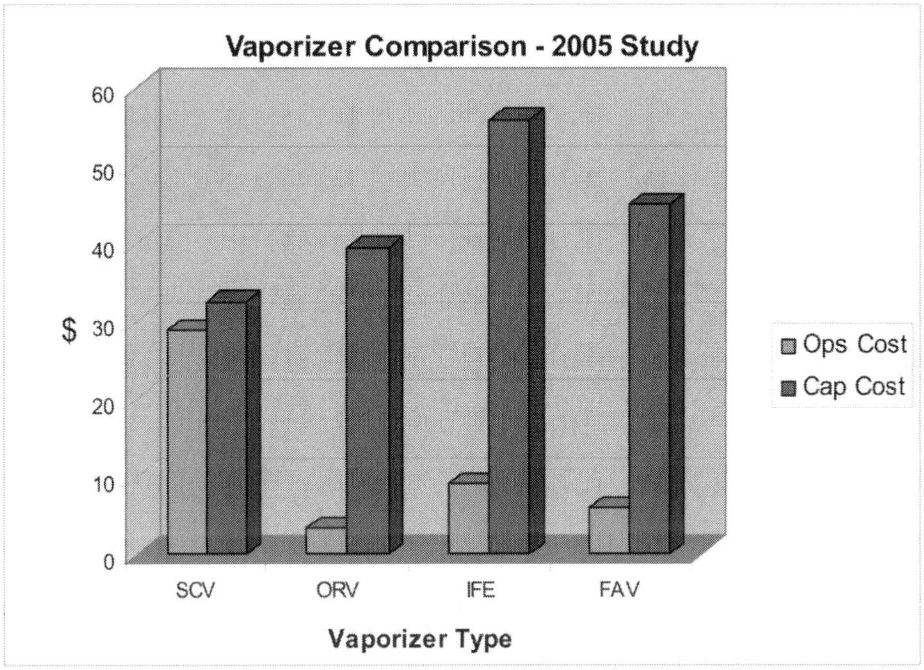

Figure 2
Capital & Operations Costs

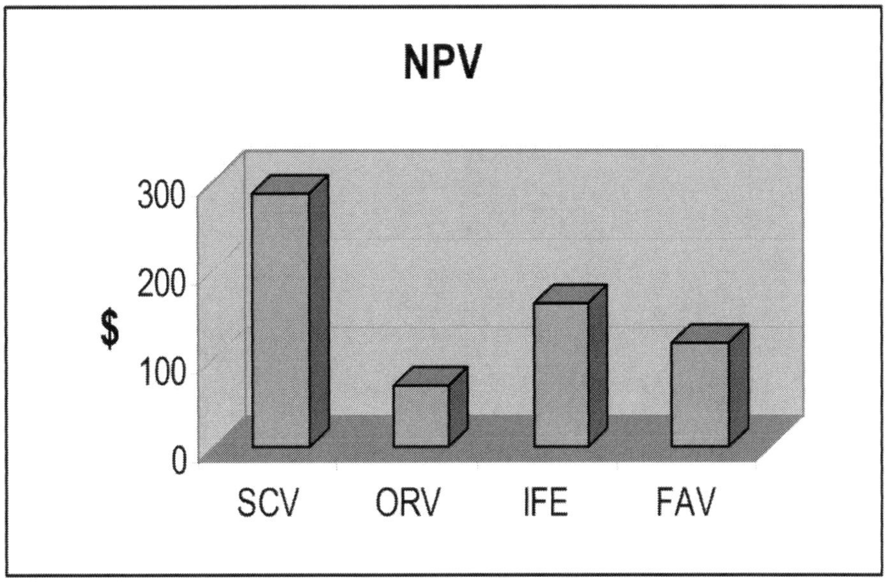

Figure 3
NPV for 20 yr @ 12 %

Figure 4

Ambient Air Vaporizer – No Fan Assist

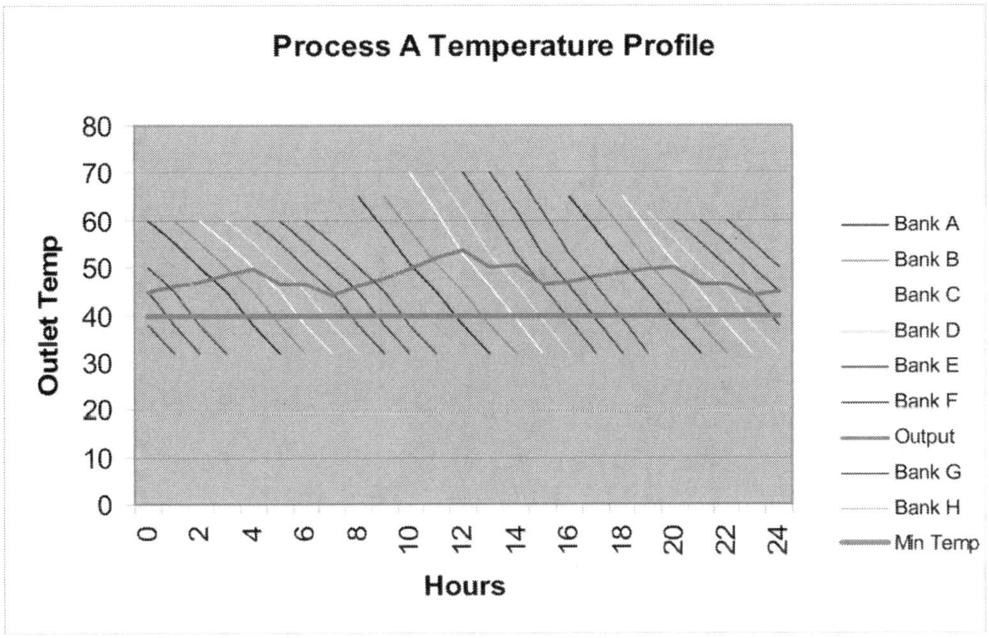

Figure 5
Average Gas Outlet Temperature – 70 °F Ambient

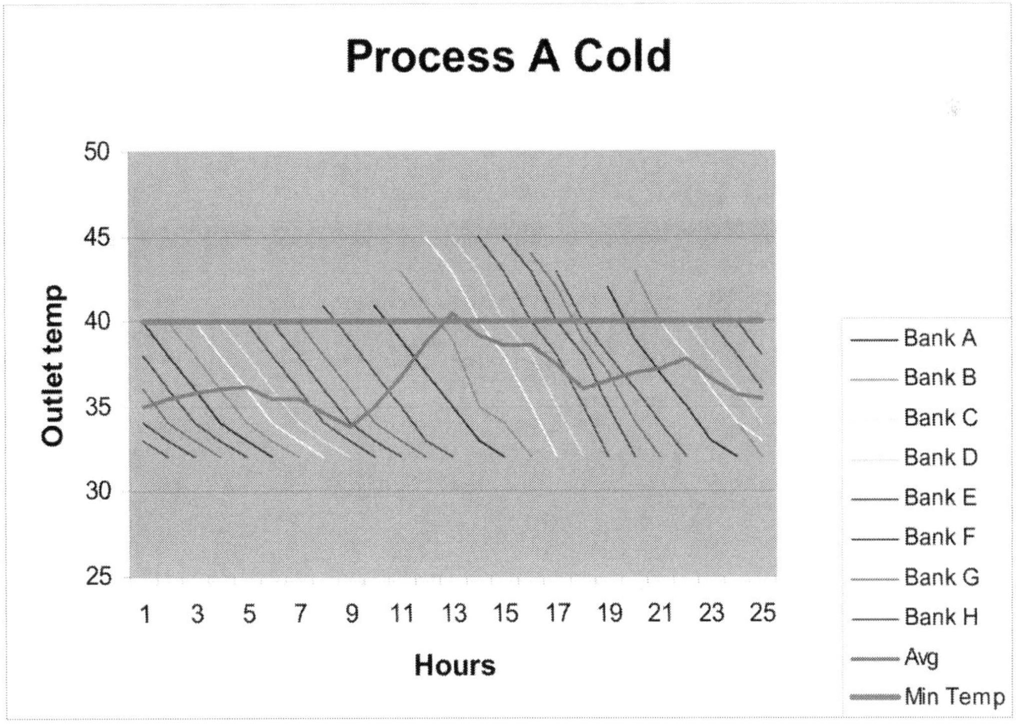

Figure 6
Average Gas Outlet Temperature – 45 °F Ambient

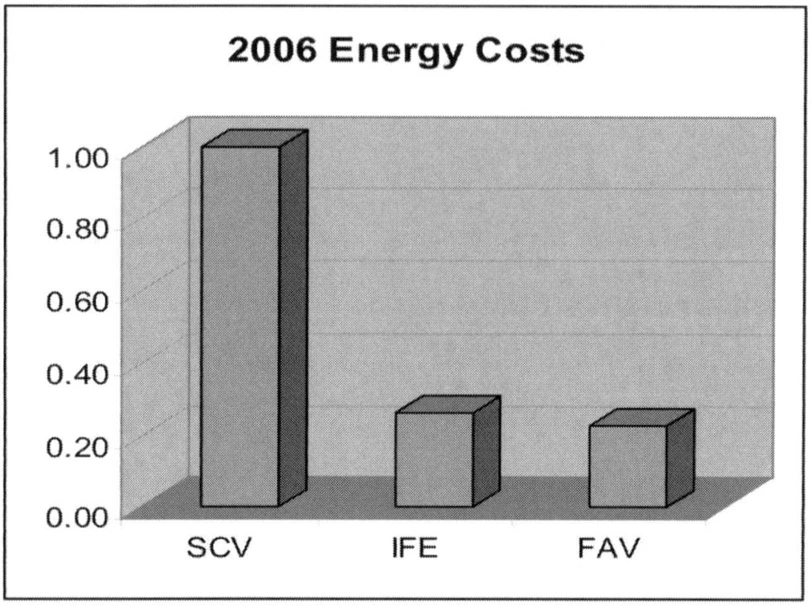

Figure 7
Normalized Energy Costs

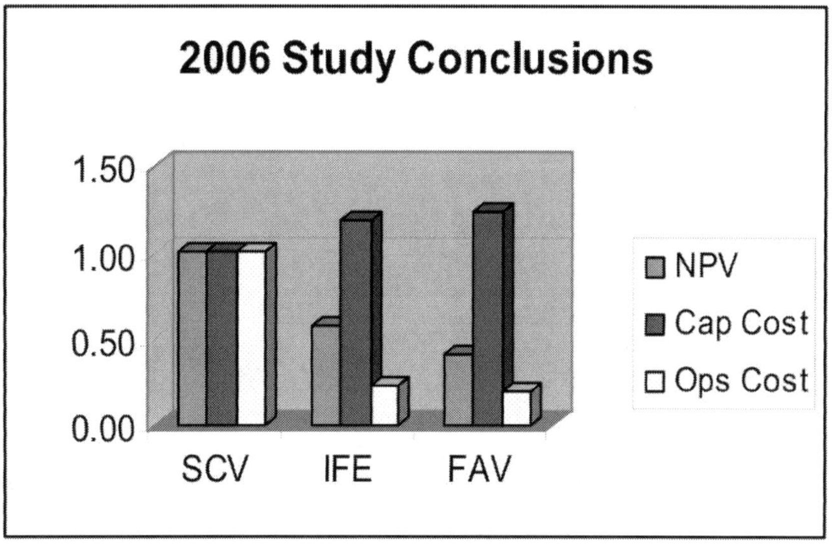

Figure 8
Normalized Costs
(NPV over 25 yr @ 12%)

Technical Challenges & Design Features of the Largest LNG Tank

Eui-Seung Park
Managing Director, Plant Division
Hong-Sung Kim
General Manger, Plant Division
Seung-Beom Hong
Manager, Plant Division
Se-Jin Jeon
Senior Researcher, Institute of Construction Technology
Daewoo E&C Co., Ltd., C.P.O. Box 8269, Seoul, Korea

Joseph. H. Cho
Technical Advisor, Ph.D., PE
Kellogg Brown and Root
601 Jefferson Avenue, Houston, TX 77002

Seung-Rim Lee
Senior Researcher, Institute of Gas Safety R&D
Korea Gas Safety Corporation, 332-1
Deaeya-Dong, Sihung-Si, Gyunggi-Do, 429-712, Korea

AIChE Spring National Meeting, April 2006
6th Topical Conference on Natural Gas Utilization
Orlando, Florida, April 23-27, 2006

Introduction

Since the first LNG cargo was imported to Korea at the end of 1986, the downstream LNG industry of Korea has seen remarkable development in each sector from receiving terminals to the nationwide trunkline network. As a result, LNG is positioned to become a major primary energy source in the domestic market.

The importance of LNG storage facilities in Korea has recently been highlighted by the country's considerable increase in consumption and the security of a stable supply against seasonal peak loads in winter, which have resulted from sizeable investments to address the issue in a timely manner.

This paper discusses past major activities in the various fields of the LNG tank industry (such as engineering, construction, technology development, safety study, etc.) starting from the construction of the first receiving terminal at Pyeongtaek in 1986 to the design of the world's largest above-ground LNG tank which is scheduled to begin construction in early 2006. The design of full containment LNG tank of this 200,000m^3 has been developed with an emphasis on the cost saving and land use efficiency under the Korean geopolitical specifics. This paper also discusses the technical challenges and design features of the largest full containment.

Increase in LNG Tank Capacity

Currently, Korea is the world's 2nd largest LNG importer after Japan. The Korean LNG industry has been developing correspondingly with an increase in LNG consumption by approximately 16% per annum since the first importation at the end of 1986 to the Pyeongtaek terminal from Indonesia.

In terms of market development, 22.2 Million Tones per Annum (MTPA) of LNG were imported and supplied in 2004 through the nation-wide natural gas trunk line network (about 2,500km in length), which was constructed and operated by Korea Gas Corporation (KOGAS). Particularly, the annual gas demand growth rate is 9.7% from 1997 to 2004, which was more than 4 times greater than of global growth rate of 2.3% in natural gas consumption. Besides, its supply rate of city gas to households was 68.8%. The demand from the city gas sector, which shows a big seasonal fluctuation, takes a considerably large portion of total demand (see Table 1). However, the demand for LNG from the power sector shows a comparatively stable demand pattern around 35%.

Korea has executed a long-term supply and demand program increasing the storage rate from 9.6% in 2005 to 12.7% by 2015 in order to make preparations against the annually constant increase of demand and seasonal peak loads. The Korean government expects that large storage volumes contribute the stability of supply and demand. This can also meet the governmental energy-reserving policy in natural gas.

Table-1 shows the increase in LNG demand by around 4% makes the expansion of storage and supply facilities inevitable, which requires sizeable investment and the enlargement of LNG tanks considering the economy of scale.

Table 1 – Perspective of LNG demand and expansion plan for storage capacity

		1997	2000	2003	2005*	2007*	2011*	2014*	2017*	Annual Growth Rate (%, 03~)
LNG Demand (10^3 Ton)	City Gas	5,770	9,528	11,979	13,658	15,386	18,852	21,700	24,893	5.36
	Power	5,377	4,689	6,468	7,612	8,572	6,992	6,014	6,764	0.32
	Total	11,147	14,217	18,447	21,270	23,958	25,844	27,714	31,637	3.93
Storage Capacity (10^6m^3)	Newly	-	7.0	14.4	10.2	7.0	20.8	8.4	5.6	
	Accumulated	1.3	2.0	3.4	4.5	5.2	7.2	8.1	8.6	
Storage Rate	(%)	5.2	6.3	8.4	9.6	10.2	13.2	13.7	12.7	

Note) * is planning data from 'The 7th Long-term Supply and Demand Program, by MOCIE (Ministry of Commerce, Industry and Energy), December 2004'

Development of LNG Tank

1st Generation: Above-ground membrane LNG tank with dike

The first LNG tank introduced in Korea was the membrane type at the Pyeongtaek terminal. In 1983, the membrane technology developed by Technigaz, France, was adopted for the first three units (100,000m^3 x 3) and the same technology was applied to the terminal expansion, resulting in construction of total 10 storage tanks. These tanks can be called the 1st Generation LNG tank' for convenience.

2nd Generation: Above-ground 9% Ni full containment LNG tank with dike

After construction of Pyeongtaek terminal, the 2nd LNG terminal was built from 1990 to 1996 at Incheon in order to meet abrupt increase in LNG demand and supply natural gas to Seoul Metropolitan area. This comprised 9%-Ni full containment LNG tanks (100,000m^3 x 3) so-called the 2nd Generation LNG tank. This type of tank was state-of-the-art at the time of construction since only about 15 units[1)] had been built in a few countries in the world. Traditionally, the single containment type of LNG tank with 9% Ni inner tank was preferred from an

1) In the basis of countries operating import terminals (Source : A study on the containment type of LNG tank, KOGAS, Aug. 1998)

economic point of view and dominated the market with over 100 units in the world. However, the geopolitical specifics of the armistice as a divided country after the Korean War influenced KOGAS to select the full containment type of LNG tank even though it required high capital investment because of the safety concerns of the tank. Ten (10) units of above-ground LNG tanks (#1~10) had been constructed at Incheon terminal by 2000.

3rd Generation: In-ground membrane LNG tank

Then came the so-called 3rd Generation LNG tank, in-ground membrane type. These tanks were constructed for the first time at Incheon terminal from 1997 to 2004. A total of 8 units of in-ground tanks (#11~18) were constructed through 4 individual projects using the Japanese membrane technology. Two tanks (#19~20) that have adopted the KOGAS membrane technology are under construction. These have resulted from a wide range of research and development of membrane and its components such as insulation material. Besides, their individual storage capacity of 200,000m^3 deserves attention as the largest volume in the world for 8 tanks (#13~20) out of 10 units.

4th Generation: Above-ground 9% Ni full containment LNG tank with no-dike

From 1999 to Oct. 2002, the 3rd LNG terminal in Korea was built at Tongyeong and the so-called 4th Generation LNG tank has been installed, no-dike 9%-Ni full containment LNG tank (140,000m^3 x 3, #1~3) with the largest diameter of the LNG tank (86m in O.D.) in the world. This terminal has made a significant contribution to the supply of natural gas in the southern area of Korea. After 2 more tanks (#4~5) were installed, a total of 7 tanks (#6~12) with the same capacity have been constructed based on the KOGAS design since Jan. 2002 (#6~10 tanks in operation). The super-size above-ground LNG tank with a 200,000m^3 capacity will be constructed at this terminal in early 2006, which will be the first and biggest full containment LNG tankl in the world.

During two decades of LNG history, a very dynamic market for the LNG tanks has been developed with a remarkable increase in LNG demands. The market contributes development of the large storage tanks, adopted state-of-the-art technology. It can be said that this results from devotional efforts by the Korean government led by KOGAS and some companies involving in LNG tank design and construction.

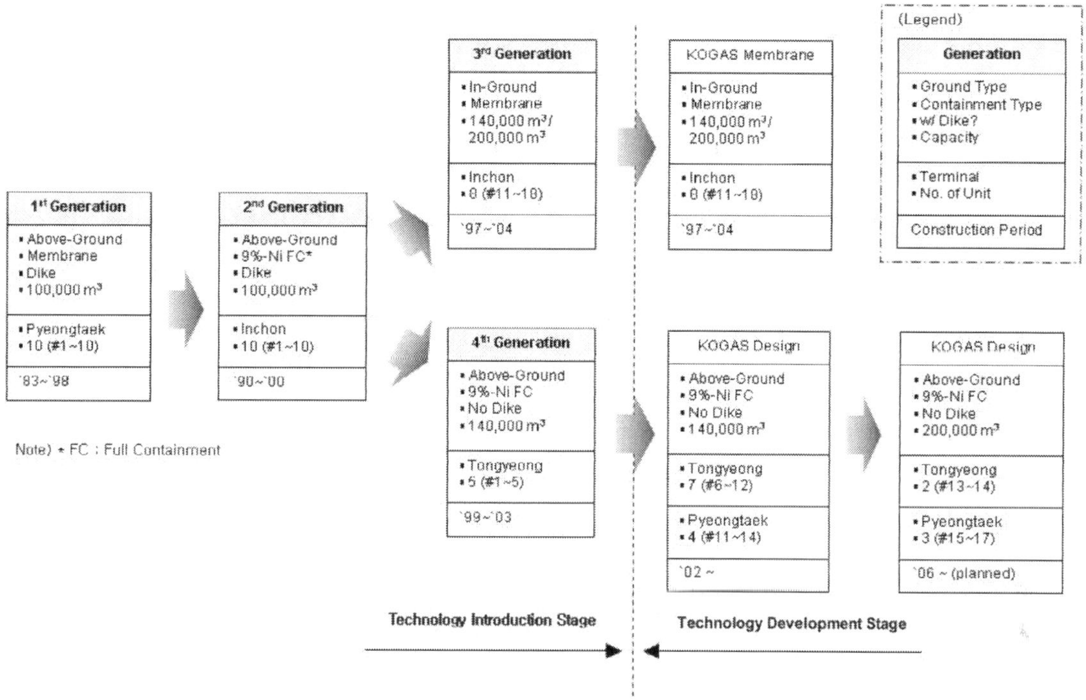

Fig. 1 - Development of LNG Tanks in Korea

Security of Safety for LNG Tank

Even though the Korean LNG industry has only about running for 20 years, there are considerable efforts to establish a Korean regulatory system for enhancing LNG tank safety. This was led by Korea Gas Safety Corporation (KGS) and supported by KOGAS and government agency. The High Pressure Gas Safety Control Law was enacted in 1996 and then KGS began to implement verification of safety for LNG tanks. KGS completed 'A Study on Inspection Standards of LNG Storage Tanks' in 1998 and this has guided storage system design for Incheon and Pyeongtaek terminals.

KGS also performed 'A study on Inspection Technology and International Standards for LNG Storage Tanks' in 2002, which results in a standard for the no-dike LNG tanks (7 tanks in operation) at Tongyeong for the first time in Korea. This detailed study on the functions of a dike for LNG tank was the basis of this guideline, which regards the outer wall of a full containment tank as a dike (or impounding wall). The guideline does not require the dike since full containment outer wall is considered having the same function and safety of the dike. After this guideline has been enacted, no-dike was installed for the full containment type of

LNG tanks. As a result, the terminal operator has benefited from enhanced land use efficiency.

In 2004, KGS implemented the 'Comparative Risk Assessment of LNG Tank Design' with the help of the British consulting firm, AEA Technology, and the Korea Institute of Gas (KIGAS). They compared and evaluated the safety of the ordinary full containment LNG tank and the KOGAS membrane tank through structural analysis and quantitative risk assessment. The latter KOGAS membrane tank was found to have the same level of safety as the full containment LNG tank installed at Tongyeong.

KGS is developing Korean Industrial Standard (KS) codes similar to the international codes and standards such as BS, API and EN code with governmental agency. The efforts of KGS towards the establishment of proper safety standards in time for Korean LNG tank industry were successfully achieved during a short period and eliminated unnecessary debate over safety issues. As a result, KGS has contributed greatly to the national plan for LNG storage facilities and the development of its related industries.

Design of the Largest Aboveground LNG Tanks

The Korean LNG tank industry has developed technically and its market was activated along with 9%-Ni LNG tank's installation at Incheon terminal. At the same time, the market competition has been serious as well, which led to a reduction in construction costs per unit volume of tank capacity. For example, the LNG tanks at the Tongyeong terminal were constructed at the cost of 70% of the tanks at Inchon[2]. Keeping pace with the increase in LNG demands, the government policy of enlargement of LNG tanks has been aimed at reducing the national budget. It has also led to competitiveness in the LNG tank construction market and expedited development of the technology in engineering and construction areas.

In terms of above-ground LNG tank capacity, economy of scale was clearly achieved as shown in Fig. 2. The data used were collected from the various projects at Incheon and Tongyeong terminal and analyzed with adjustment for the identical timeframe taking into account for annual inflation by 3%.

---

[2] Estimated by the contract amount and storage capacity taking account of inflation

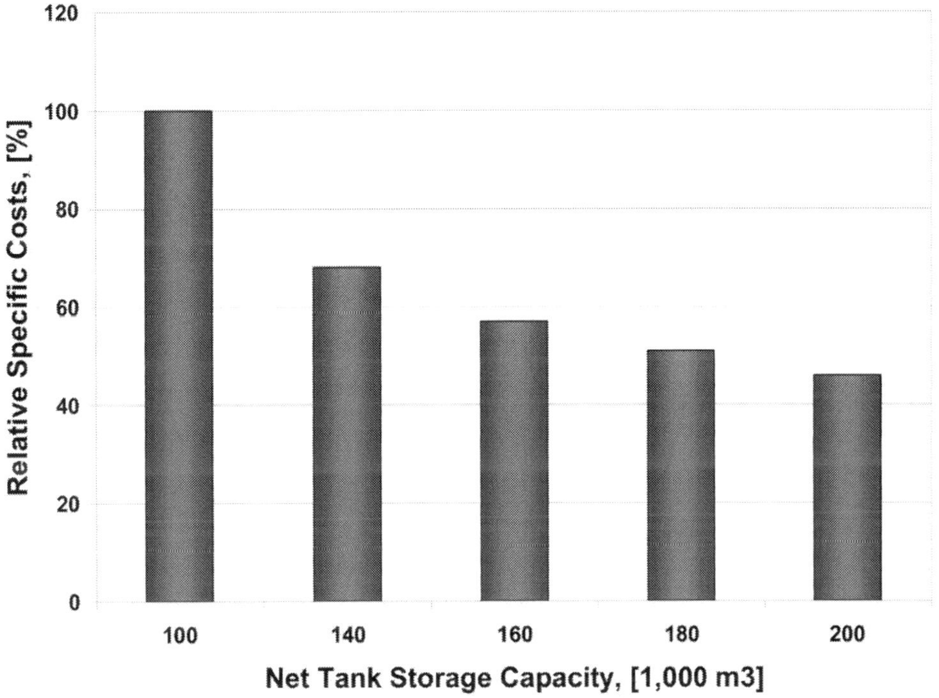

Fig. 2 - Relative Specific Tank Construction Costs

* Notes)

- Specific tank construction costs were obtained from total construction costs divided by net storage volumes (m^3).
- 100,000 m^3 tank has dike, while others are no-dike type.
- 200,000m^3 capacity LNG tank costs is based on Kogas' planned Tongyeong project

Daewoo Engineering and Construction Co., Ltd. (DWC) has participated in the LNG tank industry since 1995. DWC has independently started the development of 200,000m^3 above-ground LNG tank based on design and construction experiences at Incheon and Tongyeong terminals, as a strategy to be a market leader at the end of year 2000, while the world's largest tank was an 180,000m^3 above-ground LNG tank at the Senboku and Himeji terminals in Japan. DWC targeted the following objectives in the development of the 200,000m^3 tank.

- Cost reduction per unit volume
- Enlargement of land use efficiency
- Technology development of structural analysis and materials
- Reduction of construction period

In February 2005, DWC succeeded to develop the design and structural analysis technology for inner and outer tanks on the basis of various codes and standards, first hand experience from previous LNG projects and the design and analysis technology acquired from nuclear containment. The various LNG tanks have been installed around the world including Korea principally in accordance with BS7777 and API 620 codes. However, each code has its own strict and different provisions for the thickness of plate for the inner tank and the hydrostatic test level, which in fact limited the maximum capacity up to 160,000m^3 for the LNG tanks. But now the newly established European code prEN14620 replacing BS7777 can allow the inner tank design with over 50mm thickness and partial hydro-test level up to 125% of the maximum liquid loads. Based on this new regulation, DWC implemented a study on various sizes of LNG tanks with the optimum height to diameter(H/D) ratio and finally selected the most economical H/D ratio securing the safety in the seismic loads of OBE(0.1g) and SSE(0.2g) without anchor strap. Then, DWC designed and verified the tank with structural analysis having the following specifications.

- Capacity: 200,000m^3 / 90m (ID) X 35.3m (inner tank height)
- Internal pressure: 290mmbar / thickness of 1st plate: 31.2mm
- Radius of curvature for dome roof: 0.8d
  (Safety check with allowable stress & buckling)
- Ring beam: designed to reduce thrust transmitted from the dome roof & internal tendon

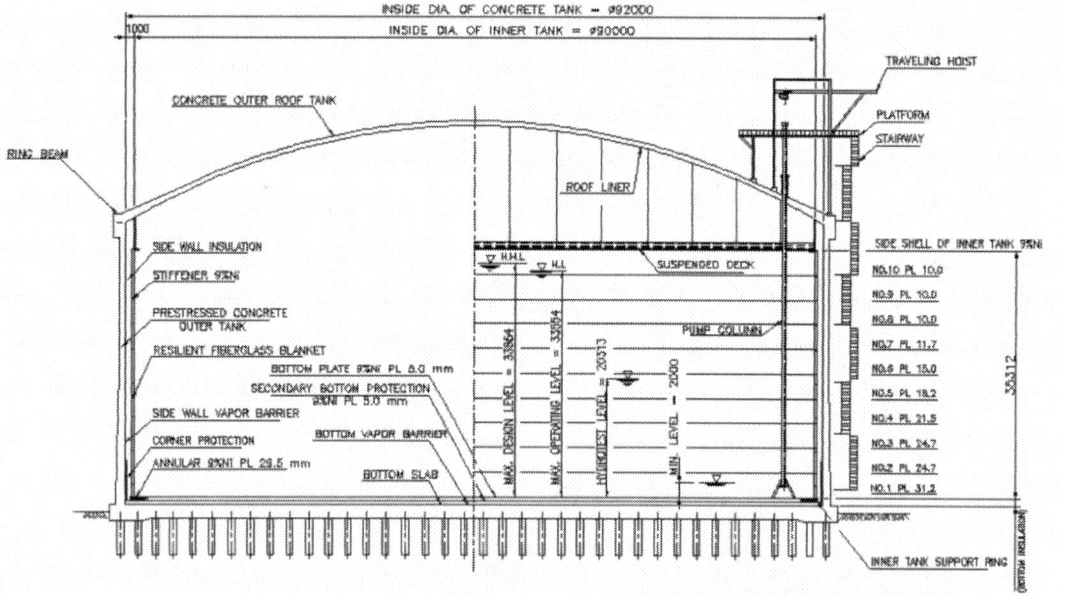

Fig. 3 - Sectional View of 200,000m^3 LNG Tank

Conclusion

As the second largest LNG importer in the world, the Korean LNG industry has grown remarkably in a relatively short period of 20 years. Along with this, LNG storage facilities have also been steadily expanding as an increase in LNG demands to secure the supply of peak loads. This also meets the Korean government energy-reserving policy. Additionally, thanks to the constant efforts of LNG-related parties, such as Korea Gas Safety Corporation, KOGAS, and Engineering and construction companies, it was possible to develop the various state-of-the-art LNG tanks.

These efforts gave birth to the Korean membrane and 9% Ni LNG tanks overcoming many difficulties. Through these efforts, DWC could secure engineering and analysis technology for the largest 200,000m^3 full containment LNG tanks of 9% Ni type based on its own advanced technology as a leading contractor in the Korean market. These also lay the foundation to leap into the global market, where the trend of expansion of LNG facilities is very apparent due to an increase in LNG demands and oil prices.

References

1. Se-Jin Jeon, "Design Basis for Large Above-Ground Tank", GASEX2002, May 2002.

2. MOCIE (Ministry of Commerce, Industry and Energy), "The 7th Long-term Supply and Demand Program", December 2004.

3. KOGAS (Korea Gas Corporation), "A study on the containment type of LNG tank", August 1998.

4. KOGAS R&D Division, "Design Information on the Largest LNG Storage Tank", May 2005.

Biography of Speaker

Eui-Seung Park is a Managing Director and a Technical Fellow of Program Committee B of International Gas Union (IGU). He has worked for Daewoo E&C for around 30 years and has been involved in many gas monetization and LNG projects. He graduated from Seoul National University with a BS degree in electrical engineering. He also completed Advanced Managerial Program (AMP). He is a head of Plant Division and currently in charge of managing LNG projects in the division. E-mail: espark@dwconst.co.kr

Maintenance and Inspection Methodologies for LNG Liquefaction Plants and Terminals

Dominik Uznanski
Senior Research Engineer, R&D Division, Gaz de France, France
Sture Angelsen
Service Area Leader, DNV Consulting, Det Norske Veritas, Norway
Gjermund Våge
Senior Consultant, DNV Consulting, Det Norske Veritas, Norway
Jens Petter Tronskar
Project Manager, DNV Technology Services, Det Norske Veritas, China

AIChE Spring Meeting, April 2006
6th Topical Conference on Natural Gas Utilization
Orlando, Florida, April 23-27, 2006

1. Introduction

Increasing demand for LNG world-wide has led to the development of a number of new gas fields and construction of LNG liquefaction plants and receiving terminals to serve the global trade of LNG. The global trade of LNG was 70 million tons per year in 1995, 111 million tons in 2002 and is expected to reach 190 million tons in 2010. The pressure to reduce the cost of LNG has resulted in lower unit costs for gas carriers and liquefaction plants, through introduction of new technology and larger capacity plants.

The planned increase in LNG production, transport and consumption, raises public concern about the safety, health, and environmental risks associated with use of LNG. This is related to design, operation and maintenance of the facilities.

There are very few records of LNG incidents and accidents and the overall safety records of LNG plants and terminals in general are impressive compared to those of other industries. However, some incidents have occurred such as in the Malaysian MLNG Plant Tiga, train 7 in 2003[1] and the gas explosion in the LNG plant at Skikda in Algeria in 2004[2].

Today, there is a growing number of LNG liquefaction plants and terminals. At the same time, the LNG market is a buyers' market and competition forces plants to

---

[1] On August 16, 2003 the Propane Compressor Gas Turbine (C3 GT) exhaust system of Train 7 of MLNG Tiga was involved in a fire incident. Prior to the incident, an unknown crack had developed in the joint between the tube and header of the regeneration gas coil. In connection with a compressor trip, fuel gas escaping from the regeneration coil mixed with very high temperature air in the waste heat recovery unit. The gases ignited, causing extensive equipment damage. (Extract from Petronas' accident report.)

[2] On January 19, 2004, in Skikda, Algeria, a steam boiler that was part of an LNG production plant exploded, triggering a second, more massive vapor-cloud explosion and fire that took eight hours to extinguish. The explosions and fire destroyed a portion of the LNG plant and caused death, injury, and material damage outside the plant's boundaries. (http://www.ferc.gov/industries/lng/safety/safety-record.asp)

reduce the production cost and improve production reliability of the facilities due the large contractual obligations involved. The increasing competition in the LNG business puts extremely strict requirements on the LNG plants and terminal reliability to maintain the buyers' trust and confidence.

2. Modern inspection and maintenance methodology

Modern methods for inspection and maintenance planning safely improve production availability by preventing trips due to equipment failure and incidents. This is done by identifying the risk drivers and root causes, introducing mitigating actions, improving inspection and maintenance, ensuring implementation and incorporating activities in the workflow. The importance of improving production availability for the plant competitiveness and profitability, warrants use of advanced technology for inspection and maintenance optimisation and systems to ensure the activities are included in the work flow, timely executed and reported.

It is generally recognized that the most effective way to achieve this and to optimize the utilization of resources, is to introduce risk based methods for inspection and maintenance strategy development. A number of different methods have been developed by different organisations to achieve this:

- Risk based inspection (RBI) for fixed or static equipment
- Reliability or risk centred maintenance (RCM) for rotating equipment
- Reliability based methods can be used to monitor performance and determine test intervals for safety critical systems, such as pressure protection systems, gas detection systems and so on.

Within the framework of the European Risk Based Inspection and Maintenance Procedures for the European Industry (RIMAP) project, DNV as the overall project manager, participated in developing the technology for a holistic risk based inspection/function/testing and maintenance planning methodology. The project also developed detailed application workbooks for selected industries. The RIMAP project developed a common approach for all equipment types (static, rotating, and safety systems) based on application of risk based methods.

Successful implementation of risk based of techniques relies on the following aspects:

- Framework
- Work process
- Detailed procedures

We briefly discuss these separately below.

2.1. Framework

In order to put the idea of risk based maintenance and inspection into action it is necessary to install proper procedures and tools within an adequate framework to ensure the required quality, transparency, and documentation.
The framework consists of the

- working process (management related issues) and the
- detailed procedures (technical related issues).

2.2. Work Process

To implement and manage a system for risk based inspection and maintenance management sets requirements to a plant's (maintenance) management system. Figure 1 illustrates the work processes involved in developing a risk based maintenance and inspection plan, implementing and managing this plan at a plant or facility.

RIMAP defines the working processes and provides requirements to the personnel that shall execute them. Implementation of risk based inspection and maintenance planning also requires an active management that focuses on issues such as:

- Management of change
- Operating procedures
- Safe work practices
- Pre-start-up reviews
- Emergency response and controls
- Investigation of incidents
- Training
- Quality assurance

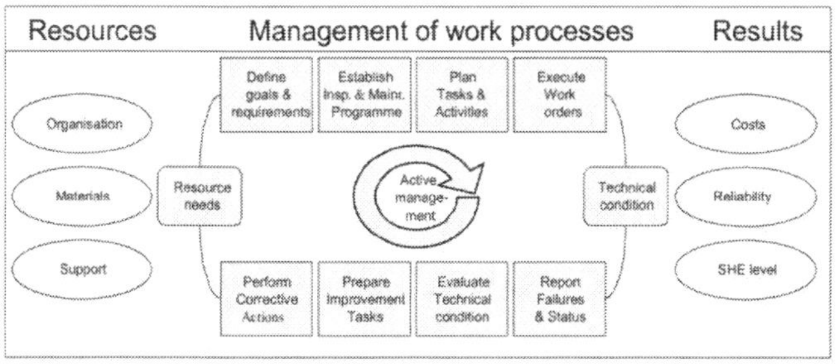

Figure 1 Work process

As risk based inspection and maintenance planning is a multidisciplinary task, it requires a team, where all necessary competencies are represented and expertise are available (e. g. EN 45004). Another important issue is the recurrent evaluation at a certain interval to assess all the modifications and new data available to make sure the entire process runs into a circle.

The steps in the process are the same for all industry sectors although the level of maturity may vary.

2.3. The RIMAP description of Risk, PoF, CoF

The RIMAP project provides guidelines on how to perform risk based inspection and maintenance planning for all types of equipment: active components, static components, and safety critical equipment.

Figure 2 illustrates how different methods of analysis are applied to each type of equipment.

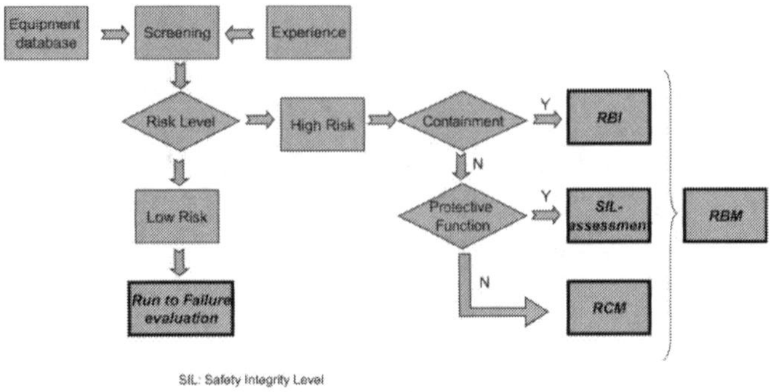

Figure 2 Structure of analysis

The steps required to perform maintenance and inspection planning are similar for each type of equipment.

The RIMAP project provides guidelines on how to perform risk based inspection and maintenance planning for all types of equipment: active components, static components, and safety critical equipment. The steps required to perform maintenance and inspection planning are similar for each type of equipment. The steps in the analysis are similar for all equipment classes:

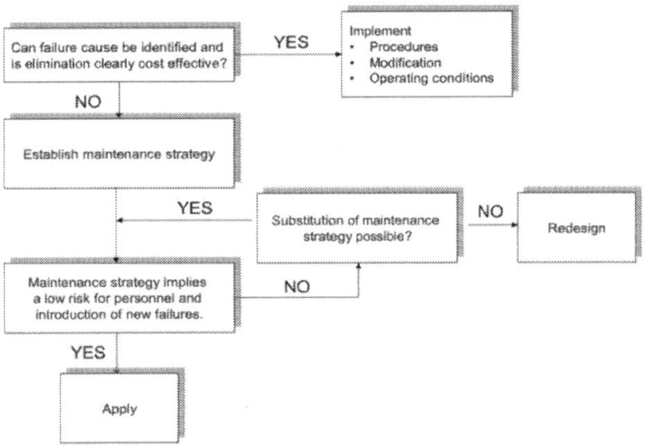

Figure 3 Procedure to identify mitigating activities

3. Challenges

The inspection and maintenance methodologies mentioned above must be based on data obtained from investigations linked to the various industries concerned (namely LNG). Such data exists, including data obtained by Gaz de France. There are several challenges that need to be addressed to implement such an optimised inspection and maintenance planning program in the LNG industry. However, before detailing these, a short review of past Gaz de France efforts to improve the understanding of aging of LNG installations and materials will be presented.

3.1. Previous Gaz de France R&D work on the aging of cryogenic materials

In the past, Gaz de France, in collaboration with other oil and gas companies and engineering firms, has led a joint study to assess the extent of aging of the most widely used materials in the LNG industry.

Gaz de France had been operating its Liquefied Natural Gas (LNG) testing station, located in Nantes (West of France) near the Montoir methane terminal, for more than 40 years. This testing station, which was launched with the construction in 1959 and commissioning in 1960 of a 500 m^3 9% nickel steel self-sustaining above ground LNG tank, allowed the qualification of a number of cryogenic installations and equipment, including the 9% nickel steel inner wall LNG containment technique, an open-rack aluminum Marston type LNG vaporization unit and other equipment such as valves and flanges.

Gaz de France decided to close down the testing station in Nantes at the end of 2001 and the testing station was dismantled from September 2002 to March 2003. The dismantling of the testing station represented an opportunity to study the aging extent of various cryogenic materials commonly used in the LNG industry, some of which had been operated for more than 40 years at LNG temperatures. In particular, the 9% nickel steel inner walls of the 500 m^3 storage tank, the tank's insulation material (perlite) and concrete base foundation and the aluminum alloy of the open rack Marston LNG vaporization unit (ORV) were analyzed in detail, using a combination of non-destructive and destructive tests.

This work was carried out between 2002 and 2004 within the framework of an international partnership, in view of sharing the results and promoting enhanced exchanges on the aging analysis and longevity determination of cryogenic materials. The Korea Gas Corporation, Osaka Gas Company, Total, Tractebel, Fluxys and BP joined Gaz de France as industrial partners in this project.

The non-destructive examinations (visualization tests, paint dye penetrant tests and radiographic tests) which were carried out on the 9% nickel steel inner walls of the Nantes 500 m^3 LNG storage tank, which had been continuously operated for more than 40 years, did not reveal any significant defects and in particular no defects which were due to the aging of the tank. This finding contributes to proving that the operating lifetime of such a tank is at least 40 years.

Comprehensive destructive test campaigns were carried out by Kogas and Gaz de France on the base metals and welds of selected zones of the shell and bottom walls. These tests consisted of chemical analysis examinations, micro-hardness tests, macro and micro-structure analysis, tensile tests (yield strength, tensile strength and elongation), Charpy impact tests and Crack Tip Opening Displacement (CTOD) tests.

The preliminary destructive characterizations results obtained using chemical analysis, macro and micro-structure analysis and micro-hardness tests on the base metal and welds confirmed the properties expected of these metals.

The tensile and Charpy impact tests performed by both Kogas and Gaz de France confirmed the consistency and repeatability of the results and compared well to the original data dating back to the construction of the tank in 1960.

Further metallographic examinations of some of the main defects revealed by the radiographic examinations confirmed that all of the examined defects originated during the tank construction period and were not due to any aging process of the tank during its service life. These results also confirm that the detected construction defects did not increase in size, despite the numerous thermal and mechanical cycles they endured. With all of these results, it was possible to determine a residual operating lifetime for this tank.

Concrete samples extracted from the support structure of the 500 m^3 tank concrete base slab were examined for compressive strength, unit volume weight, carbonation depth and amount of chloride ion. Although more than 40 years has passed since completion of the concrete structure, no deterioration caused by chloride ion diffusion and carbonation has occurred.

The Nantes ORV unit was examined by non-destructive examinations. Visualization and paint dye penetrant tests were conducted on the outer and inner surfaces of the header and tube junctions, as well as the inner surfaces of the tubes. No cracks caused by aging deterioration from the ORV unit operation were found. In addition, it has been confirmed that the structure of this ORV unit had little anxiety for stress concentration.

Thus, by analyzing the aging extent of the materials through both non-destructive and destructive tests, the longevity and durability of these materials was determined.

However, it is not enough to gain knowledge about the life expectancy of various cryogenic installations and materials. The next step is to develop a complete methodology for the diagnosis of the aging of such materials and references or guides indicating defect acceptability and repair criteria, allowing optimized inspection and maintenance planning. The major steps in this upcoming work planed by Gaz de France are detailed hereafter.

3.2. Upcoming Gaz de France R&D work on the aging of cryogenic materials

In the past, Gaz de France has developed a Defect Assessment and Repair (DAR) code for its transmission pipeline network, which spans more than 30 000 km in France. This DAR is recognized and accredited by the competent authorities. The idea today is to develop an equivalent framework for the diagnosis (inspection), assessment of defects and repair directed towards various LNG installations and materials. This work is divided into five main steps, as detailed hereafter.

1. The first step is to list the eventual degradations which occurred in LNG plants and terminals and identify the corresponding degradation mechanisms.
2. The second step consists in identifying the installations and materials to be monitored and to carry out an inventory of the diagnostic tools (methods and measuring means) available allowing such monitoring its accuracy and practical use. The materials to be studied will be essentially non-accessible and critical to safe operation, such as cryogenic steels used for LNG containment (9% nickel steel and corrugated stainless steel), LNG transfer pipes (stainless steel) and the insulation used with these materials (perlite, glasswool, polyurethane foam, ...).
3. The third step consists in developing a methodology for the management of the aging of cryogenic materials, similar to the framework already developed

for the gas transmission network. This methodology should consider the risk and uncertainty involved using the concepts from RIMAP as described above.

4. The fourth step consists in developing a referential for defect acceptability criteria. This will allow the definition of various levels of defect acceptability, based on parameters such as residual wall thickness, defect sizes and shapes and non-destructive test results and uncertainty herein. The knowledge gained from the Nantes LNG testing station cryogenic materials investigations will help in establishing such criteria.
5. The fifth step will consist in developing a referential for the repair criteria of the defects. Such a referential will allow an assessment of the possibility to repair defects and will be of great help in the decision making process for optimized installation aging management.

4. Conclusion

The

- growing number of LNG plants and terminals and the need to build new installations,
- pressure to reduce production costs and improve production availability of LNG plants and terminals and
- the need to guarantee a very high level of safety and environmental respect for all the stakeholders

warrants development of effective methods for asset management for LNG plants and terminals. The introduction of risk based methodologies for optimization of inspection and maintenance planning have shown large benefits for refineries, petrochemical, chemical and power plants : overall savings on inspection and maintenance have been quoted as typically 10-25% and increased availability by 0.5-1.0% in addition to an improvement in safety performance. Authorities in several countries embraced the methodology and have granted approval for deviation from the prescriptive rules for inspection interval for pressure vessels. Insurance companies have realized the benefits and have granted significant premium reductions to the companies implementing inspection management systems and risk based methodologies in the inspection and maintenance planning.

To address these challenges DNV and Gaz de France therefore propose to initiate a JIP to develop risk based inspection and maintenance procedures for LNG plants and terminals. This will be based on generally accepted principles for risk based inspection and maintenance and on cryogenic properties:

1. Identification of degradation mechanisms relevant for LNG plants and terminals,
2. Identifying the installations and materials to be monitored and to carry out an inventory of the diagnostic tools,
3. Develop a methodology for management of ageing of cryogenic materials,
4. Develop acceptance criteria such as residual wall thickness, sizes of defects, etc.,

5. Develop repair criteria of the defects. Such a referential will allow an assessment of the possibility to repair defects and will be of great help in the decision making process for optimized installation aging management.

Combining new understanding of cryogenic effects with methods from modern inspection and maintenance management will provide an effective tool for asset management for LNG plants and terminals.

Gaz de France, as an LNG operator, considers these issues as important and launched this idea together with DNV to develop these concepts.

5. Selected References

RIMAP RTD or RIMAP Demo project: http://research.dnv.com/rimap

RIMAP TN: http://www.mpa-lifetech.de/rimap

Uittenbroek D., Blaauw A., Wahab H., Hirose H., Hirai S. (1998) Demolition, inspection and tests of disused LNG tank in Brunei. LNG 12 Conference, May 4-7, 1998, paper 5.2.

Uznanski D. et al. (2004) Investigations into the aging extent of various installations of the Nantes cryogenic testing station. International Gas Research Conference, November 1-4, 2004, paper 1151-A.

6. Biography of speaker

Mr. Dominik UZNANSKI joined Gaz de France's Research and Development Division in 1993 as a research engineer at the Cryogenic Studies Section. Since then, he has been involved in numerous studies on the behavior of LNG in storage tanks (including research on stratification behavior and tank filling operations), on LNG vapor heavy gas dispersion and on water curtain mitigation methods and has gained extensive experience in these fields. He is also involved in the development of GDF's LNG tank management software named LNG MASTER and its delivery to end users, including Gaz de France's two LNG receiving terminals and external clients. He has managed several projects and is currently a Senior Research Engineer and dealing with LNG Technologies. In this respect, he is involved in particular in studies on LNG offshore installations and in developing a methodology for the diagnosis of the aging of LNG installations.

Mr. Uznanski holds Bachelor of Science degrees in chemical engineering and mathematics from the University of Maryland, U.S.A. (1986) and a Master of Science degree in fluid dynamics and heat transfer from the Ecole Centrale Nantes engineering school, France (1991).

An Overview of Mustang Engineering's LNG Smart® Vaporization Technology

Randy M. Sharp
Process Engineering Specialist
Ned Baudat, P.E.
Process Inventor and Project Manager
F. Pat Ross, P.E.
LNG Smart® Vaporization Technical Expert
Midstream Group, Mustang Engineering, L.P., Houston, Texas, USA

AIChE Spring Meeting, April 2006
6th Topical Conference on Natural Gas Utilization
Orlando, Florida, April 23-27, 2006

Introduction

With global oil and gas prices near record highs, opportunities for the Liquefied Natural Gas (LNG) industry have never been better. Rather than waste a precious natural resource, gas-rich nations are increasingly exercising their gas monetization options by creating facilities to liquefy natural gas and transport it via specialized marine carriers. Dozens of revaporization facilities, large and small, exist worldwide, with more being planned and constructed. Existing revaporization facilities receive, store, pressurize, vaporize, and distribute the vaporized LNG. In order to vaporize LNG from its cryogenic liquid state at -260 °F to typical pipeline temperature (35-40°F) for distribution to market, two predominant vaporization technologies have been used in the past; the Submerged Combustion Vaporizer (SCV) and the Open Rack Vaporizer (ORV).

SCV technology has been used principally onshore, but can be also used on some offshore revaporization facilities. SCVs operate by the use of a submerged burner to heat a water bath containing a set of vaporization tubes. High pressure LNG is vaporized in the tubes that are submerged in the heated water and is then sent to the gas pipeline. Typical SCV facilities use approximately 1.3% to 2% of the sendout gas as fuel for the SCV burners.

ORVs use seawater as the heat source for LNG vaporization. For this reason, ORVs are best suited for near shore or fixed offshore applications. ORVs do not lend themselves to floating applications. The inlet seawater is filtered through mesh screens, treated with a biocide, pumped to the top of the ORV, then flows downward over a vertical sheet of heat exchange tubes, and is cooled by the vaporizing LNG within the tubes before collecting in an open basin at the base of the ORV. The cooled seawater is then discharged back to the source body of water by gravity flow or by pumping. If the seawater flow rate is sufficient, ice formation on the outer surface of the tubes can be prevented.

For an ORV facility designed for 1 billion standard cubic feet per day (BCFD) gas sendout, approximately 125 to 155 million gallons of seawater intake is required daily, depending on seawater temperature change and return temperature limitations. The seawater intake must be mechanically screened and treated with biocidal chemicals, typically sodium hypochlorite, to prevent fouling on the vaporization tubes and in the associated piping. Typically, one spare ORV is installed to allow for periodic cleaning of the basin and tube surfaces while maintaining full send out capacity. In the United States, ORV systems are meeting with stiff environmental opposition. There are indications that ORVs may negatively impact the marine life in their vicinity due to mechanical, thermal, and chemical effects. Of special concern are the cumulative effects of two or more terminals located in close proximity. Despite numerous environmental impact studies, there is not an agreement to what extent the ORVs impact the environment and marine life.

Mustang Develops a New Vaporization Technology

In an effort to develop a vaporization system which uses less fuel than the SCV and has less actual or perceived environmental impact than the ORV, Mustang Engineering has developed an alternative vaporization process. The LNG Smart® Vaporization System uses forced-draft air heat exchangers coupled with an intermediary fluid loop. The combined technology provides several significant benefits. First, it is energy efficient. Electrical loads are similar to those of ORV systems. Depending on the vaporization facility site's historic climatic conditions, the LNG Smart® System can provide up to 100 percent reduction in fuel gas consumption and associated combustion emissions products when compared to SCV systems. Typical SCV vaporization systems require 1.3% to 2% of the facility's send-out as fuel. In locations where the ambient conditions do not allow for full air vaporization duty throughout the year, thermally efficient backup heaters can be installed and the fuel gas consumption and corresponding emissions are then determined by the site-specific climatic conditions. Table 1 illustrates selected global site-specific potential fuel savings.

Table 1: LNG Smart® Vaporization System Fuel Savings vs. SCV for Selected Vaporization Facility Locations

Location	Annual Average Air Temperature (°F)	Average Relative Humidity (%)	Fuel Gas Savings (%)
Thailand	82	80	100
Jamaica	81	75	100
Miami, FL, USA	76	71	99
Lake Charles, LA, USA	68	68	89
Shanghai, China	61	72	77
Rotterdam, Holland	52	68	61
Boston, MA, USA	51	64	60
Caribou, ME USA	39	70	41

LNG Smart® Vaporization System Description (Patents Pending)

The LNG Smart® Vaporization System utilizes a proprietary air-warmed heat exchanger (the KF Air Heater) to recover sensible heat from the ambient air and latent heat by condensing water vapor from the air. This heat is used to warm a solution of Potassium Formate and water (KF Fluid). The KF Fluid can be considered analogous to the seawater in an ORV. Each KF Air Heater contains layers of finned and/or bare tubes, and has top-mounted fans forcing ambient air downwards through the tube bundle. The KF Air Heaters are elevated above grade to prevent recirculation of cold air from the bottom of the unit to the fan inlet. The tubes contain flowing KF Fluid, which is cool due to the previous cross exchange with the vaporizing LNG. As the ambient air passes over the surface of the cold tubes, the air is cooled and water vapor condenses from the air onto the outer tube surfaces. The condensed water continuously forms on, and drips from, the tube surfaces. This falling water washes over the tubes below and finally exits the bottom of the exchanger with the flow of cooled air. The exiting air falls to the ground and is driven away from the unit by the downdraft. The condensed water is collected in a basin beneath the KF Air Heaters, and can be utilized in the plant for fire water, filtered and chemically treated for use as potable water, or discarded as fresh water. The rate of water condensation depends on the ambient temperature and relative humidity, but averages approximately 500 GPM per BCFD of LNG vaporized at typical design ambient conditions.

Once heated by the KF Air Heaters, the KF Fluid is sent to the KF Surge Tanks. The KF Surge Tanks provide surge capacity for the KF Fluid and constant suction pressure for the KF Pumps. The KF Pumps provide the driving pressure for the KF Fluid circulation. The pressurized flow of warmed KF Fluid is then sent to the KF-LNG Vaporizers where, acting as a heating medium, it transfers its heat into the LNG, and is simultaneously cooled. The cooled KF Fluid is then returned to the KF Air Heaters for warming, and the KF circulation cycle is completed.

The LNG Smart® system can be installed as a standalone vaporization system or as an existing system upgrade. For example, if SCVs are already in place, then the SCVs can be utilized for full or partial backup heat during ambient temperature and relative humidity conditions which are below the design conditions. If installed as a standalone system, backup heat, if necessary, is installed in the form of boiler style fired glycol heaters. If the glycol heaters are used, the heated glycol solution is circulated to the KF-Glycol Heater to provide trim heat to the KF Fluid. The decision to install partial or full backup heating capacity depends on the site climatic conditions and the risk tolerance of the client.

Figure 1 shows a simple schematic of the LNG Smart® Air Vaporization process with backup heating installed.

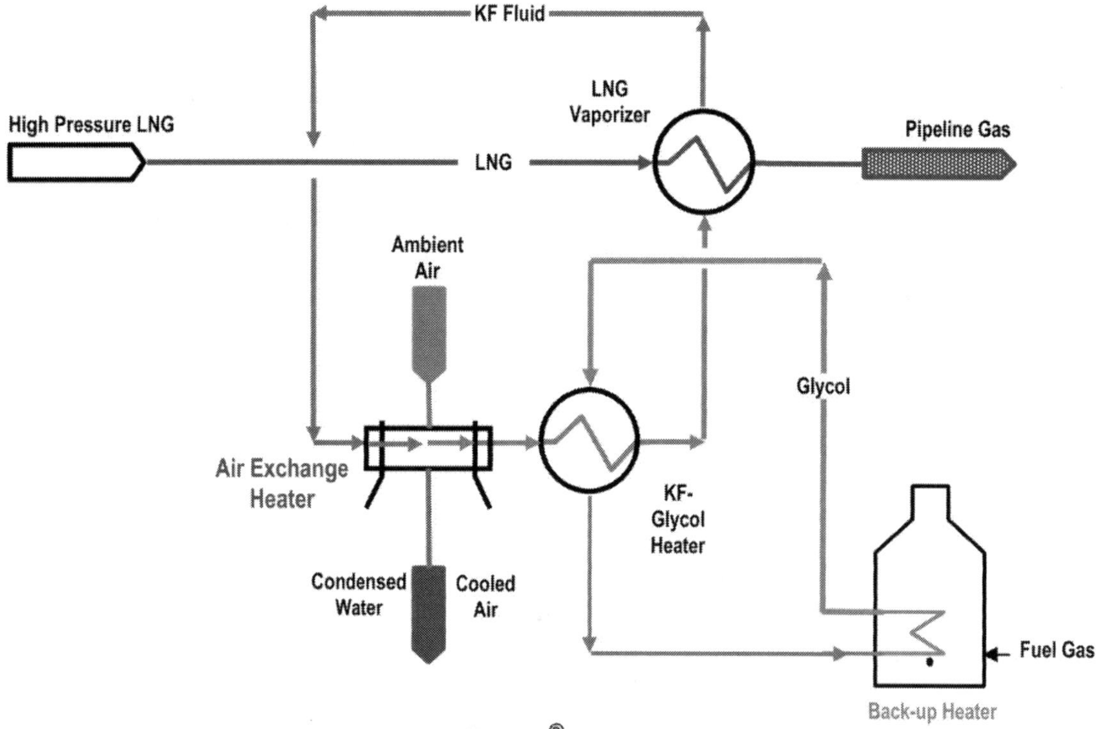

Figure 1: Schematic of the LNG Smart® Vaporization System with Backup Heat (Patent Pending)

LNG Smart® Vaporization System Operating Philosophy

At ambient temperatures at or above the design temperature, the LNG Smart® Vaporization System will provide all of the vaporization duty required for the design sendout rate. However, the pipeline gas sendout temperature will increase as the ambient temperature increases above the design point due to the excess surface area provided by the KF Air Heaters. During warmer than design ambient conditions, the return temperature of the cooled KF from the KF LNG Vaporizers will also increase above the design temperature.

Due to the design of the Air Vaporization System, warm weather conditions provide the opportunity to either increase pipeline sendout or the reduce facility power use while maintaining design gas sendout capacity and temperature. An option for power savings is to shut down a portion of the KF Air Heater fan motors. This method is especially suited for large facilities. This method can only be applied if the air discharge temperature remains high enough to prevent excessive tube surface icing, and if the LNG sendout temperature can also be maintained at or above the required sendout temperature. The additional available ambient heat also provides the option to increase the gas sendout rate to take advantage of the additional available duty. This option does not save power, but does make full use of the excess system heating capacity if the additional pipeline gas can be utilized. Spare pumping capacity, line sizing, and pipeline packing capacity must be considered for this option.

If ambient conditions fall below the design temperature, then the application of backup heat through the KF-Glycol Heaters is initiated. The boiler-type glycol heaters have very low minimum turndown capacity, so the backup heat can be applied on an as-needed basis. Backup heat requirements and approximate annual utilization of backup heat can be determined by a site-specific climatic study.

Construction, Operation, and Performance Testing of a Pilot-Scale KF Air Heater

In 2005, Mustang constructed a pilot-scale unit to test the performance characteristics and refine the design of a scaled-down KF Air Heater. The test unit uses a commercial refrigeration unit to chill KF Fluid, which is then fed to the Test KF Air Heater. The tube configuration, KF inlet flowrate, KF inlet temperature, and air flow can be modified to provide a variety of operating configurations. Each configuration can then be tested at a variety of temperature and relative humidity conditions to determine heat transfer performance and operating parameters. From the data collected to date, Mustang has determined that some tube configurations are not practical, and has decided on a standard KF Air Heater tube configuration. While the detailed results are confidential, in general, the empirical data collected from the operation of the test unit have validated the theory that heat obtained from ambient air can be transferred into an intermediate fluid, which can then be used to vaporize LNG. Operation of the test unit has also helped to refine the design of the KF Air Heaters and the total air vaporization system, has confirmed the thermodynamic properties of the KF fluid, and has helped to refine and verify the thermodynamic simulations used to design and predict the overall system performance. In addition, the test unit provides a unique opportunity for clients to view and understand how the LNG Smart® Vaporization System works.

Although Mustang believes that sufficient empirical data has been collected to confidently scale the test unit operating characteristics to a full-scale LNG vaporization facility, the pilot unit is still being modified and tested. Among the parameters currently being tested are a revised tube bundle design and a protective tube coating for offshore and other harsh applications. As opportunities arise for additional applications of the air heating process, the test unit will continue to be modified with design changes aimed at gaining additional insight into the system operating parameters, construction material viability, and improvements in efficiency and performance.

Advantages of the LNG Smart® Vaporization System

Reduced Energy Consumption – The proprietary air exchange vaporization process offers a significant reduction in fuel gas consumption when compared to an SCV vaporization system, up to 100% depending on the climatic conditions at the facility location. Reduced fuel consumption translates to significant monetary savings for the vaporization facility operator. In a Gulf of Mexico case study based on a 1 BCFD gas sendout rate facility using this technology and natural gas prices at $9/MMBTU, the value of a 90% annual fuel gas savings is approximately $38 million.

Reduced Emissions and Easier Environmental Permit Acquisition – Reduced combustion of fuel translates to reduction of combustion-related emissions. Compared to an SCV facility, the emissions reductions are in proportion to the reduction in fuel gas use. The proven reduction in emissions facilitates the permitting process. Regulatory approvals from FERC and local authorities could be expedited because of the improved environmental conditions offered by this technology.

Standardized Components – While the LNG Smart® Vaporization System and its configuration is proprietary to Mustang Engineering, virtually all operating components are readily available, and are common to traditional LNG vaporization or petrochemical facilities. This shortens facility construction schedules. The proprietary KF Air Heater is not a standard air-cooled heat exchanger, but the components used to manufacture it are essentially the same. Mustang has worked diligently to standardize the design of the KF Air Heater such that any air cooled heat exchanger manufacturer can duplicate our standard design.

Fresh Water Product – The LNG Smart® Vaporization System generates cool, fresh water as a by-product of the vaporization process. The water can be made potable for consumption, retained in a basin below the KF Air Heaters for use as plant firewater, used for irrigation, or simply discarded as fresh water runoff.

Disadvantages of the LNG Smart® Vaporization System

Need for Backup Heat In Some Locations – The need for backup heat depends mainly on the site climatic conditions. The LNG Smart® system is best suited for warm, humid climates such as the US Gulf Coast, the Caribbean, Mexico, Thailand, India, Northern Australia, and locations with similar climates. However, economic calculations indicate that the system is economically viable for climates where it can be used at least 60% of the time, on average.

Plot Area Requirements – The KF Air Heaters for 1.0 BCFD of vaporization have a footprint of about 0.75 acres. Additional equipment brings the vaporization equipment footprint up to about one acre. This is about a 50% larger footprint than an SCV system with similar capacity. This is more likely to be a factor in facility revamps than in new stand-alone construction, as the area for the vaporization equipment is small in comparison to the total plant size.

Efficiency and Behavior during Low Ambient Temperature Operation – Despite the test unit performance data, there are still questions regarding KF Air Heater performance at ambient temperatures below approximately 40 °F. This is near the lower limit of application for the KF Air Heaters, and only limited data has been collected at these temperatures. Further, the data collected have only been with tube bundle configurations that have been discarded as inadequate. Further testing of the low temperature operating regime will be performed as ambient conditions allow. Uncertainty at low temperatures is more important in mild climates where more annual hours are spent below 50 °F.

Conclusions

Mustang has developed the LNG Smart® Vaporization System to provide a fuel efficient, environmentally friendly, and easy to permit vaporization option, when compared to existing vaporization technologies. The LNG Smart® Vaporization System is best-suited for warm climate applications, but can still provide significant fuel gas savings in cooler climates when compared to a traditional SCV or other fuel burning vaporization systems. The performance characteristics of the heart of the LNG Smart® system, the KF Air Heater, have been tested and verified at the pilot scale. Mustang has developed simulations and techniques that allow it to design and predict the system performance for any site with sufficient climate data.

Speaker Biography

Randy Sharp has 5 years experience with Mustang Engineering as a Process Engineering Specialist. Randy has worked for the last three years in Mustang's Midstream Group to help create, test, and refine the LNG Smart® suite of products. Randy has a BS degree and a MS degree in Chemical Engineering from Oklahoma State University.

Ned Baudat is the inventor of the LNG Smart® Vaporization System as well as the other LNG Smart® products. Ned has more than 40 years experience as a process engineer. Most of Ned's experience is in LNG, Natural Gas Liquids recovery, and petrochemicals. Ned has worked at Mustang since 2003. Ned has a BS in Chemical Engineering from Rice University and is a registered PE in Texas. Ned holds several patents.

F. Pat Ross is a subject matter expert for the LNG Smart® Vaporization process in that he has helped to refine the design of the KF Air Heaters, has helped to design, construct, and operate the Air Vaporization Test Unit, and has helped to develop the process guarantees and contracting agreements. Pat has 40 years of engineering experience. Pat has a BS in Chemical Engineering from the University of Houston and an MS in Environmental Management from the University of Houston, Clear Lake.

Optimization of Ambient Air Vaporization Processes

Joseph H. Cho, Ph.D., P.E.
Technology Advisor-Process
Robert Femat
Process Engineer
Heinz Kotzot
Section Leader, LNG & Gas Processing
Charles Durr
Director LNG, GTL and Gas Processing
Kellogg Brown & Root, LLC

AIChE Spring Meeting, April 2006
6th Topical Conference on Natural Gas Utilization
Orlando, Florida, April 23-27, 2006

ABSTRACT

In order to comply with the LNG industry's requirements for minimizing life cycle costs, many LNG vaporization processes have been proposed and developed. However, the selection of a feasible LNG vaporization scheme is often constrained by economic justification and/or site-specific environmental issues, such as air emissions (NOx, CO, CO_2, etc) and the impact of seawater effluent on the ecosystem.

This paper presents the development of ambient air-heated vaporizers: natural draft vaporizers and fan-assisted forced draft ambient air vaporizers. Air emissions of this process are compared with that of conventional Submerged Combustion Vaporizers based on Best Available Control Technology. Technical features and economic advantages of each vaporization process are addressed. This paper also discusses the optimization methodology of vaporization processes.

INTRODUCTION

New and expanding LNG terminals face several engineering design challenges, one of which is environmental impact due to emissions concerns. A LNG receiving terminal's main source of emissions generally comes from its LNG vaporization process. In order to comply with the LNG industry's requirements for minimizing life cycle costs, many vaporization processes have been proposed and developed. However, the selection of a feasible LNG vaporization scheme is often constrained by economic justification and/or site-specific environmental issues, such as air emissions (NOx, CO, CO_2, etc) and the impact of seawater effluent on the ecosystem.

Optimum design of an LNG vaporization process directly affects an LNG terminal's operability, availability, capital expenditure, and operating costs. The selected process scheme also influences project permitting. Environmental constraints are some of the most important issues in selection of the vaporization process, and as such, alternatives to the more common ORV and SCV processes must be considered for future applications.

One such alternative is LNG vaporization by way of ambient air vaporization. The ambient air vaporization process offers the advantage of reduced energy consumption by application of air as a reusable heat/energy source.

This paper presents the development of ambient air-heated vaporizers of which there are two types: natural draft air vaporizers and fan-assisted forced draft air vaporizers. Air emissions of these processes are compared with that of conventional Submerged Combustion Vaporizers based on Best Available Control Technology. Technical features and economic advantages of each vaporization process are addressed. This paper also discusses the optimization methodology of the forced draft ambient air vaporization process.

ENVIRONMENTAL CONSIDERATION

Air Emission

Air quality is primarily determined by the type and amount of contaminants, the topography of the air basin, and meteorological conditions. An emission estimate should include discharges from the vaporizer, emergency generator, gas turbines, fire water pump engines, vent/flare system, and any fugitive reactive organic gas sources.

The air quality mainly depends on the source of emissions due to dispersion, atmospheric reactions, and transport inside of the project site. Since the vaporization system is the dominant source of terminal emissions, the NO_2, SO_2 and CO discharge levels must be considered in the selection of the vaporization process. The expected total emissions for the selected vaporization system should be lower than the stipulated Federal and State ambient air quality standards.

Liquid Discharge

The potential sources of liquid effluent at terminals include the following:

- Process wastewater such as water blowdown from SCV's, leakage from heat transfer fluid, etc.
- Cold seawater from ORV's
- Condensate from the ambient air heaters
- Potentially contaminated storm water
- Sanitary wastewater or treated effluent

Parameters to be included in the wastewater discharge permit are invariably site specific, taking into consideration wastewater composition and intake water quality. The effluent outfall is critical due to the stringent limitation on effluent discharge and the concern for recirculation of colder seawater at the seawater intake pumps [1].

When using seawater for cooling purposes, the World Bank Guidelines state: "The effluent should result in a temperature increase of no more than 3°C at the edge of the zone where initial mixing and dilution take place. Where the zone is not defined, a 100 meter distance from the point of discharge can be used [2]".

When providing seawater for LNG vaporization the discharge temperature of water decreases. Although the regulations for allowable seawater temperature change were initially developed for seawater temperature rise, they are now also used for cold water return, since no regulations currently exist for discharging cooled water. The seawater discharge from an ORV is typically 5-7°C cooler, which can be easily blended into a large body of water. However, other environmental issues have to be addressed.

Environmental Issues on Seawater

The environmental assessment on seawater usage is a sensitive issue for domestic terminals in the U.S. [3]. A project that uses seawater as a heat source must consider the effects of cold water on Ichthyoplankton, which consists of fish larvae and eggs. These are free floating and easily pulled into the seawater circulating system [4]. The impact on changes to the ecosystem has to be properly investigated to mitigate the loss of fish species.

Micro-organisms present in the seawater system can be affected and destroyed by the temperature drop and strong turbulent water circulation. The seawater organisms are also affected by the residual sodium hypochlorite that is used as an anti-fouling chemical for protection against bio-fouling. Providing an alternative method to control bio-fouling of the seawater supply system could be a favorable approach in persuading the permitting authority, as well as the environmentalist.

Many seawater intake designs have been developed to minimize the impact on the environment and to protect marine life from entering the seawater intake. This includes a dual barrier system, which is constructed of a passive marine life exclusion system, consisting of a series of fine mesh fabric filter or wedge-wire screen panel. The size of the screen can be selected to reduce the impact on marine life. At this time there are no clear environmental findings or studies available that clearly identify the impact of ORV on the sealife that would specify the filtration system or clearly disallow ORVs.

AMBIENT AIR HEATED VAPORIZATIONS

Natural Draft Ambient Air Vaporizer (AAV)

The natural draft Ambient Air Vaporizer (AAV) uses air as the heating medium as a free heat source. Such vaporizers are in service for peak shaving plants at locations where the LNG vaporization duty and flow rate is relatively small (Fig. 1). At this smaller scale, plot space is not a major consideration. These systems are based on using single or multiple units in banks with common interconnecting pipes. Traditionally, these vaporizers have been designed with half of the parallel units on-line while the other half are off-line for defrosting.

If no supplemental heating is required, exclusive use of ambient air can totally eliminate emission and noise. In this process the water formed from condensation and melting ice must be collected. However, this water is environmentally neutral and can be easily disposed. These vaporizers will also require construction of an LNG containment sump built under the vaporizers, resulting in additional cost and potentially raising some other environmental issues.

Great care must be taken to assure continuous vaporization during low ambient temperature conditions and foggy days with no wind. In order to avoid the dense icing on the surface of the heat exchanger tubes, frequent defrosting with a 4-8 hour switching cycle is required. The intervals will vary, depending on temperature, wind conditions and humidity. Long cycle times lead to dense ice on the surface of the heat exchange tubes, requiring long defrosting time. Some of the technical features of AAV are summarized below.

- A long heat exchange tube facilitates air draft as reverse chimney effect because the top air density is lighter than the cooled air at the bottom.
- In order to avoid the dense icing on the surface of the heat exchange, frequent defrosting is required. A service cyclic design is required for deicing.
- The recommended minimum space between units is 6 x 10 ft [5].
- A special heat exchange tube with a stainless steel tube inserted in the aluminum star-finned exchanger tube can be used for the cyclic service.

Fig. 1 – Ambient Air Vaporizer
(Courtesy of Cryoquip)

Some other issues that must be addressed are:

- A large number of small units in operation will have much more fog generation than fewer large units. Therefore, dense fog dispersion must be studied. When the air is calm, the fog will tend to remain in the vaporization area;
- The large plot space demand has to be included in the life cycle cost;
- Experience and risk have to be included in the evaluation because these ambient air draft vaporizers have only been used for peak shaving facilities (scale up issue);
- No public information is available for heat capacity vs. ambient temperature to determine vaporizer performance;
- Expected defrosting performance needs to be confirmed in various operating conditions such as low temperature, high humidity, long operation, fog, no wind, etc. Optimum cycling time has not yet been established;
- Fog and no-wind conditions control the number of vaporizer trains. Back-up facilities need to be included, depending on the specific meteorological conditions and terminal availability requirements.
-

Forced Draft Air Heater Vaporizer (AHV)

This scheme uses an intermediate fluid between a shell-and-tube LNG vaporizer and forced draft air heater, which is intrinsically the same as the conventional air coolers to reheat the Heat Transfer Fluid (HTF) by ambient air (Fig 2). The HTF is provided to the vaporizer after it is heated by the air heater. Air flow direction from top to bottom is recommended to minimize cold air recirculation. The system is fully backed up by fired heaters, which can vaporize the LNG when the air is too cold to provide heat at the required temperature. The air heater with intermediate fluid option was in operation since early 2004 at Dahej LNG Terminal, India, and no problems have been reported (Fig. 3).

The forced draft ambient air vaporizer would have significantly lower operating costs because the recovered heat from ambient air would save on fuel consumption for LNG vaporization. The operation and utilization rate of the force draft ambient air heaters are site specific.

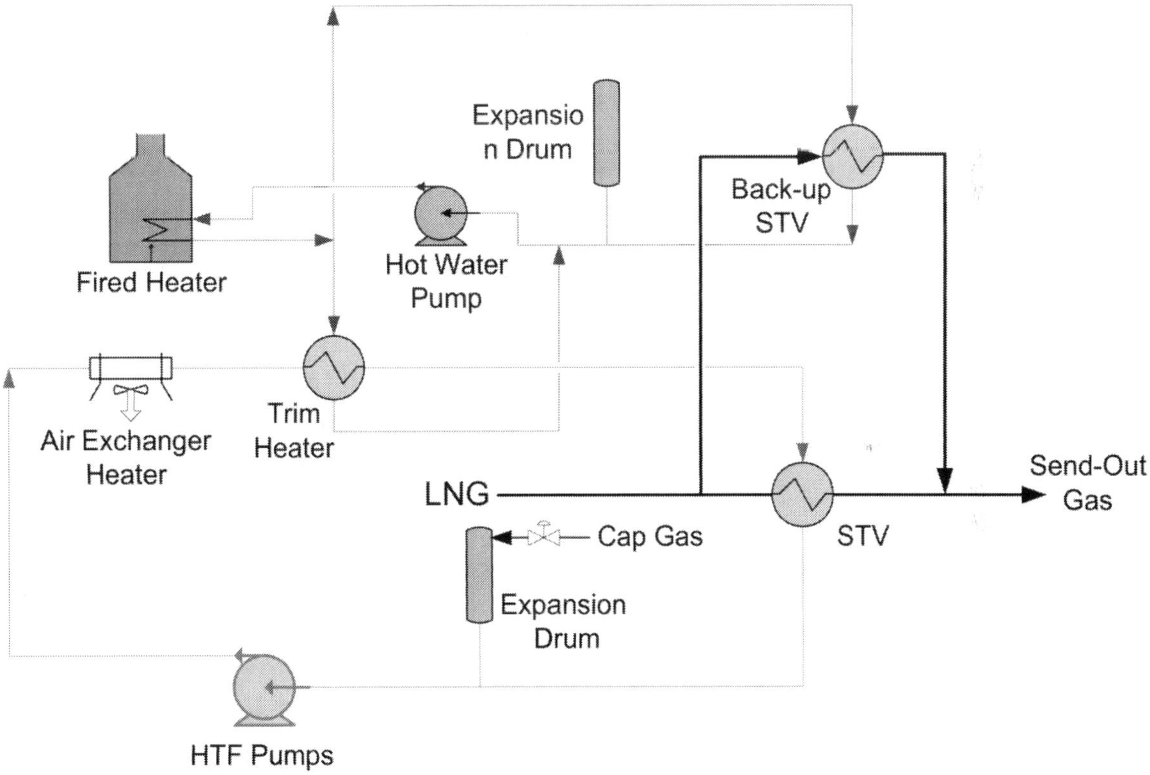

Fig. 2 – Forced Draft Air Heater Vaporizer

The selection of a HTF for LNG vaporization and the design temperature directly affects the equipment sizing, reliability of the vaporization system, operating cost and environmental impact. The HTF needs to be chosen based on operating temperature range, low freezing points, and non-flammability. Other physical properties that should be considered are those that would affect operation and heat exchanger design such as freezing point, density, thermal conductivity, viscosity and specific heat. A low-viscosity HTF at low temperature decreases both heat exchange surface area of the air heater and the vaporizer such that the total installed cost can be reduced. The low-viscosity HTF also reduces HTF pumping power, resulting in low operating costs.

The same concept of the indirect air-to-LNG vaporization process with propane as a HTF has been developed by Mustang Engineering Co. under the trade name "LNG SMART." In this process, propane is vaporized in air-fin heaters and in turn condenses to vaporize LNG in shell and tube heat exchangers.

Fig. 3 – Air Heaters (Courtesy of Dahej Terminal)

Since the HTF outlet temperature from the air heater is lower than that of the firing vaporization option, the selection of the vaporizer is an important parameter. Although the selected HTF should have a low freezing point, freezing inside the vaporizer remains a concern because the LNG inlet temperature is as cold as -160°C. To overcome such freezing concerns, a novel design arrangement of the shell and tube exchangers in a three-shell assembly was developed (Fig. 4).

The operation mechanism of this vaporizer is such that incoming LNG introduced to the first exchanger, 2401MC, is vaporized by the warmed LNG vapor from the exchanger 2402MC. During this process, the warmed vapor is chilled by heat exchange with cold LNG and is warmed up again by the exchanger 2403MC by a HTF media. This novel vaporization scheme significantly reduces the possibility of HTF freezing inside the vaporizer. This allows a HTF with a higher freezing point than that of a conventional STV. It also reduces viscosity of the HTF and improves the flow scheme in the heat exchanger such that AH design can be more compact than designs using the conventional STV.

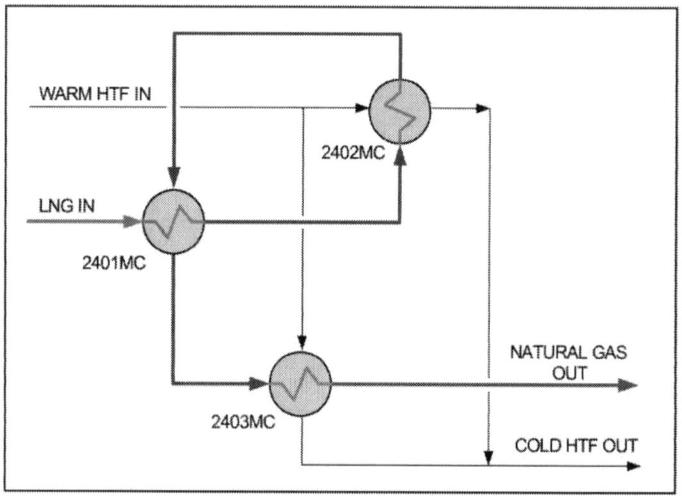

Fig. 4 – Schematic of the Three-Shell Vaporizer

OPTIMIZATION OF THE AIR HEATER VAPORIZER

Optimization Process

Once the feasibility of air heater units has been established for particular site conditions, preliminary engineering may begin in conjunction with evaluation of site meteorological data. After preliminary engineering of the units has been completed, CFD modeling should be developed based on the selected plot plan layout. At this point, further development of the air heater units becomes an iterative process with the CFD modeling playing a key role in the optimization of the air heater design.

Entrainment of cold fan exhaust into ambient air heater fan intakes (also referred to as cold air recirculation) can reduce the efficiency and operability of air heater units and must be taken into consideration in the air heater design. The colder air exhausted by air heaters has a higher density than that of warmer surrounding air and tends to settle downwards. This settling effect often has an impact on the performance of the air heater unit. Wind conditions also have an impact on cold air recirculation effects, as both wind direction and wind velocity can have a significant influence on the efficiency and operability of ambient air heaters.

The projection of recovered heat as a function of site conditions requires developing correlations of the re-circulated air temperatures with wind direction, wind velocity, and associated air paths around surrounding structures (LNG tanks, building, etc). The recirculated air temperatures under a given wind velocity and direction should be taken into account in estimating the recovered heat. High ambient air temperatures are usually not cause for concern and in most caves are favorable. However, low ambient air temperatures must be considered in order to ensure that the required duty is achievable with the projected air heater size.

Two variables affect the amount of heat that can be recovered from ambient air for a particular air heater design: ambient air temperature and dew point temperature. Optimization of the air heater design must be performed by varying these two variables in order to determine a practical design with a lowest life cycle cost.

At the beginning of the optimization process, the air heater design temperature should be optimized without consideration of relative humidity. A higher design ambient temperature results in the design of a smaller air heater size, resulting in low capital cost, but requires a large supplementary heat supply throughout the entire year. Alternatively, a lower design ambient temperature results in the design of a larger air heater size, which in turn will result in high capital cost. Figure 5 illustrates the relationship between the ambient air and recovered heat with the different ambient air design temperatures.

Wet bulb temperature taken in conjunction with dry bulb temperature is an indication of dew point temperature. Dew point temperature is an important consideration in the design of air heater units because it can be related to the amount of heat that an air heater can recover from latent heat as a result of condensation of water from the air. The amount of heat recovered from latent heat is significant in that it comes as an additional benefit over the amount of heat recovered from sensible heat. The relationship between design relative humidity and capital cost of the air heater is shown in Fig. 6. A higher design relative humidity decreases capital costs, but frequently requires operation of the supplementary heat source. Determining the appropriate design relative humidity value also requires optimization based on the life cycle costs along with frequency analysis of relative humidity data.

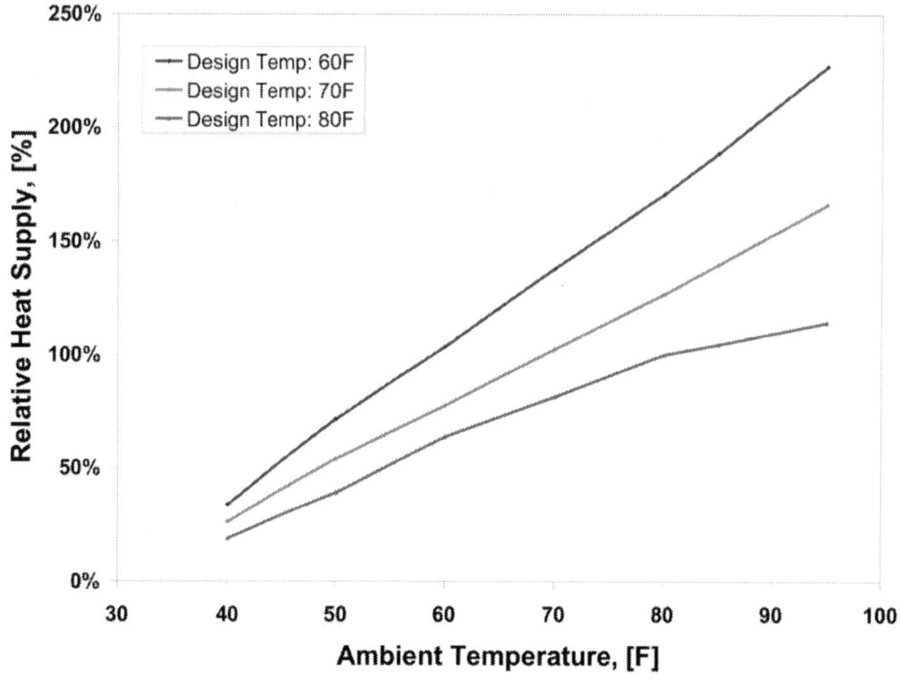

Fig. 5 – Recovered Heat from Different Design Ambient Air Temperature

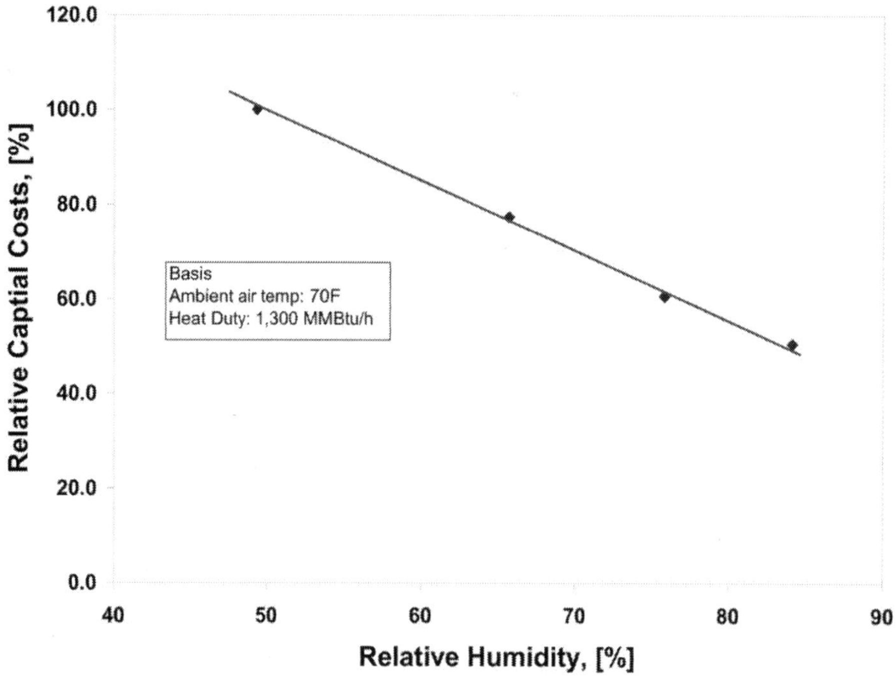

Fig. 6 – Relationship between Relative Humidity and Capital Costs

The minimum threshold temperature to stop operation of the air heater should also be determined. The power consumption and an even point of heat gain from the air heaters can be used to determine the turn-off temperature of the air fan. Continuous fan operation is recommended unless ambient air temperature drops below 45°F. During winter seasons, it is expected that fans may be turned off for a considerable duration depending on the site conditions.

Figure 7 presents a typical heat recovery with 70% relative humidity in the Gulf of Mexico region. During the winter season (December to February), supplementary heat is required, such that a considerable amount of fuel gas is required. The average heat recovery rate is a strong function of the ambient air design temperature and actual meteorological data.

In certain conditions, condensation of water from the air will occur as a result of the cooling effect on the air from the air heaters. Condensed water must be properly disposed of, which includes heating of the condensed water prior to disposal in order to minimize temperature pollution.

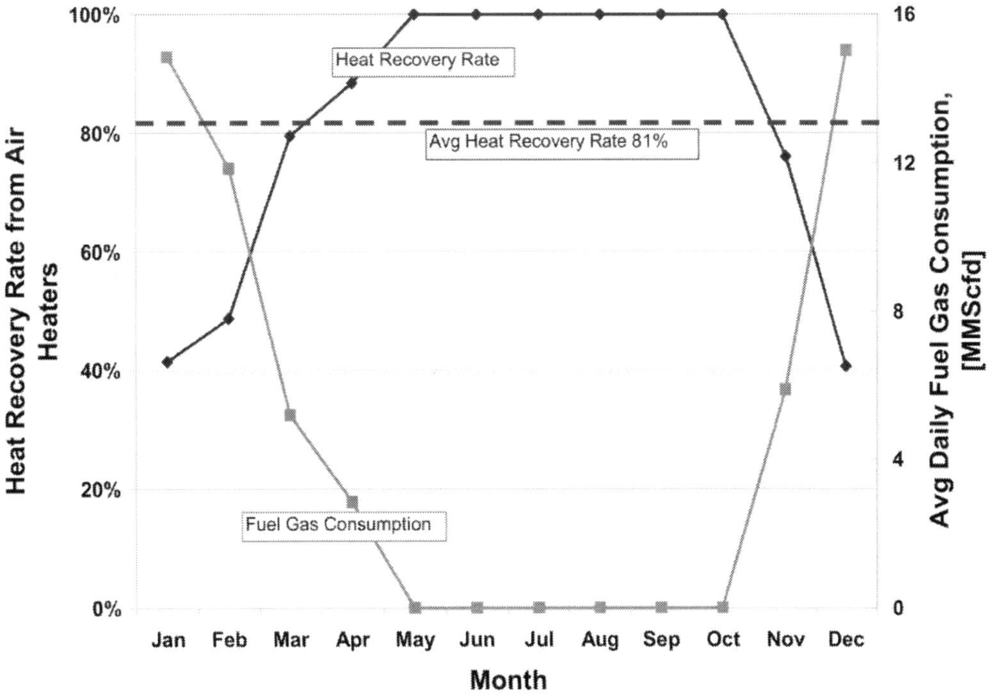

Fig. 7 – Typical Heat Recovered from AHV in Gulf of Mexico Area

ECONOMIC COMPARISON

Plot Space

Total plot space required for ambient air vaporizers is an important issue to consider due to the relatively large amount of area needed for installation. The natural-draft AAVs typically require more plot space due to equipment size in comparison to SCVs and forced draft AHVs. The necessary plot space is increased due to spacing requirements between units in order to avoid cold air recirculation. Additional plot space is required for the supplemental heat source and Vaporizers. The forced-draft AHVs require the longest plot space for the air heater unit. Table 1 below shows a comparison of total required plot space for each of three options based on an example LNG terminal with a natural gas throughput of 1.5 BScfd.

Table 1 – Plot space Comparison

		Installation	SCV	AAV	AHV
Plot Dimension	SCV	10 SCV's	250 x 250	-	-
	Fired Heaters	3 x 350 MMBtu/h	-	250 x 250	250 x 250
	STV	375 MmScfd x 5	-	100 x 100	100 x 100
	Ambient Air Heater	1350 MMBtu/h	-	275 x 1,100	90 x 1,330
Total Plot Area		ft^2	62,500	375,000	192,200
		Acres	**1.44**	**8.62**	**4.42**

Note: The indicated plot space for the AAV process is based on projected values and requires pilot tests to be conducted since scale-up vaporizers have not yet been developed.

ECONOMIC ANALYSIS

Economic evaluations for each of three vaporization schemes have been evaluated with a pre-tax life cycle cost model. The expenses (capital, maintenance, and operating) for each scheme were calculated and used to determine an expense based Net Present Value (NPV). No tax credit was considered in this model because of revenue exclusion.

It was assumed that the terminal is located on the coast of the Gulf of Mexico. The basic vaporization conditions are as follows:

- Vaporization Rate: 1,500 MMScfd
- Send-out pressure: 1,300 psig
- Send-out temperature: 34°F (at the battery limit)
- Vaporization heat duty: 900 MMBtu/hr

Table 2 summarizes the relative economic comparison of the two ambient air vaporization processes using the SCV option as a basis. The AAV requires higher capital costs than the AHV option. In addition to the capital expenditure of each vaporization scheme, there are important parameters affecting economic evaluation, such as fuel cost, electric power cost, and thermal efficiency of the supplementary heat supply equipment. Since the AAV process does not require the electrical power for fan motors and the HTF transfer pumps, annual operating cost is 26% of the basis SCV option, while the AHV process requires 37% of the SCV option. The overall expense-based Net Present Value (NPV) shows 43% and 53% for the AAV process and the AHV process, respectively.

Table 2 – Relative Economic Comparison

Description	SCV	AAV	AHV
Estimated Capital Costs *1	100	380	350
Annual Operating Costs *2	100	26	37
Expense Based NPV *3	100	43	53

Notes
1. Supplementary heat sources and vaporizers included.
2. $6/MMBtu fuel gas costs.
3. 25-year life cycle with a 12% discount rate.

The overall economic evaluation is primarily affected by fuel gas cost and capital cost. In order to evaluate the economic effect of the fuel gas costs, a sensitivity analysis was performed. Fig. 8 presents the results of a sensitivity study on NPV based on varying fuel gas costs. This sensitivity study assumes a fixed electricity cost ($0.08/kWh) and a 25 year life cycle with a 12% discount rate.

As fuel gas cost decreases, the expense NPV of ambient air vaporization becomes nearer to that of the SCV option; whereas the relative expense NPV decreases (becomes more economical) as fuel gas cost increases. The AAV process proves to be slightly more economical than the AHV process because of power savings on fan motors and HTF pumps.

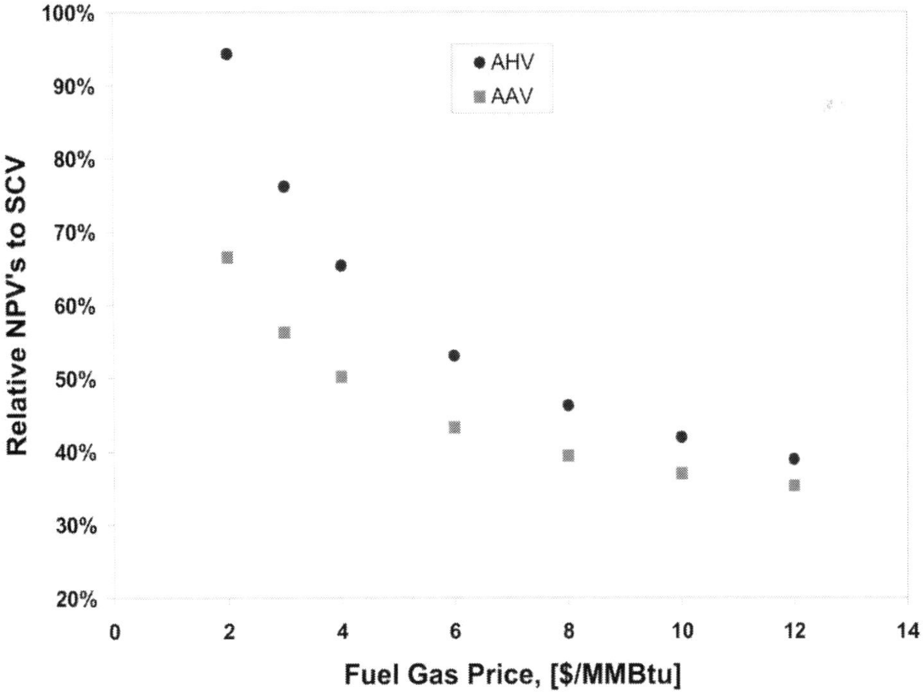

Fig. 8 – Sensitivity Study of NPV with Fuel Gas Costs

CONCLUSION

An LNG receiving terminal's main source of emissions generally comes from its LNG vaporization process. Site-specific environmental constraints and permitting issues must be carefully reviewed in the selection of an appropriate vaporization process for a given LNG terminal. Tighter environmental constraints have become a catalyst for alternatives to the more common Submerged Combustion Vaporizer (SCV) processes and such alternatives must be considered for future applications.

Reusable energy schemes such as those using Ambient Air Vaporizers (AAV) or Forced Draft Air Heater Vaporizers (AHV) offer some advantages, but require high capital costs in return for lower operating costs. Design parameters for ambient air vaporization are also heavily dependent upon site climate conditions. The two key variables that affect the amount of heat that can be recovered from ambient air for a particular air heater design are ambient air temperature and dew point temperature. Optimization of the air heater design must be performed by varying these two variables in order to determine a practical design with an attractive net present value. Further optimization of the air heater design will take air recirculation into account as a function of wind speed and direction.

An economic comparison of the ambient air vaporization processes indicates a lower life cycle costs in comparison to that of SCV vaporizers. However, plot availability is a major consideration for any ambient air vaporization option and the attractiveness of ambient air vaporization fluctuates as a function of the price of fuel gas. Lastly, the AAV option is unproven in LNG terminals and the projected cost savings must be evaluated against the operational risks.

REFERENCES

1. Cho, J.H., Patel, H., Coyle, D.A., and Durr, C.A. : "A Case Study for Integrated LNG-to-Power : Technical Comparison of Available Cost Reduction Methods," Proceedings, presented at the AIChE 2004 National Meeting, New Orleans, 25-29 April 2004.
2. World Bank Group: "Table 4, Liquid Effluent from Onshore Oil and Gas Production, Oil and Gas Development," Pollution Prevention and Abatement Handbook (1998).
3. "Coast Guard Suspends Offshore 2004 Terminal Applications, "LNG in World Markets, Vol. 16, No.9, p.14-16, October 2004.
4. "Sinking Feeling : Offshore U.S. Terminal Developers Run into Cost, Environmental Challenges," LNG Express, Vol. XIV, No. 10, p. 2-3, 2004.
5. Himmelberger, L and Billman, P.: "LNG Vaporization Technology: The Environment vs. Energy Source", Proceedings, presented at the AIChE 2005 National Meeting, Atlanta, GA, 10-14 April 2005.

BIOGRAPHY of SPEAKER

Joseph H. Cho is a Technical Advisor in the LNG, GTL and Gas Processing Division at KBR. He is a registered professional engineer and has been in the LNG industry for sixteen years. He was involved in process design of LNG liquefaction plants and import terminals for various regions as a senior engineer. He graduated from the University of Oklahoma with a Ph.D. degree in Petroleum Engineering.

Email : Joseph.cho@halliburton.com

Practical Optimization of LNG Terminal Process

In-Soo Chun
Senior Executive Vice President
Chang-Hag Kim
General Manager
Hyundai Engineering Co., Ltd.

AIChE Spring Meeting, April 2006
6th Topical Conference on Natural Gas Utilization
Orlando, FL, April 23-27, 2006

ABSTRACT

Incheon LNG Terminal is one of the largest LNG Receiving terminals in the world. It provided the Seoul metropolitan area with 9 million tonne per annum (Mtpa) of natural gas in 2005. The terminal began operations with three storage tanks (100,000m^3 each) and 630 t/h of regasification capacity in 1996. In order to comply with the abrupt increase in natural gas demand, the terminal performed continuous expansion. The terminal currently operates 18 LNG storage tanks (10 above-ground, full containment tanks and 8 in-ground tanks: total storage capacity 2.5 million m^3) and 3,300t/h of regasification capacity. Two additional 200,000m^3 inground tanks are under construction.

The development of a terminal of this size posed many technical challenges. However, this terminal has achieved a high standard of reliability and a well-optimized LNG regasification process through relentless practical design improvement and value improving practices by the engineering company with the cooperation of the terminal operator.

This paper discusses detailed practical optimizations for the following issues:

- Design LNG composition envelope with weathering impact
- Vapor handling system with two train-dual recondenser system
- Operational flexibility with different tank types (in-ground and above-ground tanks)
- Hot working and tie-in procedure near the operating LNG facility
- Optimum vaporization with an unfavorable site condition (low seawater temperature in winter)
- Process verification and optimization with dynamic simulations

INTRODUCTION

After successful operation of Pyeong-Taek LNG terminal in 1986, the first LNG import terminal in Korea, Korea Gas Corporation (Kogas) took a step forward to build the second LNG import terminal (Incheon LNG terminal) in 1991. The terminal was built on a reclaimed island west of Incheon city, which is a main transportation hub to the Seoul metropolitan area. The total reclaimed area for the terminal was about 99 ha (245 acres), which could accommodate sequential expansions.

The first phase of the terminal consists of 3 LNG tanks (100,000m^3 each, 9% Ni full containment), 360t/h of regasification capacity, and 270t/h of back-up vaporizer capacity. As a result of continuous expansion, the terminal operates 18 LNG storage tanks (10 above-ground full containment and 2 in-ground membrane tanks with two under construction), and total storage capacity reaches 2.88 million m^3. The terminal's regasification capacity increased to 3,300t/h (equivalent to 3.8BScfd) through its consecutive expansion.

There were many technical challenges associated with the development of the biggest terminal. However, this terminal has achieved a high standard of reliability and a well-optimized LNG regasification process with the cooperation of the terminal operator through relentless practical design improvements and value improving practice by the engineering firm.

This paper describes the technical challenges and process optimization during Front End Engineering Design and Detail Engineering Design.

PROJECT GENERAL

After the oil crisis in the 80's, the Korean government responded by diversifying its energy import sources. Since Korea did not have enough energy resources, most of its required energy had to be imported from foreign countries. Diversified energy sources and various import sources have been considered essential factors securing Korea's economic development. The government also needed to mitigate the environmental damages that usually accompanied industrialized development activities.

This was a driver in the decision to import LNG to Korea. The first LNG cargo arrived at Pyeong-taek terminal at the end of November 1986. Regasified natural gas was distributed to power plants (Pyeong-taek and Incheon), and the city gas supply was limited to the Seoul metropolitan area. As the living standard improved and consumption pattern moved up, the gas demands abruptly increased accordingly. Before the nationwide trunk line was completed, the gas supply was limited to only the Seoul metropolitan area.

The second terminal was located in Incheon city, the western part of the Seoul metropolitan area. The terminal Front End Engineering Design was completed in 1993 and construction began in 1994 [1]. Offloading from the first LNG cargo was successful in late 1996. Figure 1 shows the terminal locations currently operated and trunk lines. Currently, four (4) LNG import terminals are under operation in Korea: Pyeong-taek,

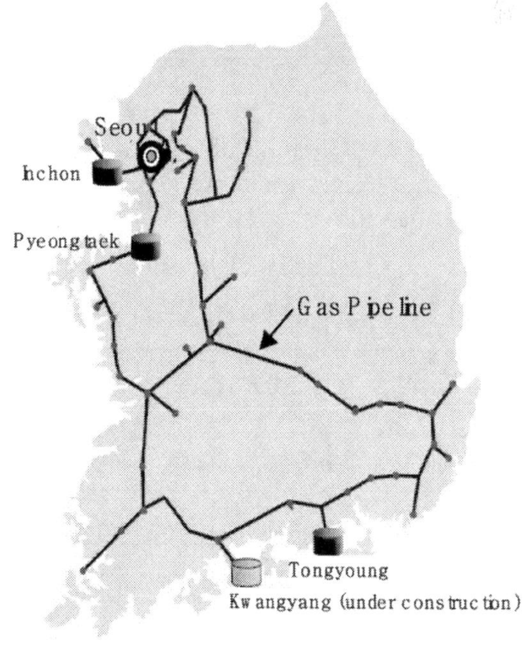

Fig. 1 – Korean Natural Gas Infrastructure [2]

Incheon, Tongyoung, and Kwangyang terminals. Pipeline grids cover cities nationwide.

Traditionally, Korean's heating demand is high because of severe winter conditions. The gas send-out load highly weighs on winter season. The annual gas send-out pattern is shown in Fig. 2. In order to meet high gas demand in the winter, it is inevitable to have a large storage capacity. This requires high construction costs of a terminal since it is the largest single cost factor. However, the large storage volume allows the terminal operator to have an initiative in LNG purchase negotiations. This also allows early spot purchases and provides great operation flexibility.

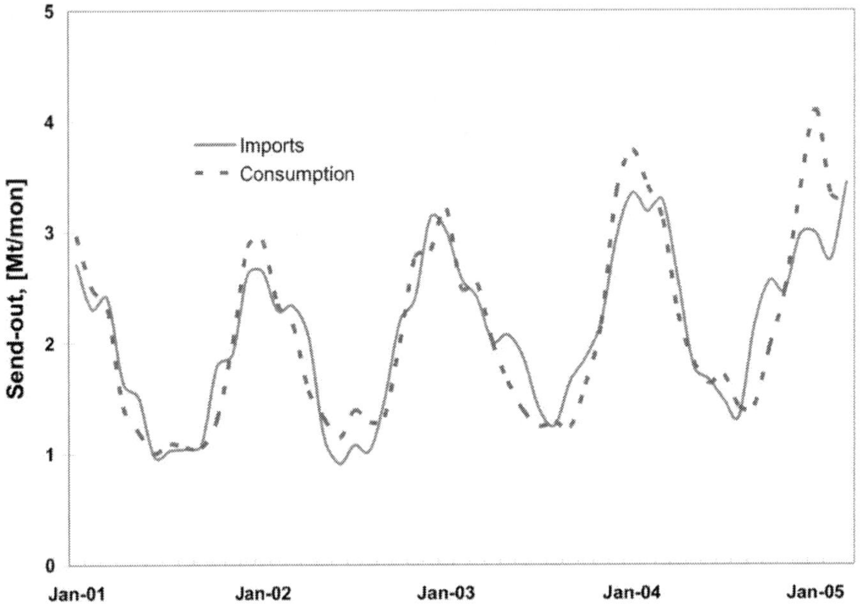

Fig. 2 – Natural Gas Demand Pattern [3]

The Incheon terminal operates as a base-load terminal. The terminal imported about 9 Mtpa of LNG in 2005. The initial regasification capacity (630t/h) and storage capacity have been increased through the continuous expansion projects. The following Table summarizes these expansion activities.

Table 1 – Summary of Incheon Terminal Step Expansion

Description	Base Plant	1st Expansion	2nd Expansion	3rd Expansion	4th Expansion	Total *2
	Oct-96	Dec-98	Dec-02	Sep-04	6-2009 *1	
LNG Tanks	3	3	8	4	2	20
Storage Capacity (1,000 m3)	300	300	1,080	800	400	2,880
ORV (t/h)	2 x 180	3 x 180	3 x 180	-	-	2,250
SCV (t/h)	3 x 90	5 x 90	10 x 90	3 x 90	3 x 120	3,690
Total Vaporization (t/h)	630	990	1440	270	360	

OUTLINE OF INCHEON TERMINAL PROCESS

Figure 3 illustrates the simplified process scheme of Incheon LNG terminal. LNG is transported by LNG tankers from different import sources. LNG is unloaded to the land storage tanks through unloading arms and unloading lines. The design unloading rate is 11,000 m^3/hr with three 16-in. full bore liquid unloading arms. At the initial facility design, return gas blowers were relocated from other terminals. Later it turned out that the return gas to the LNG ship was sent back by a pressure difference between the tank and ship without using blowers.

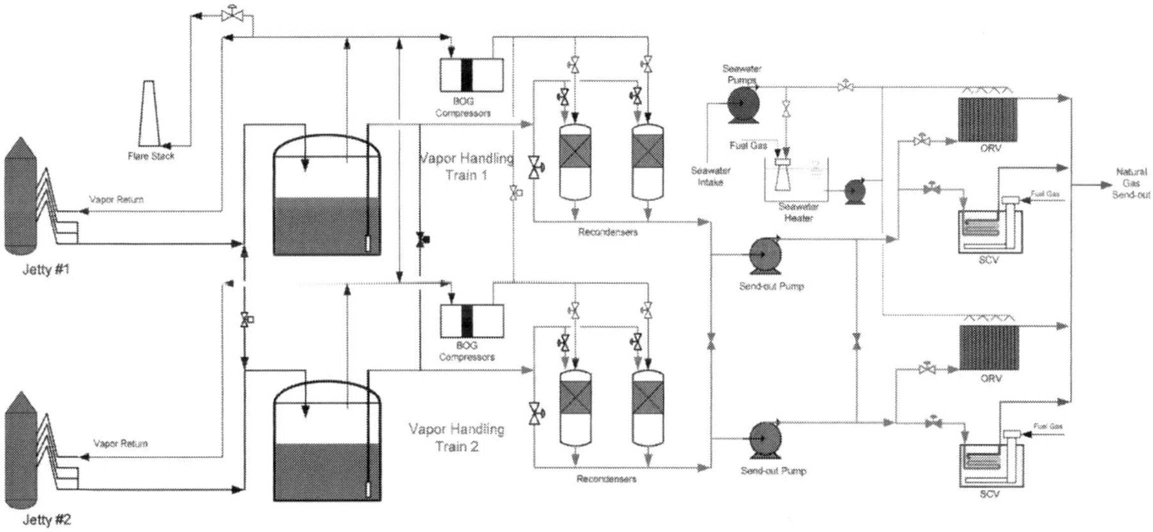

Fig. 3 – Incheon Terminal Process Scheme

The terminal has two berths allowing dual unloading or single unloading from either berth. This configuration also allows one ship unloading while another berth is on stand-by. The stand-by ship can perform vapor breathing operations to prevent over pressuring by extruding high pressure of vapor to onshore BOG header. It minimizes methane emission during unloading operation. Two berth systems can provide storage area zoning operation: LNG unloaded from one jetty can be stored in the above-ground tanks (8 tanks 100,000 m^3 each), and LNG unloaded from the other jetty can also be stored in 2 sets of Above-ground tanks and all in-ground tanks (10 tanks, total 1.7 million m^3 including the 4th expansion as shown in Table 1). Intentionally, all stored LNG can be blended to minimize density difference between tanks and achieve even calorific values of send-out gas.

It is generally considered that one large unloading line will be cheaper than two equal-sized smaller lines. However, there is a very real advantage in dual unloading lines. In case one line were to have operational problems, then the other line is still available for unloading the ships at half capacity or higher. Also, most terminals require the recirculation of LNG in the unloading line during the period a ship is not unloading. This is normally required so that the recirculated LNG picks up the heat out of the terminal via the send-out. Since the terminal has two unloading lines (2 x 32-in.) from each berth, this configuration allows recirculation to be accomplished without the addition of a smaller second line.

At the initial stage, the terminal planned to build above-ground full containment tanks. As the terminal expanded, it needed more storage capacity to cover the demand gap between summer and winter as shown in Fig. 2. As tankage area extends, the required hydraulics increases at the ship manifold. Significant factors in determining hydraulics are tank static head and frictional losses across the unloading arms. Underground storage tank could alleviate the hydraulic limitations. A large in-ground storage tank makes use of high cost reclaimed island more efficiently than conventional smaller above-ground tanks. In-ground tanks allow adherent positioning of tanks compared to above ground tanks. This also saves plot space because of elimination of the dike area.

Boil-off gas (BOG) from storage tanks is gathered in a BOG header and pressurized by BOG compressors. The pressurized BOG is then routed to the recondenser where it is recondensed with sub-cooled LNG discharged from in-tank pumps. The recondensed liquid is further pressurized by the send-out pumps and re-gasified by vaporizers.

The BOG rates are 0.075%/day and 0.1%/day for above-ground tank and in-ground tank, respectively. The large storage volume makes considerably higher BOG rates than other terminals. Ten (10) BOG compressors are installed to handle the BOG during unloading operation.

Open Rack Vaporizers (ORV's) are the base-load vaporizers, and Submerged Combustion Vaporizers (SCV's) are used to back up ORV's during peak shaving operation. However, the available seawater temperature drops below the design points during winter. The ORV performance then decreases considerably as seawater temperature decreases. In order to compensate the seawater temperature drops, twelve (12) seawater heaters can be operated.

TECHNICAL CHALLENGES AND OPTIMIZATION OF PROCESS

Design LNG Composition Envelope

Design LNG compositions vary depending on LNG import sources. In order for the terminal to have high operation flexibility, a design composition envelope was established. The LNG compositions from expected LNG sources were investigated and voyage distances were also estimated. LNG specifications of each import source may have a range of composition data. The composition data were analyzed statistically with weight values of each import source. This analysis resulted in projected LNG calorific value ranges of blended LNG at the terminal.

The nitrogen in LNG easily evaporates during sea transportation. For long voyages, some LNG export plants artificially inject liquefied nitrogen in the LNG in order to prevent methane evaporation from the cargo. The evaporation of nitrogen changes the arrival specifications, which deviates the LNG specifications from export plants.

Since nitrogen and methane have higher "K" values than other components in LNG, a special simulation program was developed to estimate the arrival LNG compositions because of weathering effects. Figure 4 illustrates the weathering impact of two different BOG rates. BOG rates and voyage time are main parameters in the weathering imports. The lower BOG rate of the LNG cargo mitigates the N_2 evaporation. A narrow LNG design composition envelope was developed, which resulted in reliable equipment design.

For small changes in the LNG composition, the effect on the quantity of boil-off gas is negligible, but the effect on the nitrogen composition of the boil-off gas is quite significant. An increase in heat ingress to the LNG tankers will generate more vapor during the voyage. This will influence the composition of the boil-off vapors and affect the quantity of boil-off gas generated, depending on the length of the voyage. Since nitrogen is the more volatile component, the longer the ship spends in transit, the less the nitrogen content of the boil-off gas. This is also a function of the heat leakage of the LNG tanker, designed according to the latest literature for about .15 to .25% day.

This design tool reasonably narrows down the design composition envelope, which results in allowing the prevention of oversizing main equipment. From the actual operation data feedback, this design tool has been confirmed its value in designing main equipment, such as vapor handling system and pumps.

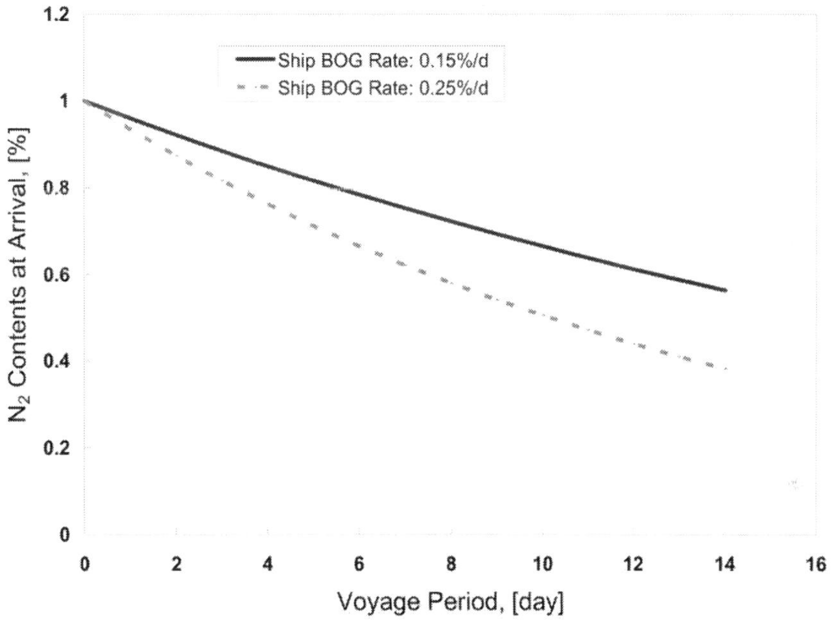

Fig. 4 – LNG Weathering Process Impact on N2 Composition

Operational Flexibility with Different Tank Types

The terminal utilizes both above ground (A/G) tanks and in-ground (I/G) tanks. These two types of tanks are zoning with its tank type. In order to maximize the unloading operation flexibility, the LNG unloaded from Jetty # 1 can be loaded to either A/G or I/G tanks. By the same token, the LNG unloaded from Jetty # 2 can also be loaded to either tank zones.

The unloading lines and double jetty configuration allow single or double unloading (simultaneous unloading from each jetty). Unloaded LNG will then fill one tank or be distributed among several tanks, depending on LNG tank stock levels.

For the dual unloading operation, the following were considered in systems engineering:

- Hydraulic availability for a single tank filling (either A/G tank or I/G tank);
- Minimization of transient pressure impact due to cargo pump failure;

- Minimization of transient pressure impact due to the sudden closures of emergency shut down valves;
- Hydraulic head balance between two types of storage tanks.

Piping design must consider the impact of fluid hydraulic hammer which occurs whenever there is a sudden change in fluid velocity, resulting from numerous causes such as block valve operation, valve failure, pump failure or closure of a check valve when flow is reversed. Sudden velocity reduction changes the kinetic energy, compressing the fluid and stressing the pipe walls and supports. The resulting pressure surge can exceed pipe line design pressure and cause a line rupture.

When the indicated surge pressure sets the design pressure for the unloading line, the pipe must be selected with a wall thickness to safely withstand this pressure. However, since economics suggest use of as thin a pipe wall as possible, a means of reducing the pressure surge is preferred. In order to achieve economic design as well as system safety, valve closure time is set to exceed the pipe time period. Since fluid velocity changes are more pronounced as valves approach shutoff, the most useful values in minimizing the pressure surges are those with the slowest closure rate. Another consideration was valve selection since some valves exhibit less pressure surge for the same closure time. Gate valves were least beneficial here since both butterfly and globe valves create a lesser pressure buildup. Butterfly valves were used because they were capable of dissipating kinetic energy over the full-closure time. In fact, using a butterfly rather than a gate valve could reduce the maximum surge pressure by half or more.

Hydraulics required for unloading consists mainly of frictional losses across the unloading arms and lines and static head difference between the ship manifold and tank top filling line. The elevation difference between A/G and I/G tank creates a hydraulic imbalance. This is mitigated by tank zoning. In order to provide operation flexibility, simultaneous filling to two types of tanks was designed.

Hot Working and Tie-in Procedure

Ensuring safe operation of the terminal, the effective expansion construction was placed in the first place during about 10 years of continuous expansion. Constructability of expansion work is generally considered less than the independent project because of a limitation of access and mobility. Safe work controls are extremely important in both the operation of an existing terminal and expansion construction. In order to achieve this, as well as to ensure high constructability, a unique safe work procedure was established including a tie-in procedure ensuring the safety operation during the hot work near the operating LNG facility.

The procedure requires providing a separate access road and fence in order to prevent general workers from accessing the operational facilities without permission and supervision. After expansion work had been done, it was required to tie-in between the expanded facility and operating facility. Ensuring safety in tie-in work was the top priority. Uninterruption of the existing terminal operation was also emphasized. The details of hot working procedure are illustrated in Fig. 5.

A work scope has to be defined in the work definition stage so that the worker will do the confined work only. This eliminates the unnecessary work and enhances safe operation of an existing facility. A work procedure has to be prepared and submitted to the Safety Review Committee. The committee reviews the appropriateness of working procedures and safety plans. The details involve process review, work scope, impact of gas send-out, and safety achievement methodology. Once the work procedure is approved by the committee, the superintendent will supervise the work at the designated working date.

The normal tie-in procedure is also discussed as shown in Fig. 5. It requires blocking the designated section. Sometimes, a small cryogenic valve is found to be leaking. Then the blocking section has to be redefined. The selected block has to be warmed up. The blocked section has to be inerted by nitrogen. Before the tie-in work, the Safety Enforcement Team checks the working conditions in order to ensure the hot working. It includes general working conditions, safety devices, fire fighting devices, and methane contents. The superintendent allows the hot work if the methane density is less than 10% of Level of Explosion Level (LEL).

As the tie-in point can not be tested by hydrostatic pressure or the pneumatic closure test, it is strictly required to be the first grade of non-destructive examination on the welding area. In order to avoid any pocket of mixed gas with air, nitrogen is injected again for inerting the blocked section. The nitrogen should be injected with the same direction of LNG or gas flow. The dew point of the section blocked is measured and the acceptance criterion is -40 °C. Gradual cool-down is then applied on the tie-in section. The normal cool-down grade is 15 – 20 °C/hr. Pressure equalization is also applied if the block section is a high pressure line.

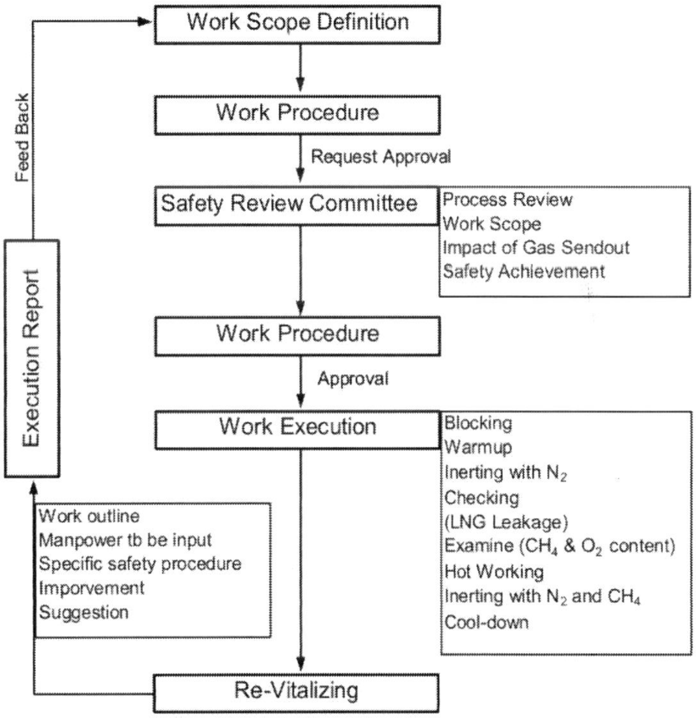

Fig. 5 – Hot working Tie-in Procedure

Parallel Recondenser Operation

During ship unloading, the LNG sent into the land storage tank physically displaces vapor out of the tank into the vapor-handling system. At the same time, the ship's vapor-space pressure will start to drop because liquid volume is being unloaded. A part of the volume is made up by boil-off in the ship. However, at the maximum ship unloading rate, the generated boil-off volume is insufficient to replace the liquid volume being transported. Therefore, vapor return from the shore is necessary to make up the volumetric differences. To accomplish this vapor return, the vapor out of the storage tank is pressurized by the return gas blower and directed through an insulated line to the ship, where it enters via a vapor loading arm. The vapor return can also be achieved by a simple pressure difference between land storage tanks and an LNG ship.

Vapors emitted from the top of LNG storage tanks are generated by three sources:

- Heat leak into the system (tank and piping systems);
- Energy input by LNG carrier pumps;
- Displacement vapor as a result of filling the LNG storage tank.

Other parameters which have an effect are: 1) Composition of liquid loaded into tanker at liquefaction plant, 2) heat leak into LNG tanker, 3) Duration of voyage, and 4) LNG tanker equilibrium pressure.

Generally, the vapor flow from the land storage tank during ship unloading is larger than the vapor-return requirements for the ship. Storage tank pressure has a major bearing upon compressor capacity requirements. By maintaining a 60 – 70 mbar pressure differential between storage tank pressure and saturation pressure of the LNG on the ship, the net energy input does not form flash vapors. Instead, this energy input manifests itself as a slight rise in saturation temperature of the stored LNG. During unloading operation, the tank operating pressure increases to 250 mbarg from about 150 – 180 mbarg of normal holding operation pressure.

A vapor handling system for the terminal is shown in Fig. 6. The excess vapor must be pressurized by the BOG compressors and routed to the recondenser, where it is recondensed by heat exchange with the sub-cooled LNG discharged from the in-tank pumps. The recondensed LNG and by-passed LNG is mingled and sent to the send-out pumps.

The major consideration in recondensing is that it does not work well at low send-out rates. Thus, if low send-out rates occur with great frequency or if there is a strict requirement that flaring (venting) never occurs (in cases of infrequent low send-out) then it may be necessary to install the pipeline compressor. The recondensing assessment study results indicate that BOG could be recondensed with the send-out rate except during dual unloading under minimum send-out conditions. Since this was a highly unlikely situation and could be properly managed, the terminal installed four sets of recondensers in parallel as shown in Fig. 6.

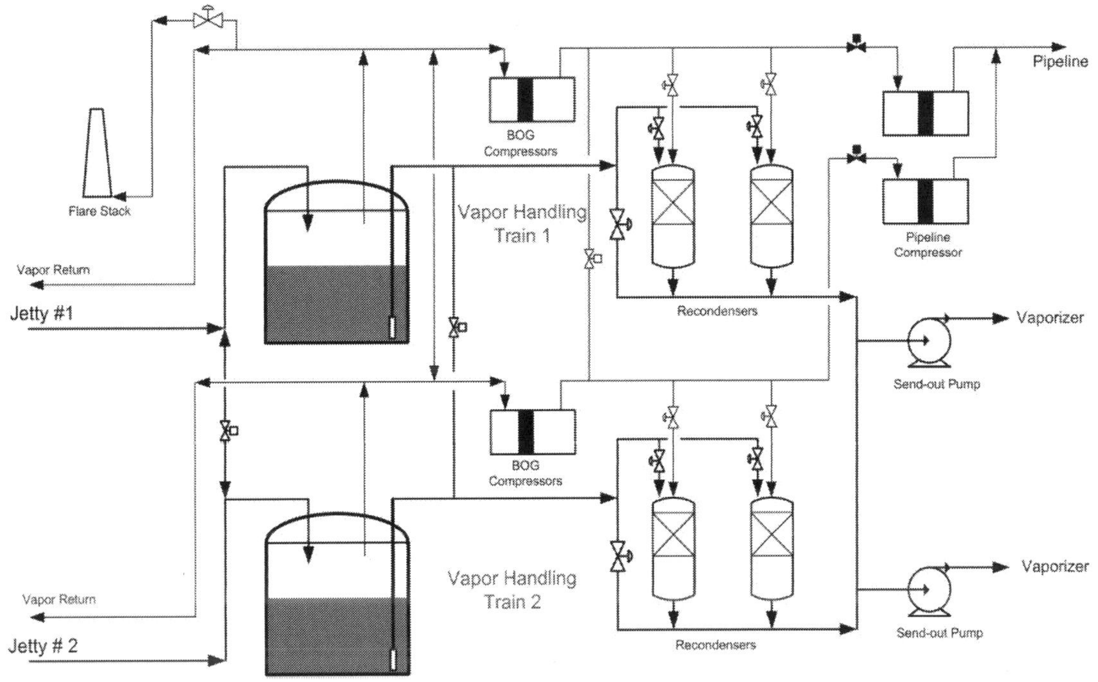

Fig. 6 – Terminal Vapor Handling System

The terminal vapor handling system adopts a two train concept. Each train consists of two recondensers, of which capacity is 75% of each train vapor handling volume. Even when one recondenser goes down, this configuration allows the 75% vapor recondensing operation in a train. The excess vapor can be forwarded to other train, where it can be recondensed.

The recondenser has the following functions:

- Condense boil-off gas (BOG) by direct contact with sub-cooled LNG on the wet exposed surface of the recondenser packed bed.
- Act as a surge drum to provide inventory for the send-out pumps.
- Provide a sufficient and controlled suction pressure for the send-out pumps.

The recondenser is a vertical, cylindrical vessel with a packed bed in the center of the upper portion. Liquid is present in the lower part of the vessel, in part of the annulus around the packed bed and in part of the packed bed. The level of liquid in the packed bed is a function of the quantity of BOG and the liquid loading of the recondenser. Liquid from the LNG storage tanks and jetty recirculation liquid enter the top of the recondenser packed bed. BOG compressor discharge vapor also enters the top of the packed bed and is condensed by the sub-cooled LNG as they flow down through the packed bed. Vapor from the send-out header enters the top of the annulus as a cap gas, which maintains the annulus pressure [4].

The terminal selected a dual volume control concept for the recondenser. In volume control, the measured annulus liquid level and the measured packed bed liquid level are used to calculate the total volume of liquid in the recondenser. The volume of liquid in the annulus plus the volume of liquid in the packed bed is kept constant. If the liquid level in the packed bed increases, the volume control reduces the level in the annulus to maintain a constant total volume. In this dual recondenser volume control configuration, the following four independent items were selected to control:

- Vapor distribution to each recondenser by BOG inlet control valves;
- The annulus pressure control by the vapor control valves;
- The ratio of liquid to BOG vapor or the liquid level in the packed bed is controlled by the control valve in the liquid LNG line to the packed bed;
- The volume of liquid in the recondenser is controlled by the control valve in the liquid LNG line to the recondenser liquid space or in the recondenser bypass.

The control valve in the LNG line to the recondenser liquid space controls the volume by opening to send more liquid to the recondenser on falling level. When in the bypass, it controls the level by opening to send more liquid around the recondenser, thus causing less liquid to leave the recondenser on falling level.

Optimized Vaporization

In order to achieve low cost vaporization and minimize impact on the environment, Open Rack Vaporizers (ORV's) and Submerged Combustion Vaporizers (SCV's) are combined as a back up and peak shaving was chosen from various vaporization technologies. A seawater facility requires a long seawater intake pipe (about 1,500m) since the tide difference is high (about 5m – M.S.L.). The available seawater temperature drops 0°C, which is generally not recommended for ORV operation during the winter. Figure 7 illustrates the vaporization scheme of the terminal. LNG from the recondensers from each train is fed to the send-out pumps and the pressurized LNG is routed to the vaporizers. Seawater is supplied to ORV's. When the temperature of seawater supplied to ORV's is lower than the design point, it is mixed with the heated seawater from the seawater heaters.

As shown in the seasonal gas demand pattern in Fig. 2, gas demand in the winter is higher than in the summer. However, the seawater temperature as a heat source is lower in the winter. Since LNG regasification is the most important parameter for lowering terminal operating costs, optimization of LNG vaporization has been performed with these unfavorable conditions.

The following parameters were investigated for the vaporization optimization:

- Performance of ORV with the lower seawater temperature;
- Optimum seawater operating temperature;
- Economic combination of ORV and SCV;
- Economic justification of seawater heating;
- Overall economic assessment of vaporization facility.

In the Open Rack Vaporizer, the relationship between the number of heat exchange panel blocks and seawater requirement has been investigated in view of optimization. However, it was found that ORV manufacturers recommended a seawater temperature difference based on the manufacturer's equipment optimization, which resulted in a less number of heat exchange panels. By analyzing the ORV performance curves with the various operating conditions, the relationship was extracted as shown in Fig. 8. The recommended seawater temperature differences are based on a 42MW heat duty (equivalent to 180t/h – 200t/h of LNG vaporization). Manufacturers confirmed that the minimum outlet temperature from the ORV was 1.5°C, which is obtained when seawater inlet temperature is 3°C.

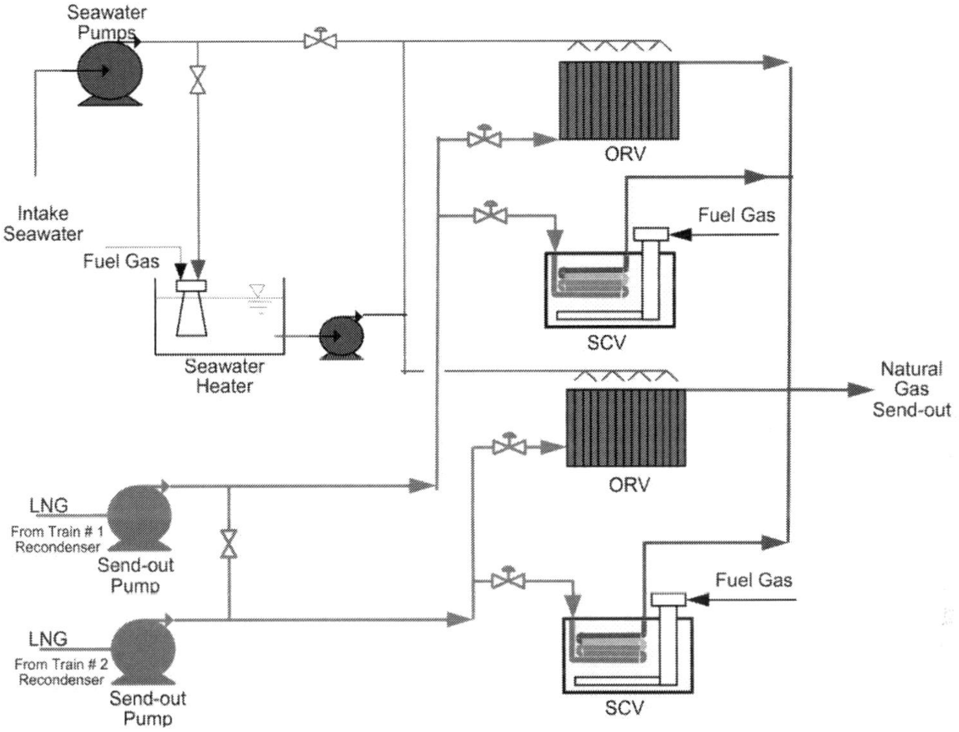

Fig. 7 – Vaporization Scheme

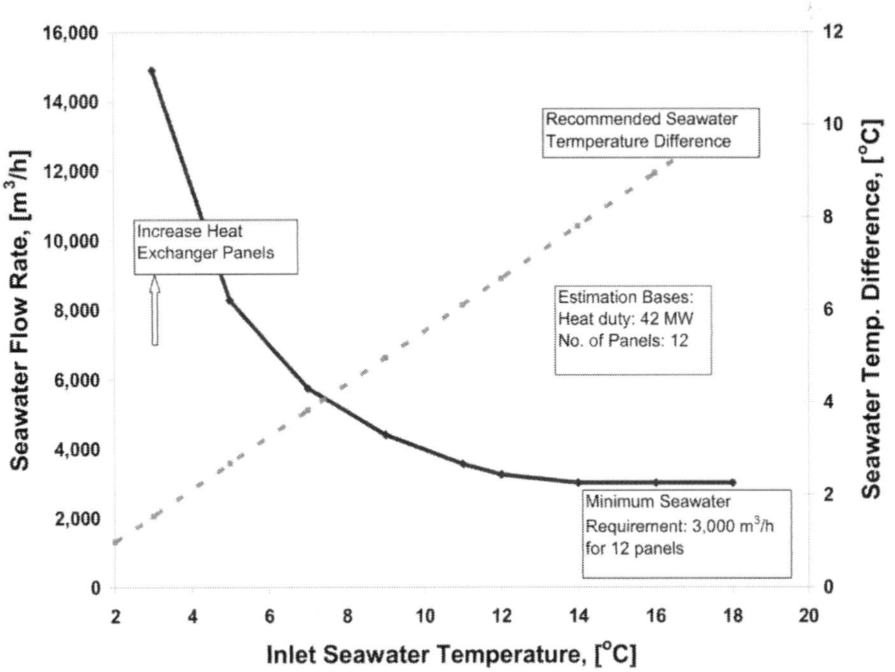

Fig. 8 - ORV Performance with Seawater Temperature

The generalized seawater requirement abruptly increases as inlet seawater temperature decreases, while seawater requirement is constant after a certain point even though its temperature increases. This is because seawater should be provided to maintain the minimum water film on the surface of the heat exchange tubes. The seawater requirement varies depending on the number of the heat exchanger panels and tubes.

The higher seawater design temperature requires a smaller ORV and seawater lift structure. This requires less capital and operation costs. However, these are only valid when the seawater temperature is high enough through the whole year. High seawater design temperature increases the SCV operation and seawater heating process, resulting in an increase in operating costs as seawater temperature decreases. During the winter, the fuel gas requirements for the operation of SCV's and seawater heaters increase exponentially as the recovered heat from seawater decreases (seawater temperature drops). The total economic analysis results are shown in Fig.9.

In order to optimize the design seawater temperature, expense-based Net Present Value (NPV) was calculated. From the economic analysis results, design seawater temperature was set at 5°C. The number of tube panels required was also determined based on design seawater temperature and expected delta temperature. The parameters used in economic analysis are summarized in Table 2. It was assumed that fuel cost was $3/MMBtu. Considering current fuel cost, the optimum design temperature may be shifted slightly to the lower point.

Table 2 – Economic Parameters

Description	Parameters
Facility life time	25 years
Discount rate	12%
Electricity cost	5¢/kWh
Fuel cost	3 $/MMBtu
Inflation rate	3%
Tax rate	Not considered

Application of Dynamic Simulation

Building an LNG facility requires capital investment. The terminal operator wants to ensure, prior to construction of the terminal, that latest industry standards for safety and efficiency can easily be met while reducing the necessary capital investment. Dynamic simulation can address many operational issues while reducing capital costs through the reduction of over design. Dynamic simulations were performed to simulate the start-up, unexpected shutdown, and a variety of trip scenarios of the main facility, such as reconderser, vaporization, and unloading system.

Because typical and popular steady state simulators cannot easily solve the pressure driven system, processes are often simulated with estimated parameters to simplify the system. However, simplifying the system introduces error, which can only be overcome by employing a design margin [5]. The customized dynamic simulation package was developed based on "OTISS" by ASPEN Tech. This simulator was integrated with the design and operation parameters verified by actual operations at previously designed terminal.

The dynamic tool used on this terminal project provides capabilities that are far superior to what is available from traditional process simulators. This dynamic tool allows operational confirmation of the dual recondenser system, increasing the effectiveness of control system and optimization of the vaporization system. The enhanced capabilities allow the design engineer to analyze margins and design factors of the sized equipment, thereby optimizing the owner's return on investments.

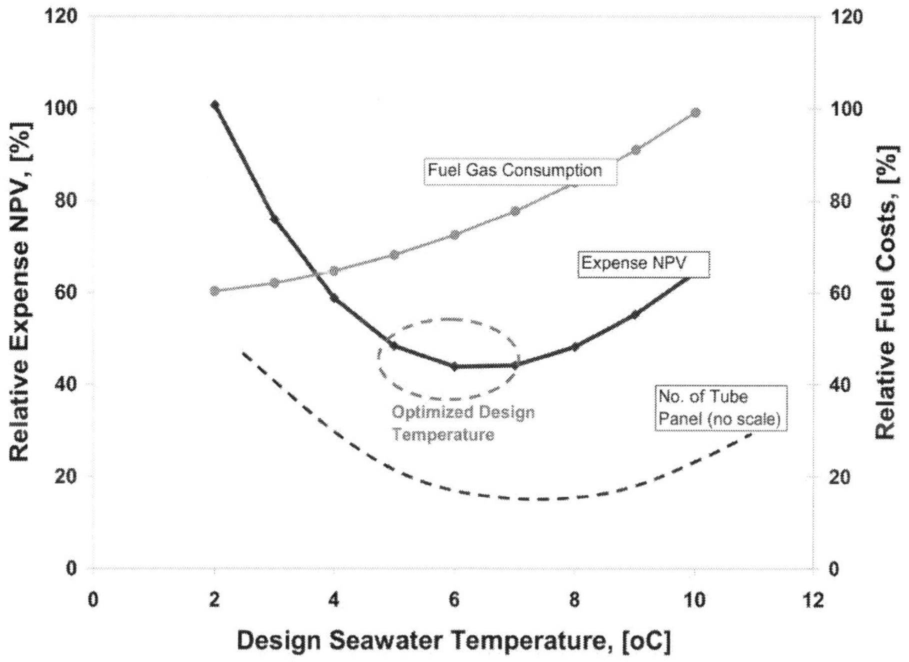

Fig. 9 – Economic Analysis of Different Seawater Temperature

The primary objectives for using dynamic simulation were three-fold. First, the terminal design was improved through the use of dynamic simulations, which can calculate the flare load, thus reducing the size of the flare. Second, it will aide in reducing the risk of excess operating pressure due to unexpected surge during certain scenarios, such as sudden LNG cargo pump failure, power outages, and emergency shut down valve closure. Dynamic simulation allows the valves to be properly sized and controllers properly tuned to avoid pressure surge may be avoided. Third, accurate dynamic simulations can give the owner the confidence necessary to operate the advanced control system, such as two train recondensings with a dual recondenser system. The dual recondensers linked with each train have been verified by the dynamic simulation during detail design.

CONCLUSIONS

The simulation tool for LNG weathering effect reasonably narrows down the design composition envelope, which results in allowing to prevent over sizing main equipment. From the actual operation data feedback, this design tool has confirmed its value in designing main equipment, such as vapor handling system and pumps.

Successful unloading operations with two different tank types: above-ground tank and in-ground tank, have been achieved by hydraulic equalization with the storage tank zoning. This results in high operation flexibility in LNG filling and blending.

The terminal vapor handling system adopts a two train concept. Each train consists of two recondensers, of which capacity is 75% of each train vapor handling volume. Even when one recondenser goes down, this configuration allows the 75% vapor recondensing operation in a train. The excess vapor can be forwarded to the other train, where it can be recondensed. Stable operation of the dual recondenser system has been achieved by volume controls.

Through the evaluation of the start-up, shutdown, and various trip scenarios, via dynamic simulation, proper operational procedures have been developed. An inherently safe design, as well as a thoroughly analyzed safety system, can be in place in the event of an incident.

ACKNOWLEDGEMENT

The authors wish to thank Korea Gas Corporation (Kogas) for permission to prepare and present this paper. Special thanks are also extended particularly to Dr. J. H. Cho, Kellogg Brown and Roots, for his kind assistance and valuable discussions.

REFERENCE CITED

1. Kwon, Y.J.: "Commissioning of Korea Gas' New Receiving Terminal at Incheon, proceeding of Gas Tech 1996, Vienna, Austria, 3-6 December 1996.
2. Source of Korea Infrastructure of Natural Gas
3. "Korea Monthly LNG Import and Demand", Korea Energy Economics Institute, http://www.keei.re.kr/keei/frame/d_results.html
4. Cho, H., Shah, SN, and Lim, K.H.: "Case History for Gas Re-Liquefaction System at the Pyeong-Taek LNG Terminal, South Korea", Proceeding of Gas tech 2000, Houston, 14-17 November 2000.
5. Perez, V., Hauche, W.E., Freels, J.A, Davis, K., and Durr, C.A.: "The 4.5 MMTPA Train – A Liquefied Natural Gas", Twelfth International Conference on LNG, Perth, Australia, 4-7 May 1998.
6. Kim, H.Y., Lee, J.H., Kim, D.H., Baek, Y.S., Kim, K.W., Park, K.H.: "A Plan for Operating Cost Reduction and Technology Development of LNG Vaporizing System", Proceeding, Presented at the Fourteenth International Conference on Liquefied Natural Gas, Doha, Qatar, 21-24 March 2004.

BIOGRAPHY of SPEAKER

In-Soo Cheon is a Chemical Engineer with a B.S.ChE degree from Seoul national University, Korea. He is currently senior executive vice president of Hyundai Engineering Co., Ltd.

He has worked as project manager for LNG distribution system in Seoul Metropolitan Area from 1983 to 1989. He has also worked as Project Director for Inchon LNG Receiving Terminal Project in Korea from 1991 to present. His career in LNG and gas processing spans 35 years involved in various gas monetization projects from conceptual design to commissioning. Through the LNG projects, he has gained extensive knowledge in efficient and practical design towards safe operation of the LNG terminal. Email: ischun@hec.co.kr

SCV with Low Emission – Flameless Burner Technology

Peter W. Falcone, P.E.
Product Development Manager
Frank M. Maupay
Product Manager, LNG Vaporizers
Selas Fluid Processing Corporation
Five Sentry Parkway East
Blue Bell, PA 19422

AIChE Spring Meeting, April 2006
6th Topical Conference on Natural Gas Utilization
Orlando, Florida, April 23-27, 2006

Abstract:

When designing today's LNG terminals, great care is given to selection of the vaporization technology best suited for the terminal's location and ambient conditions. A number of factors are taken into consideration: footprint, reliability, ambient conditions, capital expenses, operating expenses, and emissions. Increasingly, environmental issues such as air quality standards and impact on local marine life and ecosystems are influencing the selection process. Over the past five years, all four US terminals have undertaken significant expansions and in each case the terminals opted to continue with SCV as the preferred method for regasification. With these expansions, particular care has been taken to reduce vaporizer emissions to meet new and tightening air quality regulations while maintaining the original design's high efficiency. During this period, emission control abatement technologies have been at the forefront of SFPC research and development efforts.

A relatively new technology employing a flameless oxidation process is being pursued as the most effective abatement technology yet. This in house flameless oxidation technology has been used to destroy organic waste streams. Patented in 1989 by a subsidiary of SFPC, the process entails introduction of a lean, homogenous air fuel mixture via a specially designed dip tube into a bed of preheated inert media. A critical velocity profile stabilizes a reaction wave and provides the driver for the combustion process which will heat the water bath. CFD modeling software was used to develop the first prototype Flameless Thermal Oxidizer (FTO) heater, designed and tested at the SFPC Technology Center in August 2005. Results from this testing was positive and has led to scale up of a one half commercial scale pilot unit. The prototype unit is expected to enter the testing phase later this summer. Expectations are that the FTO burner technology will lower both NOx and CO emissions into the single digit range without post-combustion treatment and without compromising the simplicity, efficiency, and robustness that SCVs are known for.

Introduction:

With increasing demand for LNG imports, over 50 baseload terminals have been proposed in North America. Environmental issues are driving many of the permits for these new terminals, and emissions are a main concern. Submerged combustion has been the mainstay of LNG vaporization for the existing terminals and continues to be the primary choice for many of the newly proposed sites.

Much has been done over the past 10 years to reduce the SCV emissions to comply with ever stricter regulations. The first approach to lowering emissions was to install a water injection system in the SCV fuel delivery system, designed to lower the burner flame temperature. This resulted in a 50% reduction of NOx emissions and was deemed to be adequate with the relatively low cost of implementation. New installations are required to further lower NOx emissions to an overall reduction of 70 to 90 percent based on the uncontrolled burner emission level.

Technologies capable of meeting these requirements are complex and expensive. At present, proven combustion controls can achieve 70 to 80 percent reduction, and post-combustion controls are needed to reach 90 percent reduction in NOx emissions. Another component of the more stringent SCV emission limits is carbon monoxide. As a general rule of thumb, existing combustion control technologies are capable of reducing NOx emissions at the expense of increased carbon monoxide emissions. In other words, NOx and carbon monoxide emissions are inversely related to each other.

Post-combustion control technologies may avoid this NOx/carbon monoxide relationship but require additional chemical reagents to remove these contaminants from the gas stream. Chemical reagents sometimes slip through the process to the environment, becoming a separate emission problem to resolve. In the case of LNG vaporizers, some post-combustion emission reduction technologies require specialized catalysts to function and must be replaced periodically.

The GH Flameless Thermal Oxidizer (FTO) is a combustion control technology capable of reducing NOx emissions by 90 percent when compared to uncontrolled SCV burner emissions. It also avoids the NOx/carbon monoxide inverse emissions level relationship by utilizing a novel packed-bed reactor design incorporating a non-catalytic, inert porous medium. Consequently, the GH FTO features the emission reduction capabilities for both NOx and carbon monoxide, proven by post-combustion technologies, without the use of special reagents and catalysts.

SFPC proposes to integrate the GH FTO ultra low emission technology as the heat source required for vaporization within the SCV. The GH FTO flameless technology has been used successfully for several years to oxidize waste vent streams. A number of design changes will be required to successfully adapt the GH FTO to the traditional SCV design. The operating conditions, pressure drops and temperatures will be much more aggressive. Considerations will also be made to maintain the SCV's inherit design advantages including safety, efficiency, low maintenance, and ease of operation.

To better understand the requirements necessary to integrate the two technologies a review of both commercially proven processes is in order.

Submerged Combustion Vaporizers:

In the conventional single burner SCV process shown in Figure1, combustion air is introduced into the burner at two locations. Most of the air enters the upper volute section, with the remainder of the total air supplied to the region around the main fuel gas injector. With this arrangement, the main burner fires upward into the central section between the two volutes, where the combustion gases are contacted with the main supply of combustion air from the upper volute. The main combustion air enters the upper volute via a tangential inlet, and this imparts a swirling motion to the air, resulting in intimate mixing with the combustion gases rising from the burner. Gas subsequently recirculates back along the axis of the burner. The fuel gases then discharge into a distributor duct and set of sparger tubes located under the process tube bundle.

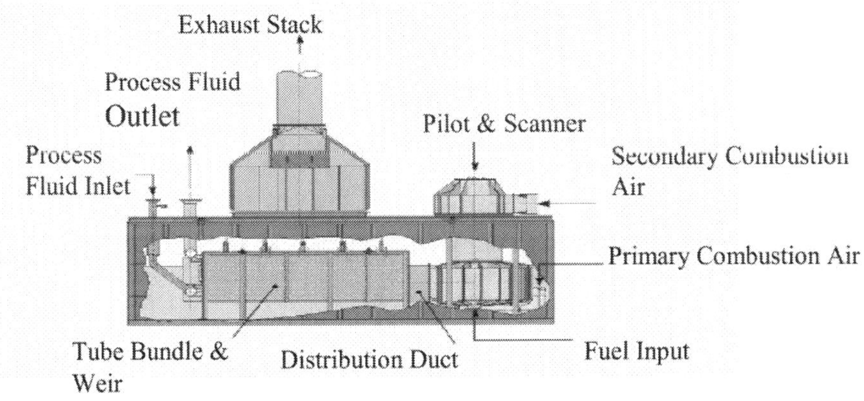

Figure 1. Sub-X Vaporizer

Injection of the combustion products into the water bath forms a frothing two-phase mixture containing many small gas bubbles. The reduced density of this two-phase mixture creates a lifting action, causing the mixture to rise. The froth flows up through the tube bundle (vaporizing the LNG) and over the weir, where disengagement of the combustion product gases occurs. The water falls back over the weir and once again re-circulates up through the heat exchanger tube bundle. Maintenance of the frothing action is necessary, even at low heat duties, to maintain uniformity and to prevent icing in the tank. The water bath maintains a very low operating temperature between 55-65 °F and the products of combustion are condensed in the bath, providing an average thermal efficiency reaching over 99%. On average, a typical single burner SCV installation will require approximately 1.3% of the vaporized natural gas send out as fuel.

GH FTO Flameless Thermal Oxidizers:

The patented Thermatrix GH Flameless Thermal Oxidizer (FTO) consists of a refractory-lined oxidation vessel with a centered diptube (Figure 2). The oxidizer is approximately half full of randomly packed inert ceramic pieces forming a uniform matrix maintained at operating temperature by the constant oxidation of the fume. Fume, air, and supplemental "trim" fuel enter the system at the top of the diptube and flow down through High Efficiency Vortex mixing tabs to pre-mix all feeds to 99.9% Coefficient of Variability prior to entering the ceramic matrix. There are no moving parts in the entire oxidizer.

Depending on fume composition, dilution air is added to provide the oxygen needed for oxidation and to ensure that the diptube mixture never exceeds 85% of the Lower Flammability Limit (LFL). Operation with non-flammable mixtures is a key safety feature of the Thermatrix technology. As the pre-mixed reactants enter the ceramic matrix, they flow between the ceramic pieces and slow down as they expand radially outward from the end of the diptube.

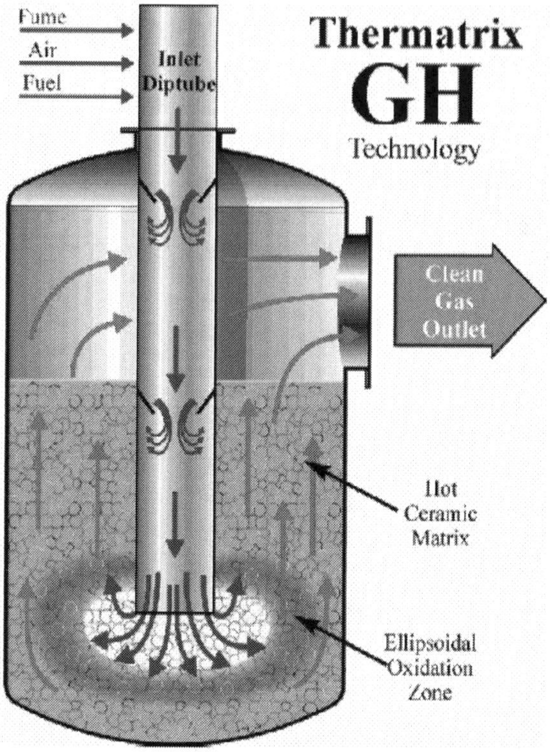

Figure 2. Flameless Thermal Oxidizer

Each fume, air, and fuel mixture has a particular oxidation reaction zone speed (typically ~1 ft/sec) based on composition and temperature. As the slowing velocity of the pre-mixed reactants reaches the oxidation speed, free radical oxidation chemistry begins. Oxidation occurs in a 10 inch thick, approximately 1800°F flameless oxidation zone deep inside a non-catalytic, hot ceramic matrix as illustrated by the diagram to the left. The oxidation zone automatically locks-in at the location where the flowing velocity of the reactants matches the oxidation speed. The supplemental "trim" fuel adjusts for variations in the fume composition and flow rate to maintain a constant composite reaction speed and keep the oxidation zone in a fixed position in the ceramic bed. This unique Thermatrix flameless oxidation zone shape and flow pattern ensures that 100% of the fume passes through a uniform zone of chemically aggressive free radicals destroying virtually every molecule of chlorinated VOC to form CO_2, H_2O, and HCl. The Thermatrix oxidation zone forms a uniform 10" thick protective barrier completely encapsulating the fume to ensure complete destruction at all times by eliminating the potential for incomplete initial combustion or "partial bypassing".

Although the VOCs fully react with oxygen to form oxidation products (CO_2, H_2O, and HCl), there is never a "flame" in the oxidizer. The reactants oxidize under relatively mild thermal conditions (1600°F to 1900°F) as they flow through the oxidation zone releasing heat which is transferred back and forth to the surrounding ceramic matrix in a "thermal flywheel" effect. The 1800°F ceramic radiates its stored thermal energy back to the flowing oxidation products. The oxidizer temperature is controlled by adjusting the air to fuel ratio. This relatively mild operating temperature is high enough to ensure complete oxidation and low enough to avoid significant NO_x production.

Submerged Combustion Vaporizer with GH FTO burner:

In the conceptual view of the integrated GH FTO design shown in Figure 3, the premixed fuel enters the refractory lined vessel through the centered diptube. Once in contact with the preheated inert media, a critical velocity profile stabilizes the reaction wave and provides the driver for the combustion process. The heated flue gas then leaves the ceramic bed and is piped into the SCV water bath. The flue gas then discharges through the sparger tubes located under the heat exchanger (tube bundle), similar to a conventional SCV.

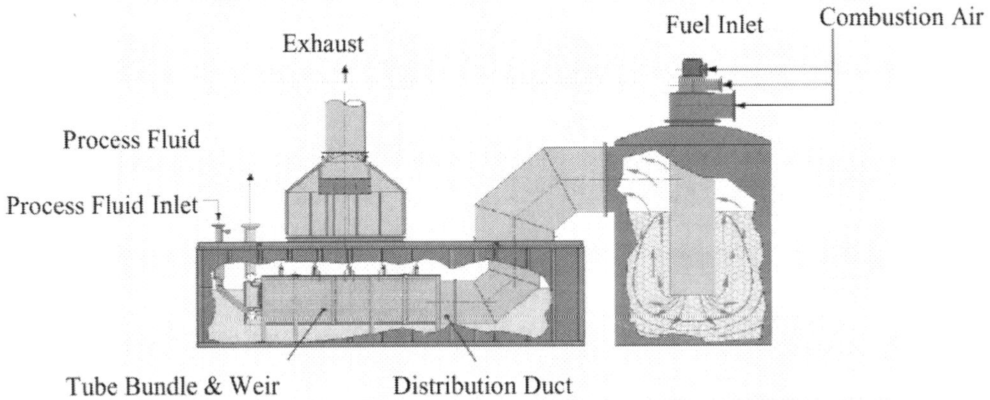

Figure 3. SCV with GH FTO Burner (GH FTO enlarged for clarity)

When modifying the conventional GH FTO for use as the SCV heater, the following considerations must be addressed:

- Higher flue gas temperatures required for vaporization
- Higher pressure required to overcome water head in SCV bath
- Operation inside the flammability limits
- Maintenance of SCV high thermal operating efficiencies
- Vaporizer process turn down

The FTO technology has demonstrated a successful track record operating in applications where the fuel/air mixture input to the unit is outside the flammability range; specifically, less than the lower flammability limit. For natural gas and air mixtures, the flammability limit is typically bounded by a range of 4.9 to 15 percent volume natural gas. In other words, the mixture will not sustain normal combustion with evidence of a visible flame even in the presence of an ignition source when outside the flammability limits.

In most demonstrated applications utilizing the FTO technology, operating temperatures have ranged from a minimum of 1600°F to a maximum of 1900°F. Operating pressure within the FTO has usually been limited to 30 inches of water column gauge. Under these conditions, the FTO has generated extremely low emissions of carbon monoxide and oxides of nitrogen with maximum values less than 5 ppmvd corrected to 3 volume percent oxygen (dry basis).

Two significant operational conditions must be satisfied in order for the FTO technology to be applied to SCV units. The temperature of the FTO exhaust gases must be increased to deliver the required thermal input to the SCV. Without such a temperature increase, the SCV would need to be enlarged significantly, adding to its capital cost. The back pressure exerted on any combustion device associated with the SCV is minimally 50 inches of water column gauge. Additional back pressure may be produced by geometrical configuration of the combustion device itself.

During the initial study phase completed to determine the feasibility of FTO technology application to SCVs, computational fluid dynamic (CFD) models were prepared to simulate the temperature and pressure profiles in the FTO technology under conditions representative of typical SCV operations. An upper temperature limit of 2600°F was imposed to allow use of reasonably available construction materials and maintain single digit exhaust gas concentrations for NOx emission performance.

Producing an exhaust gas with this temperature requires an equivalence ratio of approximately 0.62 based on a natural gas and air mixture. Additionally, the natural gas and air must be thoroughly premixed prior to entering the FTO. Since an equivalence ratio of 0.62 is inside the flammability range, this situation generated a new problem for the FTO technology since it has previously only been applied under conditions where the fuel and air mixture is lower than its flammability limit.

The most obvious concern was how to prevent a catastrophic flashback into the premixed feed system since a source of ignition would be present once it reached the FTO (reaction vessel). The solution was a special design of the feed system, acknowledging that a fuel and air mixture inside the flammability region has a characteristic flame propagation speed dependent on several properties. By carefully analyzing these properties and considering past experiences when unintentional transitions were made into the flammability region, a velocity profile map was generated which would permit safe delivery of a flammable mixture to the FTO without risk of flashbacks or other uncontrolled ignition.

Figure 4. Small scale prototype GH FTO

Having overcome these initial design challenges in theory, a small prototype FTO (Figure 4), was constructed in our testing facility to empirically verify CFD simulations prepared using process conditions directly applicable to commercial SCV operations. The test FTO unit successfully confirmed the expected temperature profile, low emissions of carbon monoxide and oxides of nitrogen, and safe continuous operation without evidence of a flashback into the feed system.

A larger prototype FTO is currently being erected for confirmation of equipment scaling factors as the technology is designed for commercial size applications (Figure 5). At this point in product development, the FTO is a stand-alone vessel which exhibits certain heat losses to the environment. These heat losses detract somewhat from the overall thermal efficiency of the process contributing to a lower thermal efficiency than the present burner design integral to the SCV. To resolve this issue, the FTO must ultimately be designed to integrate directly into the SCV so that even small environmental heat losses may be eliminated. Thermal mapping data generated during the next phase of testing will be used to confirm that FTO physical size reduction may be facilitated without negative impacts on the SCV process. The reduction in FTO physical size will permit desired integration of the technology within existing boundary limits of the SCV equipment.

The final SCV process issue is that of turndown capability. Usage of the FTO technology is designed for base-load LNG vaporization facilities which have a high level of uptime and operate under nearly steady-state conditions. Thermal output of the FTO technology may be controlled with a typical 3:1 turndown ratio. However, the time step rate of turndown is slower than a conventional burner system. Emissions do not change as the turndown is applied. Design of the FTO technology employs the use of a large ceramic media bed storing a significant amount of thermal energy. Changes in the amount of stored thermal energy take time to reach equilibrium. Adjustments to flow rate of the fuel and air mixture may be made quickly, but the process response takes some time. This is the main reason why turndown takes longer than a conventional burner system.

Figure 5. Large scale prototype GHTO

It is envisioned that an LNG import terminal facility utilizing multiple vaporizers would specify FTO technology for the full-time portion of the facility and specify burner technology for use in trimming the facility output as demand fluctuates. This combination will help to minimize total environmental emissions while retaining a high degree of flexibility in the vaporized LNG output capacity of the facility.

Results:

A CFD model was prepared to simulate the expected temperature profile within the FTO technology when operated with a natural gas and air mixture having an equivalence ratio of approximately 0.62. Figure 6 represents the graphical CFD model output. The oxidation reaction wave is very compact due to the relatively high flame speed of the natural gas and air mixture. Static temperature contours indicate that a maximum value is reached as the reaction wave expands outwardly from the end of the mixture feed pipe, which is indicated by the dark blue coloration.

System Model:

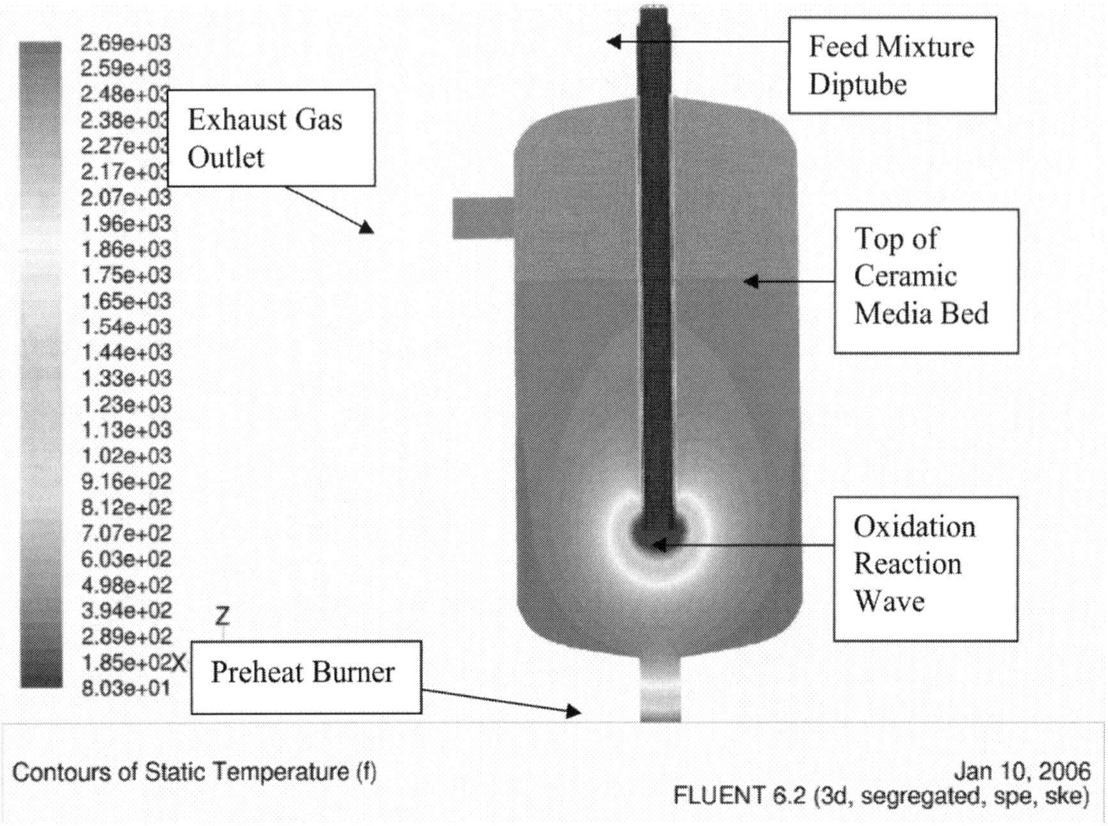

Figure 6. CFD Model

If this temperature profile is confirmed in the scaled-up prototype FTO, it will be a significant indication that the physical size of the FTO may be reduced to allow integration directly within the SCV equipment as mentioned earlier in this presentation. The temperature profile also indicates that the oxidation reaction is completed well before the product gases are exhausted from the FTO. This will ensure very low emissions performance for carbon monoxide.

Assumptions made during preparation and execution of the CFD model were as follows:

- The fuel and air were completely premixed as they entered the feed pipe assembly located on the FTO centerline.
- Composition of the fuel and air mixture remained constant during the simulation.
- The inert ceramic media was loaded into the FTO and preheated to a minimum temperature of 1400°F prior to the premixed fuel and air stream introduction.
- Back pressure on the FTO exhaust was set to emulate expected process back pressure with an operating SCV.
- No additional air was delivered to the FTO media bed via the preheat burner port for cooling or any other purpose.

Conclusions:

Development efforts continue with the FTO technology and its application as a low emission heat source for the SCV product. Based upon the data obtained to date from prototype testing and the verification of CFD simulations, it is concluded that the FTO technology may be adapted to provide sufficient volumes of 2600°F flue gases to replace the existing burner system.

Emissions generated by the FTO technology will be less than 10 ppmvd for both carbon monoxide and oxides of nitrogen, corrected to 3 volume percent oxygen (dry basis). This will be facilitated by combustion controls only and not require any NOx reduction catalyst or consumable reagents.

Data has indicated a strong potential to reduce physical size of the FTO technology, further allowing integration of the heat source directly into the SCV equipment. This will save plot space at the end user site and improve overall thermal efficiency by directing all of the heat released into the SCV and avoiding environmental heat losses associated with a standalone FTO.

References:

1. Huebner, Will (2002) "NOx Performance for the Thermatrix Flameless Thermal Oxidation Technology"
2. Huebner, Will, (2001) "Dioxin Performance of Thermatrix Flameless Thermal Oxidation vs. Flame-based Thermal Oxidation Technologies"
3. Vogel, E.R., Balfoort, J., Cobb, D., French, G. (1983) "TX LNG Vaporizer Design and Operation", in *AGA Transmission Conference*

Biography of Speaker:

Peter Falcone is the Product Development Manager at SFPC. Responsible for improving existing product performance and bringing new products to commercialization in support of market opportunities, Pete has been with SFPC for 11 years. He holds both BS and MS degrees in Chemical Engineering from Villanova University and is a registered professional engineer in Pennsylvania and Delaware.
pete.falcone@selasfluid.com

Safety Management Activities on LNG Industry in Korea

Kyoshik Park, Dal-Young Park, Ji-Yoon Kim, and Young-Do Jo
Korea Gas Safety Corp.
Daeya-dong, Shiheung-shi, Gyeonggi-do
South Korea

AIChE Spring Meeting, April 2006
6th Topical Conference on Natural Gas Utilization
Orlando, Florida, April 23-27, 2006

I. Introduction

Korea is one of the largest natural gas importing countries in the world, having the tenth large energy consumption in the world[1]. To meet the nation's ever increasing demand, Korea imported 22.2 million tons of natural gas via LNG in 2004.

The LPG started to be used in Korea as fuel since 1900s by the royal family of the Chosun Dynasty. Only small amount of gas has been used by the rich about 50 years since then. The gas needs for fuel expanded with economic development started from the early 1960s. The Korea Ministry of Energy and Resources launched in 1978, announcing gas period in Korea. At the beginning, liquefied petroleum gas (LPG) as well as natural gas was used as city gas. LPG had been replaced rapidly by natural gas since 1995, after a disaster at Daegu which recorded 101 fatalities and more than 200 injuries because of LPG city gas explosion.

As shown in Fig. 1, gas consumption as well as pipeline network increases very rapidly. Among the primary energy consumption in Korea, LNG took 3.7% in 1987 and 11.2% in 2004.

II. Gas Safety Law

Korea has experienced an intensified process of introducing and applying LNG for the last couple of decades, which makes Korea one of the highly advanced countries having LNG handling technology including safety management. In order to achieve the highest safety standards in the LNG facility Korea has developed safety management system of LNG processing and applications.

Before 1966, the gas-related business was controlled by the Korea Ministry of Domestic Affairs (MODA), i.e., the police station is responsible for safety of gas business. With some severe gas incidents, the jurisdiction was moved to the Korea Ministry of Commerce and Industry (MOCI) being aware that gas business and its safety is rather technical than administrative. The Korea Gas Safety Corporation was

[1] Energy Statistics and Balances of OECD Countries 2004 (IEA)

established in 1974 and became an exclusive organization responsible for safety of gas industries.

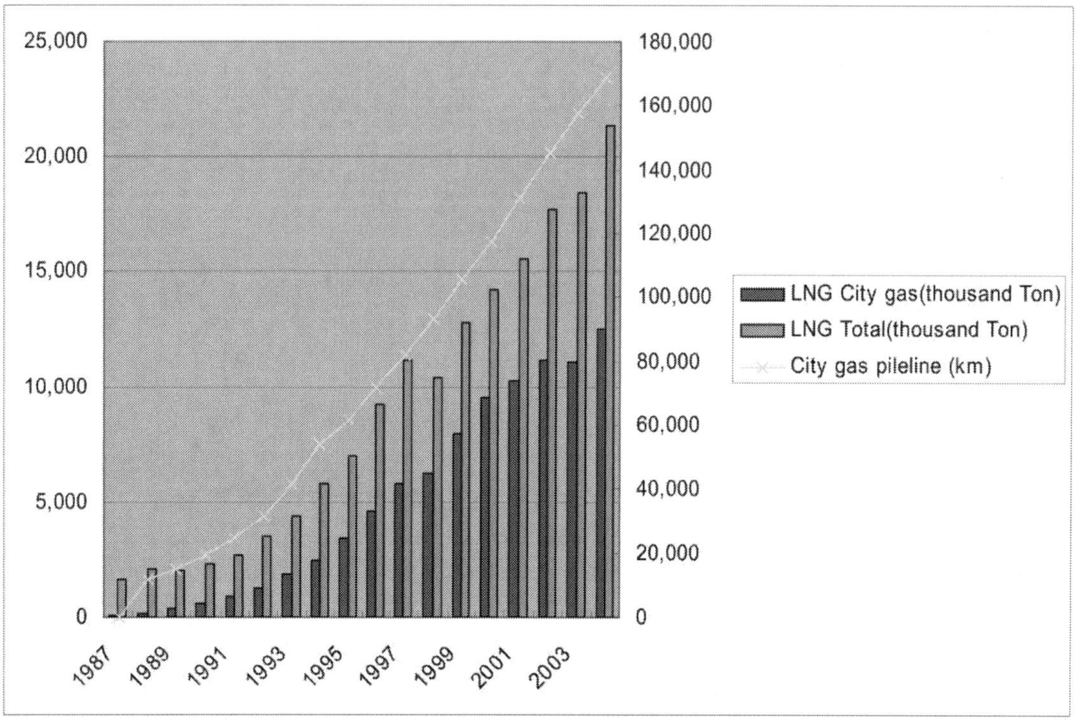

Fig. 1. Market Expansion of Korea Gas Industries

It is an irony that gas safety law became tougher with major accident. We had major accident related to gas business in 1971 and 1994. The High Pressure Gas Safety Law became effective 1973. The main content is that the gas business is under control of MOCI instead of MODA. The safety law experienced a major change in 1995 containing systematic management on gas facilities as well as its operating system, quite similar to the PSM (Process safety management) program in USA. The program work effective by reducing gas incident from 577 in 1995 to 396 in 1998. By 2005, Korea has only 104 incidents related to gas industries. The structure of natural gas industry is illustrated in Fig. 2. As shown in the figure, the KOGAS (Korea Gas Corp.) import natural gas via tanker in liquefied gas and store and supply it to power plants, large-scale consumer, industry consumer, or city gas companies, i.e., whole sale it. The KGS (Korea Gas Safety Corp.) is responsible for maintaining safety of the facility as well as its operating system.

In order to meet the future's need, the KGS is starting a program to promote to reduce gas incidents drastically and make advance in safety technology. The program covers hardware improvement such as IT (Information technology) based technology and safety devices as well as software innovation such as law and regulation and collaboration with insurance company. With this program, the KGS hope to reduce gas incidents by half and obtain zero major accident in 10 years.

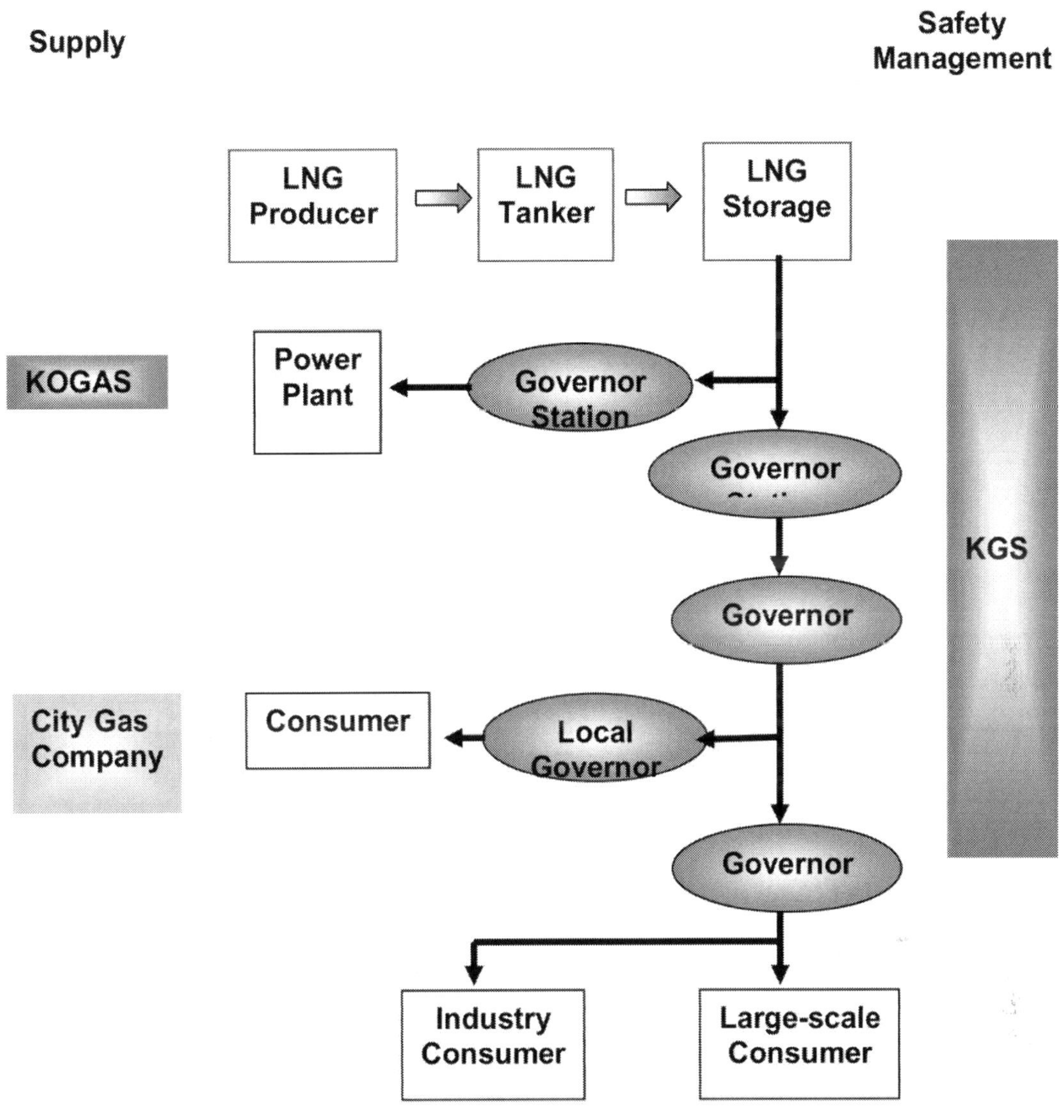

Fig. 2 Structure of Gas Industry and Safety Management in Korea.

Mailing Address

Mr. Kyoshik Park
General Manager
Korea Gas Safety Corp.
332-1, Daeya-dong, Shiheung-shi, Gyeonggi-do
South Korea, 429-712

Phone: 82-31-310-1410
E-mail: kspark@kgs.or.kr

LNG Liquefaction Plant Risk Management

Bob Dimitroff
Manager of LNG Project Support
Chevron Energy Technology Company
Houston, Texas

AIChE Spring Meeting, April 2006
6th Topical Conference on Natural Gas Utilization
Orlando, Florida, April 23-27, 2006

Abstract

This paper provides a guide and insight into how to prepare a Risk Management Plan for a Liquefied Natural Gas Liquefaction Plant (LNG) Project. LNG projects are mega-projects (>$1 Billion) and are typically located in remote international locations. In today's highly competitive business environment, properly managing risks will differentiate successful projects from those that endure hardships. The template presented in this paper is meant to help a LNG project team organize thoughts about what can possibly go wrong, to assess such risks and how to determine the appropriate response. Examples of risks for a LNG Project are used to illustrate the process.

Executive Summary

This paper describes a process for managing risks associated with a LNG Liquefaction Plant. A background of the LNG business is first provided for a basic understanding. These plants require a keen understanding of risks in order to support a project funding decision. Some aspects of these unique projects (and drivers of potential risks) include:

- Large capital expenditures - typically greater than $1 billion US Dollars
- International "greenfield" projects with significant logistical and cultural components
- Many project stakeholders – Local Governments, Partnerships, LNG Buyers, etc.
- Extensive schedules of 54+ months from feasibility engineering to plant start-up
- Current LNG industry boom is creating substantial competition for resources

The first step in the proposed Risk Management Process is Risk Identification. A template is provided with a suggested methodology and includes a proposed high level categorization of LNG Plant risks. Risk Assessment is next discussed where it is recommended that a project team performs an analysis based on the unique aspects of each project. A template is provided for teams to follow along with an explanation of the process. Last in the proposed methodology is Risk Response. The focus here is on procedures as the actual response will be dictated by the project.

Examples of key risks in a LNG plant are then provided to help illustrate how the methodology works. Currently, LNG Projects are faced with engineering and construction resource shortages. Thus Resource risks are a focus point for this section with a brief Risk Assessment and potential Risk Responses for each identified risk. Other examples of identified risks in each Risk Category are also included. Additional risks identified during research of this paper are shown in the Appendix.

Liquefied Natural Gas Business Background

Significant natural gas reserves exist in many parts of the world. Unfortunately, most of these reserves are not close to a market for the gas. Liquefied Natural Gas (LNG) is a way to link these stranded reserves with a market. Demand for Natural Gas is growing in both the US and Worldwide as shown below in the graphs below.

Figure 1: U.S./Worldwide Gas Demand

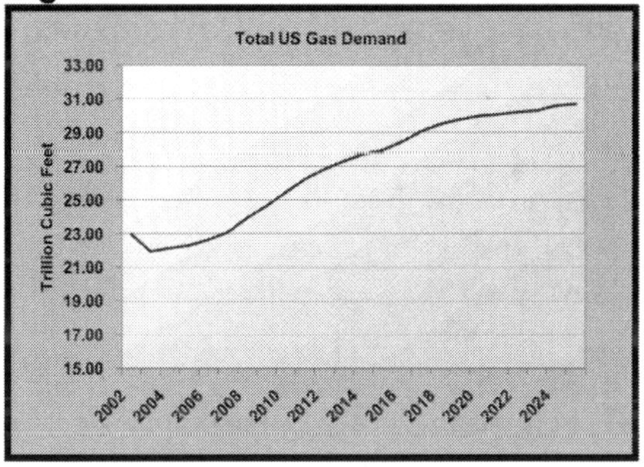

Source:
EIA–DOE http://www.eia.doe.gov/oiaf/petgas.html

Figure 35. World Natural Gas Consumption, 1970-2025

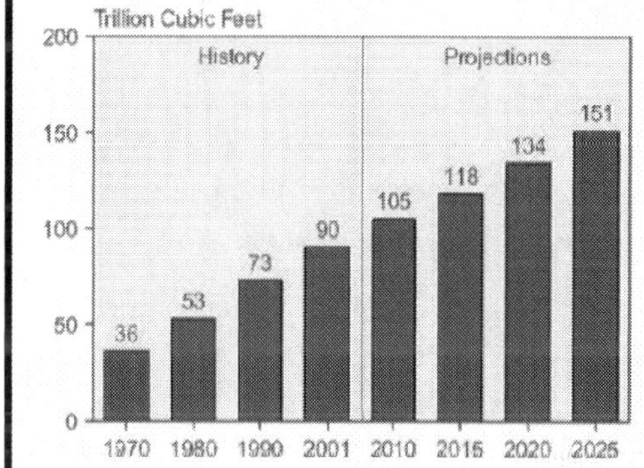

Sources: **History:** Energy Information Administration (EIA), *International Energy Annual 2001*, DOE/EIA-0219(2001) (Washington, DC, February 2003), web site www.eia.doe.gov/iea/. **Projections:** EIA, System for the Analysis of Global Energy Markets (2004).

Natural Gas is produced from the ground and transported through a pipeline to the LNG Liquefaction Plant. The gas is first processed to remove impurities such as carbon dioxide, hydrogen sulfide, mercury, and water. The remaining purified stream of hydrocarbons is then fed to the liquefaction section of the plant. The gas is liquefied by chilling it to approximately -260°F to create a near pure form of methane (CH_4) with small amounts (6-10%) of ethane (C_2H_6) and trace amounts (<2%) propane (C_3H_8). The LNG is then stored and transferred into large LNG Carriers for transport to market. Upon arriving at their destinations, the LNG is pumped through Regasification terminals where heat is added to turn the liquefied natural gas back into a gaseous state. The gas then is routed through local pipeline distribution systems to the end user. The linkage between the well head and end user is typically referred to as the “value chain”.

The investment required to bring the gas from wellhead to market is a multi-billion dollar proposition. Partnerships are common, given such large funding requirements. Host governments, typically via national oil companies, are usually involved in the project approval process – both financially and regulatory wise. Historically, long term supply contracts are established to provide pricing certainty. Currently, the business is booming and many LNG Liquefaction projects are in various project phases from Inception to Development and on into Implementation. High natural gas prices are at the root of this expansion. If the gas demand grows as expected and the pricing trend continues, it can be expected that the LNG business will remain a highly competitive environment for many years to come. As a result of this boom, and competing businesses – such as Gas-to-Liquids – key resources are in high demand, thus causing price increases. Going forward, LNG developers must recognize these risks, identify potential impacts to their projects and develop the appropriate risk response.

Risk Identification

Risk Identification is the first step of the Risk Management process as described by Smith and Merritt (2002). Risk Identification in this case has been subdivided into Risk Categorization and Risk Description. Below is an *example* list of risks for a generic LNG project. An actual project's risk listing would be expected to be more extensive and complex. Given the number of potential risks, this paper will focus on only one area of risks, Resources.

Risk Categorization

Many types of risks exist in the LNG business and have been categorized at a high level as shown in Figure 2. This categorization, or a similarly fashioned list, is recommended to organize thoughts about where risks may occur. These categories are a mix of scope based and functional based categories. Individual projects may need to go further by developing broader categories based on the unique characteristics of that project. To identify possible sources of risk, it is recommended to examine the project Work Breakdown Structure (WBS) and Organizational Breakdown Structure (OBS). This approach to categorizing risks is in alignment with the concept of developing a Risk Breakdown Structure (RBS) as described by Hillson (2002):

> "Each of these RBS structures is different, reflecting the range of possible sources of risk exposure for projects in various sectors and industries. It is therefore necessary for any organization wishing to use the RBS as an aid to its risk management to develop its own tailored RBS."

The main point for teams to remember is that there is no cookbook recipe for identifying risks. It does however help to utilize project management tools (i.e. WBS) as a method for organizing thoughts and creating discussion about where risks may exist.

Figure 2: LNG Project High Level Risk Categorization

A	B	C
#	**RISK CATEGORY**	**RISK DESCRIPTION**
1	Commercial/Marketing	
2	Cultural	
3	Environmental/Safety/Health	
4	Financing	
5	Gas Supply	
6	Other	
7	Political	
8	Project Management/Controls	
9	Resources	
10	Technical	

Risk Description

Risk Description is a key to the entire process. Risks should be fully described. By working up a detailed description, the full extent of the nature of the risk is more likely to be uncovered. In the examples below, for brevity's sake, the descriptions used are abbreviations of the full description. In practice, it is recommended to develop a unique sheet to define and track each risk including the full risk description.

Risk Assessment

The project team must next assess each risk. Risk Assessment includes identifying Risk Timing, Risk Source, Probability of Occurrence and Impact of Occurrence. Given the unique nature of each project, risk assessment must be done by the project teams.

Figure 3: Risk Assessment

A	B	D	E	F	G
#	RISK CATEGORY	TIMING FOR WHEN RISK MAY OCCUR	INTERNAL OR EXTERNAL	PROBABILITY OF OCCURRENCE	COST/SCHEDULE SCOPE/QUALITY RESOURCES IMPACT
1	Commercial/Marketing				
2	Cultural				
3	Environmental/Safety/Health				
4	Financing				
5	Gas Supply				
6	Other				
7	Political				
8	Project Management/Controls				
9	Resources				
10	Technical				

Risk Timing

Risk Timing helps the team identify when a risk may occur. Due to the extraordinary length of an LNG project (Cycle Time ~ 5 years), it is important for the team to consider those risks that are in the near term, while not losing sight of those that may occur later in the schedule. Also, a discussion about the project schedule may further yield possible risks. For instance, site preparation work may require significant soil movement and thus may be significantly affected by annual rainy seasons. Thus, in this case, project plans must recognize regional weather patterns when setting schedule milestones.

Risk Source – Internal or External

Risks can originate from within the project and/or organization and they can also come from external sources. Focusing on the potential origin of risks can help the project team document items that deserve attention. For a LNG project, many sources of external risks exist. Political risks associated with the host country are an example of such an external risk. Internally, a risk can originate from a company's organization. For example, a key specialty resource, such as a Rotating Equipment Engineer, may not be available to a project if such resources are controlled by a central organization and appropriate resource planning has not occurred.

Probability of Occurrence

Assessing the probability of occurrence is not an exact science. It is recommended that project teams use available benchmark data to understand the reasonable range of probabilities. Given the number of LNG projects in various development stages, 'real time' data is available regarding some risks. Mega-projects typically attract a lot of attention and much of what is written ultimately finds its way into the public domain. For risks that may have a large impact, a sensitivity analysis is warranted to understand magnitude of an occurrence.

Impact of Occurrence – Cost/Schedule/Scope/Quality/Resources

Estimating the impact of a risk occurrence can be done using information from the original project justification as well as from other historical data bases. For example, information on Shell's Sakhalin LNG project costs was recently made public. According to Blagov (2005),

> "Although the official statement was short on specific details, there was a major reason for concerns relative to the Sakhalin-2: the project's cost estimates have doubled to $20 billion."

Most project managers focus on Cost and Schedule. A risk occurrence can also impact Scope, Quality and/or Resources. The project team should consider how a risk impacts these other project knowledge areas. While impacts on cost and/or schedule are relatively easy to estimate, impacts on quality and/or resources may lead to other risks which have additional impacts. In today's environment where contractors and company personnel are scarce, a situation that impacts resources could have a domino affect on other tasks with more far reaching circumstances than if adequate resources are available.

Risk Response

Risk response requires the project team to identify potential actions to either eliminate or minimize the risk. These mitigation measures can either reduce the probability of occurrence and/or reduce the impact of occurrence. The team should consider whether one action is sufficient or if multiple efforts are required to mitigate the risk. The steps leading up to choosing how to respond to a risk are briefly described below.

Risk Mitigation Measures

Risk mitigation measures are actions that can be taken by the team to either minimize (or eliminate) the risk from occurring and/or that minimize (or eliminate) the impact. As with the risk description, the team should be specific about what tasks will be employed to address each risk. The template also prompts the team to consider mitigation success with the three columns: *Mitigated Probability of Occurrence, Mitigated Risk Impact and Probability Measures Work.* Shown here is a simplified version of the risk management process. As previously recommended, a separate tracking sheet for each risk should be used to capture the detailed risk description and associated mitigation information.

Figure 4: Risk Response

A	B	H	I	J	K	L	M
#	RISK CATEGORY	RISK MITIGATION MEASURES	MITIGATED PROBABILITY OF OCCURRENCE	MITIGATED RISK IMPACT	PROBABILITY MEASURES WORK	$ COST OF MITIGATION MEASURES	RISK RESPONSE TYPE
1	Commercial/Marketing						
2	Cultural						
3	Environmental/Safety/Health						
4	Financing						
5	Gas Supply						
6	Other						
7	Political						
8	Project Management/Controls						
9	Resources						
10	Technical						

Cost of Mitigation Measures

The team must also consider the costs of implementing mitigation measures to help decide how to respond. While some actions may simply require a few man hours to implement, other mitigations may be very costly and could sway the team to implement other measures.

Risk Response Type

The last step is determining how to respond to the risk. There are several different responses a team can take:

1. **Avoidance** – Completely avoid the risk. Prevent the risk from occurring. This response should be considered if the impact of occurrence is devastating to the project or company.
2. **Mitigation/Reduction** – Take actions to reduce impact and/or probability of risk.
3. **Transfer** – Transfer impact of risk to others.
4. **Insure** – Buy insurance to protect against risk.
5. **Accept/Assume** – Accept risk and react as necessary. Develop contingency plans for if the risk occurs.
6. **Do Nothing/Ignore** – Accept the risk and accept impact. Do not develop contingency plans as the impact is not of consequence.

Examples of LNG Project Key Risks

Below are some examples of key risks in each of the major risk categories that have been identified. With each risk is a brief description. For those risks in the Resource category, additional commentary is provided about Risk Assessment and potential response recommendations. A focus is provided on Resource related risks as these have grown recently with the boom in the business. It is recommended that project teams perform more in-depth assessments based on the actual circumstances.

Resources: Resources are expected to be a key risk for LNG Projects in the next several years due to increased competition – not only within the LNG business – but also with other construction activities. Each risk can potentially drive up costs and extend schedule. Quality can also be an issue with some of these risks.

- **Raw Materials** - Such as carbon steel and specialty steel (i.e. 9% Nickel) – The worldwide economic expansion has also contributed greatly to steel price increase. While steel prices have dropped off slightly from the end of 2004, prices are still substantially higher (80%) than 2003 pricing. According to Gas Monetizer (2005):

> "The only hindrance to greater world market development for natural gas is a shortage of construction materials and management talent. Steel, aluminum and nickel prices are soaring. Nickel prices have nearly tripled since the 2002 recession."

- **Assessment:** In addition to cost impacts, raw material shortages require close attention by the project team in order to prevent schedule delays. Quality concerns may also crop up since many supply sources are running at or above full capacity, creating quality control problems and the specter of substandard substitutes being provided.
- **Response:** Mitigation against raw materials shortages, delivery delays and/or price increases may include:
  - *Utilizing alternative materials* – This is not a solution for all circumstances and any alternative must be technically acceptable. Engineering design must consider performance of alternatives.
  - *Early procurement of materials* – This option requires commitment to materials earlier than originally scheduled which may also introduce additional risks. If materials are procured early and changes are later made in the design, these early procured materials may go to waste.

- **Procured Equipment** - Items such as LNG Carriers, Heat Exchangers, Turbines, Compressors, etc. are in high demand/short supply. For instance, growth in Gas-To-Liquids is expected to create shortages in braised aluminum heat exchangers.

  - **Assessment:** High demand has created both cost and schedule pressures.
  - **Response:** Mitigation against shortages, delivery delays and/or price increases may include:
    - *Utilizing alternative designs* – This is not a solution for many circumstances and any alternative must be technically acceptable.
    - *Early procurement of materials* – This option may require a substantial financial commitment earlier than originally scheduled, and prior to the Final Investment Decision. This may also introduce additional risks. If materials are procured early and changes are then made in the design, or if the project does not gain approvals, these early procured materials may end up as surplus.

- **Construction Equipment** – Some locations may require world class sized construction equipment such as heavy lift barges and large dredging equipment.

  - **Assessment:** Due to the level of project activity, such equipment may be in short supply again pressuring both costs and schedule.
  - **Response:** Mitigation against shortages, delivery delays and/or price increases may include:
    - *Modify engineering design to eliminate need for large equipment* – This may not be feasible for all items. It is recommended to hold a Constructability session in conjunction with the engineering design so that plans can be scrutinized by construction experts to ensure they can realistically be built.

- *Secure contracts early for equipment* – This may be costly but probably a better alternative than not having the equipment when needed or paying a premium at a later date.

- **Skilled Construction Labor** – Many prospective LNG projects will be located in under-developed countries where the number of skilled laborers is limited. Training can be time consuming and still may not eliminate quality concerns. However, employing local content can be a major commitment for plant owners. From the June 2004 issue of Remote Gas Strategies:

  "Exporting governments want to educate, train and employ as many of its citizens as possible. In Sakhalin II, Shell and its partners have agreed to use 70-perccent local content and employ 2,400 Russians full time."

  - **Assessment:** Lack of skilled labor will result in high training costs and significant time commitments for training plus potential delays if labor is not available.
  - **Response:** There is no easy mitigation strategy for this risk.
    - *Conduct early and ongoing training* – This must be started early enough to provide the right skills at the right time. An Operations Planning session should be held early to identify training requirements.
    - *Import labor* – This may not be a viable option depending on the requirements of the host government. If available as an option, arrangements must be made for work visas and living arrangements.

- **Major EPC (Engineering, Procurement, and Construction) Contractors** – Work load has increased such that some contractors are turning down work. Only a small, elite community of large contractors with LNG Plant experience exists. A workshop held August 10, 2005 offered by Zeus Development entitled "EPC Contractor Selection and Management in a Full-Capacity Construction Environment" was sold out, indicating the level of interest industry has in this subject. According to a study by Transmar (2005)

  "The quality and size of the process-oriented contractors' LNG staffs is in decline. There are only 5 or 6 top-level teams ("A teams") available to handle multi-billion dollar liquefaction projects. ... The reasons for the decline are demographic – "A" level LNG staff members are in their late fifties or sixties in age and retiring. The other reason for the decline is that contractor teams are being depleted by headhunters poaching key personnel for operators and other contractors trying to enter the LNG industry"

  - **Assessment:** Securing a contract for Engineering Design, Procurement and Construction of these facilities requires consideration of contract price, ability to meet schedules and resulting quality concerns, both engineering and construction.
  - **Response:** Mitigation against lack of contractors may include:
    - *Consider alternative contracting strategies* – It is now a "buyer's" market. Many different strategies have been employed within the industry. It is recommended to conduct a contracting strategy workshop early in the project to determine which strategies align best with host country and owner needs.

- **Individuals experienced in LNG** – Investor companies need some personnel who are experienced in LNG to help manage their interests. As with the EPC firms, only a limited number of such resources exist. Competition for these people has led to unsolicited offers of employment with large pay increases. According to the Gas Monetizer (2005):

  > "Demand for qualified engineers is also high. Royal Dutch Shell, for example, is reported to be in the market for 1,000 engineers. Headhunters have expressed frustrations that retired engineers no longer take their calls, illustrating that demand for capacity had reached beyond plant gates and into the halls of academia."

  Additionally, The Globe and Mail (2005) reports:

  > "Inflation is an issue, but the real difficulty is probably more sensitive. It is a problem that is troubling every major energy company: people. To be precise, a shortage of people; a chronic lack of engineers, geophysicists, cost estimators and project managers. There is a critical shortage worldwide of the skilled professionals who can erect mammoth pieces of energy infrastructure to order, on time and keep them running like clockwork."

  - **Assessment:** Shortage of experienced personnel will have the tendency to reduce quality and increase schedules. Cost impacts may also be felt.
  - **Response:** This risk needs to be mitigated since a reduction in quality can further lead to additional risks including safety concerns which are not acceptable. Potential actions include:
    - *Hiring LNG experienced individuals* – Competition for experienced personnel is fierce and costly.
    - *Mentoring non-LNG experienced individuals* – The level of expertise required makes this a long term commitment.
    - *Accepting the longer schedules* – If resources are not available, companies may need to accept longer schedules to ensure adequate quality.

- **Commercial/Marketing: *LNG sales agreement.*** LNG projects are based on supplying gas via a long term contract to particular markets. Prior to finalizing sales and purchase agreements, there is significant risk that those market(s) won't materialize in the appropriate time frame to match the LNG project investment profile. It is typical for negotiations to take time and thus needs to be worked in parallel with plant front end engineering design.

- **Cultural: *Compatibility of Partners.*** Most LNG projects consist of multiple organizations in a partnership. Problems can arise if these partners are not aligned in their goals as well as approach to the project. For instance, some companies may be aggressive and utilize shortcuts while others may be more conservative requiring full reviews of project plans.

- **Environmental/Safety/Health/Social: Environmental concerns may arise due to construction activities or impacts of long term operations.** While site selection typically takes into consideration both flora and fauna, selection should also include the potential impacts of construction activities can result in the following types of pollution: soil, water, air, noise and/or light. Additionally, similar concerns about waste streams associated with ongoing operation must be considered during design. For instance, as described by Heiersted, et al (2004), for the Snohvit LNG Project

> "In the development of the Snohvit project, Statoil corporate philosophy on HSE issues is ranked very high when setting criteria for technology selection...There is a very high environmental attention on the impacts regarding emissions to air and sea."

- **Financing: Large funding requirements/Credit worthiness of partners**. Given the large capital expenses, there is a risk that one or more of the partners will not have the funds to proceed. According to Inniss (2005):

> "An LNG project is still very capital intensive and requires investors with have deep pockets. All the major petroleum companies are investing through the chain to ensure the security of their investment and to jump-start the globalization of LNG trading. They have moved out of their traditional areas and are investing all the way into the market."

- **Gas Supply: Supply composition – Hydrocarbons, Impurities (H_20, H_2S, CO_2, Hg), Calorific Value.** Projects proceed based on an assumed, or "as tested", condition for the gas supply. A change in reservoir conditions can result in a modified composition. This could significantly impact LNG plant output. According to Madouri (2004):

> "During recent years, the quality of the feed natural gas to the liquefied natural gas plant GL1Z has become leaner in heavy hydrocarbons. Leaner feed gas makes it increasingly difficult to obtain the required propane and butane to be recycled to the scrub tower. This has led to long start up times to stabilize the scrub tower operation and sometimes causes certain instabilities even during normal operation."

- **Other/Construction: Lack of Infrastructure.** The lack of infrastructure in the nearby area may dictate that the project considering building an airport, roads, marine port, lodging for workers and other logistical support.

- **Political: Civil War/Disagreements amongst local communities.** Turmoil in local government and communities can devastate a project. Local labor will be tough to find and hostilities could endanger project personnel.

- **Project Management/Controls: Multiple international locations - Engineering & Construction.** Due to the size and international nature of these projects, it is likely that most projects will have work going on in multiple locations. Coordination amongst these various locations is paramount to having a well-run project.

- **Technical: Use of new Technology - Use of unproven equipment and/or processes.** As the industry continues to grow, so do the size of the plants. With this growth in throughput is a correlating increase in the size of equipment. Cryogenic compressors required for the refrigerant cycles are pushing the envelope for operating conditions and must undergo extensive testing prior to placing larger machines into service.

Conclusion/Recommendations

Managing a LNG Liquefaction Project requires close attention to the multitude of risks that may occur. What has been presented here is a methodology that can be used to identify, assess and respond to such risks. It is recommended that project teams begin risk management early in the project since understanding the risks will help the team establish appropriate estimates and expectations.

Competition in the business has increased substantially in the last few years. Resources required to plan, design, construct and operate these project are becoming critically scarce. This paper has provided some focus on the possible Resource risks that can occur on a LNG project and what may be done to mitigate against these risks. To be successful in the future, companies will need to proactively manage these Resource risks.

References

1. Bell Globemedia Publishing Inc., "Shell's latest fiasco rooted in critical skills shortage", The Globe and Mail, July 21, 2005
2. Blagov, S., "Sakhalin Energy Projects face Reality Check", Available from http://www.jamestown.org/edm/article.php?article_id=2370064 2005, Internet, 6-Aug-05
Energy Information Administration, "US Gas Demand", Available from http://www.eia.doe.gov/oiaf/petgas.html; Internet, 6-Aug-05
3. Energy Information Administration, "Worldwide Gas Demand", Available from http://www.eia.doe.gov/oiaf/ieo/figure_35.html, Internet, 6-Aug-05
4. Gas Monetizer, "World Gas Construction Boom Here to Stay, Oil Prices to Remain High, Some Analysts Say", Gas Monetizer, Vol. I, No. 2, February 2005, Pg. 3
5. Heiersted, R.S., S. Lillesund, S. Nordhasli, G. Owren, and K. Tangvik, "The Snohvit Design Reflects a Sustainable Environmental Strategy", 2004, LNG 14 Conference Proceedings, Paper No. PS5-4
6. Hillson, David, "Use a Risk Breakdown Structure (RBS) to Understand Your Risks" Proceedings of the Project Management Institute Annual Seminars & Symposium October 3–10, 2002, San Antonio, Texas, USA ", Available from http://www.risk-doctor.com/pdf-files/rbs1002.pdf
7. Iniss, Helena, "Energy Economics Operational Vertical Integration in the Gas Industry, with Emphasis on LNG: Is it necessary to ensure Viability?", Date unknown, Available http://www.dundee.ac.uk/cepmlp/car/html/car8_article22.pdf, Internet, 7-Aug-05
8. Madouri, Abdelouahab, "Improvement of High Heating Value of Commercialised Liquefied Natural Gaz of GL1Z Plant by Optimising the LPG Extraction", 2004, LNG 14 Conference Proceedings, Paper No. PS2-7
9. Remote Gas Strategies, "New Era of Mega Projects Will Create New Mega Managers", Vol. VIII, No. 6, June 2004, Pg 5
10. Smith, Preston G. and Merritt, Guy M., "Proactive Risk Management: Controlling Uncertainty in Product Development", 2002, Productivity Press, Page 29
11. Transmar, "Gauging the Available Technical Resources in the Global LNG Industry: Do We have the Cabability to build the Future?" a Multi-Client Study Sponsored by ChevronTexaco, March 2005
12. Zeus Development, LNG Express, "EPC Contractor Selection and Management in a

Full-Capacity Construction Environment" Available from http://www.lngexpress.com/mc/, Internet, 6-Aug-05

Appendix 1 – Additional LNG Project Risks

Commercial/Marketing:

- Competing Projects – This is a prime example of how one risk can impact other areas. Competing projects pose problems in two ways – 1) Competition for resources (see below) and 2) Competition for Markets
- Alternative Energy Sources – Alternative energy sources such as Coal, Nuclear and Renewable Sources - Water, Wind, and Solar - are competing with LNG to supply energy to markets. A risk exists that other sources can replace possible market options for LNG. Rising costs of LNG projects will improve the competitive position of these other sources.
- Gas Pricing – Due to the significant capital investments required along the entire value chain, LNG projects need sufficient price support to maintain positive economics.

Cultural: This risk category is often overlooked by some in considering risk management plans. While these risks are not typically project killers, they can be big enough to create significant problems for the project team.

- Host Country – The project team must understand and recognize cultural norms and practices within the host country. Risks can arise by failing to recognize such differences.
- Contractors and Partners – EPC Contractors and partners can also potentially have cultural differences. For instance, some contractors are very amenable to requests by the owner while others will point out all the issues with such requests prior to proceeding.

Environmental/Safety/Health:

- Archaeological – Site selection must also consider the potential for finding archaeological or culturally important artifacts.
- Safety/Health Concerns – Safety and Health are similarly concerns during all phases of the project.
- Socio-Economic Issues – Given the size of these projects, considerations must be given to the impacts that such construction will have on the local community. For example, a large influx of workers may put a strain on local infrastructure

Financing:

- Currency Fluctuations - Many projects must deal with multiple currencies. Changes in exchange rates can cause substantial changes in project costs.

Gas Supply:

- Supply composition – HC, Impurities, BTU Content – Projects proceed based on an assumed or "as tested" condition for the gas supply. A change in reservoir conditions can result in a modified composition. This could significantly impact LNG plant output.
- Supply rates/pressure – As with the composition, the actual conditions for the plant may vary from design conditions. If pressures or rates, especially rates, are less than expected, plant economics will be negatively impacted.
- Reserves volumes – Volumes need to be sufficient to supply the plant for the expected long term (20+ years) operation of the plant.

Other/Construction: During the execution phase of the project,

- Lack of Infrastructure – Logistical requirements of a LNG are immense. A plant may require construction of an airport, roads, port and accommodations to support the large volume of materials and people needed for building the plant.
- Substandard performance by contractor(s) – Cost and schedule estimates are made assuming certain work productivity levels and specified quality requirements. Failure to achieve these targets will negatively impact overall project measures.
- Force Majeure Events – Earthquakes, Hurricanes, Tsunami's, etc. – A plant will be designed for the "100 year event" based on historical information. However an occurrence during construction can be catastrophic.

Political:

- Boundary disputes – Some areas of the world do not have accurate property ownership records. Disputes about land ownership near proposed plant locations should be avoided.
- National companies – Special interests – National Oil Companies involved as a partner may also have special interests that need to be addressed that are not necessarily in the best interests of other partners. Partnerships need to work together to understand the goals and needs of all parties so that a common win-win solution can be found.
- Requirements of National/Local Governments – All projects need to keep government entities high on the list of key stakeholders. Special requirements and "wish lists" can be added to the project scope destroying overall project value for the partnership.
- Changes in Government – Changes in government or key personnel can result in changes for the project.
- Local Content – Most host countries require a substantial percentage of the labor and/or materials to be provided locally. Achieving these requirements can present a logistical problem at best for the owner companies.
- Objectives of Partners – Since many partners can be involved in a project, the goals and objectives of each partner must be compatible for the project to move forward. Decision making can severely slow down if there is incompatibility.
- Public Opposition – Opposition by public groups or individuals to the project can result in scope changes and delays.

Project Management/Controls

- Project Interfaces – Due to the size of these projects, there may be multiple contractors in various locations working on different parts of the project scope. Risks lie in managing the interfaces between the various work groups and making sure their deliverables flange up.
- Extremely long schedules (> 5 years for project life cycle) with many activities – While a long schedule in itself is not a risk, there is a risk of managing a project with many activities and the potential for multiple critical paths. Due to the length of these projects and the number of activities building in schedule float is recommended.
- Schedule Delays/Pressures – A schedule delay in itself is not a risk as are the pressures that accompany delays. However a delay can create multiple critical paths and divert attention of the team creating additional risks.
- Cost Overruns/Pressures – Similar to the schedule delays, a cost overrun typically diverts attention of the team if required to provide documentation/justification. A loss of focus can help create additional risks.

- Scope Changes – Any change must be analyzed in order to ensure a safe facility still exists given the change. Modifications to scope will also potentially impact schedule, cost, quality and resources required.

Technical: Many risks can arise technically. This typically is a key area of focus for project teams, as it should be. An example of the various types of technical risk is:

- LNG Production - Actual vs. planned – Meeting planned production capacity is a risk due to commercial considerations. Equipment Availability and Auto-Consumption (Fuel Use) are key factors in achieving planned output.
- Train Size - Matching throughput to Reserves and Commercial considerations – While the size of the train is not a risk, the economics of the project will be at risk if a proper match of reserves, product sales and equipment sizing is not achieved.
- Geotechnical - Soil stability – Given the large foundations required for the rotating equipment, any weakness in the soil can cause many headaches.
- Marine - Ability to berth LNG Carriers – Various environmental factors must be accounted for in order to achieve good berth availability. These factors include: wind, waves, currents, water depth, and sedimentation
- Marine - Breakwater – The ability to berth LNG Carriers is also affected by the size of breakwater required. Ocean bottom soil conditions must be understood to build a stable breakwater.

Biography

Bob Dimitroff received a Bachelor of Science degree in Mining Engineering in 1980 from the University of Pittsburgh and a Masters of Business Administration in 1985 from California State University, Bakersfield. He is currently working towards a Masters of Science in Project Management from The George Washington University with an expected graduation date of May 2006. Bob obtained his Professional Engineers license in Mechanical Engineering from the State of California in 1993. He also earned Project Management Professional Certification from the Project Management Institute in 1998.

He has over 25 years experience in the oil and gas business for Chevron in a variety of Facilities Engineering and Project Management positions, domestically and internationally. He is a Manager of LNG Project Support in Chevron Energy Technology Company.

Risk Comparison of Offshore vs. Onshore LNG Receiving Terminals

Cynthia Spitzenberger
Consultant
DNV Consulting, Det Norske Veritas, Houston, Texas

AIChE Spring Meeting, April 2006
6th Topical Conference on Natural Gas Utilization
Orlando April 23-27, 2006

Introduction

More than 50 new LNG terminals are proposed for North America. There have been concerns about the potential hazards posed by these facilities, especially onshore LNG terminals. Some believe that the solution lies offshore through the development of floating or gravity based LNG terminals. The processes and technologies involved in typical onshore and offshore terminals are reviewed within this paper. A risk assessment methodology using a risk matrix approach facilitates the discussion. Resulting from the risk analysis, main contributors of the facility hazards can be identified and a risk profile is generated and compared for each terminal type.

Typical Onshore Receiving LNG Terminal

An onshore import terminal consists of the following main components that contribute to its function and risk picture:

- LNG carrier ship
- Unloading dock facility
- Storage facilities
- Re-gasification unit
- Export facility

An LNG carrier, which can range in size from 120,000m^3 up to the projected future capacity of 250,000m^3, travels from the LNG liquefaction facility to the unloading jetty of the onshore terminal. Depending on the size of the terminal and the carrier, an unloading operation may be scheduled as often as once every few days. The LNG carrier is moored to the unloading jetty and the carrier pumps offload and transfer the LNG to the onshore LNG storage facilities. The LNG is transferred to the onshore terminal via multiple offloading arms and an unloading line that traverses the jetty to the storage tanks. Depending on the size of the carrier, the offloading process may take between 10 and 16 hours.
The onshore storage tanks are usually of full containment construction, meaning there is a steel alloy inner tank surrounded by a concrete outer tank. Depending on cost and land availability, single or double containment tanks may be used

instead. In addition, the storage tanks are sometimes surrounded by a bund that would also contain the volume of the tank. Usually the terminal will consist of 2 or 3 storage tanks. The storage tanks are usually over 200ft in diameter and 150ft in height and are prominent features of any terminal. The tanks operate at atmospheric pressure yet have enough insulation to maintain the cold temperature required to maintain the liquefied state of the LNG.

In-tank sendout pumps transfer LNG from the tanks to vaporizers that re-gasify the LNG for export through a pipeline. LNG may also be transported by trucks which load at the terminal and deliver LNG to satellite storage locations. Several vaporizer technologies are possible, including the following:

- Submerged combustion vaporizers (SCV) – burn fuel gas to heat a water bath that vaporizes the LNG,
- Open-rack vaporizers (ORV) – use sea water for vaporization of LNG, and
- Shell and tube type vaporizers (STV) – using different fluids as the heat source.

Vapor lines remove excess "boil-off" gas from the storage tanks and use the vapor to equalize the pressure in the carrier tanks during unloading. Boil-off gas compressors and re-condensers may be used to re-liquefy the vapor for storage and export. During times of no unloading, the terminal is in "holding" mode, and LNG is circulated through the unloading pipelines to maintain the terminal's readiness for the next carrier visit.

In the US, FERC approves or denies applications for siting, construction, operation or expansion of onshore LNG terminals. A new onshore terminal built on existing designs is expected to cost $400 million or more (for a 180-360 Bcf per year facility).[1]

Figure 1 Example Onshore Terminal [2]

Typical Offshore Receiving LNG Terminal

An offshore import terminal consists of the following main components that contribute to its function and risk picture, which are similar to those for an onshore facility:

- LNG carrier ship
- Unloading platform facility
- Storage facilities
- Re-gasification unit
- Export facility
- Structural system

The offshore facility is classified as a deepwater port if it is located outside of state boundaries, and in federal waters. Thus these offshore terminals may be offshore at a minimum distance of 9 nautical miles off the coast of Florida and in the Gulf of Mexico (GOM) and 3 nautical miles along the rest of the US coast. Offshore terminals fall under the jurisdiction of the US Coast Guard (USCG) and the US Maritime Administration (MARAD) according to the Deepwater Port Act. The cost of implementing an offshore facility may be up to twice that required for an onshore facility.

An offshore facility must carefully consider its structural design. The main types of configurations are:

- Gravity Based System (GBS) - consists of concrete caissons that rest on the seafloor and support the terminal topsides.
- Floating system – permanently moored or floating, similar in design to FPSO (floating production, storage, and operating) or FSRU (floating storage, re-gasification unit) facilities.

The same LNG carriers could offload their cargo at the offshore terminal. Depending on the design of the terminal, there are a variety of offloading solutions. Offloading could be conducted side-by-side with the carrier alongside the terminal; the carrier could dock at a jetty facility extended from the terminal; or offloading could be performed through a buoy system.

The LNG on the terminal would be stored in storage tanks similar to those in existing LNG carriers or even in independent tank configurations similar to that of onshore facilities. The offshore facility requires the same process components as in an onshore facility; except that novel technologies and alternative equipment are favored given the offshore environment. Space, equipment footprints and weight are critical factors in an offshore facility. Depending on the layout, the process equipment may be placed on top of the storage tanks or located on a different module connected to the tanks.

Seven of the eight proposed or licensed offshore terminals in the GOM use some form of open loop vaporization (OLV) technology – also referred to as open rack vaporizers (ORV) – which utilizes sea water as the heat medium to vaporize the LNG.[3] The large amount of sea water required – on average 100 million gallons per day per billion cubic feet of capacity – has prompted environmental concerns regarding impact to fish populations due to entrainment and impingement that would occur from the seawater intakes and the potential impact on fishery stocks.[3] The possible environmental impact created from such vaporization technology would occur on a daily basis; a recent study indicates that the previous calculations for environmental impact of OLVs have been overly conservative.[3]

The natural gas is exported from the terminal via underwater pipelines that tie-in with existing pipelines or connect and tie-in onshore to the natural gas distribution pipeline system.

a) GBS Offshore Terminal [4] b) Floating Offshore Terminal [5]

Figure 2 Example Offshore Terminals

Risk Assessment Approach

A qualitative risk analysis approach was utilized to compare the risks associated with onshore and offshore LNG receiving terminals.[6] The following steps were used to setup the evaluation method:

- Develop a risk criteria basis
- Perform a hazard identification exercise to identify the potential risks for each type of facility
- Evaluate the likelihood of occurrence for each hazard scenario
- Evaluate the severity of each hazard scenario according to the various possible consequence categories

Risk Criteria Basis

A risk matrix approach satisfied the evaluation goals and provided a succinct presentation of the risk ranked results. Likelihood and severity levels were defined as presented in Tables 1 and 2. The combination of the likelihood and severity levels represents the risk level and the corresponding location in the risk matrix as presented in Table 3.

Table 1 Definition of Likelihood Levels

Level	Description	Indicative Frequency (per year)	Definition
A	Frequent	> 0.5	Occurs frequently
B	Probable	0.5 - 0.05	Occurs several times
C	Occasional	0.05 - 0.005	Likely to occur during lifetime
D	Remote	0.005 - 0.0005	Unlikely to occur during lifetime
E	Improbable	< 0.0005	Unlikely event, may not be experienced

Table 2 Definition of Severity Levels

Level			Code	1	2	3	4
Description				Minor	Major	Critical	Catastrophic
Consequence Category	Human / Personnel	Crew	C	Minor injury to crew	Serious injury to crew	One fatality to crew	More than one fatality to crew
		Third party	T	No injury to third party	Minor injury to third party	Serious injury to third party	Third party fatalities
	Environmental		E	Negligible pollution	Reportable pollution Minor release; no long term effect	Reportable pollution Major release; limited effects	Reportable pollution Uncontrolled pollution; long term effect
	Material / Assets	Company Property	P	Minor damage	Damage	Major damage	Loss of facility
		Down time	D	Negligible down time	Down time up to one day	Down time up to one week	Down time more than one week
		Reputation	R	Negligible or no loss of reputation	Reputation affected locally	Reputation affected with authorities, noted in industry	Major public interest; loss of reputation in the industry
		Third party assets	A	No effect on third party	Minor damage to third party assets; short repair duration	Major damage to third party assets; long repair duration	Extensive damage to third party assets

Table 3 Risk Matrix

		SEVERITY			
		1	2	3	4
LIKELIHOOD	A	M	H	H	H
	B	M	M	H	H
	C	L	M	M	H
	D	L	L	M	M
	E	L	L	L	L

The combination of the severity and likelihood levels creates three risk levels – high (H), medium (M), and low (L). The high risk level is normally considered as unacceptable risk and risk reducing (mitigating) measures are required to reduce the risk level. Medium risk levels are within the ALARP (As Low As Reasonably Practicable) region and risk reducing measures should be implemented if cost-benefit evaluations are favorable.

Risk Evaluation

The risk assessment was performed for the various sections of the LNG receiving terminals and the possible hazardous events that may occur. Tables 4 and 5 present the overall results of the risk evaluation. The classifying of likelihood and severity levels was based on previous qualitative analyses of LNG releases.

LNG Carrier at Berth

Depending on the unloading rate, the LNG carrier may be at berth for up to a day at the terminal facility. During the time at berth, a high or low energy collision may occur between the carrier and another vessel. The low energy collision may penetrate the outer hull of the carrier, but is unlikely to breach the inner tank, and thus there would be no release of LNG. However the high energy collision is assumed to cause a release by penetrating both the inner and outer hulls. Due to the nature of a high energy collision, it is assumed that ignition of the release would occur. The consequences would be similar between the onshore and offshore terminals. Depending on the size of release, there could be multiple fatalities of onsite personnel and the colliding boat's crew may be injured.

For the onshore facility, the LNG carrier will be at berth for up to a day at the terminal jetty. The likelihood of this event occurring at an onshore facility is remote for a high energy collision and occasional for a low energy collision. Enforcement of safety zones around the carrier should lower the chance of a large vessel colliding with the docked carrier. Smaller vessels, such as tugs or pleasure craft, have an increased chance of avoiding security forces and colliding with the carrier.

At an offshore facility the likelihood of a high energy collision is increased due to the undefined traffic lanes in open waters for ocean-going vessels. The offshore terminal may be inappropriately used as a navigational waypoint increasing the chance that if there is a failure onboard an ocean-going vessel, it may be on collision course with the terminal. However, there are several actions that would lower this probability of occurrence, such as proper identification on nautical charts, and adequate radar and warning systems.

Even though the severity ratings are the same, the higher likelihood class category for the offshore facility has some high risk rankings. Each terminal location, onshore or offshore, has unique vessel traffic patterns and densities that would adjust the likelihood for a collision event.

Unloading Dock

A release from an unloading arm or rupture of an unloading hose is considered likely since both onshore and offshore unloading will involve movement of the unloading carrier. However the release from the arm or hose will be minimized through the use of quick closing control valves that will activate if the connection between the carrier and facility is broken or there is excessive relative movement between the two.

Although not as likely to occur, a release from the unloading pipeline that transfers the LNG to storage could be of significant size to cause property damage and would result in critical downtime. A differentiation between onshore and offshore is that third party assets will unlikely be impacted by such a release offshore.

Storage Facilities

Although the release of an entire tank's inventory could result in multiple fatalities and catastrophic property damage and downtime, the likelihood of such an event is improbable. The event would have to cause the failure of both the inner and outer containment systems found both onshore and offshore. The offshore terminal would not impact third party lives, as the onshore tank would, but the large release to the water may impact local marine life and may pose a greater threat environmentally due to the large release area and cold release temperature.

Re-gasification Process

Three release sizes were considered resulting from the re-gasification process unit – large, medium, and small. The likelihood of occurrence for the three events on both facilities is the same, which small releases more probable than the large and medium releases.

A large release is likely to impact offsite personnel and third party assets for the onshore terminal, while the offshore terminal is expected not to have third party impacts. The property damage and down time severity categories are considered to be more severely impacted on the offshore facility for the medium and small releases than if they occurred onshore. Depending on temperature, LNG vapors will quickly disperse. However, if there are areas of congestion where the vapors may collect and reach flammable concentrations, there could be explosive consequences. The offshore design unless reviewed carefully, may increase congested areas and thus provide circumstances for more powerful explosions and consequences.

Export Pipeline

The frequency of an export pipeline release is considered to be occasional for both onshore and offshore. Onshore export pipelines are usually buried, while offshore export pipelines are subsea. Depending on the size of the release, the consequences may result in a crew fatality and some downtime. The offshore facility's export pipeline will have little impact on third party interests, while the on land export pipeline will most likely impact third parties in some fashion, depending on the developments surrounding the pipeline.

Extreme Weather

Extreme weather, such as a hurricane would impact an onshore facility with extreme wind, storm surge, and flooding conditions. Depending on the strength of the storm and the design level of the facility, the weather could cause failure of a storage tank or other damage to the facility. Offshore, extreme weather would also impact the terminal with extreme wind and wave loads. Operating procedure in the event of a hurricane would be to evacuate any personnel from the offshore facility. Thus the effects of a hurricane on an offshore facility will not impact personnel or third parties, which will occur at an onshore facility. For both facility types, extreme weather could cause catastrophic damage and facility downtime. If the offshore terminal is a floating structure, then it may be feasible to move the facility in advance of the extreme weather and thus avoid damage.

The likelihood of extreme weather impacting a terminal is occasional for both facility types.

Earthquake

An earthquake can equally impact onshore and offshore facilities. Both would need to be designed for the seismic activity levels within their respective areas according to the design standards. As with extreme weather, hazardous consequences at an offshore terminal would not impact third party personnel and assets. The likelihood of a damaging earthquake is considered remote for both facility types.

Sabotage

Depending on the act of sabotage, critical damage may be done to the facility with multiple injuries or even fatalities of personnel resulting. Again any hazardous event resulting from a terrorist act at the offshore facility would not impact third parties. The impact to third parties onshore is dependent on the resulting magnitude of the hazard event.

Adequate security measures to limit access to relevant personnel would naturally be in place for both facilities. The offshore facility is more vulnerable due to its isolation and limited manning.[7] Although the likelihood for successful implementation of an act of sabotage for both is remote, the offshore facility was evaluated at a lower likelihood than the onshore facility. Targeting an offshore facility would not serve the terrorist purpose in terms of third party injury and damage and thus onshore facilities would be the more likely targets for such acts.

Offshore Specific

The following hazards have been assessed as only applying to an offshore LNG receiving terminal:

- Helicopter collision
- High energy collision with terminal
- Low energy collision with terminal

Transport of personnel to the offshore facility is assumed to be via helicopter, which poses the hazard of crashing at the facility and causing a release and/or fire event. The damage may be major, and the personnel within the helicopter would retain some injuries.

Similar to the berthed LNG carrier, the offshore terminal is at risk from high and low energy collisions. Only the high energy collisions are expected to cause a release event and the consequences may be catastrophic in terms of damage and downtime, with the possibility of crew fatalities.

The likelihood of each of these offshore specific hazards is evaluated to be occasional, possible during the facilities lifetime.

Table 4 Onshore and Offshore LNG Receiving Terminal Risk Evaluation

ID	System	Event	ONSHORE															OFFSHORE														
			Likeli-hood	Severity							Risk Category							Likeli-hood	Severity							Risk Category						
				C	T	E	P	D	R	A	C	T	E	P	D	R	A		C	T	E	P	D	R	A	C	T	E	P	D	R	A
1	LNG Carrier	High energy collision with carrier at berth	D	4	4	3	3	4	3	3	M	M	M	M	M	M	M	C	4	4	3	3	4	3	3	H	H	M	M	H	M	M
2		Low energy collision with carrier at berth	C	1	2	1	1	1	1	2	L	M	L	L	L	L	M	C	1	2	1	1	1	1	2	L	M	L	L	L	L	M
3	Unloading Dock	Unloading arm / hose release	B	1	1	2	1	2	1	1	M	M	M	M	M	M	M	B	1	1	2	1	2	1	1	M	M	M	M	M	M	M
4		Unloading pipeline release	C	2	1	3	2	3	1	2	M	L	M	M	M	L	M	C	2	1	3	2	3	1	1	M	L	M	M	M	L	L
5	Storage Facilities	Release of entire tank inventory	E	4	3	3	4	4	3	2	L	L	L	L	L	L	L	E	4	1	4	4	4	3	1	L	L	L	L	L	L	L
6	Re-gasification Unit	Large release from process equipment	D	3	2	3	4	4	3	2	M	L	M	M	M	M	L	D	3	1	3	4	4	3	1	M	L	M	M	M	M	L
7		Medium release from process equipment	C	2	1	2	2	3	2	1	M	L	M	M	M	M	L	C	2	1	2	3	4	2	1	M	L	M	M	H	M	L
8		Small release from process equipment	B	1	1	1	1	1	1	1	M	M	M	M	M	M	M	B	1	1	1	2	2	1	1	M	M	M	M	M	M	M
9	Export Pipeline	Export pipeline release	C	2	2	3	2	3	2	2	M	M	M	M	M	M	M	C	2	1	3	2	3	2	1	M	L	M	M	M	M	L
10	Extreme Weather	Damage to facility / structural supports	C	2	2	3	4	4	3	3	M	M	M	H	H	M	M	C	1	1	3	4	4	3	1	L	L	M	H	H	M	L
11	Earthquake	Damage to facility / structural supports	D	2	2	3	3	3	3	3	L	L	M	M	M	M	M	D	2	1	3	3	3	3	1	L	L	M	M	M	M	L
12	Sabotage	Damage to facility / structural supports	D	3	3	3	3	3	3	3	M	M	M	M	M	M	M	E	3	1	3	3	3	3	1	L	L	L	L	L	L	L
13	Specific to Offshore Facility	Helicopter collision																C	3	1	2	2	3	2	1	M	L	M	M	M	M	L
14		High energy collision with terminal																C	4	4	3	3	4	3	3	H	H	M	M	H	M	M
15		Low energy collision with terminal																C	1	2	1	1	1	1	2	L	M	L	L	L	L	M

Table 5 Onshore and Offshore LNG Receiving Terminal Risk Results

ONSHORE

		Severity			
		1	2	3	4
Likelihood	A				
	B	3C, 3T, 3P, 3R, 3A 8C, 8T, 8E, 8P, 8D, 8R, 8A	3E, 3D		
	C	2C, 2E, 2P, 2D, 2R 4T, 4R 7T, 7A	2T, 2A 4C, 4P, 4A 7C, 7E, 7P, 7R 9C, 9T, 9P, 9R, 9A 10C, 10T	4E, 4D 7D 9E, 9D 10E, 10R, 10A	10P, 10D
	D		6T, 6A 11C, 11T	1E, 1P, 1R, 1A 6C, 6E, 6R 11E, 11P, 11D, 11R, 11A 12C, 12T, 12E, 12P, 12D, 12R, 12A	1C, 1T, 1D 6P, 6D
	E		5A	5T, 5E, 5R	5C, 5P, 5D

OFFSHORE

		Severity			
		1	2	3	4
Likelihood	A				
	B	3C, 3T, 3P, 3R, 3A 8C, 8T, 8E, 8R, 8A	3E, 3D 8P, 8D		
	C	2C, 2E, 2P, 2D, 2R 4T, 4R, 4A 7T, 7A 9T, 9A 10C, 10T, 10A 13T, 13A 15C, 15E, 15P, 15D, 15R	2T, 2A 4C, 4P 7C, 7E, 7R 9C, 9P, 9R 13E, 13P, 13R 15T, 15A	1E, 1P, 1R, 1A 4E, 4D 7P 9E, 9D 10E, 10R 13C, 13D 14E, 14P, 14R, 14A	1C, 1T, 1D 7D 10P, 10D 14C, 14T, 14D
	D	6T, 6A 11T, 11A	11C	6C, 6E, 6R 11E, 11P, 11D, 11R	6P, 6D
	E	5T, 5A 12T, 12A		5R 12C, 12E, 12P, 12D, 12R	5C, 5E, 5P, 5D

As demonstrated in Table 4 and Table 5, the offshore terminal is exposed to three additional hazardous events – helicopter collisions and high and low energy collisions with the terminal – and thus has more entries in the risk matrix. Yet the distributions of the offshore and onshore risks fall within the same matrix areas. Some of the risks present at both facility types receive a lower risk level in the offshore terminal than in the onshore terminal, such as risks for third party personnel and property.

In Table 5 for the onshore terminal, there are two entries for high risk (severity 4, likelihood C); they both correspond to extreme weather and the property damage and downtime that may result from such an event. The offshore terminal has nine entries at the same high risk level; including risks resulting from extreme weather, high energy collision with carrier at berth or with the terminal, and medium process releases. Most of these events may be investigated and with careful design and safety measures bring the risk levels into the ALARP region.

Conclusions

The onshore and offshore terminals pose common and distinctive hazards. Offshore LNG receiving terminals have been favored recently for decreased exposure to third parties and to avoid community opposition and permitting obstacles. Although attractive, offshore terminals present a similar risk picture as that of onshore terminals, with the addition of several hazards inherent to the offshore industry.

Each terminal design – whether onshore or offshore – has a unique risk picture considering the possible hazards and the corresponding frequency. Modern terminal design carefully considers siting issues which incorporate the implementation of safety and security measures that mitigate the hazard severities and to bring risks within the ALARP region of the risk matrix.

References

1 National Gas Intelligence, Intelligence Press, Inc. October 28, 2004.

2 Cheniere. Corpus Christi Receiving Terminal. (http://www.cheniere.com/LNG/LNG_corpus_artist_rendition.shtml).

3 Exponent, "An Evaluation of the Approaches Used To Predict Potential Impacts of Open Loop LNG Vaporization Systems on Fishery Resources of the Gulf of Mexico," Center for Liquefied Natural Gas, November 2005.

4 Shell. Gulf Landing Terminal (http://www.shell-usgp.com/downloads/GL.Brochure.final.pdf)

5 BHP Billiton Cabrilo Port (http://lngsolutions.bhpbilliton.com/pdf/BHP%20DW%20Fact%20Sheet.pdf)

6 Skramstad, E. et.al. "Use of Risk Analysis for Emergency Planning of LNG Carriers," Det Norske Veritas, 2000.

7 Parfomak, P., "Liquefied Natural Gas (LNG) Infrastructure and Security: Background and Issues for Congress," CRS Report for Congress, September 9, 2003.

8 Coyle, D. and Patel, V., "Processes and Pump Services in the LNG Industry" Proceedings of the 22nd International Pump Users Symposium, 2005.

9 Yang, C.C. and Huang, Z., "Lower Emission LNG Vaporization" LNG Journal, November/December 2004, p24-26.

10 Napier, D.H. and Roopchand, D.R., "An approach to hazard analysis of LNG spills", Journal of Occupational Accidents, 7: p251-272, 1986.

Biography of Speaker

Cynthia Spitzenberger is a Consultant for DNV Consulting in Houston, Texas. She provides technical risk assessment services to both the onshore and offshore sectors. Cynthia has participated in numerous LNG consequence and risk assessments regarding LNG transport and terminal facilities. She attended the University of Texas at Austin and received a BS in Chemical Engineering. (cynthia.spitzenbeger@dnv.com)

Details of 35 m Diameter LNG Fire Tests Conducted in Montoir, France in 1987, and Analysis of Fire Spectral and other Data

Hugues Malvos
Gaz de France
Research & Development Division
361, avenue du Président Wilson - BP 33
93211 Saint-Denis La Plaine cedex, France
Phani K. Raj
Technology & Management Systems, Inc
102 Drake Road
Burlington, MA 01803

AIChE Spring National Meeting 2006
6th Topical Conference on Natural Gas Utilization
Orlando, Florida, April 23-27, 2006.

Abstract

A series of three 35 m diameter LNG fire tests was conducted on an insulated concrete dike in July & September of 1987. Only limited data from these tests has been published before. This paper presents the details of the tests conducted, types of instruments used, data measured and important results obtained. The fire emissive power obtained from the wide angle and narrow angle radiometer data are compared with similar results from other previous field tests. The importance of these results in modeling the behavior of large LNG fires and predicting the thermal emissions from fires postulated to occur from large LNG releases are discussed. Also indicated in this paper are the data on the spectral emission from the 35 m diameter LNG fires. Analysis and comparison with similar data from the China Lake LNG fire on water tests are provided.

Introduction

A series of three tests involving an outdoor fire on a pool of LNG confined in a specially constructed, flat, 35 m diameter insulated concrete dike was conducted in July – September 1987 in the field test facility of Gaz de France (GdF) at the Montoir de Bretagne methane terminal in France. This test series was performed as a joint program by a consortium of six companies comprising Shell, British Gas (BG), British Petroleum (BP), Elf Aquitaine, Total CFF and Gaz de France.

The objective of the test program was to understand the burning characteristics of and quantify the radiant thermal emission from large LNG pool fires in which the dominant heat for the vaporization of the liquid pool comes from the fire. GdF carried out the construction of the special test facilities during 1986 and 1987. Three fire tests were performed under different wind conditions. During the tests, Gaz de France (Direction des Etudes et Techniques Nouvelles) had overall control of the test facility, while comprehensive measurements were made by British Gas (Midlands Research Station, UK), Shell Research Ltd., (Thornton Research Center, UK) and Gaz de France. This paper presents some of the salient data obtained during these tests including those presented previously in European conferences [Nédelka, et al., 1989, and Nédelka, 1991][(1)] and additional information that have not been published before.

The principal components of the test facility consisted of the following:

1 A 35 m diameter circular dike constructed on graded earth and topped with ordinary concrete and a 100 mm (≈ 4 inch) thick colloidal, liquid impervious, insulation quality concrete spread and cured to be as flat as possible. The overall depth of the dike from the top of the encircling wall to the top of the insulated concrete layer was about 1 m. [See Figure 1].

2 A 75 □m (≈ 0.003 inch) thick high strength, flammable, polyethylene ("Valeron") cover stretched over the entire dike to prevent solar and convective heat input to the LNG pool during the filling process and before ignition. Eight (8) holes, each 1 m in diameter, were cut in the cover to allow vapors to escape and not build pressure inside the "tent."

3 Six (6) LNG road transport tankers connected to the dike through two parallel, insulated pipelines each of 90 mm (≈ 4 in) diameter and about 250 m long. [See Figure 2]. LNG was fed into the dike at 4 points equally spaced along the circumference of the dike.

4 A number of measuring instruments (listed later) to measure different aspects of the fire and radiant thermal output.

Salient Features of Various Instruments

The experimental setup consisted of a number of instruments both inside the dike and outside the dike to measure various parameters. Most instruments were monitored remotely from an instrument building located about 235 m from the center of the dike and protected by a water spray curtain. General locations of the instruments are indicated in Figure 2. The various instruments used are described below:

Instruments outside the dike

Wide angle radiometers (WAR): A total of fifty WARs were used in these tests. Of these, 43 were slower-response WARs with 120° total angle field of view and located at different radial and azimuthal positions indicated in Figure 2. These WARs were located such that their axes were horizontal and "looking" to a point 1 m above the dike floor center.

1 Dominique Nédelka of Gaz de France was the overall project manager for this series of tests.

Five fast response (< 20 ms time constant) WARs were deployed at positions nominally upwind, crosswind and down-wind of fire. Cine cameras at the locations of the WARs provided a synchronous record of the flame as viewed by each radiometer. Using the view factor derived from the cine recordings and the instantaneous readings from the radiometers, average SEP (Surface Emissive Power) values were calculated for the visible flame.

Two WARs with a 50° view angle were placed relatively close to the dike behind aluminum shields and were so oriented as to see the radiation only from the part of the flame above 33.6 m (measured from the dike floor). That is, radiation from only those layers of the fire with substantial smoke was seen by these radiometers.

Narrow angle radiometers (NARs): Three Pyroelectric detector type NARs were located at 155 m, 155m and 215 m from the edge of the dike in three different directions. These fast response (< 30 ms time constant) NARs had apertures adjusted to a fully illuminated spot of 1.55 m diameter at a distance of 155 m and the optics were such that they had essentially a transmissivity of 1in the 0.8 □m to 40 □m spectral range and a maximum flux capacity of 380 kW/m^2. For values lower than 20 kW/m^2 the NAR readings were assumed to be unreliable. The NARs at these locations were steerable to facilitate taking of the surface emissive power readings from different parts of the flame.

Band-pass NARs: Three NAR spectrometers with different IR band-pass filters (0.62 □m, 2.25 □m and 4.02 □m) were located at a distance of 155 m from the dike rim and fixed on a movable platform operated by a person located throughout the experiment at this location. On this movable platform was located a narrow angle video camera whose vision coincided with the spot "seen" by the narrow angle radiometer.

Spectrometers: Three different types of spectrometers were used. Firstly, a circular variable filter instrument was used at a distance of 122 m from the bund edge (nominally crosswind). This instrument made scans every 2 s over a wavelength range of 1.6 □m to 8.0 □m and was also steered to view different positions on the flame surface; again a video camera was used to determine the aiming point. A nitrogen-purged grating monochromator spectrometer was placed in a well-insulated box at 20 m from the edge of the dike and looking at the fire with an angle of 11.5° elevation. This obtained repeated scans from a fixed position on the flame surface, taking a total of 40 s for each scan from 2.0 □m to 7.8 □m. A second grating spectrometer was located at the same position and scanned the wavelength region 0.4 □m to 1.05 □m i.e., the visible and near infrared regions.

Instruments inside the dike

Thermocouple array: Two vertical thermocouples arrays, each of 20 thermocouples, located at heights above the dike floor varying from 20 mm to 200 mm were placed inside the dike to measure the liquid-vapor interface location at different times. Also, the temperature of the liquid was used to infer the composition of the liquid.

Helium purged dip tubes: LNG vaporization rate was also determined directly by two helium purged dip tubes which measured the pressure required to force a small flow of helium gas through the liquid. In this way the instantaneous level of LNG was calculated and hence over time the mass-burning rate was calculated.

Calorimeters: Six calorimeters were provided inside the dike just above the liquid surface to measure the heat flux input into the liquid due to radiant and convective heat transfers from the fire. These instruments were distributed along a diameter of the dike to measure the local variation in the heat flux to the pool.

Gas grab sampling: Two sampling lines were floated on the liquid to withdraw gas samples from points just above the surface. Remotely controlled valves routed the gas to different storage bottles as time progressed. The samples were used to infer the composition of the vapor actually burning at any given time.

Other Instruments

Other instruments used in the tests were to measure wind speed, wind direction and atmospheric conditions (temperature, pressure and relative humidity). In addition, numerous still photographs were also taken from different locations from and around the dike during the fire tests.

Test Conditions

Three tests were conducted over a span of about 8 weeks, under different wind and atmospheric temperature conditions. Table 1 provides data on volume and composition of LNG used and other test conditions. The LNG used has almost 90% methane content and the volume of LNG spilled in each test was approximately 200 m^3, whereas the mean volume in the dike prior to ignition was about 93 m^3. The variation, among the three tests, of the LNG depth in the dike just prior to ignition resulted from the different LNG volumes of spill, the variation in the duration of spill and the evaporation of LNG due to heat flow from the concrete floor and the surrounding atmosphere. In fact, to minimize the total evaporation before ignition a thin polyethylene canopy was provided over the dike. The duration of spill was progressively reduced by a factor of 2 for each test, (from a 4 hour spill in test # 1 to 2 hr release in test # 2 to one hour release in test #3). Two tests (# 1 and # 3) were conducted under relatively low wind speeds and one test (# 2) was conducted under a moderately windy condition.

Summary of Measured Data

Table 2 indicates a summary of parameters measured to characterize the fire. In general, mean values are shown except as otherwise indicated. Height-wise measurements of narrow angle radiometer data (time averaged high and low values, respectively) are also indicated. Plate 1 shows a typical photograph of the LNG fire during the steady state burning of methane content of LNG in test 1. Plate 2 shows the methane burning period fire in test 2 in high wind conditions. Plate 3 is a sample of the fire geometry during the steady state burning of methane in test 3.

Burning time and Liquid Evaporation Rate: It is seen that, in general, the total burning period of the fire (400 ± 50 s) is about the same for all three fires, the longer duration representing the case of larger quantity of liquid in the dike. Also noticed is the fact that the measured heat flux from the fire to the liquid (at the surface) is the same for test 1 and test 2, whereas for test 3 it is higher. The flame was bent significantly (about 60°) in test 2 because of higher wind speed, while in test 1 and test 3 the angle of bending of the flame was about the same ($\sim 45^{o}$). One expects the heat flux feed back in test # 2 to be less because of the significantly larger bending angle of the fire and the corresponding lower "view factor" between the fire and the dike area; however, the calorimeter data indicate the opposite trend. This is perhaps due to better mixing of LNG vapor with air due to higher wind speed (and, hence, higher turbulence in the lower heights of the fire) resulting in more efficient combustion, and thus higher fire temperature, at the bottom portions of the fire. In Table 2 the calculated liquid evaporation rates are indicated based on the heat flux data measured by the calorimeters located just above the liquid surface. This calculation assumes that all heat received close to the liquid surface is absorbed by the liquid. Also indicated are the directly measured evaporation rates obtained from the helium purged dip tubes. It is seen that the two results do not agree both in magnitude and trend-wise.

The calculated burning rate from the fire-heat-feed-back data is almost 2.5 times the liquid evaporation rate measured directly by the helium purged dip tubes! One can speculate several reasons to explain this discrepancy. The first is that not all heat flux "aimed at" the liquid surface is absorbed by the liquid pool because the liquid is boiling violently resulting in the formation of liquid droplets and vapor mixture just above the liquid surface which absorb and scatter a substantial part of the heat flux incident just above the liquid surface. It should also be noted that methane absorbs IR ("band absorption") in the 3.3 □m, 6.5 □m and 7.6 □m bands. The second reason is that LNG (liquid) is a poor absorber of infrared radiation and much of the heat received is reflected back to the fire both from the liquid surface and the bottom of the pool (which is only ~ 100 mm in depth, initially). These data may indicate, therefore, the substantial influence of methane vapor/liquid droplets in the regions just above the liquid surface in preventing the heat feed back from the fire to the boiling liquid. Another reason that should be considered is the presence of soot at low temperature close to the liquid surface, in the low part of the flame: this soot and LNG Vapor mixture is a powerful absorber of infrared radiation (continuous band) and constitute therefore an efficient screen in preventing the heat feed back from the fire to the boiling liquid. While the helium purged dip tube data and the calorimeter data do not indicate a reduction in the heat feed back to the pool with increased bending of the fire, it is possible to speculate the effect of a deep layer of methane vapor above the liquid surface (in reducing the heat feed back) in the burning of LNG on top of a LNG tank under the conditions of high wind and when the liquid surface is at a significant depth below the lip of a LNG tank.

Radiant Heat Output from the Fire: The radiant heat out put from the fires was measured with both narrow angle radiometers (NARs) and wide-angle radiometers (WARs). Figure 3 shows, schematically, the protocol for test # 2 for spot emissive power data collection using the NAR mounted on a movable platform and located 155 m from the edge of the dike. In this scaled diagram, the location of the dike is shown as well as the aim points of the NAR. The numbers next to the dots indicate the sequence of aim. The sequence was initiated at about 32 s after ignition and measurements were taken for duration of about 5 s at each point. Similar measurements were taken during test # 1 and test # 3, although the exact locations of the aim were different because of different fire plume inclinations to the vertical.

Figure 4, Figure 5 and Figure 6 show the NAR data for test 1, test 2 and test 3, respectively. These are as measured data and have not been corrected for atmospheric absorption. The data indicate the mean value and standard deviation for the 5 s measurements of the apparent surface (spot) emissive power (SEP) at each NAR aim point, recorded from a NAR position at 155 m from the edge of the dike and uncorrected for atmospheric attenuation. The horizontal axis in these figures represents the measured SEP in kW/m^2. The Y-axis of the figures represents the distance along the tilted flame axis to a horizontal plane containing the NAR "aim point.". Unfortunately, no data are available for test #1 above 10 m plume axial length (visible fire plume length in this test was 75 m). In Figure 7 are shown the statistical distributions of SEP measurements made within a period of 5.6 s at three representative locations, namely, the bottom of the fire, the middle of the fire plume axis and at the "top" of the visible fire.

The data from Figure 4 through Figure 6 indicate considerable scatter. However, the scatter increases with increasing axial distance from the bottom of the fire. One can attribute this feature to the intermittent visibility of the burning zone ("visible fire") due to smoke obscuration, the obscuration increasing with increasing height (or length) from the bottom of the fire. In spite of the data scatter, one can discern some trends; there seems to be a region at the bottom (extending about 10 m in axial length) wherein, the SEP is reasonably constant and beyond which the SEP decreases. The 10 m length value for the "constant SEP" region is very subjective. It is also noticed that while in test 3 the decrease with height is more pronounced, in test 2 this decrease is not so prominent even though the trend for the decrease shows clearly. This may be ascribed to the higher wind speed in test 2 that reduces the incidence of (or the frequency of) smoke obscuration of the burning zone. The maximum values (uncorrected for atmospheric absorption effects) in the NAR measured apparent SEP data range from 201 kW/m^2 to 210 kW/m^2 (neglecting three NAR values over the entire test data that are suspicious, such as a single measurement of 297 kW/m^2 shown in Figure 6). If these apparent SEPs are converted to maximum flame surface

SEPs, using the estimated atmospheric transmissivity, the values are in the range 308 kW/m^2 to 316 kW/m^2

An examination of the data presented in Figure 7 (for test 2) indicates that, on a statistical basis, the decrease in apparent SEP with height is dramatic. That is, within the averaging period, the effect of smoke obscuration is significant as one goes from the bottom of the fire to the top of the fire. This fact is also shown by the increase in the standard deviation in the data from the bottom of the fire to the top of the fire. This is also due to the increase in the intermittency of smoke obscuration; that is, the fraction of the time during which the smoke obscures the burning region increases as one goes from bottom to top. This observation is indeed supported by visual observations of the fire in the movie films.

Figure 8 shows the calculated emissive power averaged over the entire fire and the three tests. The calculations are based on measured wide-angle radiometer data corrected for atmospheric attenuation of the IR heat flux from the fires. Two types of averaging are shown for the 35 m diameter Montoir tests. In the first case, the actual visible fire plume areas are used (obtained from painstaking measurements from movie films and synchronizing the areas with the WAR heat flux readings at the same "clock" time). The view factors to the WARs from these individual radiation-emitting areas on the fires are calculated and the mean surface emissive power is calculated. This result is shown with its attendant error bars in Figure 8. The mean of this value is about 265 kW/m^2. In the second calculation, the same WAR data are used but with the flame shape being considered as an ideal circular tilted cylinder of base diameter equal to the dike diameter and axial length equal to that obtained from Thomas' correlation for the visible plume with the measured burning rate (helium purged dip tube data). This average value is 165 kW/m^2. Also shown in Figure 8 are, the reported values reported in the literature for mean emissive powers for LNG fires of different diameters. These values from the literature are based on the description of the pool fire as having an ideal cylindrical shape with the height of the fire being determined by Thomas' correlation. The results presented in Figure 8 clearly indicate the importance of specifying the flame shape and using the corresponding value for the mean SEP when making predictive calculations of the hazard zone around a LNG fire.

The distribution of incident heat flux on a vertical plane element at 1 m above ground at locations all around the fire in test 2 is shown in Figure 9. The 2.5 kW/m^2, 5 kW/m^2 and 7.5 kW/m^2 contours are plotted in this figure. As expected, the distances on the downwind side to the different flux levels are larger than the corresponding distances on the upwind side due to the wind induced tilting of the fire plume. It should be noted that the downwind distance to 5 kW/m^2 is also increased because of the extension of the firebase in the downwind direction due to wind drag. In the case of test 2 this drag ratio was found to be 1.2 (see Table 2). The upwind, cross-wind and down-wind distances from the center of the dike to the 5 kW/m^2 contour are, respectively, 4.3, 7.1, and 8.6 times the dike radius. If the effect of flame drag is discounted (as may be the case in windless condition or fire on top of a LNG tank), the actual distance downwind to 5 kW/m^2 will be reduced, perhaps closer to 7 times the radius. This is considerably smaller than the normal value of 10 times the fire base radius often quoted in the literature for the distance to 5 kW/m^2 in smaller LNG fires under no-wind conditions. This means that, overall, larger fires are less radiative than smaller fires, which result is also shown quantitatively in Figure 8.

The Fire Spectral Emission data obtained by two spectrometers located at 20 m from the edge of the dike in test 2 are shown in Figure 10. These heavily insulated and nitrogen purged spectrometers were designed to capture the emission in the IR range (2 μm to 8 μm) and in the visible-to-near-IR wavelength range (0.3 μm to 1.05 μm). The spectrometers were at a height of 2.25 m above the dike floor and were aimed at an elevation angle of 11.5° with horizontal. That is, the field of view of the spectrometer was aimed at a point about 10 m on the dike vertical center line above the floor (or 6.3 m above the dike floor, on the fire "surface," if one assumes the fire surface to be at the dike edge). This region of the fire represents the intensely hot burning zone.

The spectral data measured by both the "visible and near IR" spectrometer and the IR spectrometer are shown in Figure 10. The maximum spectral radiance measured is 96.4 kW/m^2 μm and this occurs at a wavelength of 2.26 μm. Also plotted in the same figure is the emission spectral radiance from a "black body" at 1516 K. The temperature of the black body is chosen to have the same spectral emission radiance at 2.26 μm wavelength as was measured. It is known that there is very little atmospheric absorption at this wavelength.

The results presented in Figure 10 indicate that the emission from the fire essentially tracks the black body curve, except in the 2.75 μm H_2O band and in the 4.3 μm CO_2 band. The absorption in the 2.75 μm H_2O band and in the 4.3 μm CO_2 band is significant. In these bands there is near total absorption indicating that even over a relatively short path length of 21 m from the fire surface, there is substantial absorption by the atmospheric water vapor and CO_2. The data in Figure 10 further indicates that the luminous soot has a high emissivity; and the combined emissivity of the luminous soot and the band emissions from H_2O and CO_2 in the fire is slightly less than but close to unity. This is in contrast to the emission from a 13 m LNG fire in which, based on spectral data, the luminous soot emissivity was estimated to be 0.14 and the overall fire emissivity was calculated to be 0.61 [Raj, 2005]. This implies that the Montoir fire (35 m diameter) was essentially an optically thick emitter whereas the China Lake fire (13 m diameter) was not optically thick. It is postulated that the optical thickness of LNG fires can be assumed to be close to 20 m. For the 13 m diameter China Lake LNG fire the temperature for the fire was calculated indirectly from the spectral data to be 1500 K [Raj, 2005].

It is interesting to note that the equivalent blackbody radiative temperatures of the fire determined from the 13 m diameter China Lake LNG pool fire on water and from the 35 m diameter Montoir LNG pool fire on land are practically the same (1500 K and 1516 K, respectively) within the estimated uncertainty in determining the fire emission after correcting for the absorption in the intervening atmosphere. The black body emissive power corresponding to 1516 K temperature is 300 kW/m^2. The maximum emissive power estimated from the NAR data (see Figure 6) after applying the atmospheric absorption correction factor is 300±10 kW/m^2. This is very close to the black body emissive power indicated above. Considering the slight uncertainty in the estimated value of the atmospheric transmissivity (0.665) this is a remarkable agreement between NAR data and spectrometer data.

From the above results we may be able to conclude that irrespective of the size of the LNG fire, for diameters larger than 35 m, the maximum emissive power cannot exceed that of a 1516 K black body (or 300 kW/m^2). It is emphasized that this does not mean that all of the flame surface, even the visible spots, have emissive powers of 300 kW/m^2; In fact, even for the 35 m Montoir fires the calculated mean surface emissive power (SEP) over the actual visible fire surfaces is 265 kW/m^2. It should be further noted that the mean SEP for emission from the total surface of an idealized geometry fire is 165 kW/m^2 (see Figure 8). Therefore, the use of such numbers as 300$^+$ kW/m^2 over an entire idealized surface area of a LNG fire to calculate the hazard distances from accidental release of LNG on either water or land is incorrect.

Principal Findings

The principal findings from the 35 m diameter LNG fire tests conducted in Montoir are:

1 The mean liquid evaporation rate due entirely to the heat feed back from the fire is 0.14 kg/m^2 s. This value obtained from the measurement of the actual liquid depth variation with time is considerably lower (by about a factor of 2.5) than the value that can be calculated using the data on the heat flux (from the fire) into the liquid pool measured inside the dike and close to the liquid surface. The discrepancy may be explained as due to (i) methane vapor absorption of heat radiation at the lower levels of the fire, (ii) absorption and scattering by any liquid droplets thrown up due to the violent boiling inside the pool, and (iii) absorption by low temperature soot present close to the liquid surface. Note that the heat input into the pool from the dike floor during the fire test is very small because of the insulated concrete as well as the fact that the floor was pre-cooled.

2 The NAR readings at the bottom of the fires, time averaged over 5 s, are generally in the range 180 kW/m^2 to 200 kW/m^2. The maximum values recorded range from 201 kW/m^2 to 210 kW/m^2. Using the calculated atmospheric transmissivity values (0.65 to 0.665) consistent with the atmospheric conditions and the location distance to the NAR, the maximum spot emissive powers range from 308 kW/m^2 to 316 kW/m^2. It should be noted that these SEP values occur only at one location and do not represent the overall SEPs even in the lower regions of the fire.

3 The fires in all three tests were very smoky beyond a certain height (or length along the tilted axis). The narrow angle radiometer readings confirm that the time averaged emissive power of the fire decreases with height (or

length along the axis). The mean emissive power calculated from actually visible burning area spots over the entire length of the fire plume is about 265 kW/m^2. The mean emissive power value, based on idealized fire geometry (tilted circular cylinder with axial length given by Thomas' correlation), is 165 kW/m^2.

4 The spectral data indicate that the 35 m LNG fire is optically thick and is essentially a black body emitter at an equivalent temperature of 1500 K. This emission applies only to those parts of the fire that are luminous. However, considerable smoke at higher regions masks the high emission rate lowering the apparent emissive power averaged over time and fire size.

Conclusions

From the data gathered in these tests, analyses of the data, and observations from photographs and video films we conclude that:

1 It is incorrect to estimate the burning rate using the radiative characteristics of the lower parts of the fire near the pool and the heat of vaporization of the liquid. In fact, it is entirely possible that 0.14 kg/m^2 s may constitute the highest burning rate consistent with a heat feed back from the fire. That is, in larger diameter fires one could expect the burning rate to be lower. This burning rate must be added to the burning rate due to heat from the substrate to get the overall burning rate.

2 Large LNG fires (greater than 35 m diameter) should be even sootier and produce significantly more smoke mass per unit mass of fuel burned. This conclusion is based on the understanding that black smoke formation is the result of partial combustion of methane due to lack of oxygen in and near the core of the fire.

3 Care must be exercised in using the results of these tests, especially the mean emissive power value quoted (265 kW/m^2 vs. 165 kW/m^2). These values are associated with a particular flame area and should therefore be used only in conjunction with corresponding geometrical representations of the flame. Consistency in the use of the numbers with the models on which they are based is very important. The 165 kW/m^2 value constitutes an example specific to a particular descriptive model of the fire geometry.

4 The maximum emissive power from a LNG fire is 300 kW/m^2 corresponding to a black body emission at a temperature of 1516 K. While greater than 300 kW/m^2 SEP values have been measured in one or two locations in the lower regions of the fire, these values do not represent the average SEPs for the lower fire region.. SEP values equal to or greater than 300 kW/m^2 have been used in some reports and literature.Such values are incorrect to use for

representing the emissive power values over the entire idealized fire geometry.

Note: The tests reported were conducted on an insulated concrete dike, which essentially cut off heat input from the substrate (ground). Even then copious amount of black smoke was noticed; this has been attributed to the reduction of oxygen concentration in the fire center affecting the combustion reactions. The effect of black smoke production is to reduce the overall thermal emission from the fire to the surrounding. One can speculate that large LNG fire on water will produce more black soot, and hence radiate at lower overall emissive power than a fire of the same size on land. This is because the evaporation rate on water is much higher and the effect of lack of oxygen is to reduce combustion efficiency even more than on an equivalent land fire.

References

Nédelka, D., J. Moorhouse and R.F. Tucker, "The Montoir 35 m Diameter LNG Pool Fire Experiments", TRCP.3148R, 9th Intl. Conf & Expo on LNG (LNG 9), Nice, France, 1989.

Nédelka, D., "Calculation of radiation effects from LNG fires", Proceedings EUROGAS90, European Applied Research Conference on Natural Gas, Trondheim, Norway, 1990 May 28-30.

Raj, P.K., "LNG Fire Spectral Data and Calculation of Emissive Power", Paper presented at the International Symposium on Process Safety, Mary Kay O'Connor Process Safety Center, Texas A & M University, October 25 & 26, 2005, College Station, TX.

Acknowledgment

The authors express their gratitude and thanks to Ms. Isabelle Serres, Mr. Michael Godeau, and Ms. Julia Mariotti of the Research & Development Division, Gaz de France for their efforts to find and provide the data from experiments conducted 19 years ago and for the technical review of the manuscript.

Table 1: Test Conditions

Test #	Test Date (1987)	Volume of LNG used (m³)	Composition of LNG (***measured in tank trucks, avg***) (% molar)	Dike filling time (s)	Initial depth of liquid pool (***from dip tube data***) (mm)	Wind speed range (m/s)	Atmospheric	
							Temperature (°C)	Relative humidity (%)
1	7/23	238	CH_4 90.33 C_2H_6 8.95 C_3H_8 0.342 N_2 0.341	235	80	2.7 – 4.8	25	53
2	9/23	198	CH_4 89.90 C_2H_6 8.70 C_3H_8 0.80 N_2 0.42	125	90	7.0 – 10.1	21	54
3	9/25	196	CH_4 90.10 C_2H_6 8.91 C_3H_8 0.64 N_2 0.266	67	110	2.8 – 4.8	14	85

Table 2: Summary of Results

Test #	Total burning time of the Fire (s)	Steady state (***methane burning***) period (s)	Heat flux feed back from fire to the pool (kW/m^2)	Liquid burning rate: Calculated from [3] measured heat feed back (kg/m^2 s)	Liquid burning rate: Helium purged dip tube data (kg/m^2 s)	Mean visible flame length (m)	Mean flame tilt from the vertical (deg)	Maximum downwind flame drag length as a ratio of diameter (Drag Ratio)	Emissive power of the visible part of the flame [4] (kW/m^2)	"High" and "Low" Narrow angle radiometer data [5], [6] (kW/m^2)
1	400	30 -170	168.4 ± 45.7 [2]	0.330 ± 0.09	0.146	75.0 ± 7.1	47.0 ± 9.9	1.05	260.5± 6.4	201 - 79
2	450	35 - 200	188.6 ± 55.4	0.370 ± 0.11	0.142	77.8 ± 4.3	57.3 ± 3.2	1.2	264.8± 6.6	191 - 85 208 - 50 167 - 69
3	500	57-160 [1]	167.4 ± 44.7	0.328 ± 0.09	0.134	71.0 ± 1.0	44.7 ± 8.1	1.1	261.3± 2.5	297 -110 222 - 56 162 - 30

(1) Measurements lost after this time. Steady methane burning period is longer than 160 s.
(2) The number after the symbol ± is the standard deviation
(3) Based on methane latent heat of vaporization (510 kJ/kg)
(4) Calculated from WAR heat flux data, measurements of actual emitting surface areas of flame from video films of flame shape, and (corrected for) atmospheric absorption. The numbers are the mean and standard deviation of the data.
(5) Time averaged mean values of as measured NAR data (not corrected for atmospheric absorption) Numbers on the first line represent the high and low time-average values for the first 10 m along the fire plume; Second line numbers are the data for the 10 m to 35 m along fire axis; third line numbers are for the flame above 35 m.
(6) No NAR data were taken at fire axial distances beyond 10 m in test # 1.

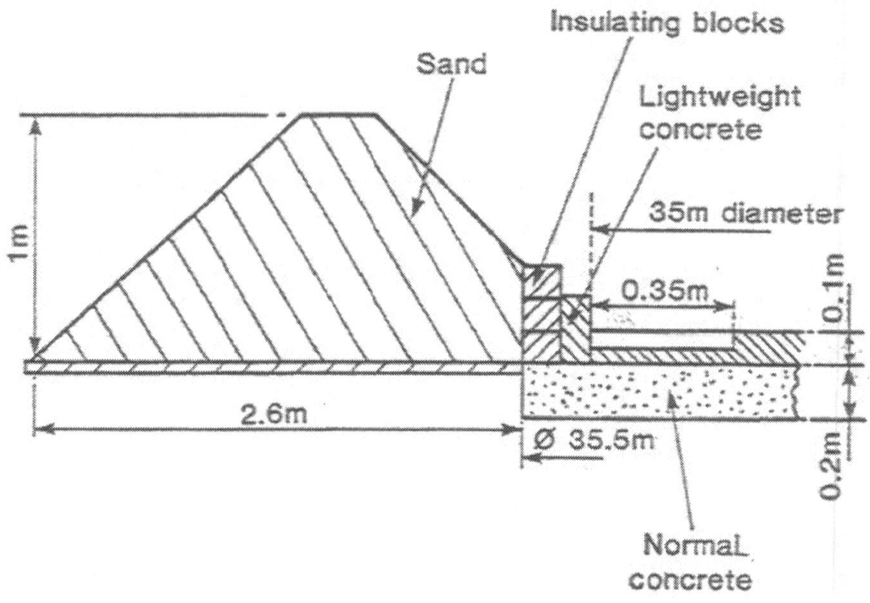

Source: Nedelka, et al.,(1989)

Figure 1: Diagram showing the details of the construction of the dike

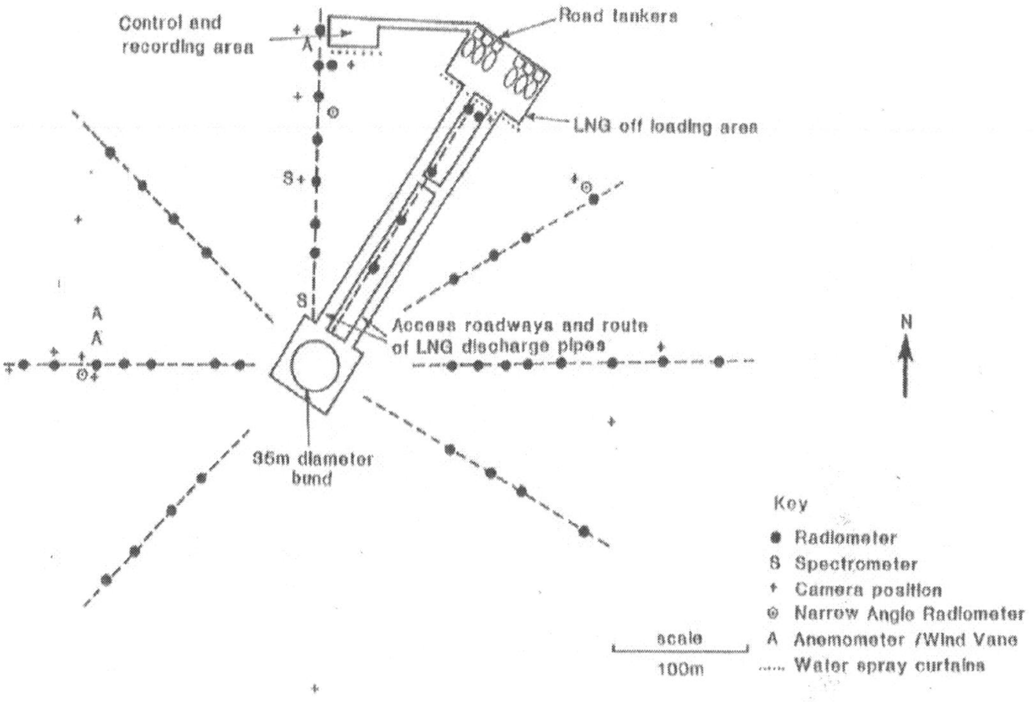

Source: Nedelka, et al.,(1989)

Figure 2: Test site layout and locations of different instruments

(Test 1 configuration shown)

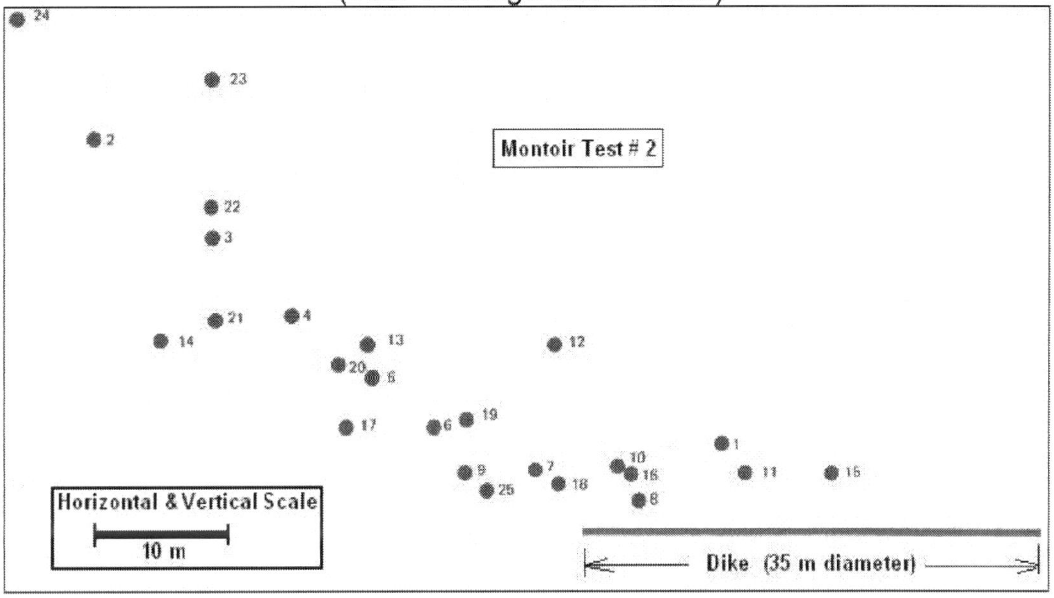

Figure 3: Scaled schematic showing the aim points on the fire and sequence of data collection from the NAR on the movable platform at 155 m from dike edge

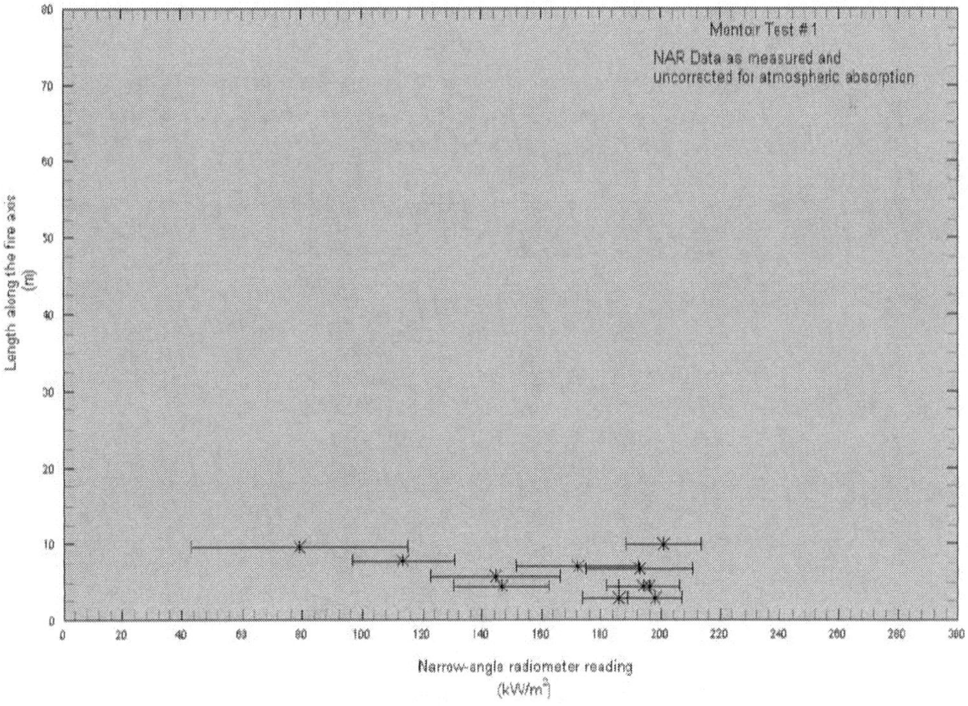

Figure 4: Narrow angle radiometer data from test #1 taken at 155 m distance from dike edge

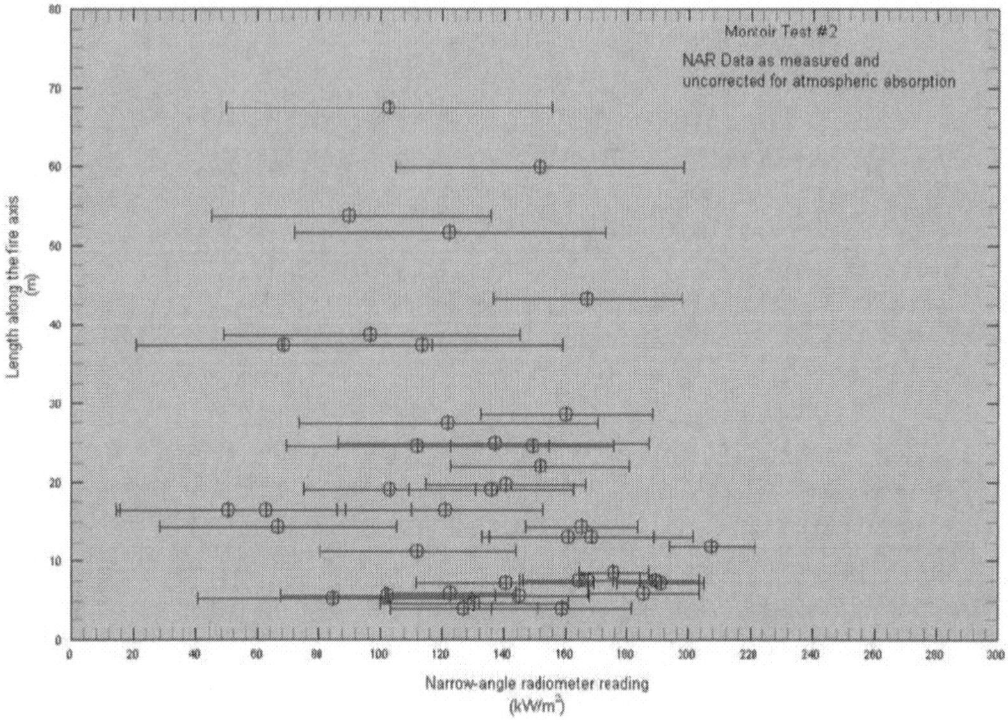

Figure 5: Narrow angle radiometer data from test # 2 taken at 155 m distance from dike edge

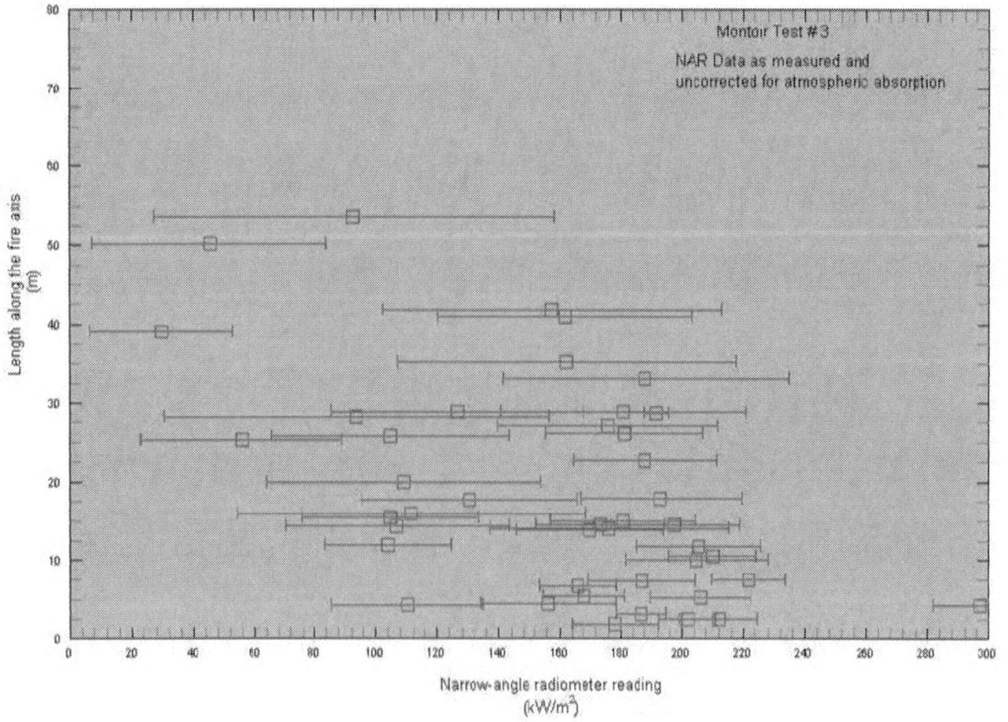

Figure 6: Narrow angle radiometer data from test # 3 taken at 155 m distance from dike edge

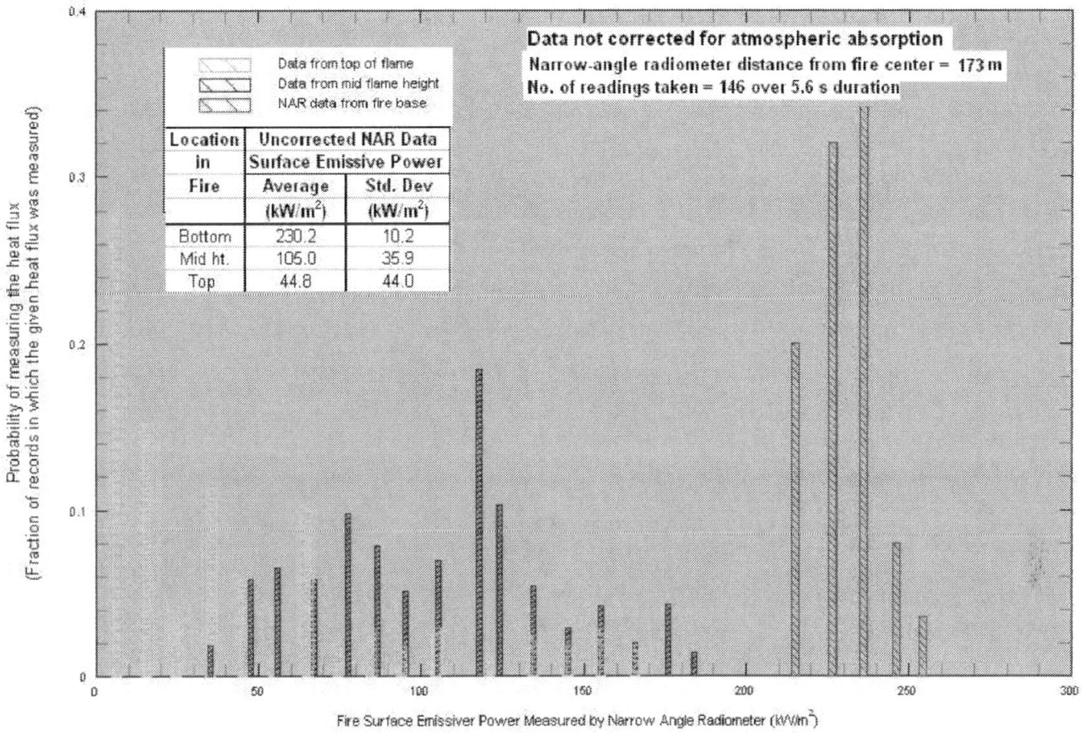

Figure 7: Statistical distribution of NAR data taken at different times from three locations on the LNG fire plume in test # 2

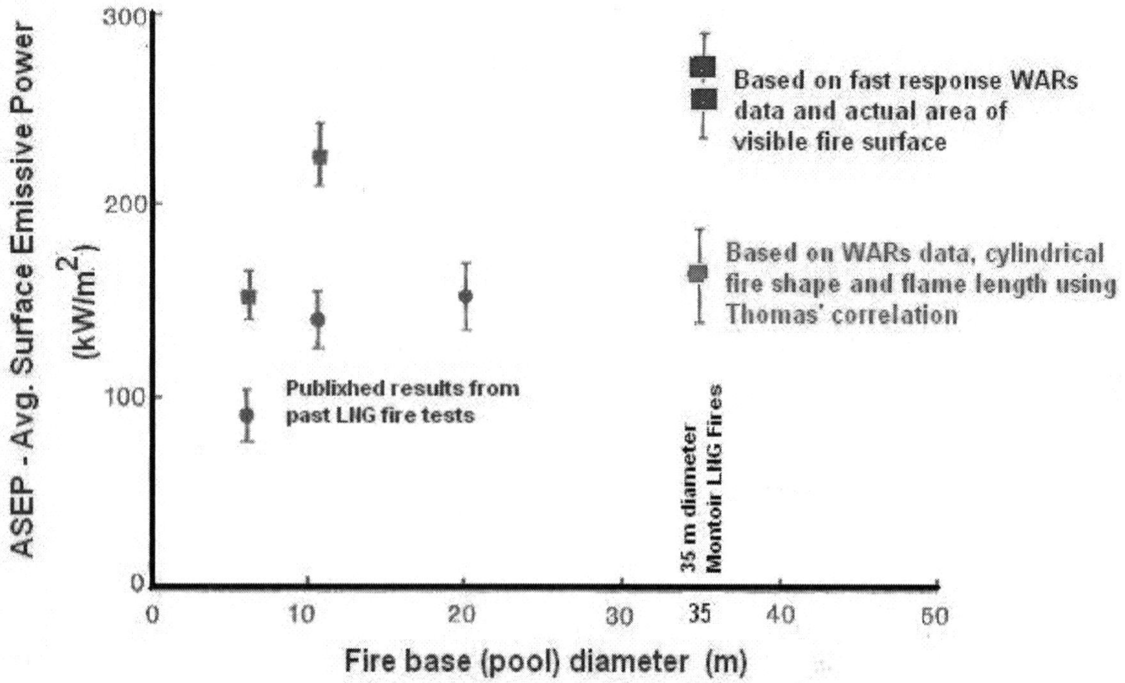

Source: Nedelka, et al. (1989)

Figure 8: Calculated emissive powers from different size LNG pool fires

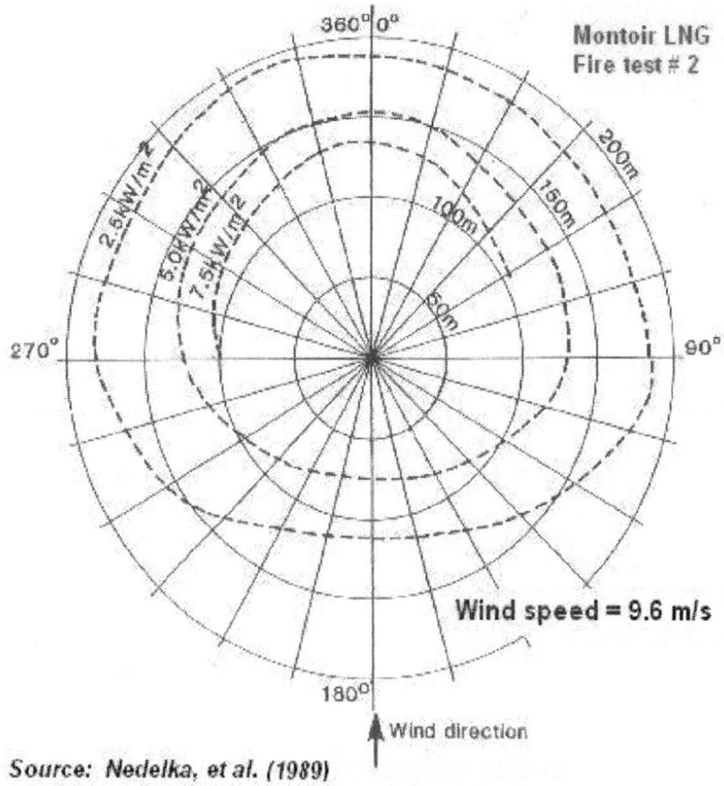

Source: Nedelka, et al. (1989)

Figure 9: Contours of radiant heat flux to vertical plane surfaces oriented to aim at 1 m above the dike center

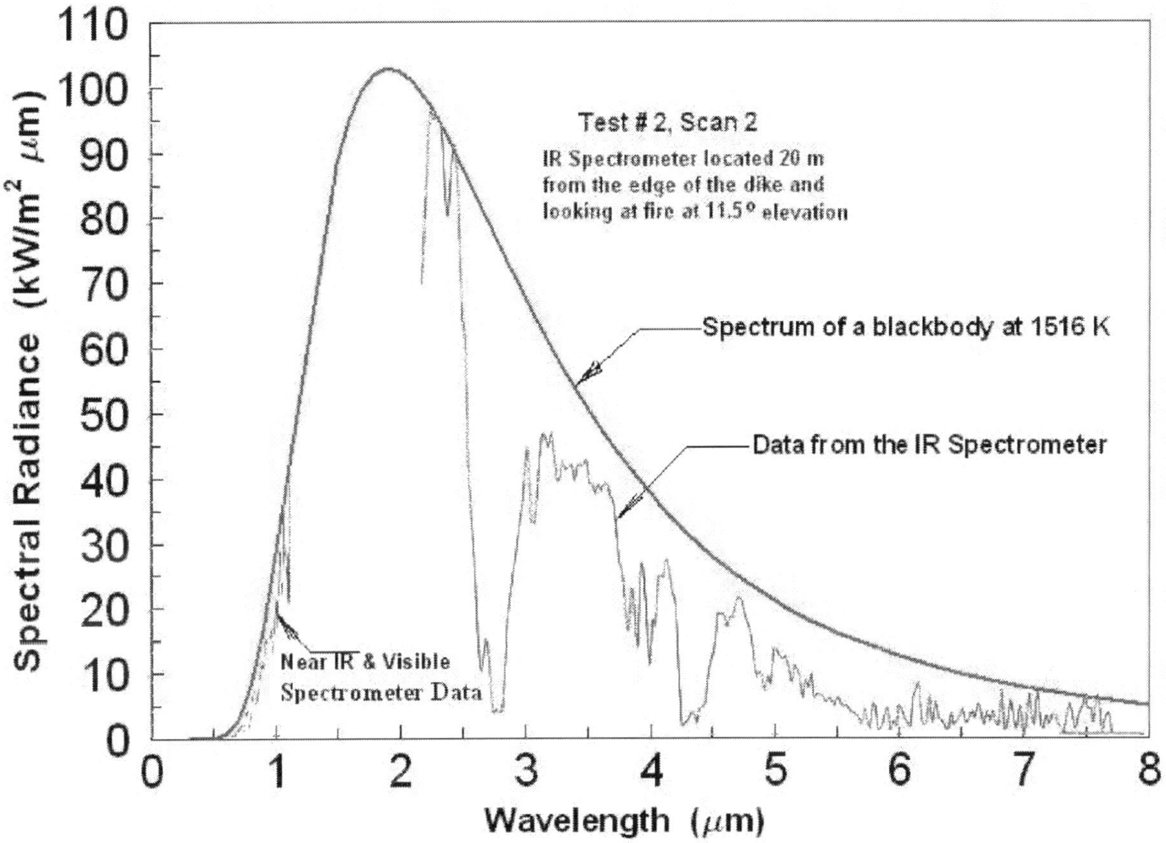

Figure 10: Comparison of the data from IR spectrometer and visible range spectrometer with black body spectrum

Plate 1: Photograph of the typical shape and characteristics of the fire in test 1 during the steady state period of methane burning

Plate 2: Photograph of the typical shape and characteristics of the fire in test 2 during the steady state period of methane burning & high wind conditions

Plate 3: Photograph of the typical shape and characteristics of the fire in test 3 during the steady state period of methane burning

A Non-Equilibrium Model for BLEVE Simulation

Wensheng Lin, Anzhong Gu and Xuesheng Lu
Institute of Refrigeration and Cryogenics
Shanghai Jiaotong University,
Shanghai, China

AIChE Spring National Meeting 2006
6th Topical Conference on Natural Gas Utilization
Orlando, Florida, April 23-27, 2006

Abstract

Boiling liquid expanding vapor explosion (BLEVE) occurs when there is a sudden loss of containment of a pressure vessel containing a superheated liquid or a liquefied gas. The resulting thermal radiation is intense and has the potential to cause severe health-damage even loss of life as well as other material damage.

When a liquefied gas vessel is depressurized rapidly, it becomes highly superheated. Consequently, a great deal of bubbles generated homogeneously in the bulk liquid, and the liquid vaporize violently. The higher the superheat, the larger the number of bubbles generated. Consequently, the pressure will increase dramatically, sometimes even surpass the initial pressure in the vessel, causing further destruction.

A new non-equilibrium model is suggested to simulate this BLEVE process. Firstly, a new hypothesis for predicting homogeneous nucleation rate in superheated liquids is proposed according to the probability distribution of energy fluctuation in statistical thermodynamic fluctuation theory. Then, the pressure decrease and re-increase process is simulated by combining a bubble growth and collapse model based upon non-equilibrium thermodynamic theory and a model for the dynamic process of vapor space.

A case study is given to illuminate the dynamic processes of pressure, temperature, mass, liquid level, outlet velocity, evaporating heat as well as homogeneous nucleation rate during release of the LPG tank. The influences of initial temperature (initial pressure), crack diameter, initial liquid level, tank volume and percent of propane in LPG on the LPG vapor explosion thermal processes are then discussed.

Bubble Generation

Introduction to Previous Work

Boiling liquid expanding vapor explosions (BLEVE) may happen during storage and transportation of liquefied gases. Due to rapid increase of temperature (e.g. ink in a jet printer) or rapid decrease of pressure (e.g. sudden release of a pressure vessel with a two-phase containment), a liquid may reach very high superheat degree. This may result in the violent and explosive vaporization of liquid, and most catastrophically, lead to BLEVE. As we all know, vaporization in liquid firstly appears as a lot of nuclei. Thus, the generation and growth of bubbles in superheated liquid are crucial to the vapor explosion process. Furthermore, because of the high speed of pressure decrease, the superheat is almost uniform in the whole liquid space, the nucleation process can be regarded as a homogeneous one. As a result, the homogeneous nucleation rate in the superheated liquid is a key factor that influences the BLEVE process.

For the homogeneous nucleation rate, much research has been made on.

Among the theoretical approaches to calculate the homogeneous nucleation rate, the kinetic theory is most widely used, and is relatively practical. With this theory, when dealing with the homogeneous nucleation and bubble generation in liquids, the superheat of liquids is generally taken as the driving force to form nuclei. Adopting the statistical thermodynamics of liquids, and assuming the formation speed of activated groups in unit volume (this speed is in fact the nucleation rate), the work needed to form a bubble with the critical radius r_c at constant temperature can be figured, and the homogeneous nucleation rate in liquids, J_k, can be decided with the superheat degree of liquid [1]:

$$J_k = N\sqrt{\frac{2\sigma}{\pi m B}}\exp\left[\frac{-16\pi\sigma^3}{3kT(p_s - p_l)^2}\right] \tag{1}$$

In Eq. (1), p_s is the saturation vapor pressure corresponding to temperature T of the superheated liquid, and p_l is the real pressure of the liquid. B is a constant. The difference between these two pressures represents the superheat degree of the liquid.

Equation (1) is by far most widely used to determine the homogeneous nucleation rate in liquid. Although some improvements have been made after Eq. (1) was proposed, it remains the most accepted one.

It is obvious that the homogeneous nucleation rate is decided by the superheat degree when calculated with the kinetic theory. Once the superheat degree is given, the homogeneous nucleation rate is decided; and vice versa.

Energy Fluctuation and Homogeneous Nucleation Rate

For a system under stable equilibrium state, fluctuation is only a kind of inevitable disturbance that does not influence the system behavior. But for a system under metastable equilibrium state, fluctuation within the system becomes important as the system goes far away from the equilibrium state. When the fluctuation is large enough, the system can be led into another state. When the energy fluctuation in a metastable liquid system with the scale *r* under certain temperature is large enough to achieve the work to form a nucleus, vaporization occurs. This liquid temperature, under which homogeneous nucleation occurs, is the thermodynamic limit of superheat liquid, and the scale of the system is at the same level of the bubble with critical radius.

The above demonstration is the staring point to determine the superheat limit of liquids with statistical thermodynamic fluctuation theory. The expressions to calculate the superheat limit of liquids in Gibbs canonical and grand canonical systems were derived from this theory in [3,4]. The superheat limits of many liquids under different pressure were also calculated. The results turned out to be satisfactory.

The successfulness of calculating liquid superheat limits with fluctuation theory implies the possibility of calculating homogeneous nucleation rate of liquids, which is closely related to the liquid superheat limit, in a similar way. For this reason, a hypothesis is advanced in this paper:

When the energy fluctuation in a metastable liquid system with the scale r_c under certain temperature is large enough to achieve the work to form a nucleus, vaporization occurs. The number of systems, in which vaporization occurs, is decided by the probability that the energy fluctuation in the ensemble achieves the work needed.

According to the hypothesis presented, the expressions about the relationship between homogeneous nucleation rate and energy fluctuation can be deduced.

First of all, the critical radius r_c of a bubble nucleus at temperature T should be determined, because only the bubble nuclei whose radii are larger than the critical radius may grow up, while the bubble nuclei with smaller radius are due to collapse.

The vapor pressure in the bubble under equilibrium is

$$p_v = p_s \exp\left[\frac{v_l(p_l - p_s)}{RT}\right] \tag{2}$$

According to the force equilibrium of the bubble,

$$p_v = p_l + \frac{2\sigma}{r_c} \tag{3}$$

Combining the above two equations, the critical radius is

$$r_c = 2\sigma\left\{p_s \exp\left[\frac{v_l(p_l - p_s)}{RT}\right] - p_l\right\}^{-1} \tag{4}$$

Then the work needed to form a bubble nucleus with critical radius can be determined. To calculate this work precisely, the change of Gibbs energy should be considered, but this can be approximated well as the sum of surface work and volume work, the later cannot be omitted at high pressure. So

$$W = \sigma A + p_l \Delta V = 4\pi r_c^2 \sigma + \frac{4}{3}\pi r_c^3 \left(1 - \frac{\rho_v}{\rho_l}\right) p_l \tag{5}$$

The probability that the energy fluctuation in the ensemble achieves the work needed can then be decided. According to the statistical thermodynamic fluctuation theory, the energy fluctuation in the ensemble conforms to the normal distribution $N(0,\sigma(E))$. The standard error of the energy fluctuation in a Gibbs canonical system with the scale r_c is

$$\sigma_1(E) = \sqrt{kT^2 C_v} = \sqrt{\frac{4}{3} kT^2 \pi r_c^3 \rho_l c_{vl}} \tag{6}$$

Thus, the probability that the energy fluctuation in the ensemble achieves work W is

$$P\big|_{\Delta E \geq W} = \int_{W/\sigma(E)}^{+\infty} \frac{1}{\sqrt{2\pi}} e^{-x^2/2} dx \tag{7}$$

The number of bubble nuclei generated in unit volume of the ensemble is expressed as

$$n = \left[1/\left(4\pi r_c^3/3\right)\right] P\big|_{\Delta E \geq W} \tag{8}$$

For liquids that becomes superheated as temperature rises or pressure drops, the number of bubble nuclei generated in unit time and in unit volume of the ensemble, which is in fact the homogeneous nucleation rate, is defined respectively as

$$J_T = \frac{\partial n}{\partial t} = \frac{\partial n}{\partial T}\frac{\partial T}{\partial t} \tag{9}$$

$$J_p = \frac{\partial n}{\partial t} = \frac{\partial n}{\partial p_l}\frac{\partial p_l}{\partial t} \tag{10}$$

Contrasting Eq. (9)(10), which calculate the homogeneous nucleation rate with fluctuation theory, with Eq. (1) adopting the kinetic method, some characteristics of the new approach may be found.

Firstly, the fluctuation approach is a dynamic one. Adopting this theory, the homogeneous nucleation rate is related not only to the liquid temperature and pressure, but also to the changing velocity of temperature or pressure at the moment. The importance of the velocity of temperature increase or pressure decrease is mentioned in many research works. In fact, due to temperature decrease or pressure increase, new bubble nuclei will not generated in the superheated liquid. Furthermore, the nuclei already formed may collapse. This kind of phenomenon cannot be explained with Eq. (1), but can be easily demonstrated with Eq. (9) and (10).

Secondly, the fluctuation approach clearly indicates that the homogeneous nucleation rate of liquids is not a thermal property of substances, nor is it a parameter of state. The superheat limit of liquids may be regarded as a thermal property, but the homogeneous nucleation rate, which is closely related to it, is a process variable. It is influenced by the temperature and pressure that are changing with time, so it cannot be regarded as a thermal property or a parameter of state. But with Eq. (1), it seems that the homogeneous nucleation rate is a thermal property or a parameter of state, and this may lead to some misunderstandings.

Bubble Growth and Collapse

In order to set up the model for bubble growth and collapse, the following assumptions are made: (1) bubbles with critical radius have already generated; (2) uncompressible liquid and ideal vapor; (3) bubbles and thermal boundary layers are spherical.

Fig.1. shows a bubble, the thermal boundary layer and the surrounding bulk liquid. The model is set up for the thermal boundary layer.

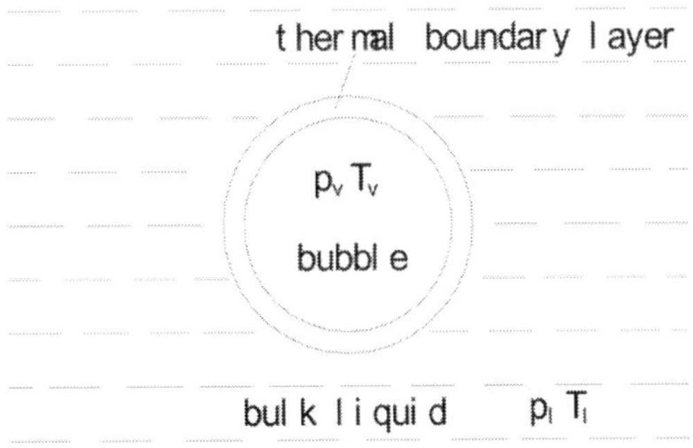

Fig.1. Bubble Model

Continuous equation

$$\frac{1}{r^2}\frac{\partial}{\partial r}\left(\rho_l r^2 u_r\right)=0 \tag{11}$$

Momentum equation

$$\rho_l \varepsilon\left[R\ddot{R}+\left(2-\frac{\varepsilon}{2}\right)\dot{R}^2\right]=p_l(R,t)-p_l(\infty,t) \tag{12}$$

where

$\varepsilon = 1-\rho_v/\rho_l$, $\dot{R}=dR/dt$, $\ddot{R}=d^2R/dt^2$

Energy equation

$$\frac{\partial T}{\partial t}+\frac{\varepsilon R^2\dot{R}}{r^2}\frac{\partial T}{\partial r}=\frac{12\eta_l\varepsilon^2}{\rho_l c_{pl}}\frac{R^4}{r^6}\dot{R}^2+a_l\frac{1}{r^2}\frac{\partial}{\partial r}\left(r^2\frac{\partial T}{\partial r}\right)+\frac{\alpha_p T}{\rho_l c_{pl}}\frac{\partial p_l}{\partial t}$$
$$-\frac{\alpha_p T\varepsilon^2 R^3\dot{R}}{c_{pl}}\left(\frac{2\dot{R}^2+R\ddot{R}}{r^4}-\frac{2\varepsilon R^3\dot{R}^2}{r^7}\right)$$
(13)

where

$$\alpha_p=-\frac{1}{\rho_l}\left(\frac{\partial\rho_l}{\partial T}\right)_p$$

The heat transfer from liquid to vapor is divided into two parts: the latent heat absorbed by liquid and the heat directly absorbed by vapor in the bubble. Thus, the energy equation for vapor-liquid interface is

$$\lambda_l\left.\frac{\partial T}{\partial r}\right|_R=\rho_v\Delta h_{l,v}\dot{R}+\frac{R}{3}\Delta h_{l,v}\frac{d\rho_v}{dt}+\frac{R}{3}\rho_v\frac{n-\kappa}{n-1}c_{Vv}\frac{dT_v}{dt}$$
(14)

Initial conditions and boundary conditions:

$$t=0,\quad T(r,0)=T_\infty,\quad p_l(r,0)=p_{l0},\quad p_{v0}=p_{l0}+\frac{2\sigma}{R},\quad \dot{R}(r,0)=0$$
(15)

$$r=R,\quad T_l(R,t)=T_v$$
(16)

$$r=R,\quad p_l(R,t)=p_v(R,t)-\frac{2\sigma}{R}+2\eta_l\left.\frac{\partial u_r}{\partial r}\right|_R$$
(17)

$$r=R,\quad \lambda_l\left.\frac{\partial T}{\partial r}\right|_R=\rho_v\Delta h_{l,v}\dot{R}+\frac{R}{3}\Delta h_{l,v}\frac{d\rho_v}{dt}+\frac{R}{3}\rho_v\frac{n-\kappa}{n-1}c_{Vv}\frac{dT_v}{dt}$$
(18)

$$r\to\infty,\quad T_l(r,t)=T_\infty,\quad u_r=0,\quad p_l(r,t)=p_{sys}(t)$$
(19)

These are control equations based on non-equilibrium thermodynamics, which describe the growth and collapse of bubble under changing pressure.

Dynamic Process in Vapor Space

Mass Flow for Vapor Space

(1) Discharge at the crack

When $\frac{p_\infty}{p_v} \le \left(\frac{2}{\kappa+1}\right)^{\frac{\kappa}{\kappa-1}}$,

the flow rate at the crack is

$$\dot{m}_1 = C_D A \sqrt{p_v \rho_v \kappa \left(\frac{2}{\kappa+1}\right)^{\frac{\kappa+1}{\kappa-1}}} \tag{20}$$

Otherwise,

$$\dot{m}_1 = C_D A \sqrt{p_v \rho_v \frac{2\kappa}{\kappa-1}\left[\left(\frac{p_\infty}{p_v}\right)^{\frac{2}{\kappa}} - \left(\frac{p_\infty}{p_v}\right)^{\frac{\kappa+1}{\kappa}}\right]} \tag{21}$$

The average velocity of discharging vapor is

$$u = \frac{\dot{m}_1}{\rho_v A} \tag{22}$$

(2) Bubbles entering vapor space
The rising velocity of bubbles is

$$\dot{u}_b = \left(\frac{\rho_l}{\rho_v} - 1\right) g - \frac{3 u_b \eta_l}{R_b^2 \rho_v} \frac{\eta_l + \frac{3}{2}\eta_v}{\eta_l + \eta_v} \tag{23}$$

The time for a bubble to reach the surface is

$$t = \frac{H}{2u_b} \tag{24}$$

where *H* is the depth of liquid.

Thus, the average radius of bubbles when entering vapor space $R_{b,max}$ can be found out with the method introduced in the last section. The bubble mass flowing into vapor space is

$$\dot{m}_2 = \frac{4}{3}\pi R_{b,\max}^3 \rho_v J_{t=n-k} V_l \tag{25}$$

where $J_{t=n-k}$ is nucleation rate at time step *n-k*, and V_l is the volume of liquid.

(3) Vaporization at interface

The mass entering vapor space by surface vaporization is

$$\dot{m}_3 = AJ_v = -AL_v \frac{\partial}{\partial z}\left(\frac{\mu_v}{T}\right) \tag{26}$$

where μ is the chemical potential.

Model for Dynamic Process of Vapor Space

Consider a LPG storage system as that in Fig.2. At the beginning, a crack suddenly appears on top of the tank.

The following assumptions are adopted to set up the model: (1) LPG is stored in a vertical cylindrical tank; (2) The crack is a circle at the center of the tank top; (3) Gravity is negligible; (4) The vapor is incompressible; (5) Constant properties except density.

Continuous equation:

$$\frac{\partial \rho_v}{\partial t} + \frac{\partial u_r}{\partial r} + \frac{\partial u_z}{\partial z} + \frac{u_r}{r} = 0 \tag{27}$$

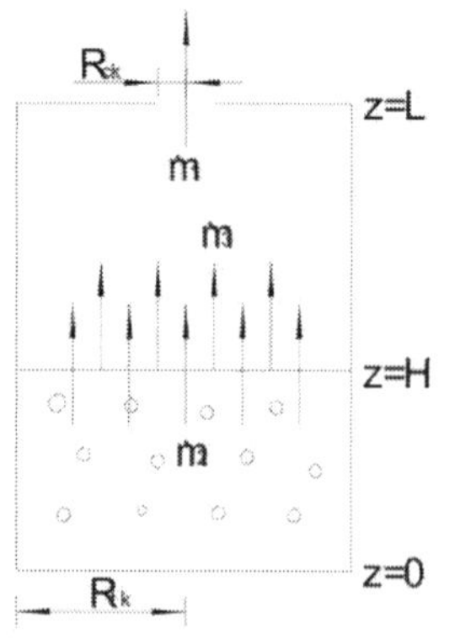

Fig.2. LPG Storage System

Momentum equations:

$$\rho_v\left(\frac{\partial u_r}{\partial t}+u_r\frac{\partial u_r}{\partial r}+u_z\frac{\partial u_r}{\partial z}\right)$$

$$=\frac{1}{r}\left[\frac{\partial(r\tau_{rr})}{\partial r}+\frac{\partial(r\tau_{zr})}{\partial z}\right]=-\frac{\partial p}{\partial r}+\eta_v\left(\frac{\partial^2 u_r}{\partial r^2}+\frac{1}{r}\frac{\partial u}{\partial r}+\frac{\partial^2 u_r}{\partial z^2}-\frac{u_r}{r^2}\right)$$

(28a)

$$\rho_v\left(\frac{\partial u_z}{\partial t}+u_r\frac{\partial u_z}{\partial r}+u_z\frac{\partial u_z}{\partial z}\right)$$

$$=\frac{1}{r}\left[\frac{\partial(r\tau_{rz})}{\partial r}+\frac{\partial(r\tau_{zz})}{\partial z}\right]=-\frac{\partial p}{\partial z}+\eta_v\left(\frac{\partial^2 u_z}{\partial r^2}+\frac{1}{r}\frac{\partial u_z}{\partial r}+\frac{\partial^2 u_z}{\partial z^2}\right)$$

(28b)

Energy equation is not introduced here, assuming no heat exchange at the short period of BLEVE.

The initial conditions and boundary conditions are:

$$t=0,\quad T(r,z,0)=T_\infty,\quad \mathbf{u}(r,z,0)=\mathbf{0},$$
$$p_v(r,z,0)=p_{v0}(H\le z<L),\quad p_v(r,L,0)=p_{v0}(R_{ck}\le z\le R_{tk}),$$
$$p_v(r,L,0)=p_\infty(0\le r\le R_{ck})$$

(29)

$$r=R_{tk},\quad \mathbf{u}(R_{tk},z,t)=\mathbf{0}(H\le z\le L)$$

(30)

$$z=L,\quad \mathbf{u}(r,L,t)=\mathbf{0}(R_{ck}\le r\le R_{tk})$$

(31)

$$z=L,\quad p_v(r,L,t)=0(0\le r\le R_{ck})$$

(32)

$$z=H,\quad u_z(r,H,t)=u_2(t)(0\le r\le R_{tk}),\quad u_r(r,H,t)=0(0\le r\le R_{tk}),$$
$$T(r,H,t)=T_l(0\le r\le R_{tk})$$

(33)

where $u_2(t)$ is the moving velocity of vapor-liquid interface.

Case Study

To solve equations (11)-(14), system pressure is needed. On the other hand, system pressure is greatly influenced by bubble generation and growth. That is to say, to solve equations (27)-(28), the moving velocity of interface, which is caused by bubbles, is needed. So, all these equations should be put together in order to get a solution.

As a basis, some specific conditions of the LPG system are given as:

Inner diameter of tank: 0.4m;
Tank height: 0.6m;
Crack diameter: 0.03m;
Initial liquid level: 0.3m;
Initial temperature: 320K (corresponding to 1.6027MPa);
Content: 100% C_3H_8.

Fig.3. shows the pressure response at different initial temperatures, when all other conditions are the same as the basis condition. It can be seen that initial temperature (pressure) has substantial influences on the pressure response. When initial temperature is low enough, no pressure rebound is found.

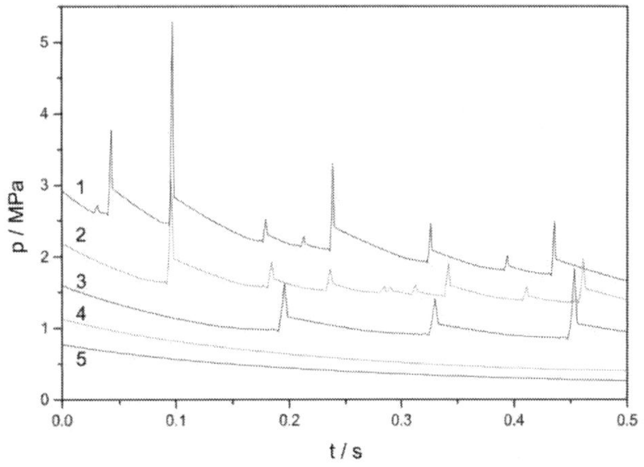

1 – T_0=350K ; 2 – T_0=335K ; 3 – T_0=320K ; 4 – T_0=305K ; 5 – T_0=290K

Fig.3. Effects of Initial Temperature on LPG Vapor Explosion Processes

Fig.4. shows the pressure response at different crack diameter, when all other conditions are the same as the basis condition. It can be seen that crack diameter has great influences on the pressure response. When crack is small enough, no pressure rebound is found.

Fig.5. shows the pressure response at different initial liquid level, when all other conditions are the same as the basis condition. It can be seen that initial liquid level greatly influences the pressure response. When initial liquid level is low enough, no pressure rebound is found.

Fig.6. shows the pressure response at different tank volume, when all other conditions are the same as the basis condition. The tank volume has obvious influences on the pressure response. Smaller tank results in more violent pressure rebound.

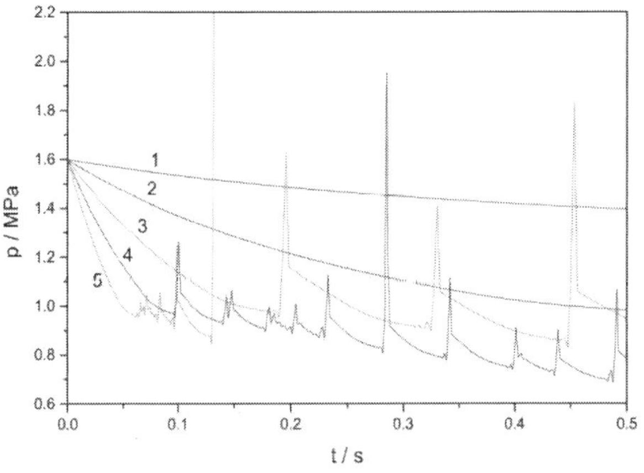

1 – D=10mm ; 2 – D=20mm ; 3 – D=30mm ; 4 – D=40mm ; 5 – D=50mm

Fig.4. Effects of Crack Diameter on LPG Vapor Explosion Processes

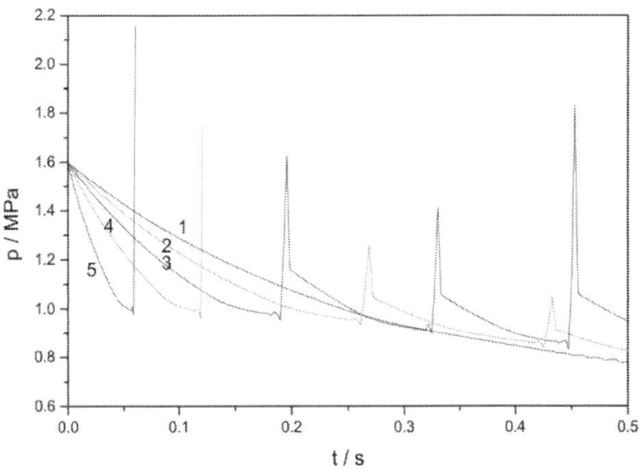

1 – H_0=100mm ; 2 – H_0=200mm ; 3 – H_0=300mm ; 4 – H_0=400mm ; 5 – H_0=500mm

Fig.5. Effects of Initial Liquid Level on LPG Vapor Explosion Processes

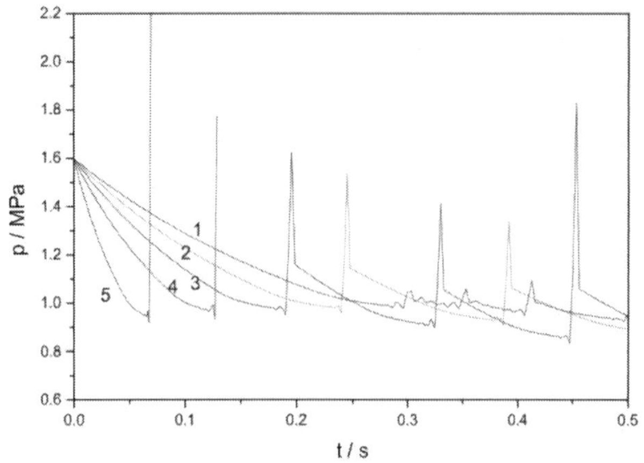

1 – H=1000mm ; 2 – H=800mm ; 3 – H=600mm ; 4 – H=400mm ; 5 – H=200mm

Fig.6. Effects of Tank Volume on LPG Vapor Explosion Processes

Fig.7. shows the pressure response at different liquid contents, when all other conditions are the same as the basis condition. The tank volume has obvious influences on the pressure response. When the proportion of n-C4H10 increases, pressure response becomes gentler.

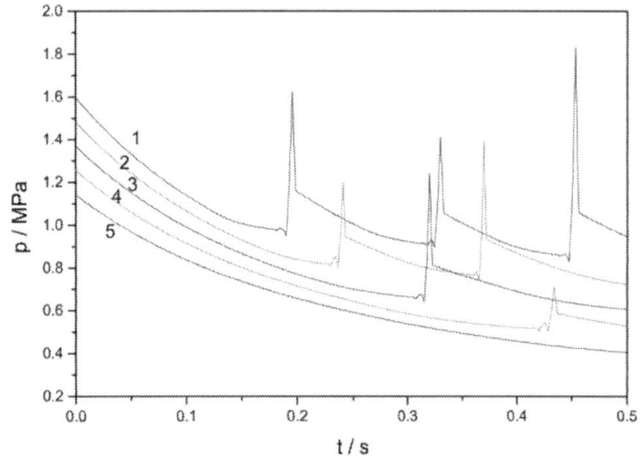

1 – 0%C_4H_{10} ; 2 – 10%C_4H_{10} ; 3 – 20%C_4H_{10} ; 4 – 30%C_4H_{10} ; 5 – 40%C_4H_{10}

Fig.7. Effects of Contaminant Contents on LPG Vapor Explosion Processes

Conclusions

A new approach for predicting homogeneous nucleation rate in superheated liquids is proposed according to the probability distribution of energy fluctuation in statistical thermo-dynamic fluctuation theory. The number of systems that nucleation takes place in an ensemble is supposed to be decided by the probability that the energy fluctuation in the ensemble achieve the work needed to form a stable nucleus. The new method takes temperature increase or pressure decrease speed into account, reflecting that the homogeneous nucleation process is dynamic, and the nucleation rate is not a physical property of materials.

A bubble growth and collapse model based upon non-equilibrium thermodynamic theory is set up. Different from the common models that suppose the bubble expands along the saturate vapor line and thus obeys the Clausius-Clapeyron Equation, the bubble expands (or compresses) polytropically in the present model. This avoids the theoretical drawback that the bubble expansion is propelled by negative chemical potential difference, which violates the Second Thermodynamic Law.

A model for the dynamic process of vapor space is set up. The mass flows of the vapor space are analyzed, the method for deciding each mass flow is given, and a model for bubble rise is presented. Subsequently, a dynamic mathematical model for describing the flow and heat transfer in the vapor space is proposed, and the solution method is also given.

The above models are then combined together to obtain numerical solutions. A case study is given to illuminate the dynamic processes of pressure, temperature, mass, liquid level, outlet velocity, evaporating heat as well as homogeneous nucleation rate during release of the LPG tank. The influences of initial temperature (initial pressure), crack diameter, initial liquid level, tank volume and percent of propane in LPG on the LPG vapor explosion thermal processes are then discussed.

Acknowledgements

The authors are grateful to China's National Natural Science Fund (No. 50076024) for the financial support to the research.

References

1. Blander, M., Katz, J. L., (1975), "Bubble nucleation in liquids," *AIChE Journal*, Vol.21, No.5, pp.833-848.
2. Avedisian, C. T., (1985), "The homogeneous nucleation limits of liquids," *Journal of Physical and Chemical Reference Data*, Vol.14, No.3, pp.695-729.
3. Zeng, D. L., Jing, C. J., (1995), "Determination of superheat limit of liquids using fluctuation theory," *Science in China (Series A)*, Vol.39, No.3, pp.301-308.
4. Liu, C., Ming, X. J., Zeng, D. L., (1997), "Determination of theoretical superheat limit of liquid," *Journal of Engineering Thermophsics*, (In Chinese), Vol.18, No.3, pp.265-269.
5. Venart, J. E. S., Sollows, K. F., Sumathipala, K., et al, (1993), "Boiling liquid compressed bubble explosions: experiments/models," *ASME FED*, Vol.165, pp.55-60.
6. Kielec, D. J., Birk, A. M., (1997), "Analysis of fire-induced ruptures of 400-L propane tanks," *Journal of Pressure Vessel Technology*, Vol.119, No.3, pp.365-373.

Biography

Xuesheng Lu

Institute of Refrigeration and Cryogenics
Shanghai Jiaotong University
1954 Huashan Road
Shanghai, 200030
China

Phone Number: +86-21-62932602
Fax Number: +86-21-62932602
Email: luxs@sjtu.edu.cn

Born 1949

B.Sc. 1977, Shanghai Jiaotong University, Refrigeration engineering

1999-current, Professor of Refrigeration and Cryogenics,
Shanghai Jiaotong University, Institute of Refrigeration and Cryogenics
Director of Laboratory of Refrigeration and Cryogenics

Principal Research, Education and Consulting Activities:
Liquefaction, separation, storage and transportation of gases; LNG technology; Adsorption of natural gas and hydrogen; Refrigeration and air-conditioning engineering; New refrigerants; Cryogenic measurement.

Published about 100 refereed papers, chapters in books and books.

Uncertainties in modeling the Physical Behavior of Large LNG Releases upon Sea-water

F. Kootstra[1]
Netherlands Organisation for Applied Scientific Research
TNO Built Environment en Geosciences
The Netherlands

AIChE Spring Meeting, April 2006
6th Topical Conference on Natural Gas Utilization
Orlando, Florida, April 23-27, 2006

Abstract

A detailed investigation of recently applied modeling approaches and used methodologies for LNG fire and dispersion modeling after an accidental or intentional release has shown that the currently applied (computer) models still contain many assumptions. Some of these modeling assumptions are due to the unknown physical behavior of LNG after releases in such large amounts, others due to the limited validity range of the applied physical effects models, the so-called model limitations. Most of these modeling assumptions will have a large influence upon the final results of e.g. a quantitative risk assessment for an LNG terminal.

In this paper an overview will be given of the applied assumptions after an LNG release. The assumptions are categorized after the unknown physical behavior or model limitations. The assumptions are made throughout the complete chain of subsequent models that describe a release of LNG.

A typical LNG release scenario includes:

- LNG outflow from a hole in one side of a vessel after e.g. a collision,
- Evaporation of a violently boiling and quickly spreading LNG pool upon sea-water,
- A possible LNG pool fire upon sea-water close to the LNG carrier,
- Dispersing of the cold LNG vapor evolving from the evaporating pool, which initially is denser than air,
- A possible explosion (after late ignition) of the dispersing LNG vapor from the evaporating pool.

---

[1] Telephone: +31555493208; Fax: +31555493390; E-mail address: freddie.kootstra@tno.nl, Department of Industrial and External Safety, Laan van Westenenk 50, P.O. Box 342, 7300 AH Apeldoorn, The Netherlands

Finally some recommendations will be given to achieve (in the near future) a more realistic way of modeling LNG releases from LNG carriers.

Introduction

Liquefied natural gas, or LNG, is natural gas in its liquid form. When natural gas is cooled to minus -161 °C, it becomes a clear, colorless and odorless liquid. LNG is neither corrosive nor toxic. Natural gas is primarily methane, with only small amounts of other hydrocarbons. LNG weighs less than half the weight of water so it will float if spilled on water. The LNG production chain consists of several steps before actually arriving in residential areas for domestic consumption e.g. heating and/or cooking. The LNG chain is: production - LNG liquefaction terminal – transportation - LNG terminal – distribution – consumption. This paper will focus upon the 'transportation' and the 'LNG terminal' of the LNG production chain in relation to safety issues for public and environment. More precise: upon the uncertainties in modeling the physical effects behavior and consequences of large LNG releases upon sea-water e.g. at an LNG terminal. Natural gas is the cleanest burning fossil fuel. It produces less emissions and pollutants than e.g. coal or oil. Since LNG occupies only a fraction (1/600) of the volume of natural gas, and takes up less space, it is more economical to transport across large distances and can be stored in larger quantities. LNG is a price-competitive source of energy that could help to meet the future economic needs worldwide. LNG is transported in double-hulled ships specifically designed to handle the low temperature of LNG. These carriers are insulated to limit the amount of LNG that boils off or evaporates. When LNG is received at a terminal, it is transferred into insulated storage tanks especially designed to hold LNG. These tanks can be found above or below ground and keep the liquid at a low temperature to minimize the amount of evaporation. The temperature within the tank will remain constant if the pressure is kept constant by allowing the boil off gas to escape from the tank. This is known as auto-refrigeration. The boil-off gas is collected and used as a fuel source in the facility or on the tanker transporting it. When natural gas is needed, the LNG is warmed to a point where it converts back to its gaseous state. This is accomplished using a re-gasification process involving heat exchangers.

When cold LNG comes in contact with warmer air, it becomes a visible vapor cloud. As it continues to get warmer, the vapor cloud becomes lighter than air and rises. When LNG vapor mixes with air it is only flammable if it's within 5-15% natural gas in air. If it's less than five percent natural gas in air, there is not enough natural gas in the air to burn. If it's more than 15 percent natural gas in air, there is too much gas in the air (and not enough oxygen) for it to burn. As a liquid, LNG is not explosive. LNG vapor will only explode if in an confined space. When enough LNG is spilled on water at a very fast rate, a rapid phase transition, or RPT, can occur. Heat is transferred from the water to the LNG, causing the LNG to instantly convert from its liquid phase to its gaseous phase. A large amount of energy is released during this rapid transition between phases and a physical explosion due to the vigorous evaporation can occur. While there is no combustion, this physical explosion can be hazardous to any nearby person or buildings.

Definition of the problem

In the harbor of Rotterdam (the Netherlands) large amounts of LNG will, in the nearby future, be imported by sea-going vessels. LNG is transported as a completely cooled (refrigerated) liquid at atmospheric pressure. If such a (sea-going) LNG vessel releases its cargo due to an accident with another vessel, or an intentional (terrorist) attack, this can lead to a serious leakage of the LNG from one the cargo tank of the vessel. From this leakage the LNG will flow out upon the water and from a (boiling) pool which can burn or evaporate, and eventually (when evaporating) will disperse as a (initially dense) gas in the atmosphere, from which the vapor cloud might possibly explode when the LNG concentration in the vapor cloud is within the flammable limits of methane. A precise and validated way of modeling such a scenario from outflow until dispersion is of great importance to determine the possible effects and eventual and risk. The nowadays used standard techniques in the Netherlands, as described in the colored books [1,2], cannot be followed blindly, while in this case it is seagoing navigation and refrigerated LNG and in [2] the situation is described for inland navigation and pressurized LNG. Besides the even more important fact that the physical effect models for fire and dispersion (as described in [1]) are mostly analytical models which are build, based and validated upon small-scale experiments compared to the transported LNG amounts considered here. Especially the next to be described behavior of LNG will determine the eventual concentration in the neighborhood, and is strongly dependent upon the modeling and assumptions made: At atmospheric conditions and a temperature of -162°C (directly after release), the LNG will (violently) evaporate from the pool upon sea-water and form a vapor with a density of 1.7 kg/m^3 (or even higher when droplets are included). By contrast, LNG at ambient temperature is lighter than air (density is 0.55 kg/m^3). Therefore, LNG vapor which evolves from a liquid release upon sea-water will initially stay close to the water surface until the vapor heats up (by mixing in of air or other external heat sources) until the vapor cloud reaches a temperature around -108°C, from that point the vapor will be lighter than air and raise into the air and quickly disperse. For being able to give reliable predictions concerning the LNG concentration in the neighborhood of the vessel or at shore, after an accidental or intentional release of LNG in which a sea-going vessel is involved, a validated modeling of such a scenario is an important condition. The careful and precise investigation of the outflow, (pool fire), evaporation, dispersion (and possible explosion) is thereby of great importance. For being able to predict how the dispersion of the LNG cloud in the neighborhood of the release will evolve, the above processes cannot be looked at separately but an accurate and integrated modeling approach is needed. These abovementioned processes are heavily influenced by:

- The outflow conditions (location of the leak in the vessel with respect to the waterline).
- The weather conditions (temperature, humidity and wind speed).
- The total amount of LNG that is released, under which conditions, and the outflow rate.

- The spreading behavior of, and the evaporation rate from, the LNG pool.
- The amount of LNG and water droplets (aerosols) in the vapor cloud.
- The confinement of the vapor cloud and location of ignition sources.

Objective and research method

On the basis of the above defined problem, the following objectives were formulated:

- The determination of the validity, and the uncertainties, concerning the modeling of LNG releases from a vessel.
- To indicate (if necessary) the effective possibilities to reduce the uncertainties in the LNG modeling.

To reach these objectives a concrete examination of the validity of the nowadays-common risk analysis methods (and available modeling information) for a release of LNG from a sea-going vessel will be performed. Based upon these results we will give some recommendations upon how to achieve a more realistic way of modeling large LNG releases from LNG carriers.

Potential hazards

An LNG release upon water from a sea-going vessel can result in one or more of the following potential hazards for humans and/or the environment around the release:

1. Exposure to thermal radiation resulting from a pool fire upon sea-water, after the ignition of the (evaporating) LNG pool.
2. Direct flame contact from a flash fire, after a delayed ignition of a flammable vapor cloud resulting from the evaporating LNG pool.
3. Exposure to overpressure, after a delayed ignition of a flammable vapor cloud resulting from the evaporating LNG pool.

The behavior and the modeling of LNG

The above two sections, 'definition of the problem' and 'objective and research method', will be worked out more extensive in this section by comparing the formal theoretical description with nowadays used modeling methods, from which the uncertainties can be directly deduced. First a short summary of the physical properties of LNG will be given, after which the different processes of outflow, evaporation, (pool fire) and dispersion (including explosion) will be treated.

Some typical physical properties of LNG:

- LNG is simply natural gas that has been cooled to its liquid state at atmospheric pressure: - 260°F (-162.2°C) and 14.7 psia (1.013 bara). Currently, imported LNG is commonly 95–97% methane, with the remainder a combination of ethane, propane, and other heavier gases.
- LNG is transported at ambient pressure.

- Liquefying natural gas vapor, which reduces the gas into a practical size for transportation and storage, reduces the volume that the gas occupies more than 600 times.
- LNG is considered a flammable liquid.
- LNG vapor is colorless, odorless, and non-toxic.
- LNG vapor typically appears as a visible white cloud, because its cold temperature condenses water vapor present in the atmosphere.
- The normal boiling point of LNG is: -161.5 °C.
- The lower and upper flammability limits of methane are 5.5 and 14% by volume at a temperature of 25°C.

Outflow

The following four uncertainty topics were identified with respect to outflow:

1. Which release scenarios are representative and should be used?
2. Which frequencies are applicable for these release scenarios and where or what are they based upon?
3. What are the causes of these (accidental or intentional) releases?
4. How are the release rates for the scenarios determined / calculated?

After an accidental (or intentional) release of LNG from one or more compartments of a sea-going vessel, it is first of all of great importance to carefully analyze the way in which the LNG flows out on the sea-water surface. This so-called 'source-term' will strongly influence the results of the subsequent physical effects models in a specific scenario.

Several (very important) choices are made in this 'initial' stage:

- What is the cause of the LNG release from a sea-going vessel?
- Which loss of containment (LOC) scenarios are (or will be) linked to this specific cause of the LNG release?
- Which frequencies are applicable for these LOC scenarios?
- Where (or what) are the frequencies for these LOC scenarios based upon?
- Which specific choices for the sizes of the holes or release rates (and locations of the hole / release) in a compartment are consequently made to model the outflow for these specific LOC scenarios?
- After the above choices made: How is the LNG outflow consequently modeled?

First of all the cause of the LNG release from a sea-going vessel is ranked as: *accidental* or *intentional*. Accidental releases are e.g. caused by collision, grounding, striking or operational errors. When a release is ranked as intentional, this is mostly referred to as a terrorist attack. In several recent (for the public) available LNG studies [3,4] the accidental and intentional releases are linked to different types of LOC scenarios. Sometimes the LOC scenarios are skipped (not specified) and the hole sizes are directly porposed, grouped after accidental and intentional releases. For which kind of the two above methodologies is chosen, the LOC scenarios or the directly proposed hole sizes, depends upon the applicable risk approach, the *probabilistic* or the *deterministic* risk approach. Some possible examples of LOC scenarios for *accidental* LNG releases from a sea-going vessel at transit could be:

- A grounding on, or collision with the outer breakwater.
- A collision with another vessel outside the breakwater.

Some possible examples of LOC scenarios for *intentional* LNG releases from a sea-going vessel at berth could be:

- A projectile, such as an airplane or missile, striking the ship.
- An explosive charge in a small boat detonated adjacent to the ship.

The ranges of the hole size differ already in e.g. the two publicly accepted LNG studies in the USA/Canada [3,4] by ABS/Ferc and Sandia, see Table 1.

	Diameter of the hole [m]	
LNG risk study	**Accidental**	**Intentional**
ABS/Ferc [3]	1	5
Sandia [4]	0.8 – 1.4	1.6 – 4.0

Table 1 : Ranges of the hole sizes as used in [3] and [4]

In case of a probabilistic risk approach, risk (R) is expressed as R=P*E and determined by the magnitude of a certain physical effect (E) of an incident (or accident) and its likelihood (or probability) (P).

The determination of the probability (P) of a certain physical effect (E), resulting from a specific LOC scenario as caused by either an accidental or intentional release, is rather complicated due to several reasons in case of LNG.

1. The probability of occurrence of an accident is normally based upon historical records of occurrence of such an accident. The LNG shipping industry has an excellent safety record in terms of cargo loss. To date no loss of containment was recorded due to an accident in which a sea-going vessel was involved. In the late 1970s there have been two serious grounding (*El Paso Paul Kayser, LNG Taurus*) but neither of them resulted in a cargo loss. Therefore, the safety record for LNG shipping makes the determination of the probability (P) very complicated.
2. The probability of occurrence of an intentional release (a terrorist attack) is also very complicated and rather arbitrary due to obvious reasons, e.g. what are the criteria for ranking a LNG release as caused by a terrorist attack?

The final choice made, about the inclusion of specific kind of LOC scenarios, will mainly depend upon the appropriate risk approach (for the location under consideration) to determine the hazard ranges between the hazardous activity and the nearby people (or vulnerable objects).

Examples are:

1. worst-case consequence approach (deterministic)
2. maximum credible event approach (deterministic)
3. risk assessment approach (probabilistic)

The worst-case consequence approach considers only the outcome of the most severe incidents amongst the ones identified. Because the outcome of whatever happens cannot be worse than this, makes it a very attractive decision tool. The maximum credible event approach considers also (just like the worst-case consequence approach) only the outcome of the most severe incidents amongst the ones identified, but only those incidents which are considered as plausible and/or reasonably believed. This opens the possibility to take safeguards into account to reduce the likelihood (probability) of an event or reduce its potential outcome (mitigation of the physical effect). In a risk assessment approach the entire range from frequently small events, through infrequent but credible events, to the rare worst case events are included. The likelihood (probability) of every event is combined with all its possible outcomes (consequences)

Next to the size of the hole, another point of uncertainty is the location of the hole in the ships compartment. Is the hole located above or below the waterline? As indicated (and quantitatively indicated) in [12,19] this has a considerable influence upon the outflow rate and duration of the outflow from the ships compartment. Furthermore, in case of a hole above the waterline, experiments [20] have indicated that the outflow would be much slower than predicted by most models, because these models do not take into account the so-called "glug-glug" effect of the vacuum that is created in sealed tanks when a liquid is spilled.

Evaporation

The following three uncertainty topics were identified with respect to evaporation:

1. Modeling of the physical behavior of a boiling LNG pool upon sea-water.
2. Which boiling regime is applicable for the evaporation of LNG?
3. Whether or not to include the phenomena of ice-formation

After the release (outflow) of the LNG from a hole in a compartment of a sea-going vessel upon sea-water (i.e. an unconfined water surface) a spreading and (violently) boiling LNG pool upon sea-water is formed. In the past there have never been done large scale LNG release experiments upon sea-water which are comparable (in size) to the amount of LNG in which we are interested here. This causes certain implications for the validity and usefulness of the theoretically applied evaporation models because there is still quite some uncertainty around specific topics concerning the behavior of LNG upon sea-water. In other words the exact physical behavior of huge LNG releases upon sea-water is still an area which is not fully understood. A general theoretical description of the process of a boiling liquid upon a water surface can be found in [5]. Experimental measurements [6,7], showed that the evaporation flux of a spreading LNG pool on water deviates substantially from the evaporation flux as calculated theoretically within a pool evaporation calculation in the film boiling regime [5]. Theoretically a heat flux from the sea-water to the LNG pool of ≈ 42 kW/m^2 is to be expected, corresponding to an evaporation flux around 0.085 kg/m^{2}s, however in many experiments evaporation rates in the order of 0.18 kg/m^{2}s were measured. Consequently, various studies recommend the use of a higher evaporation flux in LNG pool evaporation calculations than theoretically calculated in the film boiling regime [6-9]. There is no unambiguous explanation for the deviating behavior of LNG. To be able to interpret the deviating behavior of LNG better, it is highly recommendable to perform additional (large-scale) experiments after the possible physical phenomena occurring during the evaporation of LNG. The following research questions are applicable in these experiments:

- Why is the evaporation rate of LNG higher than expected from the film boiling regime and compared to other cryogenics?
- Is there any ice-formation recorded during the LNG evaporation upon an open water surface?
- What is the influence of different types of outflow (slow, fast, from a height above the water surface) upon LNG evaporation?
- How does the LNG interact with the water after outflow?

Another topic of uncertainty is whether or not ice-formation occurs during the outflow of LNG upon sea-water. In several (small-scale) experiments for LPG, LNG and ethane [8,10-11] the formation of a solid layer of ice was recorded in the case of a spreading liquid over a confined water surface. In a recent study [12] after the outflow of refrigerated NH_3 upon sea-water (the Westerschelde river) the formation of ice was neglected. This decision was not based upon experimental finding but a comparison a one-year average wave height and wave frequency in comparison with the experimentally found [8] time which is needed to grow a solid layer of ice.

Pool Fire

The following three uncertainty topics were identified with respect to pool fire:

1. Uncertainties between experimental results and theoretical model.
2. Determination of the maximum pool size upon water.
3. Geometric modeling of the pool fire.

After the release (outflow) of the LNG from a hole in a compartment of a sea-going vessel upon sea-water a spreading and (after ignition) burning LNG pool upon sea-water can be formed. In the past there have never been done large scale LNG pool fire experiments, comparable (in size) to the amount of LNG which will be released after an (accidental or intentional) release of LNG from a sea-going vessel in which we are interested here.

This causes several sources of uncertainty while:

1. The theoretical pool fire model [1], which is used to estimate the risk associated with an LNG pool fire arising from a release from a sea-going vessel, is an empirical model which is based upon the in the past performed small-scale experiments.
2. The physical phenomena occurring during large scale LNG pool fires are an area of uncertainty.

For a pool fire important parameters are e.g. surface emissive power, pool burning rate, flame dimensions, flame tilt, atmospheric transmissivity and view factor. A comparison of the surface emissive power and pool burning rate (See Table 2) show considerable differences between the experimental available data and used/calculated parameter values in theoretical studies.

Reference	Exp. vs. Lit. Study	Pool diameter [m]	Surface Emissive Power [kW/m^2]
Raj [13]	Exp.	8.5 – 15	170 – 280
Blackmore [14]	Exp.	29	172 – 234
Minzer [15]	Exp.	30	178-248
ABS/Ferc [3]	Lit.		265
Sandia [4]	Lit.		220
			Burning Rate [kg/m^{2}s]
Raj [13]	Exp.	10 - 15	0.17 - 0.42
Mizner [15]	Exp.	20	0.10
Rew [16]	Exp.		0.282
ABS/Ferc [3]	Lit.		0.282
Yellow Book [1]	Lit.	Up to 14 17 20 35	0.94 0.255 0.106 0.14
Fay [17]	Lit.		0.34
Sandia [4]	Lit.		0.14

Table 2 : Experimental and literature values for the surface emissive power [kW/m^2] and burning rate [kg/m^{2}s]

Several small-scale LNG experiments showed that LNG burns relatively clean, with a low soot fraction. Literature (e.g. the Yellow book [1]) assumes (on average) a soot fraction (ζ) of 80%. On the other hand there is the expectation that large-scale LNG pool fires will burn less clean and therefore with a higher soot fraction. This is due to the fact that:

- There is insufficient oxygen available around such large pool fires, and
- The huge amount of oxygen which is needed for clean burning can simply physically not be transported towards these kind of large burning pools.

Next to the above described discrepancies between the experimental pool fire results and the theoretically applied pool fire model, other uncertain points in the theoretical modeling of a pool fire are:

1. The determination of the shape and maximum pool size upon water next to a sea-going vessel.
2. The calculation of the geometrical parameters in the pool fire model.

The spreading of LNG has been experimentally investigated in [6] for continuous LNG releases. The theoretical spreading model upon a water surface by TNO [5] is in good agreement with these experiments. LNG releases from a hole in a compartment of a sea-going vessel will never result in circular symmetric (spreading) pool, but rather a half-circular symmetric pool at the one side of the sea-going vessel is to be expected. The results from the spreading equations in theoretical models are all based upon free spreading circular symmetric pools. Consequently some simplifications have to be made for the shape of the pool. Another point upon which an assumption has to be made to be theoretically able to model the free spreading of a liquid upon an unconfined water surface is the use of a minimum pool thickness. When the pool thickness becomes less than its minimum value, the liquid pool will stop spreading and this moment determines the maximum size of the pool. Other points upon which theoretical pool fire models [1,3,4] differ are the equations for the calculation of the height and the tilt of the flames from the pool fire. Both are dependent upon the (characteristic) wind velocity. Differences in these parameters will finally result in heat radiation levels for certain degrees of lethality at different distances from the pool fire. Considering all the above mentioned uncertainties for large LNG pool fires, its usefulness to perform additional large scale experiments to demonstrate the effects upon the burning rate, surface emissive power and soot formation in order to reduce these uncertainties.

Dispersion

The following two uncertainty topics were identified with respect to dispersion:

1. Modeling the physical behavior of an LNG vapor cloud after evaporation.
2. Dispersion modeling uncertainties.

After the modeling of the LNG outflow from a hole in a compartment of a sea-going vessel and the evaporation of a spreading and (violently) boiling LNG pool upon sea-water, the modeling of the dispersion behavior of the LNG cloud is of great importance. Theoretically, for the exact modeling of the dispersion of the formed LNG vapor cloud, we need to answer the following question: How does the dispersion process for the released LNG vapor cloud, evolving from an evaporating pool, need to be modeled? Is it a light (buoyant), neutral (passive) or a heavy (dense) gas cloud? The answer to this question is strongly dependent upon the outflow and the evaporation of the released LNG prior to the dispersion. For the correct modeling of the dispersion it is of great importance to determine the density of the LNG cloud (initially and) during the dispersion process. The composition of the LNG cloud will consist out of several of the following compounds like e.g. the pure natural gas, LNG droplets, condensed water droplets, air (possibly humid). Important conditions which play a role in the density of the LNG cloud are:

- the initial temperature of the LNG at release.
- the fraction of LNG droplets in the vapor cloud.
- the ambient air temperature.
- the relative humidity of the ambient air.

Especially the next to be described behavior of LNG will determine the eventual concentration in the neighborhood, and is strongly dependent upon the modeling and assumptions made: At atmospheric conditions and a temperature of -162°C (directly after release), the LNG will (violently) evaporate from the pool upon sea-water and form a vapor with a density of at least 1.7 kg/m^3 (or even higher when the vapor cloud contains droplets of LNG or water). By contrast, pure LNG vapor at ambient temperature is lighter than air (density is 0.55 kg/m^3).

Therefore, LNG vapor which evolves from a liquid release upon sea-water will initially after the release stay close to the water surface until the vapor heats up and all droplets have evaporated (by mixing in sufficient air and/or other external heat sources) until the vapor cloud reaches a temperature around -108°C, from that point the vapor will be lighter than air and raise into the air and quickly disperse. For being able to give reliable predictions concerning the LNG concentration in the neighborhood of the vessel or at shore, after an accidental or intentional release of LNG in which a sea-going vessel is involved, a precisely validated modeling of such a scenario is an important condition. For the dispersion modeling of an LNG vapor cloud, different (starting) conditions can therefore be applied, depending upon the already made assumptions during the outflow and the evaporation of the released LNG prior to the dispersion. Examples of these different starting conditions are:

- Dispersion out of a spreading and (violently) boiling LNG pool;
- Dispersion out of a point source;
- Dispersion (finally) from a neutral LNG vapor cloud, which takes only the initial sagging into account;
- Dispersion of a neutral LNG vapor cloud.

The choice for one (or several) of the above conditions is, in case of uncertainty after the exact behavior of a vapor cloud, mostly based upon the type of risk approach; pessimistic or conservative. Considering all the above mentioned uncertainties (and possible modeling assumptions) prior to the LNG dispersion process have lead (among many others) to questions like: To what extent is it valid to model the LNG vapor cloud, after outflow and evaporation of the LNG out of a sea-going vessel, as a neutral gas cloud? From the previously indicated uncertainties for outflow and evaporation in combination with the theoretical behavior of the (initial) LNG cloud evolving from an evaporating pool upon sea-water, it's clear that the modeling of the to be formed LNG vapor cloud as a neutral gas cloud must be considered in most cases as a conservative approach.

Explosion

The following two uncertainty topics were identified with respect to explosion:

1. Applicability of the multi-energy-model (MEM) for large LNG vapor clouds?
2. Explosion modeling uncertainties.

Experimental research during the last decade [e.g. 18] has clearly showed that deflagrative combustion generates a blast only in those parts of a quiescent vapor cloud which are sufficiently obstructed and/or partially confined. The interaction of the expanding flow ahead of the flame with obstacles is a very efficient turbulence generator, and turbulence is the key parameter which determines the overpressure effects. The guideline for quantitative risk assessment [2], one of the commonly used standard techniques in the Netherlands, prescribes to calculate the overpressure from an explosion using the multi-energy method with the highest value of the blast strength, i.e. 10. Furthermore, the total flammable mass in the vapor cloud is equal to the mass within the LFL and the flammable mass of the cloud has to be partitioned into an obstructed and non-obstructed region. The fraction of the mass in the obstructed region is by default set to 8%, with the centre of the explosion located at the centre of the cloud.

Considering the abovementioned, the following uncertainties can be identified in the modeling of an exploding LNG vapour cloud:

- What is the moment in the MEM at which the dispersing LNG vapor cloud explodes? Is this:
    1. The moment in time at which the LNG vapor cloud contains its maximum explosive content or the maximum size of the LFL cloud.
    2. The moment in space at which an LNG vapor cloud within its flammable limits reaches an ignition point e.g. at shore.
- What is the applicability and validity of the MEM for such large LNG releases above sea-water? Because above sea-water there are no major obstacles generating turbulence.

Summary

In Table 3 a summary is given for the uncertainty topics which were described above for the modeling of large scale LNG releases categorized after outflow, evaporation, pool fire, dispersion and explosion. Indicated also is whether these uncertainties are caused by:

1. Gaps in the experimental knowledge after the exact physical behavior of such large LNG releases.
2. Limited validity of the physical effects models used to calculate the consequences
3. Some other cause.

Physical effects model	Uncertainty	Unknown physical behavior	Model limita-tions	Other
Outflow	Which release scenarios are representative and should be used?			*
	Which frequencies are applicable for these release scenarios and where or what are they based upon?			*
	What are the causes of these (accidental or intentional) releases?		*	*
	How are the release rates for the scenarios calculated / determined?	*	*	
Evaporation	Modeling of the physical behavior of a boiling LNG pool upon sea-water	*	*	*
	Which boiling regime is applicable for the evaporation of LNG?	*	*	
	Whether or not to include of the phenomena of ice-formation	*		
Pool fire	Uncertainties between experimental results and theoretical model	*	*	*
	Determination of the maximum pool size upon water	*	*	
	Geometric modeling of the pool fire		*	
Dispersion	Modeling the physical behavior of an LNG vapor cloud after evaporation	*	*	
	Dispersion modeling uncertainties		*	*
Explosion	Applicability of the multi-energy-model for large LNG vapor clouds?	*	*	
	Explosion modeling uncertainties		*	*

Table 3 : Summary of the above described uncertainties in the modeling of large LNG releases.

Conclusions

The detailed investigation of recently applied modeling approaches and used methodologies for LNG fire and dispersion modeling after an accidental or intentional release, as presented in this paper, has shown that the currently applied (computer) models still contain many assumptions. Some of these modeling assumptions are due to the unknown physical behavior of LNG after releases in such large amounts, others due to the limited validity range of the applied physical effects models. To be able to understand the physical behavior of huge LNG releases better, and for being able to extend the validity range of the applied physical effects models, it is highly recommendable to perform additional (large-scale) LNG release experiments after the possible physical phenomena occurring during the outflow, evaporation, (pool fire) and dispersion (including explosion) of large amount of LNG on sea-water from sea-going vessels.

Recommendations

Considering the performed detailed review of the till now (publicly) available LNG studies to detect the uncertainties in modeling the physical effects behavior of large LNG releases upon sea-water, the following recommendations are applicable in order to reduce the uncertainties, assumptions and shortcomings in modeling the physical effects (and consequences) of large LNG releases upon sea-water from a sea-going vessel.

The following two recommendations can be made:

1. The risk of possible large-scale LNG releases upon sea-water should be studied in more detail while using the most modern risk analysis techniques in order to give a more precise estimation of the risks associated with such large LNG releases.

2. Perform more exact, detailed and advanced modeling of
   - LNG vessel incidents and
   - The consequences of large LNG releases upon sea-water.

In order to implement the above two recommendations and reduce the uncertainties and assumptions in the physical effects modeling of large LNG releases, the consequence modeling should be validated with large-scale LNG release experiments upon sea-water.

References

1. PGS2 (CPR14E) TNO Yellow Book, third edition second print 2005, Methods for the calculation of physical effects, resulting from releases of hazardous materials (liquids and gases). Directorate-General of Social Affaires and Employment, CPR 14E third edition, The Hague 1997.
2. CPR18E, Purple Book, first edition 1999, Guidelines for quantitative risk assessment. Directorate-General of Social Affaires and Employment, CPR 18E first edition, The Hague 1999.
3. ABS Consulting Inc., 2004. Consequence assessment methods for incidents involving releases from liquefied natural gas carriers. Report prepared for Federal Energy Regulatory Commission, contract FERC04C40196. May 13, 2004.
4. Sandia National Laboratories, 2004. Guidance on Risk Analysis and Safety Implications of a Large Liquefied Natural Gas (LNG) Spill Over Water. Sandia Report SAND2004-6258. December 2004.
5. F. Kootstra, B.F.P. van het Veld and G. Opschoor, A complete description of an advanced evaporation model for non-soluble, floating liquids from water. Part IB –

Theoretical background, Proceedings of the 38th Annual Loss Prevention Symposium, AIChE 2004 Spring National Meeting, 26 – 29 April, 2004, New Orleans, Louisiana, The United States of America.
6. Feldbauer G.F., Heigl J.J., McQueen W., Whipp R.H., May W.G., Spills of LNG on water – vaporization and downwind drift of combustible mixtures, ESSO Research & Engineering Company, API-report, 1972
7. Burgess D.S., Murphy J.N., Zabetakis M.G., Hazards Associated With the Spillage of Liquefied Natural Gas on Water, Bureau of Mines, Report of Investigations 7448, 1970.
8. G. Opschoor, Investigation into the spreading and evaporation of LNG spilled on water, Cryogenics, Vol. 17, No. 11, pp. 629-633 (1977).
9. USA-FERC, Notice of availability of staff's responses to comments on the consequence assessment methods for incidents involving releases from liquefied natural gas carriers.
10. R.C. Reid, and K.A. Smith, Behavior of LPG on water, Hydrocarbon Processing, Vol. 58, No. 4 (1978).
11. E.M. Drake, A.A. Jeje, and R.C. Reid, Transient boiling of liquid cryogens on a water surface: I. Nitrogen, Methane and Ethane, Int. J. Heat Mass Transfer, Vol. 18, pp. 1361-1368 (1975).
12. F. Kootstra, A.I.van Berkel, M. Molag, Modelling refrigerated NH_3-outflow out of sea-going vessels on the Westerschelde, TNO-report R2004/334, July 2004.
13. P.K. Raj, Calculations of thermal radiation hazards from LNG fires, a review of the state of the art, A.G.A. Transmission conference, T135-148, May 1977.
14. D.R. Blackmore, J.A. Eyre and G.G. Summers, Dispersion and combustion behavior of gas clouds resulting from Large spillages of LNG and LPG on to the sea, Trans. I. Mar. E. (TM), 94, paper 29 (1982).
15. G.A. Mizner and J.A. Eyre, Radiation form liquefied gas fires on water, Combustion sciences and Technology, 35, pp. 33-57 (1983).
16. P.J. Rew and W.G. Hulbert, Development of pool fire thermal radiation model, HSE Contract Research Report No. 96-1996, Health and Safety Executive, 1996.
17. J.A. Fay, Model of spills and fires from LNG and oil tankers, J. Hazard. Mater., B96, pp 171-188 (2003).
18. C.J.M. van Wingerden, Experimental investigation into the strength of blast-waves generated by vapour cloud explosions in congested areas, 6th Int. Symp. on Loss Prevention and Safety Promotion in the Process Industries, Oslo, Norway, pp 26-1, 26-16 (1989).
19. R. Pitblado, J. Baik, G. Hughes, C. Ferro, S. Shaw, Consequences of LNG Marine Incidents, CCPS Conference, Orlando, Florida, The United States of America, June 2004.
20. Platts LNG Daily, Vacuum effect could slow LNG tanker spills: study, Volume 2 / Number 176, Tuesday September 13 2005.

Biography

Freddie Kootstra

09/1997-08/2001: Ph.D. in Mathematical and Physical Sciences at the University of Groningen, The Netherlands.
09/1992-08/1997: M.Sc. in Chemistry at the Free University of Amsterdam, The Netherlands.
09/1986-06/1992: V.W.O. at the Ichthus College, Enschede, The Netherlands.

12/2001-today: TNO Built Environment and Geosciences, Department of Industrial and External Safety, Apeldoorn, The Netherlands.
09/2001-11/2001: Software engineer / scientist at Scientific Computation and Modelling N.V. (SCM), theoretical chemistry / FEW, Free University of Amsterdam, The Netherlands.

Experience with FEM3A

Benjamin Cormier, Jaffee Suardin, Yanjun Wang
Harry H. West, and M. Sam Mannan
Mary Kay O'Connor Process Safety Center
Chemical Engineering Department
Texas A&M University
College Station, Texas

AIChE Spring Meeting, April 2006
6th Topical Conference on Natural Gas Utilization
Orlando, Florida, April 23-27, 2006

Abstract

In the 1970s and 1980s, Liquefied Natural Gas (LNG) release modeling became very important due to the potential for LNG to become an important source of energy for the U.S. Because of the cost that LNG represents in handling and storing, the LNG industry of the 1980s and 1990s became less attractive. Now due to the lack of energy, LNG is again a serious contender for energy supply. Thus the modeling of the LNG vapor cloud dispersion of Denser-than-Air gas re-emerges in its importance. Today computer power is far advanced than in the 1980s. Older vapor dispersion simulation codes such as DEGADIS, SLAB and FEM3A are reviewed and analyzed herein. New codes like CFX have become available and friendly compared to older software. In this paper, all of these LNG vapor dispersion models were tested and compared to experimental data collected during a LNG release at the Brayton Fire School at Texas A&M University. In addition this paper provides a methodology to use these emergency release models. For emergency release models, U.S. Federal Safety Standard of Liquefied Natural Gas (LNG) facilities (DOT 49 CFR 193) recommends the use of the FEM3A or DEGADIS mathematical dispersion models to estimate the concentration of 2.5% natural gas (using methane as the major component) as a critical value of LFL. This paper attempts to show how to use dispersion models to estimate the risk of a LNG release into atmosphere. DEGADIS and SLAB were used to give a first rough idea of the size of the release. Then the estimation is used to set up CFD models. In April 2006, BP is pre-testing the LNG fire school facilities before its annual workshop a week later. The test consists of poring six inches of LNG in the bottom at a 45 m^2 pit. This release can be qualified as definite continuous release of LNG in a confined environment. The originality of this setup makes the simulation unique and valuable to challenge CFD models such as FEM3A and CFX-5.7.

1 Introduction

SLAB software and DEGADIS model were studied first. These models are so-called "box-model" predictions. They only provide simulation for obstacle-free release. For simulation including obstacle and adjusting the model respect to the transport law, the use of Computational Fluid Dynamics codes (CFDs) becomes necessary. CFD models are more much intensive to prepare and requires the access to supercomputer. Early 1980s, Lawrence Livermore Stanford, and Gas Research Institute (GRI) developed one of the first versions of FEM3 in Fortran 77. For instance, FEM3 is a CFD code that can simulation more complex system for LNG release than DEGADIS. At the time when DEGADIS and FEM3 were developed, a lot of questions had risen. In order to verify the models, real releases were performed and all these models showed that they have provided good estimations of the dispersion compare to experimental data[1-4]. SLAB and DEGADIS are studied at first to identify the required inputs to run this simulation. Then this study is taken to the next level by studying FEM3A and its required inputs. In the early 2004, a new CFD method, CFX, appeared in literature to simulate LNG release and vapor dispersion cloud[5]. CFX is a much more recent CFD code that seems to be more attractive and easier to use since it employs an interface to set and run the dispersion model.

For each software, a brief description is presented and also the possibilities i of simulation options. A common scenario will be used to compare the precision of the prediction of the LNG vapor dispersion cloud.

2 Box-model prediction

Box-model prediction software refers to software that acts like a box (input in a box and the box produces the output, not really flexible). Basically, only a certain types of scenarios can be simulated.

2.1 SLAB

SLAB is available online at www.lakes-environmental.com. The initial version comes with only a two-dimensional display mode. The main purpose of the SLAB project was to create software that will be usable by IBM-compatible personal computer.

SLAB is a computer model that simulates atmospheric dispersion of denser-than-air releases. Atmospheric dispersion of chemical release is calculated by solving the conservation equations of mass, energy, momentum and species. In order to reduce the computing time that would require a three-dimensional model, the concentration component y and z calculated from the concentration respect to x. The crosswind-averaged properties are also function of x[6].

2.1.1 Possibilities

The current version of SLAB can deal with instantaneous, continuous and finite duration releases from four types of sources: A ground level evaporating pool, an elevated horizontal jet, a stack or elevated vertical jet, and a ground based instantaneous release. For the evaporating pool, it is assumed to be only vapor compare the other sources that could be either pure or mixture of vapor and liquid droplets. SLAB can also simulate the vapor cloud dispersion of neutrally buoyant release.

For each type of release, SLAB was broken into different regimes:

- For a continuous release: the release is simulated as a steady state plume release
- For a finite duration release, the release is initially treated as a steady state plume till the time is over and then it is treated as a transient puff mode.
- For an instantaneous release, the transient puff mode is used to estimate the dispersion.

2.1.2 SLAB model concept

SLAB uses basic equations including one dimension conservation equations, equation of state, width and length equations with a numerical integration to obtain initial SLAB results. The initial results include spatially averaged cloud properties, one-dimension. Then from these results, the concentration profile, cloud meander and time average are calculated to obtain the final SLAB results. The final results include time averaged, three dimensional and time dependent concentration[6].

2.1.3 Dispersion Submodels

The main code provides the dispersion model for LNG vapor cloud dispersion with no effect from surroundings. Submodels are added in order to simulate the effects of the surroundings. These submodels are source rate, height dependent wind speed, vertical and horizontal air entrainment speeds, and surface heat and momentum flow rates. These quantities are important to obtain a more realistic dispersion model[6].

2.1.4 Input file

There are 30 input parameters required to run SLAB. These parameters are categorized in six sections such as spills source type, source properties, spill parameters, field parameters, source location, and meteorological parameters.

<u>Spills source type</u>

There are four options of source types and are listed below. The rest of the input requirements will vary with source type option.

1. Evaporating pool
2. Horizontal jet release
3. Vertical jet or stack release
4. Instantaneous or short duration release

<u>**Source properties**</u>

The properties of the chemical component that would be released into atmospheric. It requires molecular weight, vapor heat capacity, boiling point temperature, heat vaporization, liquid heat capacity, and two saturation pressure constants. The data for pure component is available in SLAB. In the case of mixture, the thermodynamic data for mixture would need to be calculated and imported in SLAB

Spill parameters

This section describes the configuration of the release:

1. Evaporating pool
    a. Mass source
    b. Continuous source duration
    c. Either source area or effective evaporating rate
2. Horizontal jet release
    a. Mass source
    b. Continuous source duration
    c. Source height
    d. Material storage temperature
    e. Either source area or actual area of the opening
3. Vertical jet or stack release
    a. Mass source
    b. Continuous source duration
    c. Source height
    d. Material storage temperature
    e. Either source area or actual area of the opening
4. Instantaneous or short duration release
    a. Instantaneous source mass
    b. Source height
    c. Material storage temperature

Field parameters

This section includes concentration averaging time and maximum downwind distance. By default the concentration is measured at ground level. It is possible to choose up to four different heights to obtain the concentration profile at different levels. The height selection is useful, it gives a good idea of three dimensional vapor cloud at different concentrations.

Source location

This option is used to determine the position of the release. It is only useful when you upload the map of the facilities into SLAB

Meteorological parameters

This section includes the surrounding effect such as the surface roughness height, ambient measurement height, ambient wind speed, ambient temperature, relative humidity and stability class. You may choose different meteorological parameters scenarios and save them in your simulation.

2.1.5 Limitation

SLAB is an easy to use software for chemical dispersion model but it comes with limitations. The wind direction is constant and only one direction and seems to be very simple compared to the complexity of a real wind. It is an obstacle-free simulation. The blueprint is only there to give an idea where the vapor cloud could go during a release but SLAB does not provide simulation options for building, dike or tank obstruction.

2.2 DEGADIS

DEGADIS (Dense Gas DISpersion) model was developed in 1985, under GRI and U.S. Coast Guard sponsorship (Havens and Spicer). It was developed primarily for cryogenic dense gas dispersion. In 1988, Havens incorporated Ooms' jet model with DEGADIS and in 1989, Spicer and Havens modified Ooms' jet model to provide an elliptical plume cross-section with air entrainment[7]. DEGADIS version 2.1 is available online at www.weblakes.com. The free version requires DOS knowledge.

DEGADIS model has three more-or-less distinct fluid regimes: Buoyancy-dominated, stably stratified and passive dispersion at low momentum. In the first phase, buoyancy-dominated, the flow of gas is spreading until the kinetic energy of the buoyancy-driven flow is dissipated. This motion induces a mixing that determines the shape of the vapor could. Once all the kinetic energy is dissipated, the vapor could is leaded by the mean wind flow, this is the stably stratified phase. The final regime is the passive dispersion, induced by per-turbulence. Richardson number is used as the criteria to determine the regime. Denser-than-air gas source cloud is forming two source blankets. The primary source blanket is the growing cloud from the evaporating pool until the atmospheric take-up rate from the top[7].

2.2.1 Possibilities

DEGADIS is versatile and well-accepted model. The prediction for different component such as LNG, propane, HF, and Ammonia seems to be accurate for an obstacle-free simulation. DEGADIS assumes an unobstructed atmospheric flow field. This model can simulate only vertical jet, buoyancy-dominated, stably-stratified and passive dispersion assuming a logarithmic wind profile. It is possible to use a large spectrum of surface roughness elements.

2.2.2 Input file

The input for DEGADIS is separated in two categories: Meteorological description and release description. The meteorological description is used for see how the surrounding affects on the cloud[8].

Meteorological description

1. Windspeed at given elevation,
2. Surface roughness
3. Pasquill stability class
4. Ambient temperature
5. Ambient pressure
6. Ambient (relative or absolute) humidity
7. Ground temperature
8. Averaging time

Release description

1. Contaminant molecular weight
2. Release temperature after depressurizing to ambient pressure
3. Release density after depressurizing to ambient pressure
4. Contaminant heat capacity
5. Release rate and release area as function of time

For low-momentum, ground-level release, DEGADIS requires the mass rate and the radius of the release. In the case of transient release, these inputs can be function of time. DEGADIS uses an energy balance including the heat transfer from the ground, condensation of ambient humidity, and ideal gas law to determine the density-concentration relation.

2.2.3 Limitations

DEGADIS does not account for positive buoyancy such as plume rise. It also does not include complex meteorological flow phenomena. DEGADIS is only used for obstacle-free simulation, it does not account for terrain features, which can influence the cloud path as well as its dispersion. It is only usable with pure chemical. For transient contaminant releases where the release duration is shorter than the averaging time, DEGADIS does not provide for concentration time averaging in the wind direction.

3 Computational Fluid Dynamics (CFD)

All the previous software provide good dispersion model assuming for obstacle-free and logarithmic wind profile, thus their flexibility is limited. The introduction of computational fluid dynamics becomes necessary for consideration of terrain effects. More commonly called CFD, this type of model requires the use of more powerful computer resources than the personal computer. CFD models bring the possibility to model more complex dispersion scenarios to include the impact of dike or tank. A more complex wind profile can also be incorporated into the CFD model. FEM3A was the first CFD model for LNG release in early 1980s. A more recent CFD model is CFX which can also be used for simulating LNG vapor dispersion.

3.1 FEM3, FEM3A and FEM3C

FEM3 model was initiated in 1979 in order to develop a validated computer model for simulating gravitational spread and vapor dispersion that result from a LNG release into atmosphere. FEM3 is a three-dimensional computer model that was designed for simulating the atmospheric dispersion of heavy gas releases such as LNG. This model uses a modified Galerkin finite element method that it is able to solve the time-dependent conservation equations of mass, energy, momentum and species together with the ideal gas law for the density of the gas/air mixture. This model also includes submodels for the source, ground heat transfer and K-theory submodel for turbulence. These submodels provide a mathematical description of the physical of heavy gas dispersion including gravity spread, the effect of density stratification on turbulent mixing and ground heating into the gas cloud. FEM3 is fully running in three-dimensional, and then it can treat complex flow and dispersion scenarios over variable obstacles[9].

FEM3 model was modified to FEM3A by Lawrence Livermore National Laboratory. The code received new modifications in the submodels for turbulence and mass balance considerations. An isotropic k-□ turbulence closure was implemented. Then FEM3A received more advance changes and was renamed FEM3C. FEM3C can treat dispersion scenarios involving density variations much larger than previously and an advanced turbulence submodel based on the buoyancy-extended k-□ transport equations[9]. In this paper, the analysis was based on FEM3A only.

FEM3A accounts for a lot more atmospheric phenomena than the box-model prediction. For instance, all these software including CFD includes dense gas effects with PBL closure, non smooth surfaces, and heat transfer. But CFD also accounts for vapor holdup with dike or vapor fence with 2nd order closure, obstacles (non-uniform) with heights greater than cloud depth and for terrain effects on cloud path as well as dispersion.

3.1.1 Input file

The input requirements are more complex for FEM3A than the box-model because the configuration of the surroundings has to be specified. It is important to first use a box-model such as DEGADIS or SLAB, since it will help to provide an estimate of the required computational domain[8].

Before running the main code, FEM3A, it is required to prepare the input files. Depending on the complexity of the simulation, more input may be required. For the simple simulation with obstacle-free, only one input file is required for the FEM3A preprocessor, pfem3a. Pfem3a will generate the final input file, FEM3IN, for FEM3A. If the simulation includes obstacles, a grid with geometric shape must be prepared first before running the preprocessor. If the simulation requires using k-□ model for the wind profile, the input file for pfem3a is even longer.

The best approach is to start simple. First, run a simulation with obstacle-free and PBL profile. Then run it and check the results by plotting using TECPLOT®. Then if any obstructions are present on the ground, use GRIDGEN to generate a grid with the obstacles. Use the GRIDGEN output with pfem3a. Then run FEM3A and plot to check the results. If the simulation does not finish or the results does not seem to make sense, go back to input file for FEM3A first and then to pfem3a. When the simulation gets to a point it is complicated it is a matter of trial and error.

3.2 CFX

The latest studies show that CFX could be a great help in the dispersion model. CFX is a general-use CFD code, and it is not calibrated for the solution of specific problem like FEM3A which specifically designed to simulate LNG. However CFX has been validated in isothermal heavy gas dispersion trials [5]. So the use of CFX for this project will require more intensive work and time consuming.

CFD codes are usually composed of three parts: preprocessor, solver and post-processor for problem definition, problem solution and results processing. CFX 5.7 is composed of four subsequent[10]:

1. CFX-Build, in which the computational domain geometric is constructed.
2. CFX-Physical processor, where boundary conditions, turbulence models, domain fluids and solution scheme are defined.
3. CFX-Solver, where the code performs calculations towards the final solution.
4. CFX-Post, in which the elaboration of results takes place, with graphical capabilities.

The two first subsequent is used to define the problem then the rest of the subsequent is used in getting a solution and analyze it. For the vapor model dispersion, CFX uses the Reynolds averaged Navier-Stokes equations and is based on the finite volume method for the conversion of partial differential equations and auxiliary conditions into a discrete system of equations.

3.2.1 Turbulence modeling

A turbulence flow can be described as a flow of particles moving randomly within a flow field. In atmospheric flows, turbulence is the dominant mechanism in mixing and dilution of gaseous releases. In such release scenario, turbulence is present. Most of the turbulence models are based on the Reynolds averaged Navier-Stokes equations and can be classified as eddy-viscosity or Reynolds stress models. Eddy-viscosity model is formed with a little eddies in which the Reynolds stress is assumed. Reynolds stress model is based on transport equations for all components. According to literature, the k-□ model had shown very good performance in LNG release scenario[11]. This model is formed to a two equation eddy viscosity turbulence model into the system conservation equations.

4 LNG Experiment

The purpose of using these software models is to be able to compare their results for real releases. The LNG test will be performed at Brayton fire school Texas A&M University. The atmospheric conditions are really similar to many LNG facilities because usually these facilities are situated in a warm climate. The LNG fire school has several pools for LNG release. Primarily this facility is used to train LNG fire fighters. In April, a week before the annual BP fire school workshop, the Fire school will run tests on LNG vapor cloud dispersion and foam effectiveness as a mitigation system on LNG pool fire. For this paper, the larger pit is used, 45m^2. LNG will be spilled at the bottom of the 45 m^2 pit until LNG reaches 6 inches in height to simulate a confined continuous release. The pit is 4ft depth, by poring only 6 inches of LNG, the wall of the pit creates a barrier which affects the behavior of vapor cloud. This constraint is challenging the "box-model", so the use of CFD is inevitable.

4.1 Description of pit at Brayton Fire School

During the test at LNG Fire School, 9.9 m^3 of LNG was spilled in a 45m^2 pit. The LNG was spill the pit. The pit is 4 ft deep and 33ft by 21ft. The figure 1 shows the configuration of the pit in two-dimensional. The spill should provide about 6 inches of LNG on the bottom of the pit. The measurement of the concentration of methane was taken from the first drop of LNG till the pit is filled with about 6 inches of LNG. It would take about one hour and half. During this time, the concentration of methane was measured. During this time, the vapor cloud goes through different regime. The release is initially treated as a steady state plume till the time is over and then it is treated as a transient puff mode.

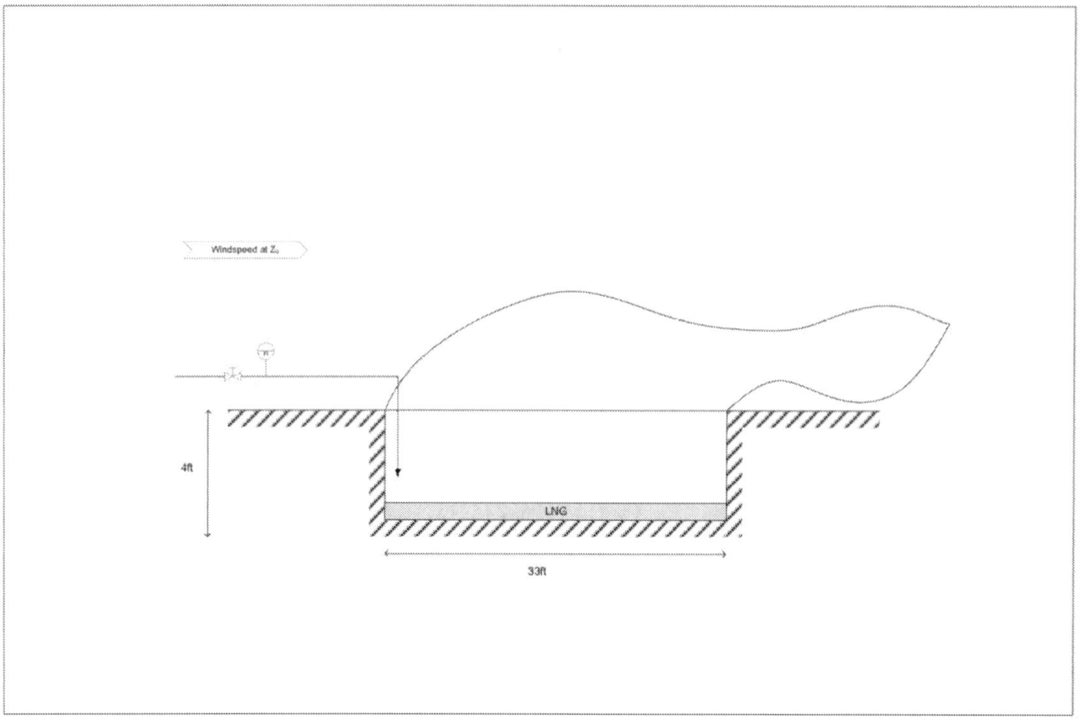

Figure 1. Scheme of the 45m^2 pit from LNG Fire School at Texas A&M University

The following parameters were assumed for the experiment:

Release description:

Evaporating pool	
Component:	100% methane
Mass source:	2.66 kg/s
Continuous source duration:	1550 s
Source area:	45 m^2
Concentration average time:	10 s
Maximum downwind distance:	80 m

Meteorological parameters:

Surface roughness:	0.03 m
Ambient measurement height (Z_0):	10 m
Windspeed at Z_0:	10 m/s
Ambient temperature:	298.15 K
Ambient pressure:	1 atm
Pasquill stability:	D
Relative humidity:	15%

The first step was to run SLAB and DEGADIS in order to give an order of magnitude to the experiment. In the figure 1, the wall of the pit represents a barrier to turbulence form the surrounding. This barrier can not be taken in account while using DEGADIS and SLAB. The use of CFD is inevitable. For this paper, we used the FEM3A and CFX-5.7 to simulate the continue release of LNG in a confined environment as described previously. Once the experiment data are collected at Brayton LNG fire school in April, the different models will be adjusted and compared.

Reference

1. Hanna, S.R., J.C. Chang, and D.G. Strimaitis, *Hazardous gas model evaluation with field observations.* Atmospheric Environment. Part A. General Topics, 1993. **27**(15): p. 2265-2285.
2. Spicer, T.O. and J.A. Havens, *Field test validation of the degadis model.* Journal of Hazardous Materials, 1987. **16**: p. 231-245.
3. Pereira, J.C.F. and X.Q. Chen, *Numerical calculations of unsteady heavy gas dispersion.* Journal of Hazardous Materials, 1996. **46**(2-3): p. 253-272.
4. Chan, S.T., *Numerical simulations of LNG vapor dispersion from a fenced storage area.* Journal of Hazardous Materials, 1992. **30**(2): p. 195-224.
5. Sklavounos, S. and F. Rigas, *Simulation of Coyote series trials--Part I: CFD estimation of non-isothermal LNG releases and comparison with box-model predictions.* Chemical Engineering Science, 2006. **61**(5): p. 1434-1443.
6. Ermak, D.L., *User's Manual for SLAB: An atmospheric dispersion model for denser-than-air releases.* 1980, Lawrence Livermore National Laboratory.
7. Spicer, T.O. and J.A. Havens, *User's guide for the DEGADIS 2.1 dense gas dispersion model* 1989, Environmental Protection Agency.
8. Spicer, T.O. and J.A. Havens, *Evaluation of mitigation methods for accidental LNG releases. Volume 5: Using FEM3A for LNG Accident consequences analyses.* 1996, Gas Research Institute.
9. Chan, S.T., *A Three-Dimensional Model for Simulation Atmospheric Dispersion of Heavy-Gases Over Complex Terrain.* 1997, Lawrence Livermore National Laboratory: Livermore, CA.
10. Sklavounos, S. and F. Rigas, *Validation of turbulence models in heavy gas dispersion over obstacles.* Journal of Hazardous Materials, 2004. **108**(1-2): p. 9-20.
11. Perdikaris, G.A. and F. Mayinger, *Numerical simulation of heavy gas cloud dispersion within topographically complex terrain.* Journal of Loss Prevention in the Process Industries, 1994. **7**(5): p. 391-396.

Biography of Speaker

Benjamin Cormier is a PhD candidate at Artie McFerrin Department of Chemical Engineering, Texas A&M University. He attended Virginia Tech where he obtained his Bachelor degree in Chemical Engineering in May 2004. Then he graduated from Texas A&M University with a Master of Science degree in Chemical Engineering in December 2005. Since the summer 2005, he has been involved with Mary Kay O'Connor Process Safety Center under the direction of Dr. Sam Mannan. Benjamin Cormier is also responsible for teaching Aspen Plus to the senior class here at Texas A&M University.

Email: Benjamin.Cormier@chemail.tamu.edu

Comparison of the Acceptability of Various Oil Shale Processes

Alan K. Burnham, Lawrence Livermore National Laboratory, Livermore, CA
James R. McConaghy, Antero Engineering, LLC, Salida, CO

Introduction

While oil shale has the potential to provide a substantial fraction of our nation's liquid fuels for many decades, cost and environmental acceptability are significant issues to be addressed. Lawrence Livermore National Laboratory (LLNL) examined a variety of oil shale processes between the mid 1960s and the mid 1990s, starting with retorting of rubble chimneys created from nuclear explosions [1] and ending with in-situ retorting of deep, large volumes of oil shale [2]. In between, it examined modified-in-situ combustion retorting of rubble blocks created by conventional mining and blasting [3,4], in-situ retorting by radio-frequency energy [5], aboveground combustion retorting [6], and aboveground processing by hot-solids recycle (HRS) [7,8]. This paper reviews various types of processes in both generic and specific forms and outlines some of the tradeoffs for large-scale development activities. Particular attention is given to hot-recycled-solids processes that maximize yield and minimize oil shale residence time during processing and true in-situ processes that generate oil over several years that is more similar to natural petroleum.

Is Oil Shale for Real This Time?

The recovery of oil from oil shale dates back centuries. Although reserves of oil shale are comparable to remaining reserves of conventional petroleum and probably greater than the amount of petroleum still to be discovered [9], it was never a significant energy source in the United States, and world-wide production has actually decreased over the past two decades due to its greater cost and environmental problems. A 1918 National Geographic article [10] proclaimed that shale oil was just about to replace crude oil due to dwindling crude supply, but new discoveries soon eliminated the need for oil shale. The real price of crude oil was nearly constant for almost 100 years until OPEC, primarily Saudi Arabia, became the supply controller in the mid 1970s, at which time the real price increased four-fold over a few years. Predictions of permanently high prices and proclamations of the national security importance of domestic energy supplies were rampant. Effort in synthetic fuels greatly increased in the United States, with oil shale being a principal player. However, OPEC could not retain control of supply and prices for long, and the sharp decline in oil prices after 1980 completely destroyed the national oil shale effort. LLNL was one of the final players, leading a CRADA to explore the HRS process with partners Amoco, Chevron-Conoco, Unocal.

The obvious question is whether the current spike will be different from the 1980 spike. Although a decrease in prices from the current high is likely, the drop probably won't be as much or as long lived. Although world-wide petroleum production has been primarily demand-limited for 125 years, it will probably reach its maximum rate within the next 10 years, while demand will continue to increase [9]. This imbalance will cause its price to rise to the level of alternatives, including conservation. In addition, the rate of production has exceeded discoveries for more than 10 years, and the gap is expected to persist. Finally, the time scale for significant shale oil production is certainly longer than the timescale for conventional petroleum meeting demand, and shale oil will probably be able only to slow the rate of production decline for conventional and non-conventional liquid fossil fuels 15 years hence.

A major difference between today and 1980 is the growing acceptance of global warming driven by fossil fuel use. The 1970-1980 boom was challenged for reasons of air emissions, water consumption, aquifer contamination, and surface disturbance. These environmental factors were all used in conjunction with the then unfavorable economics to kill the national oil shale program in 1993. They all remain, but concern about CO_2 has become substantially elevated with respect to the previous boom. These issues must be considered seriously in any process development, and the incremental cost of avoidance may be less than the economic risk of political backlash. Realistic comparisons should be made against the competition, which include biofuels and an effective electrical vehicles powered by non-fossil-derived electricity.

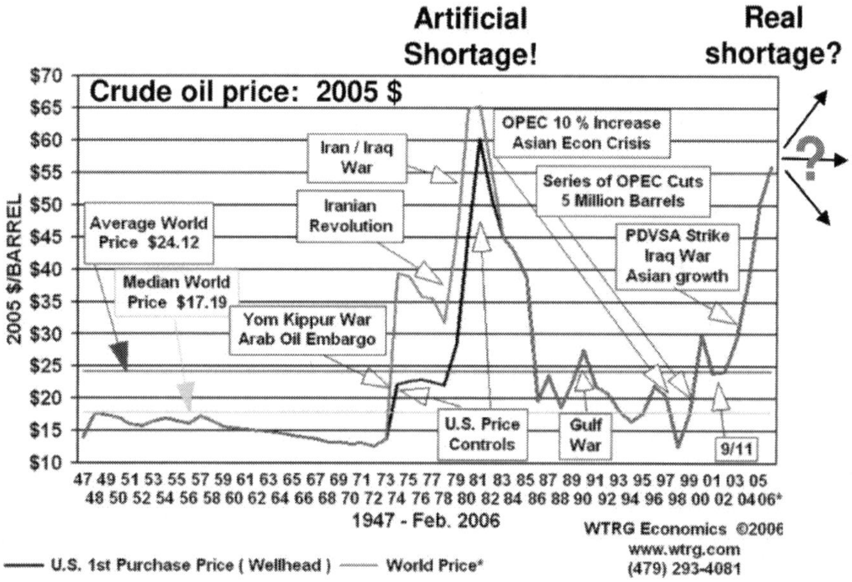

Figure 1. Oil prices since World War II showing the OPEC-inspired spike about 1980 and the current war-and-weather-related spike. Will the current spike be different from the 1980 spike?

Oil Shale Process Classification

There are many conceivable classification schemes, but the one in Figure 2 is fairly typical in terms of its distinguishing characteristics of heating method and whether retorting is done above or below ground. Belowground retorting occurs in larger "vessels" with larger particles, hence longer thermal diffusion times and slower retorting by necessity. Aboveground retorting offers the possibility of more process control (not always achieved in poorly designed processes), but with the associated higher costs of shale handling and solids processing equipment. Heating methods are often broken down into *direct* and *indirect*, where *direct* is the same as *internal combustion* in Figure 2. However, different indirect processes have substantially different characteristics depending on how the heat is delivered to the shale.

One important difference is whether the heat is delivered by a solid or gas. Roughly equal masses of each are required, and it is in principle cheaper to deliver that heat via a solid, especially if that solid contains its own heating fuel, as is the case for retorted shale. On this basis, LLNL concluded that a recycled-hot-solids process [7] had the greatest ultimate potential. Separation of combustion and retorting enables control needed for high oil yields and a concentrated pyrolysis gas stream. Well-designed retorted-shale combustors using fine shale can achieve that heat with minimal carbonate decomposition and effective SO_2 capture. Mixing of burned shale with pyrolysis products leads to effective capture of H_2S and COS, which are then converted to sulfates in the combustor. Total residence times of a few minutes for retorting and combustion combined give minimum reactor volumes and easier scaleup.

Another generic process that is currently receiving much attention is true-in-situ retorting using externally generated heat. In-situ retorting has some basic constraints. Oil shale has little native permeability, so combustion retorting can only be achieved by adding porosity by mining (MIS) [3] or explosive uplift (Geokinetics) [11]. Injecting hot fluids is not particularly effective, particularly on the time scale normally considered for enhanced oil recovery, because the highest permeability regions have the poorest oil yield and because the temperatures to be achieved require very hot fluids. If one is patient, one can achieve thermal diffusion of a few meters over a time scale of a year or more. Patience is the concept behind the primary Shell ICP method [12,13], which uses electric heaters in wells as the heat source. Of course, their patents also mention downhole burners, which would be twice as efficient. However, one must place wells very close together in such a process, and the time scale increases roughly as the square of the well spacing. Of course, the drilling costs also scale roughly as the inverse square of the well spacing, so the optimum spacing depends on the thermal and recovery efficiencies as a function spacing and the time value of money.

One possible way to improve in-situ processing is to use volumetric heating by radio waves [14]. In our view, the literature is inconclusive-to-inadequate concerning the penetration distances possible, but there are indications that it could be many meters. To be perfectly clear, the objective here is to choose the frequency with the smallest absorption coefficient to maximize the penetration distance. One pays a two-fold energy cost by using electricity in any form, but that cost is potentially recoverable from either lower drilling costs for wider well spacing or faster retorting at a given well spacing than in the basic Shell ICP conductive process. We are unaware of any analysis that has done that tradeoff carefully, and we are not convinced that available data on rf penetration makes the tradeoff even possible at this time.

Table 1. Classification of oil shale processing according to heating method and location.

Heating Method	Above Ground	Below Ground
Conduction through a wall (various fuels)	Pumpherston, Fischer assay, ATP, Oil-Tech	Shell ICP (primary method)
Externally generated hot gas	Union B, Paraho Indirect, Superior Indirect	
Internal combustion	Union A, Paraho Direct, Superior Direct, Kiviter, Petrosix	Oxy MIS, LLNL RISE, Geokinetics Horizontal, Rio Blanco*
Hot recycled solids (inert or burned shale)	Galoter, Lurgi, Chevron STB, LLNL HRS, Shell Spher	
Reactive fluids	IGT Hytort (high-pressure H_2), Donor solvent processes	Shell ICP (some embodiments)
Volumetric heating		ITTRI and LLNL radio-frequency

*This generic type has particularly challenging environmental issues related to combined poor oil yield, dilute offgas (HC and CO_2), and possibly aquifer contamination

Details of the LLNL HRS Process

LLNL investigated hot-recycled-shale processing because it appeared to have the greatest promise for speed and intrinsic control of pollutants, which would minimize processing costs. A 4-tonne/day process pilot plant, shown schematically in Figure 3, was built and operated from 1990 to 1993 to test this concept [8]. A delayed-fall combustor is used to achieve good mixing with air, plug flow, and a particle velocity roughly independent of particle size and slow enough that the combustion vessel is small compared to a lift pipe combustor. The lift pipe is still used to elevate the same, and significant combustion occurs therein, but the use of a separate delayed fall combustor gives greater control over the combustion process. A fluidized bed classifier then rejects the finest material to set the recycle ratio. A fluidized-bed mixer replaces the screw mixer in the Lurgi process, and the majority of the pyrolysis occurs in a settling-bed unit.

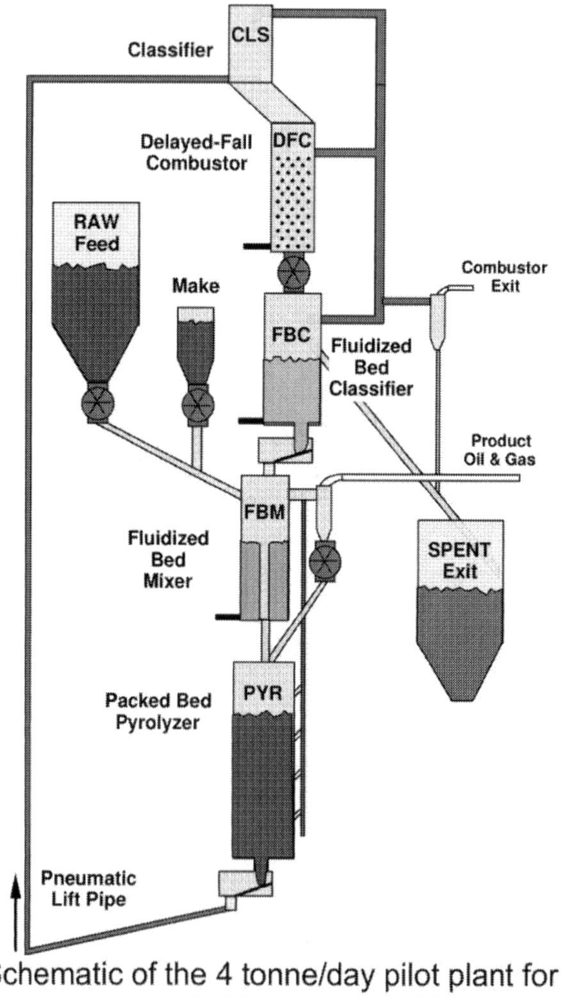

Figure 2. Schematic of the 4 tonne/day pilot plant for the HRS process operated at LLNL between 1990 and 1993.

Oil yields were typically 96% and 102% of Fischer Assay for 22 and 38 gal/ton oil shale, respectively. Clogging due to melting of rich shale was never a problem. Separation of fines from shale oil was identified as the last remaining technical challenge. Hot gas filtering and heavy-oil recycling were tested, but the results have not been publicly released. In today's environment, CO_2 mitigation must be added to the list of challenges. Carbonate decomposition ranged from 14 to 49% and correlated with combustion temperature. This implies that increased recycle ratios, which can tolerate a lower temperature in the recycled shale, may have an advantage in CO_2 mitigation costs that counterbalance the larger vessels needed for a larger recycle ratio. Also, modeling suggested that one could get improved combustor performance with an O_2-enriched combustion gas. Perhaps that would be more favorable in a situation where the flue gas was scrubbed for CO_2. These demonstrate how process optimization would be different in today's environment.

Spent shale disposal was also a major concern for environmental groups. An early study of spent shale disposal had shown that reactions between carbonate and silicate minerals under the right conditions formed significant quantities of the active components of Portland cement [15]. Although that particular study focused on injecting a grout made from spent shale into the voids of MIS retorts, later unpublished studies showed that the burned shale from the HRS process, with much less added water, could be pressed into low-permeability bricks that return the shale to much closer to its original volume. More work is needed to optimize this process, but it is easily conceivable that large blocks could be formed and moved back into the mine or cast and compacted in place, thereby drastically reducing the need for surface disposal.

True In-Situ Retorting

Starting in 1979, Mallon at LLNL [5] examined the radio-frequency approach being developed by IITRI [14]. Although a significant amount of mining was required in his concept, it was still significantly lower than MIS, since only access drifts were required. The concept was to retort the oil shale over four nights using inexpensive off-peak power. The economics depend on the assumed oil yield, and no data existed for the envisioned heating rates and pressures (7.5 ^{o}C/h and 10 atm). Nevertheless, an oil yield of 83% of FA was estimated from a variety of literature sources. Assuming an electricity cost of \$0.02/kW-h, he derived an electricity cost of \$6.7/bbl and a total cost of \$15/bbl in 1980s dollars. Doubling that value for today's costs puts it in the ballpark of what is being discussed for Shell ICP.

To refine the economic predictions, a set of experiments were done to determine oil and gas yields and composition for autogenous sweep conditions at various heating rates and pressures relevant to in-situ rf processing [16]. Initial experiments were conducted on compressed pellets having 24% porosity. It was found that the externally applied pressure delayed the vaporization of the oil and caused the oil yield to decrease due to cracking into gas and lighter oil. Oil properties at the extremes of the processing conditions are shown in Table 2. This process was later modeled (Figure 3) using the most detailed chemical kinetic model of oil shale pyrolysis at the time [17]. It included a more explicit treatment of the coking reactions in earlier atmospheric retorting models (which reduce nitrogen content) along with a pressure-dependent oil cracking model that split the oil into 11 boiling-point fractions.

Table 2. Summary of shale oil properties at the extremes of the conditions examined by Burnham and Singleton [15].

Conditions	Density, g/cm^3	H/C Ratio	Wt% N	Wt% S	90% distilled, ^{o}C
12 ^{o}C/min, 1 atm	0.906	1.61	2.7	0.66	504
1 ^{o}C/h, 27 atm	0.826	1.90	1.5	0.36	395

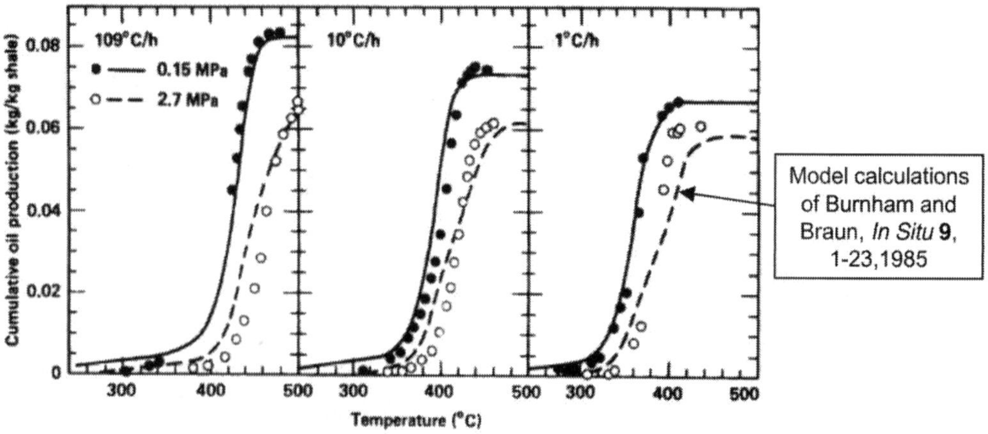

Figure 3. Ability of an LLNL kinetic model that includes liquid-vapor equilibrium calculations to match oil evolution data for conditions of shallow true-in-situ retorting.

These results become particularly important today within the context of the Shell ICP. A plot of the oil yield versus heating rate from Burnham and Singleton [16] is compared to oil yield results in the Shell patent [12]. Superficially, their yields appear to be a lot higher than expected, but further discussion is warranted. First, Burnham and Braun calculated that yields would be 5% higher for full-density shale because of reduced oil-vapor residence time at high pressure. Second, Burnham and Singleton conducted additional unpublished experiments on cores which suggested that this yield improvement might be even greater, so a yield of 80% is likely at 1 °C/h and 27 atm. Third, the oil yield relationship is almost certainly sigmoidal with heating rate. This is consistent with both distributed reactivity models that have become prominent since 1983 and hydrous pyrolysis results from the geochemical community. Fourth, there is an accounting issue of whether one considers C_4 and C_5 species in the gas as oil or gas. On the other hand, there is also an uncertainty in the Shell result related to estimating the volume and grade of shale retorted. The body of the patent says yields are 75-80% of Fischer assay. Much more information is needed to assess whether there actually is a discrepancy between the LLNL and Shell results. Independent of the precise yield estimations, the oil properties from the Shell ICP are consistent with predictions from the LLNL experiments to the extent the comparison is possible with public information.

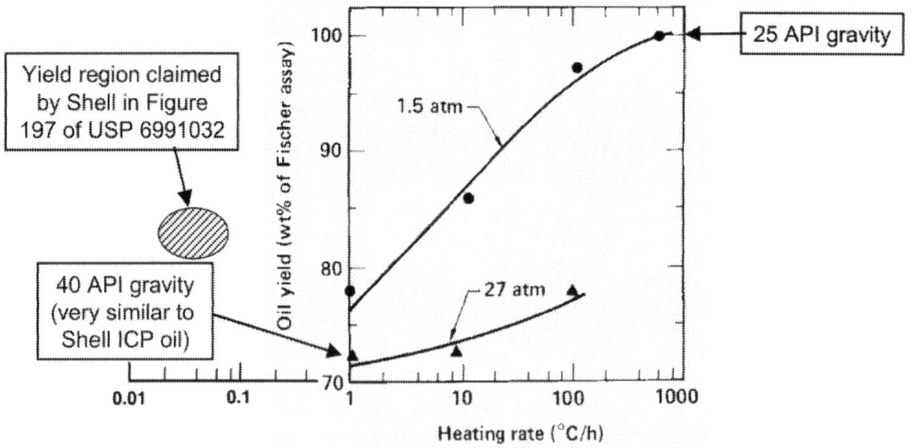

Figure 4. Comparison of yield results from Shell ICP with experiments of Burnham and Singleton [16].

With the successful application in mid 1980s of the chemical model shown in Figure 3 to petroleum formation in the Uinta Basin [18], it was both evident and commonly discussed in the organic geochemistry community that oil shale is merely a petroleum source rock that didn't get buried deep enough. When one of us [19] proposed the possibility of slowly heating large blocks of oil shale over a period of several years to convert it to and produce it as a more conventional crude oil, the common response from oil company personnel was that the time for return on investment was too long compared to new opportunities in deep water and the former Soviet Union. The lone exception was a very guarded interest by Shell employees during a January 27, 1995, presentation at their Bellaire Research Center (referenced in their recent patents), for reasons which are now obvious. Perhaps their different view was due to their long-time association with the Belridge field in California, which has very close wells over a very thick pay zone.

The LLNL concept, eventually documented [2], was to use wells spaced at tens of meters, either vertically or in some type of triplate design following bedding planes using deviated drilling, to heat cubic kilometers of deep oil shale very slowly (e.g., Figure 5). The presumption was that it would be possible to find a radio frequency at which the skin depth would be many tens of meters, thereby overcoming the very long thermal diffusion times needed for conductive heating. Targeting deeper shale than the Shell ICP has disclosed to present, the expulsion mechanism would change from vapor-driven expulsion at a pressure limited by modest lithostatic load to compaction-driven expulsion at the greater lithostatic loads. Possible contamination of aquifers becomes less important. Of course, the deep and shallow characterizations of the LLNL and Shell processes are merely two end members of a continuum of situations.

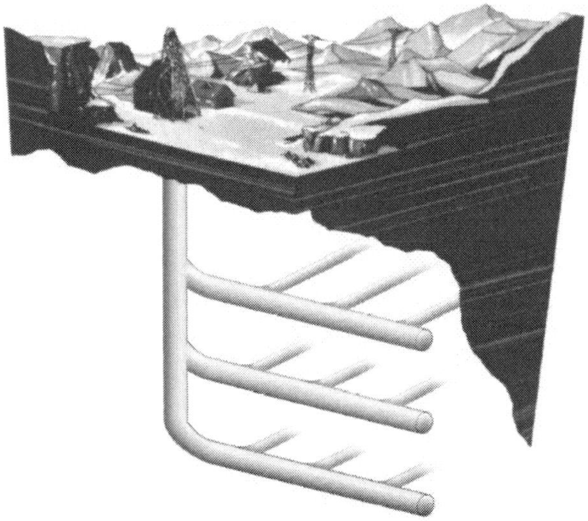

Figure 5. Artist conception of a deviated well pattern for insertion of radiofrequency antennas and withdrawal of produced oil with minimum surface disturbance.

One less-obvious advantage of slow retorting is that less energy is required. This is because the additional time at temperature enables the oil to be generated and expelled at a lower temperature. As the heating rate decreases from 3 °C/h to 3 °C/month, the completion of oil generation decreases from 400 to 300 °C [2,19]. The energy required (including losses) decreases from 124 to 87 kW-h/Mg shale. At steady state, the production rate has a correspondingly higher value for constant heat input. The additional time for thermal diffusion at the slower rates can be either good or bad. If there is preferential deposition in lean shales, the additional time enables better heat transfer to the rich shales. If there is preferential deposition into rich shales or a significant heat loss due to either fluid migration or thermal diffusion at the boundaries, the shorter time is advantageous.

There are obviously many questions about the viability of the LLNL concept. Foremost is whether the energy deposition actually can occur over a scale of several tens of meters, because the higher costs of radio-frequency energy must be more than compensated by some combination of lower drilling costs, a shorter time between energy insertion and oil recovery, or both. Published literature is not definitive on that point. If so, it would appear to be more economic than the electrical heaters currently used by Shell. It is less obvious whether rf has an advantage over downhole burners, since they would be twice as efficient thermally. Second, will large-area subsidence be tolerated by the public? If not, equivalent volumes of fluids will have to be injected. Part of that volume could be CO_2.

It is important to understand that no significant resources at LLNL or elsewhere have been used to evaluate deep, large-volume heating by radiofrequency energy. Perhaps there will be, if the Shell ICP is successful, with the objective of incremental process improvement. More generally, one might also consider heating by injection of hot fluids. Here, of course, one must consider not only how the uniformity and efficiency of the heat deposition but also the recovery efficiency of any injected non-aqueous fluid. Because of its low intrinsic permeability, it is doubtful that any injected fluid could have adequate contact with the retorting shale to significantly affect oil and gas yields.

One of the potential gains in carbon efficiency is to use non-fossil energy to generate the electricity or thermal heat used to retort the oil shale [20]. In fact, if electricity is generated from nuclear fission and then deposited by conduction, it would be even more efficient to design a nuclear fuel that could be lowered directly in the well and double the energy efficiency. Independent of the technical challenge of creating an unleachable nuclear heat source, public acceptance is an important issue. Another possibility is direct conversion of coal to electricity, which can be done at an efficiently of approximately 80% [21]. Actually, the coal is pyrolyzed to remove hydrogen-rich volatiles, and the residual char is oxidized at a ceramic electrode saturated with molten salt. Direct electrical conversion of carbon has no entropy change and therefore no loss of thermodynamic efficiency as in a Carnot cycle. The main source of inefficiency is due to internal resistance losses. In addition, the effluent gas is a nominally pure CO_2, an easily recoverable stream with small amounts of NO_x and SO_x. If successfully developed, this would make electrical heating by either conduction or radio frequency more attractive environmentally.

Conclusions

Despite the mammoth size of the US oil shale resource, oil shale has not yet been a significant energy supply because of its higher cost than conventional crude oil. The OPEC-inspired price spike starting in the early 1970s spurred a large effort in oil shale, which collapsed utterly in the early 1980s due to a retreat of crude oil prices to their historical level. The obvious question is whether oil shale is a surer investment today than then? Although energy security arguments are increasing to the levels of the 1970s, it is doubtful that energy security alone will overcome economics in the global economy. However, it is very likely that the peak in world oil production and corresponding permanent increase in price due to demand exceeding production capacity will occur sooner than significant oil shale production can be put in place, so the investment risk is lower from that perspective. On the other hand, environmental concerns are at least as important today, particularly with respect to CO_2 emissions, and it is more conceivable now than then that technology advances might actually make renewable biofuels economically competitive with fossil fuels.

In this arena of increased environmental constraints, existing and potential, it is more important to design and implement processes with lower environmental impact, including the amount of CO_2 generated per barrel of oil. This situation makes modified in-situ and any direct combustion process less attractive than in the 1970s. The HRS process, for example, produces three times less CO_2 per barrel of oil than an MIS process. In fact, the HRS process produces less CO_2 per barrel of oil than the Shell ICP process if fossil fuels are used to generate the electricity for ICP conductive heating. Even though it is much farther developed than is generally appreciated, significant effort is still required to update and demonstrate the HRS process to current standards.

When properly compared within the constraint of publicly available material, product yields and compositions from the Shell ICP process are consistent with experiments and models from LLNL in the early 1980s. In fact, those early results caused one of us [18] to independently develop a process concept similar to Shell's in the early 1990s. One difference is to use radio-frequency waves to overcome the long thermal diffusion time for conductive heating, thereby increasing the number of wells needed by an order of magnitude. Another difference is to target deeper deposits, where the expulsion mechanism changes from vaporization to compaction and where concerns of aquifer contamination are largely eliminated. The LLNL process has greater technical uncertainty, but the uncertainties could be reduced substantially with only a modest research investment.

A significant question is whether the 15-25% lower yields from an in-situ process, be it Shell ICP, LLNL radio-frequency, or some other variation, is counterbalanced by the improved oil quality? The chemistry of yield loss by its very nature reduces refining cost by rejecting heteroatoms and carbon and increasing hydrogen and saturate content. A less obvious advantage is that the resulting shale oil has a 4% greater combustion energy content per mass of carbon dioxide eventually produced. Finally, a variety of methods might improve the energy efficiency of the in-situ processes compared to using electricity generated from fossil fuels.

Acknowledgments

This work was performed under the auspices of the U.S. Department of Energy by the University of California, Lawrence Livermore National Laboratory, under Contract No. W-7405-Eng-48. We thank James L. Williams of WGRT Economics for updating and giving us permission to use his figures on historical oil prices.

References

1. D. B. Lombard, B. G. Bray, H. W. Sohns, T. S. Sterret, R. S. Brundage, and H. C. Carpenter, Design for Project Bronco, and Experiment for Nuclear Fracturing and In Situ Retorting of Oil Shale, LLNL Report UCRL-50339 (Oct. 1967).

2. A. K. Burnham, *Slow Radio-Frequency Processing of Large Oil Shale Volumes to Produce Petroleum-like Shale Oil*, LLNL Report UCRL-ID-155045 (Aug. 2003).

3. A. E. Lewis and A. J. Rothman, "Rubble In Situ Extration (RISE): A Proposed Program for Recovery of Oil from Oil Shale, LLNL Report UCRL-51768 (1975).

4. J. H. Campbell, *Modified In Situ Retorting: Results from LLNL Pilot Retorting Experiments*, LLNL Report UCRL-53168 (Aug. 1981).

5. R. G. Mallon, *Economics of Shale Oil Production by Radio Frequency Heating*, LLNL Report UCRL-52942 (May, 1980).

6. R. L. Braun, D. E. Christiansen, J. C. Diaz, and A. E. Lewis, *Fuel Processing Technology* **9**, 125-138 (1984).

7. A. E. Lewis, R. L. Braun, R. G. Mallon, O. R. Walton, Apparatus for Oil Shale Retorting, US Patent 4619738 (Oct. 1986). Assigned to USDOE.

8. R. J. Cena, Status of LLNL Hot-Recycled-Solid Oil Shale Retort, LLNL Report UCRL-; also DOE/NBM-94000080, available from www.osti.gov/bridge.

9. H. R. Johnson, P. M. Crawford, J. W. Bunger, Strategic Significance of America's Oil Shale Resource. Vol. 1. Assessment of Strategic Issues, Office of Naval Petroleum and Oil Shale Reserves, US DOE (March 2004).

10. G. E. Mitchell, "Billions of Barrels of Oil Locked up in Rocks," *National Geographic*, Feb 1918, pp. 195-205.

11. D. P. Zerga, Shale Oil Recovery Process," US Patent 4205610 (Jun. 1980).

12. I. E. Berchenko et al., *In Situ Thermal Processing of an Oil Shale Formation Using a Pattern of Heat Sources*, US Patent 6991032 (Jan. 2006), assigned to Shell Oil Co.

13. S. Mut, Testimony before the Senate Energy & Natural Resources Committee, Washington, DC, April 12, 2005: http://energy.senate.gov/public/index.cfm?FuseAction=Hearings.Testimony&Hearing_ID=1445&Witness_ID=4139.

14. J. E. Bridges, A. Taflove, and R. H. Snow, *Method for In Situ Heat Processing of Hydrocarbonaceous Resources*, US Patent 4140180 (Feb. 1979), assigned to IITRI.

15. R. G. Mallon, "Preparation and Injection of Grout from Spent Shale for Stabilization of Abandoned In-Situ Oil Shale Retorts," 54th Annual Mtg. Soc. Petr. Engr. of AIME, Sept. 1979, paper SPE 8448.

16. A. K. Burnham and M. F. Singleton, "High pressure pyrolysis of Green River oil shale," in *Geochemistry and Chemistry of Oil Shale*, F. P. Miknis and J. F. McKay, eds., ACS Symposium Series 230, American Chemical Society, Washington DC, pp 335-351 (1983).

17. A. K. Burnham and R. L. Braun, *In Situ* **9**, 1-23 (1985).

18. J. J. Sweeney, A. K. Burnham, and R. L. Braun, *AAPG Bulletin* **71**, 967-985 (1987).

19. A. K. Burnham, "Chemical kinetics and oil shale process design," LLNL Report UCRL-JC-114129 (July 1993), published in *Composition, Geochemistry and Conversion of Oil Shales*, NATO ASI Series C, Mathematical and Physical Sciences, vol. 455, pp. 263-276 (1995).

20. A. K. Burnham, *Energy* **14**, 667-674 (1989).

21. J. E. Cooper, N. J. Cherepy, R. Krueger, "Tilted Fuel Cell Apparatus," US Patent 6878479 (Jun. 2002), assigned to Regents of the Univ. of Calif.

Overview of Retorting of Eastern Oil Shale

Burtron H. Davis, University of Kentucky, Center for Applied Energy Research, Lexington, KY

Abstract

Eastern oil shale represents a major supply of oil reserves. For example, the oil shale located in just two counties in Kentucky represent two or more trillion barrels of oil when based on an overburden to shale ratio of 2.5. There are potential advantages to the production of oil from these Eastern reserves including location within 500 miles of a large fraction of the U.S. population and a ready supply of water needed for processing. A major disadvantage for the eastern shale is that it requires special retorting if the oil is to be competitive in the marketplace. One of the disadvantages of Eastern shale compared to Western U.S. shale is that a larger fraction is converted to coke under normal thermal recovery retorting processes. However, a 1.5 – 2.0 times greater yield of oil is obtained when a rapid pyrolysis such as fluid bed retorting is used. Limited data suggest that a high yield of products of higher quality can be obtained if the rapid pyrolysis is conducted in a hydrogen atmosphere. An overview of the Eastern oil shale reserves, with emphasis on the Kentucky resource,is be provided. Typical yields of product and its quality as obtained with various retorting processes are described.

Introduction

Coal, oil sands and oil shale resources now comprise the major fossil energy resources of North America. As shown in Figure 1, the resources of Canadian tar (now known as oil) sands and

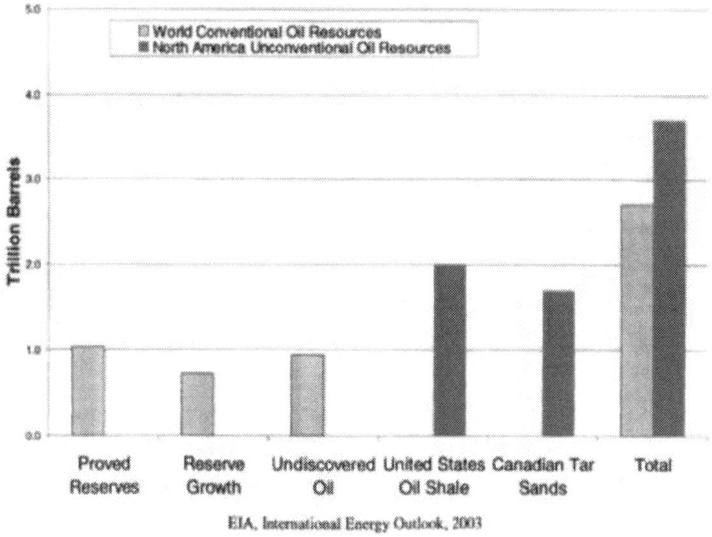

Figure 1. Unconventional oil resources exceed conventional oil resources.

the U.S. oil shale reserves exceed the proven and potential oil resources. Likewise, looking at the world's researves of coal and shale oil, each comprise more than three times that of both oil and gas resources of the world (Figure 2). Thus, the potential of oil shale for the U.S. future energy supply is enormous. In the U.S., two types of oil shale are available: those in the western and those in the eastern US. The two shales differ significantly in characteristics.

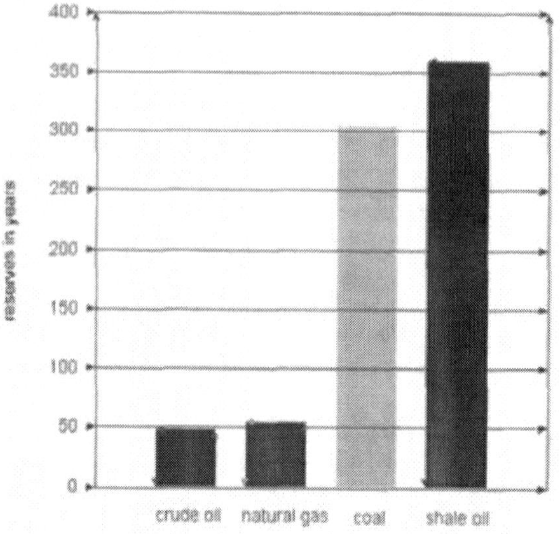

Figure 2. Carbon resources (Chemisch Magazine, 1995).

Black organic-rich marine shale and associated sediments of late Devonian and early Mississippian age underlie roughly 725,000 km^2 in eastern United States. These shales have been exploited for many years as a resource for natural gas [1]. The Eastern shales occur over a wide expanse from New York to Oklahoma and from Iowa to Alabama; however, the shales of greatest potential are those at or near the surface [2]. The most important Eastern oil shale areas are in Kentucky, Tennessee, Indiana and Ohio, where nearly flat-lying beds, averaging 9 to 12 m (30 to 40 ft) thick, are exposed at the surface in an outcrop belt nearly 16 km (1,000 miles) long. Greater than 90% of the near-surface mineable resource is in Kentucky, Ohio, Indiana and Tennessee.

The eastern shale location is shown in Figure 3 as illustrated on the cover of Proceedings of the 1985 Eastern Oil Shale Symposium, held in Lexington, KY. Recoverable resources of shale oil from the marine black shales in the eastern United States were estimated in 1980 to exceed 400 billion barrels. The eastern shales are low in carbonate content so that retorting will not cause decomposition of carbonates to produce large amounts of carbon dioxide. It is estimated that 850,000 acres of land in Kentucky is underlain by shallow deposits of oil shale. It is estimated that 30% of this shale oil is recoverable by surface mining (contour mining, not mountain top removal), assuming an overburden to economic oil shale ratio of 2.5:1 and a lower cut-off of 8.0 wt.% organic carbon. **This represents, conservatively, ten billion barrels of oil.**

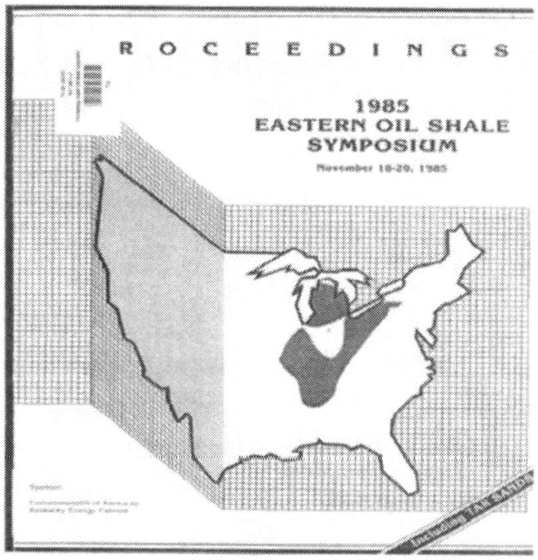

Figure 3. Eastern shale location.

Retorting

For western shale the commonly used Fischer assay provides a good estimate of the yield of oil that can be obtained during conventional retorting. However, the slow heating rate normally used for the Fischer assay leads to an underestimation of the yield obtainable from eastern oil shale. For eastern shale, fluid bed retorting provides yields of oil that exceeds that of the Fischer assay by 125 to 200 percent. An example of the impact of the rate of heating on the yield of oil is shown in Figure 4. Using fluid bed retorting, a heating rate of 100°C per minute is attainable, resulting in at least a 150% recovery based on Fischer assay.

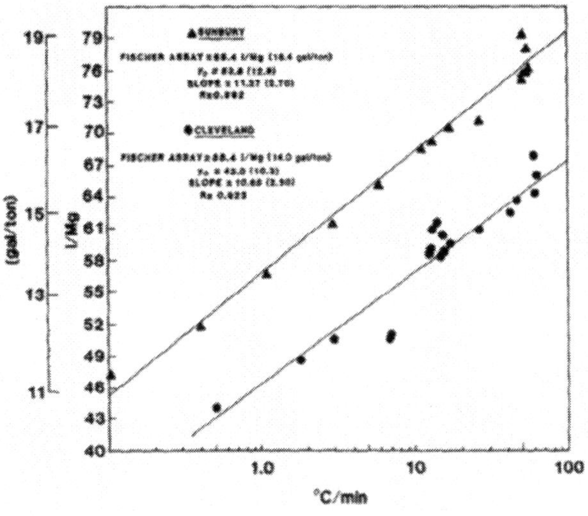

Figure 4. Semi-log plot of heat-up rate vs. oil yield.

A second difference between eastern and western oil shale is associated with the inorganic composition. Western oil shale contains significant carbonates and these decompose during retorting to release large quantities of carbon dioxide. On the other hand, very little carbon dioxide is released during retorting of eastern oil shale.

An atmospheric pressure fluid bed retorting process was developed by workers at the CAER: the KENTORT II Process (Figure 5). This process was scaled from a one inch to three inch and finally a six inch reactor; operation of these reactors demonstrated the advantage of rapidly removing the oil products from the reaction zone. The process consists of pyrolysis to produce oil in the upper zone of the reactor, gasification in the middle zone of the reactor, and combustion in the bottom zone of the reactor. Fluid bed retorting produced oil yields that were up to 150% greater than from the commonly used standard Fischer assay. KENTORT II operated at atmospheric pressure so that hydrogen was not added during the retorting.

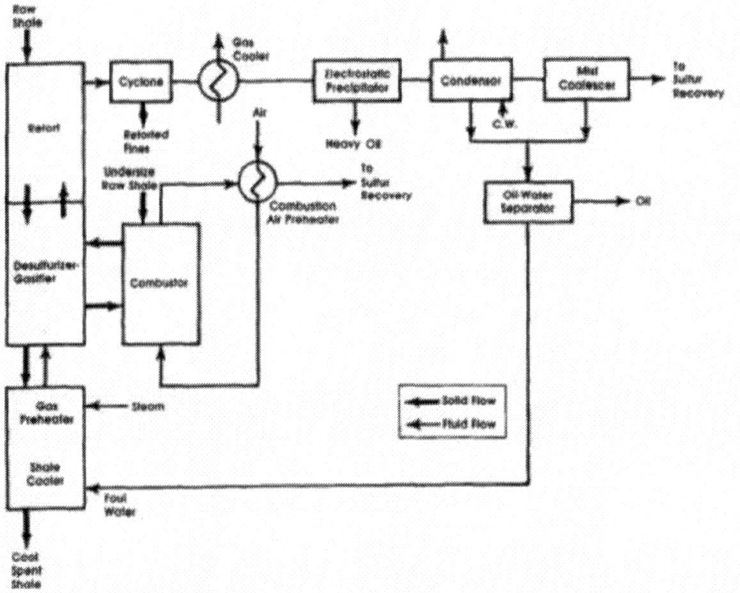

Figure 5. KENTORT II Process Concept.

The Institute of Gas Technology (IGT) developed a medium pressure oil shale retorting process that utilized a hydrogen atmosphere [3,4]. Results of the Fischer assay and hydroretorting eastern and western oil shales are compared in Table 1.

Table 1. Comparison of retorting of Eastern and Western oil shales (with similar organic carbon content) results for the standard Fischer Assay and fluid bed hydrogen retorting (HYTORT Process) (R. D. Matthews, "The Devonian-Mississippian oil hale resource of the Eastern United States," 16th Oil Shale Symposium Proceedings, (J. H. Gray, Ed.), Colorado School of Mines Press, August 1983, pp14-25.)

	Eastern	Western
Ultimate Analysis, (dry basis) wt.%		
Organic carbon	13.7	13.6
Hydrogen	1.6	2.1
Sulfur	4.7	0.5
Carbon Dioxide	0.5	15.9
Ash	78.3	66.8
Fischer Assay Analysis		
Oil Yield, wt.%	4.6	11.4
Assay, gal/ton	10.3	29.8
HYTORT Yields		
Oil Yield, wt.%	9.1	—
Assay, gal/ton	23.2	---

The Eastern oil shale compares very poorly with the Western oil shale when the comparison is based on the standard Fischer Assay. However, when a medium pressure hydrogen fluid bed retorting process is used, the yield is similar.

Based on the HYTORT Process, the potential recoverable resources are shown in Table 2. Four criteria for resource assessment were imposed in calculating rough resource estimates:

1. At least 10% organic carbon,
2. At least 3 m (10ft) thick,
3. Stripping ratio no greater than 2.5:1,
4. and overburden less than 61 m (200 ft).

On this basis, some 16,000 km^2 (6230 mi^2) were considered accessible in several states included in the study.

Table 2. Estimated resources of shale oil recoverable by the HYTORT Process in the Appalachian, Illinois, and Michigan basin areas (J. C. Janka and J. M. Dennison, "Devonian oil shale," Symposium Papers, Fuels from Oil Shale, Institute of Gas Technology, Chicago, 1980, pp. 21-116).

State	Total Area Suitable for Surface Mining, square miles	Resources Recoverable by Aboveground Hydroretorting	
		billion bbl	bbl/acre
Ohio	980	140	222,000
Kentucky	2650	190	112,000
Tennessee	1540	44	44,000
Indiana	600	40	104,000
Michigan	160	5	49,000
Alabama	300	4	21,000
TOTAL	6230	423	

In conclusion, eastern oil shale has potential for commercial operations. However, to achieve this potential it will be necessary to use retorting processes that are more sophisticated than conventional atmospheric pressure retorting.

References

1. J. R. Dyni, Oil Shale, 20 (2003) 193-252.
2. R. David Matthews, 16th Oil Shale Symposium Proceedings, 1983, pp 14-25.
3. S. A. Weil, P. B. Tarman and D. Punwani, Process for production of hydrocarbon liquid from oil shale, U.S. Patent 4,003,821, January 18, 1977.
4. D. M. Rue, F. S. Lau and D. Punwani, Hydrogen consumption during moving-bed hydroretorting of raw and beneficiated Eastern oil shales, 1988 Eastern Oil Shale Symp. Nov 30-Dec 2, 1988.

228c An Economic Perspective of above-Ground Oil Shale Retorting R&D Needs

Robert Vagnetti, U.S. DOE NETL, 3610 Collins Ferry Road, Morgantown, WV 26507-0880 and Lawrence J. Shadle, OSTA, National Energy Technology Laboratory, 3610 Collins Ferry Rd, Morgantown, WV 26507.
Technologies for extracting oil from shale were developed in the 1970s and 1980s. Based on engineering designs from this previous development era, capital costs for a 50,000 barrel per day, first-of-a-kind commercial plant built today could exceed $4 billion dollars with daily operating costs projected to be above $25 per barrel. With these costs, oil shale remains uncompetitive at current world oil prices.

In this paper these previous, publicly available, cost data are critically evaluated for both accuracy and relevance to today's economic situation. Experience shows these costs can be reduced significantly by conducting large-scale tests to reduce scale-up risks and by addressing design inefficiencies through research and development (R&D). In addition, technical advancements made over the past two decades are considered with regard to their potential to reduce costs. R&D needs are identified that improve both the economics and acceptance of new developing technologies. R&D areas of highest impact are discussed as well as those technologies which remain highly uncertain.

228d R&D Data Gaps Identified for Model Development of Novel in Situ Oil Shale Conversion Process

Lawrence J. Shadle[1], Mark Smith[2], Esmail R. Monazam[2], and Wen-Ching Yang[3]. (1) OSTA, National Energy Technology Laboratory, 3610 Collins Ferry Rd, Morgantown, WV 26507, (2) REM Engineering Services, U.S. DOE NETL, 3610 Collins Ferry Road, Morgantown, WV 26507-0880, (3) Chemical and Petroleum Engineering, University of Pittsburgh, Benedum Engineering Hall, Pittsburgh, PA 15261
The United States is endowed with 62% of world's potentially recoverable oil shale resources. Potentially recoverable oil amounts to 500 billion to 1.1 trillion barrels, more than triple the proven crude oil reserves of Saudi Arabia. Nearly 60% of this total concentrates in the relatively rich Green River formation of Colorado, Utah, and Wyoming. Almost all the R&D activities in the US were performed 20 to 30 years ago. With the rapidly increasing and sustaining high oil price due to world supply-demand imbalance, commercial development of this valuable national resource becomes an attractive alternative.

The literature of oil shale conversion processes (surface retorting and in situ retorting) was reviewed in order to formulate a model of the novel oil shale in situ process. This in situ process relies on low temperature, thermally conductive heating of oil shale over extended periods of time to produce shale oil. The purpose of this paper is to present the R&D data gaps that were identified during model formulation. In particular, the environmental issues were identified in the application of modern engineering technologies. In addition, the limitations of previous predictive models will be discussed along with specific data gaps for the modeling the new in situ process.

228e Compact, Mobile Synthetic Fuel Unit

Kai Jarosch[1], Terry Mazanec[1], Jeff McDaniel[2], Anna Lee Tonkovich[1], and Sean Fitzgerald[1]. (1) Technology Development, Velocys, Inc., 7950 Corporate Boulevard, Plain City, OH 43064, (2) Business Development, Velocys, Inc., 7950 Corporate Boulevard, Plain City, OH 43064
Modern military operations require a steady, reliable supply of liquid fuel to keep armored vehicles rolling and field generators running. Often, this fuel can not be easily sourced in the theater of action and must be brought from the elsewhere. The logistical challenge of importing millions of gallons of fuel into a hostile environment can be daunting. The U.S. Department of Defense has recognized this difficulty and they are starting to explore alternative fuel sourcing strategies, including truck transportable synthetic fuel units to produce much of the required fuel in the theater of action from locally available energy resources. Sufficiently reducing equipment size to fit in truck transportable containers requires a new approach to reactor design, which is offered by microchannel process technology.

The concept of producing synthetic fuels in compact units hinges on the ability to economically scale-down reaction hardware while maintaining sufficient capacity. Systems based on microchannel process technology have the potential to allow applications like mobile synthetic fuel units by greatly reducing the size of chemical processing hardware. Reactors using this technology are characterized by parallel arrays of microchannels, with typical dimensions in the 0.010-inch to 0.200-inch range. Processes are intensified by reducing heat and mass transfer distances, thus decreasing transfer resistance between process fluids and channel walls. Overall system volumes are typically reduced 10- to 100-fold or more compared with conventional hardware.

For the mobile fuel unit programs, microchannel technology is being applied to the two main reactor sections in a synthetic fuel process: 1) synthesis gas production from natural gas, and 2) Fischer-Tropsch reaction to convert this synthesis gas into waxes which are subsequently processed into the high quality fuels needed to run vehicles and generators. This presentation will show the compact, mobile synthetic fuel unit concept developed by Velocys and Mustang

Engineering under a congressionally funded project overseen by the U.S. Department of Defense Tank and Automotive Command (TACOM). The unit presented will be a 15,000 gallons per day facility capable of transforming locally available sources of natural gas into a synthetic fuel. In addition to the conceptual design, data from laboratory experiments on microchannel Fischer-Tropsch reactors will be included to show the viability of producing fuel in this manner.

228f Opportunities for Co-Feeding Coal and Natural Gas to a Liquids Plant

Diane. Hildebrandt[1], David Glasser[2], B. Patel[2], and B. Hausberger[2]. (1) Centre of Material and Process Synthesis, University of the Witwatersrand, Private Bag 3, WITS 2050, Johannesburg, South Africa, (2) Centre of Material and Process Synthesis, University of the Witwatersrand, Private Bag 3, WITS 2050, Johannesburg, South Africa

Mass and energy balances are typically used in the analysis of flow-sheets and equipment. They however can also be used to understand the impact of the choice of feedstock and chemistry on processes and in particular the effect on carbon efficiency in appropriate processes. This can be done at a high level, that is before the flow-sheet is even synthesized. In addition, another powerful tool that is rather underutilized in the chemical engineer's toolbox is the concept of entropy or exergy. Combining overall entropy or exergy balances with the overall process mass and energy balances allows one to look at integrating a process at the highest level. In addition it allows one to calculate targets of performance. We call this approach, ComPS, or Complete Process Synthesis. The approach has been formulated as a matrix based method (Patel et. al.).

CompS, allows one to scan all different chemistries and in effect flow-sheets at a very high level. We are therefore able to quickly scan different options before the flow-sheet even exists, and to set and compare performance targets. Previous research has then (Mukoma et. al.) allowed us to develop the flow-sheet once the chemistry had been chosen, and in effect find the target for a real process.

ComPS is used to identify opportunities in a liquids production plant. In particular we identified the opportunity for the co-feeding of coal and natural gas and looked at the various optimal ways of integrating the process. Because of the number of degrees of freedom, we can set different targets, for example a zero CO2 emissions plant or a net zero work process. We get different answers depending on the target but all show that there are large opportunities and the optimum relative amounts of the two feeds in the process are different depending on the objective chosen.

References:

Patel, Bilal, Hildebrandt, Diane, Glasser, David, and Hausberger, Brendon. (2005) Thermodynamics Analysis of Processes. 1. Implications of Work Integration. Ind. Eng. Chem. Res. Vol. 44, 3529-3537

Patel, B., Hildebrandt, D., Glasser, D. and Hausberger, B. Matrix based methods for Complete Process Synthesis, Manuscript in preparation.

Mukoma, P., Hildebrandt, D., Glasser, D. and Hausberger, B. Fischer-Tropsch Results and their Analysis for Process Synthesis. Submitted to Ind. Eng. Chem. Res.